Lecture Notes in Physics

Springer
Berlin
Heidelberg
New York
Barcelona
Hong Kong
London
Milan
Paris
Singapore
Tokyo

Physics and Astronomy

ONLINE LIBRARY

http://www.springer.de/phys/

The Editorial Policy for Proceedings

The series Lecture Notes in Physics reports new developments in physical research and teaching – quickly, informally, and at a high level. The proceedings to be considered for publication in this series should be limited to only a few areas of research, and these should be closely related to each other. The contributions should be of a high standard and should avoid lengthy redraftings of papers already published or about to be published elsewhere. As a whole, the proceedings should aim for a balanced presentation of the theme of the conference including a description of the techniques used and enough motivation for a broad readership. It should not be assumed that the published proceedings must reflect the conference in its entirety. (A listing or abstracts of papers presented at the meeting but not included in the proceedings could be added as an appendix.)

When applying for publication in the series Lecture Notes in Physics the volume's editor(s) should submit sufficient material to enable the series editors and their referees to make a fairly accurate evaluation (e.g. a complete list of speakers and titles of papers to be presented and abstracts). If, based on this information, the proceedings are (tentatively) accepted, the volume's editor(s), whose name(s) will appear on the title pages, should select the papers suitable for publication and have them refereed (as for a journal) when appropriate. As a rule discussions will not be accepted. The series editors and Springer-Verlag will normally not interfere with the detailed editing except in fairly obvious cases or on technical matters.

Final acceptance is expressed by the series editor in charge, in consultation with Springer-Verlag only after receiving the complete manuscript. It might help to send a copy of the authors' manuscripts in advance to the editor in charge to discuss possible revisions with him. As a general rule, the series editor will confirm his tentative acceptance if the final manuscript corresponds to the original concept discussed, if the quality of the contribution meets the requirements of the series, and if the final size of the manuscript does not greatly exceed the number of pages originally agreed upon. The manuscript should be forwarded to Springer-Verlag shortly after the meeting. In cases of extreme delay (more than six months after the conference) the series editors will check once more the timeliness of the papers. Therefore, the volume's editor(s) should establish strict deadlines, or collect the articles during the conference and have them revised on the spot. If a delay is unavoidable, one should encourage the authors to update their contributions if appropriate. The editors of proceedings are strongly advised to inform contributors about these points at an early stage.

The final manuscript should contain a table of contents and an informative introduction accessible also to readers not particularly familiar with the topic of the conference. The contributions should be in English. The volume's editor(s) should check the contributions for the correct use of language. At Springer-Verlag only the prefaces will be checked by a copy-editor for language and style. Grave linguistic or technical shortcomings may lead to the rejection of contributions by the series editors. A conference report should not exceed a total of 500 pages. Keeping the size within this bound should be achieved by a stricter selection of articles and not by imposing an upper limit to the length of the individual papers. Editors receive jointly 30 complimentary copies of their book. They are entitled to purchase further copies of their book at a reduced rate. As a rule no reprints of individual contributions can be supplied. No royalty is paid on Lecture Notes in Physics volumes. Commitment to publish is made by letter of interest rather than by signing a formal contract. Springer-Verlag secures the copyright for each volume.

The Production Process

The books are hardbound, and the publisher will select quality paper appropriate to the needs of the author(s). Publication time is about ten weeks. More than twenty years of experience guarantee authors the best possible service. To reach the goal of rapid publication at a low price the technique of photographic reproduction from a camera-ready manuscript was chosen. This process shifts the main responsibility for the technical quality considerably from the publisher to the authors. We therefore urge all authors and editors of proceedings to observe very carefully the essentials for the preparation of camera-ready manuscripts, which we will supply on request. This applies especially to the quality of figures and halftones submitted for publication. In addition, it might be useful to look at some of the volumes already published. As a special service, we offer free of charge LATEX and TEX macro packages to format the text according to Springer-Verlag's quality requirements. We strongly recommend that you make use of this offer, since the result will be a book of considerably improved technical quality. To avoid mistakes and time-consuming correspondence during the production period the conference editors should request special instructions from the publisher well before the beginning of the conference. Manuscripts not meeting the technical standard of the series will have to be returned for improvement.

For further information please contact Springer-Verlag, Physics Editorial Department II, Tiergartenstrasse 17, D-69121 Heidelberg, Germany

Series homepage – http://www.springer.de/phys/books/lnpp

Hugues Dreyssé   (Ed.)

Electronic Structure and Physical Properties of Solids

The Uses of the LMTO Method

Lectures of a Workshop Held at Mont Saint Odile, France, October 2–5, 1998

Springer

SEP/AE
PHYS

Editor

Prof. Hugues Dreyssé
IPCMS-GEMME
23 rue du Loess, BP 20 Cr
67037 Strasbourg, France

Library of Congress Cataloging-in-Publication Data applied for.

Die Deutsche Bibliothek - CIP-Einheitsaufnahme

Electronic structure and physical properties of solids : the uses of
the LMTO method ; lectures of a workshop held at Mont Saint Odile,
France, October 2 - 5, 1998 / Hugues Dreyssé (ed.). - Berlin ;
Heidelberg ; New York ; Barcelona ; Hong Kong ; London ; Milan ; Paris
; Singapore ; Tokyo : Springer, 2000
  (Lecture notes in physics ; 535)
  ISBN 3-540-67238-9

ISSN 0075-8450
ISBN 3-540-67238-9 Springer-Verlag Berlin Heidelberg New York

Springer-Verlag is a company in the BertelsmannSpringer publishing group.
© Springer-Verlag Berlin Heidelberg 2000
Printed in Germany

The use of general descriptive names, registered names, trademarks, etc. in this publication
does not imply, even in the absence of a specific statement, that such names are exempt
from the relevant protective laws and regulations and therefore free for general use.

Typesetting: Camera-ready by the authors/editor
Cover design: *design & production*, Heidelberg

Printed on acid-free paper
SPIN: 10720521   55/3144/du - 5 4 3 2 1 0

Preface

In recent years, computational materials science has clearly emerged as an important field of condensed matter physics. In particular, the development of new computing facilities has made it possible to study physical phenomena at the *atomic* scale by means of *ab initio* electronic structure methods. Among various approaches used, the Linear Muffin-Tin Orbitals method (LMTO) proposed in the seventies by O.K. Andersen has played a key role. In its Atomic Sphere Approximation (ASA), the LMTO method has been widely used to tackle various type of problems. In 1984, O.K. Anderson and coworkers introduced a *localized* LMTO basis set. This new approach, called Tight-Binding LMTO (TB-LMTO), has paved the way to an order-N scheme, giving new impetus to the study of numerous physical properties of systems with large number of atoms.

This book is based on selected contributions presented at a workshop, organized in October 1998 in the monastery of Mont Saint Odile (near Strasbourg, France). A large number of scientists involved in the development and the practice of the LMTO method gathered there for three days. The first part of this book is devoted to the formalisms for ground and excited states. It starts with a review, by Andersen and coworkers, of the TB-LMTO method and its generalization. The Schrödinger equation of Nth order in the energy expansion for an overlapping muffin-tin potential is solved using a minimal basis set. The aim of this third generation LMTO method is to take a further step beyond the limitations of the popular atomic-sphere approximation. The present approach uses wave functions which are accurate not only in the muffin-tin spheres but also in the interstitial region. In the conventional implementations of the LMTO-ASA method it is difficult to determine the forces on the atoms. For this reason Full-Potential LMTO approaches have been developed. In this book two different approaches are described. The first one is proposed by Methfessel and coworkers and is based on smooth Hankel functions. The use of these special functions allows the method to provides a good accuracy of the total energy with an almost minimal basis set. The second approach proposed by Wills and coworkers uses a large basis set which can describe multiple principal quantum numbers within a single *fully hybridized* basis set. This large basis set allows this method to determine the excited states to higher energies without the need for cumbersome multiple energy panel calculations like in the minimal basis set methods. However, the drawback and limitation of these FP methods remains the huge computational effort needed, which inhibits calculations of systems with large

numbers of atoms. To overcome these computational hurdles, Kollar and coworkers presented the so called *full-density* method, a method halfway between the
ASA and the FP, which combines the simplicity of the former and the accuracy
of the latter. In the initial description of the Density Functional Theory (DFT),
the eigenvalues are Lagrange multipliers, and thus no physical meaning can be
associated with them. However the success of excited states calculations based
on DFT can be understood due to the fact that the Kohn–Sham equations can
be viewed as an approximation to the quasi-particle equations where the self-
energy is local and time independent. Alouani and Wills give the basics and
some applications to the determination of optical properties and x-ray magnetic
dichroism.

One of the main success of the LMTO schemes in the last few years has
been the description of magnetic systems. A fully relativistic formalism and
the applications to spectroscopy are presented by Ebert. The key question of
the magnetic anisotropy of bulk and thin films is addressed by Eriksson and
Wills, illustrating the high level of precision reached, whereas Temmerman and
coworkers present a unified formalism to describe localized and delocalized states,
pointing out the importance of the self-interaction correction. Another spectac-
ular use of the TB-LMTO method is given by Kudrnovsky and coworkers on the
Interlayer Exchange Coupling (IEC). Ab initio formulations of the IEC between
non-collinearly aligned magnetic slabs lead to results in good agreement with
experiment.

One advantage of the TB-LMTO is its use to describe disordered systems such
as alloys with the precision of ab initio methods. A short review of the TB-LMTO
within the Coherent Potential Approximation (CPA) applied to disordered alloys
and surfaces is given by Turek and coworkers, whereas Abrikosov and coworkers
present a locally self-consistent Green's function method. This latter order-N
method is particularly interesting for systems with a large number of inequivalent
atoms.

Mathematical and numerical problems are the building blocks of ab initio
methods, and numerical algorithms for solving various parts of the formalism are
of great interest. In particular, the diagonalisation of sparse matrices, which is at
the heart of the TB-LMTO method, by efficient algorithms is highly desirable. In
this respect, Scott has given an introduction to direct methods for the solution
of large-scale linear systems, emphasizing the progress made in the development
of routines which are now available in numerical libraries. The book ends with
two contributions on the determination of the electronic structure in real space:
a real-space derivation of the TB-LMTO method by Spisak and Hafner, and the
venerable semi-empirical tight-binding method by Cornea and Stoeffler.

I would like to thank each of the sixty-seven scientists from seventeen different
countries who make this book possible, especially those who came from far away
countries, like Argentina, Brasil, India, and Japan. The meeting was held over
three days in a lively atmosphere where the most recent progress in the ab initio
methods was presented and where discussions continued well after the talks.

Besides the fourteen invited talks which form the basis of this book, nineteen shorter presentations offered the opportunity to focus on more precise points.

The "Ab-initio Calculations of Magnetic Properties of Surfaces, Interfaces, and Multilayers" TMR European network provided the first impulse for this meeting. The support of the European Science Foundation (ESF) through the program Ψ_k allowed the participation of a large number of scientists. Finally, the local support of the IPCMS (Institut de Physique et Chimie de Strasbourg) was greatly appreciated. I would also like to thank M. Alouani for constant interest and I. Galanakis, who struggled and succeeded in finalizing the electronic version of the book.

Strasbourg, August 1999   *Hugues Dreyssé*

Contents

List of Contributors

O. K. Andersen
Max-Planck-Institut FKF
D-70569 Stuttgart
Germany
andersen@and.mpi-stuttgart.mpg.de

J. Kollár
Research Institute for Solid State
Physics
H-1525 Budapest,
P.O.Box 49, Germany
jk@ppc1.szfki.kfki.hu

M. Methfessel
Institute for Semiconductor Physics
Walter-Korsing-Str. 2
D-15230 Frankfurt (Oder)
Germany
msm@th4.ihp-ffo.de

J. M. Wills
Los Alamos National Laboratory
Los Alamos, NM 87545
USA
jxw@lanl.gov

M. Alouani
Institut de Physique et Chimie des
Matériaux de Strasbourg
23, rue du Loess
F-67037 Strasbourg, France
mea@ipcms.u-strasbg.fr

H. Ebert
Institut für Phys. Chemie,
Univ. München
Butenandtstr. 5-13
D-81377 München, Germany
he@gaia.phys.chemie.uni-muenchen.de

O. Eriksson
Department of Physics
Uppsala University
Box 530
S-75112 Uppsala, Sweden
Olle.eriksson@Fysik.UU.SE

W. M. Temmerman
Daresbury Laboratory
Daresbury
Warrington, WA4 4AD
United Kingdom
W.M.TEMMERMAN@dl.ac.uk

J. Kudrnovský
Institute of Physics
Academy of Sciences of the Czech
Republic
CZ-182 21 Praha 8
Czech Republic
jk@ws7.cms.tuwien.ac.at

I. Turek
Institute of Physics of Materials
Academy of Sciences
Žižkova 22, CZ-616 62 Brno
Czech Republic
turek@ipm.cz

I. A. Abrikosov
Condensed Matter Theory Group
Physics Department
Uppsala University
S-75121 Uppsala, Sweden
Igor.Abrigosov@Fysik.UU.SE

J. A. Scott
Department for Computation and
Information
Rutherford Appleton Laboratory
Chilton, Didcot, Oxon OX11 0RA
England
J.Scott@rl.ac.uk

D. Spišák
Institut für Theoretische Physik
and Center for Computational
Materials Science
Technische Universität Wien
Wiedner Hauptstraße 8-10/136
A-1040 Vienna, Austria
spisak@hal27.cmt.tuwien.ac.at

C. Cornea
Institut de Physique et Chimie des
Matériaux de Strasbourg
Groupe d'Etude des Matériaux
Métalliques
23, rue du Loess
F-67037 Strasbourg, France
Daniel.Stoeffler@ipcms.u-strasbg.fr

Part I

Ground and Excited-State Formalisms

Developing the MTO Formalism

O. K. Andersen, T. Saha-Dasgupta, R. W. Tank, C. Arcangeli, O. Jepsen, and
G. Krier

Max-Planck-Institut FKF, D-70569 Stuttgart, FRG,
andersen@and.mpi-stuttgart.mpg.de

Abstract. The TB-LMTO-ASA method is reviewed and generalized to an accurate
and robust TB-NMTO minimal-basis method, which solves Schrödinger's equation to
Nth order in the energy expansion for an overlapping MT-potential, and which may
include any degree of downfolding. For $N = 1$, the simple TB-LMTO-ASA formalism is
preserved. For a discrete energy mesh, the NMTO basis set may be given as: $\chi^{(N)}(\mathbf{r}) = \sum_n \phi(\varepsilon_n, \mathbf{r}) L_n^{(N)}$ in terms of *kinked* partial waves, $\phi(\varepsilon, \mathbf{r})$, evaluated on the mesh,
$\varepsilon_0, ..., \varepsilon_N$. This basis solves Schrödinger's equation for the MT-potential to within an
error $\propto (\varepsilon - \varepsilon_0) ... (\varepsilon - \varepsilon_N)$. The Lagrange matrix-coefficients, $L_n^{(N)}$, as well as the
Hamiltonian and overlap matrices for the NMTO set, have simple expressions in terms
of energy derivatives on the mesh of the Green matrix, defined as the inverse of the
screened KKR matrix. The variationally determined single-electron energies have errors
$\propto (\varepsilon - \varepsilon_0)^2 ... (\varepsilon - \varepsilon_N)^2$. A method for obtaining orthonormal NMTO sets is given and
several applications are presented.

1 Overview

Muffin-tin orbitals (MTOs) have been used for a long time in *ab initio* calculations of the electronic structure of condensed matter. Over the years, several
MTO-based methods have been devised and further developed. The ultimate aim
is to find a generally applicable electronic-structure method which is *accurate*
and *robust,* as well as *intelligible.*

In order to be *intelligible,* such a method must employ a small, single-
electron basis of atom-centered, short-ranged orbitals. Moreover, the single-
electron Hamiltonian must have a simple, analytical form, which relates to a
two-center, orthogonal, tight-binding (TB) Hamiltonian.

In this sense, the conventional linear muffin-tin-orbitals method in the *atomic-
spheres approximation* (LMTO-ASA) [1,2] is intelligible, because the orbital may
be expressed as:

$$\chi_{RL}(\mathbf{r}_R) = \phi_{RL}(\mathbf{r}_R) + \sum_{R'L'} \dot{\phi}_{R'L'}(\mathbf{r}_{R'})(H_{R'L',RL} - \varepsilon_\nu \delta_{R'R}\delta_{L'L}). \quad (1)$$

Here, $\phi_{RL}(\mathbf{r}_R)$ is the solution, $\varphi_{Rl}(\varepsilon_\nu, r_R) Y_{lm}(\hat{\mathbf{r}}_R)$, at a chosen energy, ε_ν, of
Schrödinger's differential equation inside the atomic sphere at site R for the
single-particle potential, $\sum_R v_R(r_R)$, assumed to be spherically symmetric in-
side that sphere. Moreover, $\mathbf{r}_R \equiv \mathbf{r} - \mathbf{R}$ and $L \equiv lm$. The function $\varphi_{Rl}(\varepsilon, r)$

thus satisfies the one-dimensional, radial Schrödinger equation

$$\frac{\partial^2}{\partial r^2} r\varphi_{Rl}\left(\varepsilon,r\right) = -\left[\varepsilon - v_R\left(r\right) - \frac{l\left(l+1\right)}{r^2}\right] r\varphi_{Rl}\left(\varepsilon,r\right). \tag{2}$$

In (1), $\dot{\phi}_{RL}\left(\mathbf{r}\right)$ are the energy-derivative functions, $\partial\varphi_{Rl}\left(\varepsilon,r\right)/\partial\varepsilon|_{\varepsilon_\nu} Y_{lm}\left(\hat{\mathbf{r}}\right)$. The radial functions, φ and $\dot{\varphi}$, and also the potential, v, are truncated outside their own atomic sphere of radius s, and the matrix, H, is constructed in such a way that the LMTO is continuous and differentiable in all space. Equation (1) therefore expresses the LMTO at site R and (pseudo) angular momentum L as the solution of Schrödinger's equation at that site, with that angular momentum, and at the chosen energy, plus a 'smoothing cloud' of energy-derivative functions, centered mainly at the neighboring sites, and having around these, all possible angular momenta.

That a set of energy-*in*dependent orbitals must have the form (1) in order to constitute a basis for the solutions $\Psi_i\left(\mathbf{r}\right)$ – with energies ε_i in the neighborhood of ε_ν – of Schrödinger's equation for the *entire* system, is intuitively obvious, because the corresponding linear combinations, $\sum_{RL} \chi_{RL}\left(\mathbf{r}_R\right) c_{RL,i}$, will be those which locally, inside each atomic sphere and for each angular momentum, have the right amount of $\dot{\varphi}$ – provided mainly by the tails of the neighboring orbitals – added onto the central orbital's φ. Since by construction each $\varphi_{Rl}\left(\varepsilon,r\right)$ is the correct solution, this right amount is of course $\varepsilon_i - \varepsilon_\nu$. In math: since definitions can be made such that the expansion matrix $H_{R'L',RL}$ is *Hermitian*, its *eigenvectors* are the coefficients of the proper linear combinations, and its *eigenvalues* are the energies:

$$\sum_{RL} \chi_{RL}\left(\mathbf{r}_R\right) c_{RL,i} = \sum_{RL} \left[\phi_{RL}\left(\mathbf{r}_R\right) + \left(\varepsilon_i - \varepsilon_\nu\right) \dot{\phi}_{RL}\left(\mathbf{r}_R\right)\right] c_{RL,i}$$

$$\approx \sum_{RL} \phi_{RL}\left(\varepsilon_i, \mathbf{r}_R\right) c_{RL,i} = \Psi_i\left(\mathbf{r}\right). \tag{3}$$

Hence, H is a 1st-order *Hamiltonian*, delivering energies and wave functions with errors proportional to $\left(\varepsilon_i - \varepsilon_\nu\right)^2$, to leading order.

First-order energies seldom suffice, and in the conventional LMTO-ASA method use is made of the variational principle for the Hamiltonian,

$$\mathcal{H} \equiv -\nabla^2 + \sum_R v_R\left(r_R\right), \tag{4}$$

so that errors of order $\left(\varepsilon_i - \varepsilon_\nu\right)^2$ in the basis set merely give rise to errors of order $\left(\varepsilon_i - \varepsilon_\nu\right)^4$ in the energies. With that approach, the energies and eigenvectors are obtained as solutions of the generalized eigenvalue problem:

$$\sum_{RL} \left[\langle\chi_{R'L'}|\mathcal{H} - \varepsilon_\nu|\chi_{RL}\rangle - \left(\varepsilon_i - \varepsilon_\nu\right)\langle\chi_{R'L'} \mid \chi_{RL}\rangle\right] c_{RL,i} = 0, \tag{5}$$

for all $R'L'$. If we now insert (1) in (5), we see that the Hamiltonian and overlap matrices are expressed in terms of the 1st-order Hamiltonian, H, plus two *diagonal* matrices with the respective elements

$$\left\langle \phi_{RL} \mid \dot{\phi}_{RL} \right\rangle = \int_0^s \varphi_{Rl}(r)\dot{\varphi}_{Rl}(r)r^2 dr, \quad \left\langle \dot{\phi}_{RL} \mid \dot{\phi}_{RL} \right\rangle = \int_0^s \dot{\varphi}_{Rl}(r)^2 r^2 dr. \quad (6)$$

These matrices are diagonal by virtue of the ASA, which approximates integrals over space by the sum of integrals over atomic spheres. If each partial wave is normalized to unity in its sphere: $\int_0^s \varphi_{Rl}(r)^2 r^2 dr = 1$, then $\langle \phi \mid \phi \rangle$ is the unit matrix in the ASA, and the Hamiltonian and overlap matrices entering (5) take the simple forms:

$$\langle \chi | \mathcal{H} - \varepsilon_\nu | \chi \rangle = (H - \varepsilon_\nu)\left[1 + \left\langle \phi \mid \dot{\phi} \right\rangle (H - \varepsilon_\nu) \right] \qquad (7)$$

$$\langle \chi \mid \chi \rangle = \left[1 + (H - \varepsilon_\nu)\left\langle \dot{\phi} \mid \phi \right\rangle \right]\left[1 + \left\langle \phi \mid \dot{\phi} \right\rangle (H - \varepsilon_\nu) \right]$$
$$+ (H - \varepsilon_\nu)\left[\left\langle \dot{\phi} \mid \dot{\phi} \right\rangle - \left\langle \phi \mid \dot{\phi} \right\rangle^2 \right](H - \varepsilon_\nu).$$

Here and in the following we use a vector-matrix notation according to which, for example $\chi_{RL}(\mathbf{r}_R)$ and $\chi_{RL}(\mathbf{r}_R)^*$ are considered components of respectively a row-vector, $\chi(\mathbf{r})$, and a column-vector, $\chi(\mathbf{r})^\dagger$. The eigenvector, c_i, is a column vector with components $c_{RL,i}$. Moreover, 1 is the unit matrix, ε_ν is a diagonal matrix, and H is a Hermitian matrix. Vectors and diagonal matrices are denoted by lower-case Latin and Greek characters, and matrices by upper-case Latin characters. Exceptions to this rule are: $Y(\hat{\mathbf{r}})$, the vector of spherical harmonics, the site and angular-momentum indices (subscripts) R, L, I, and A, and the orders (superscripts) L, M, and N. Operators are given in calligraphic, like $\mathcal{H}$, and an omitted energy argument means that $\varepsilon = \varepsilon_\nu$.

With the $\phi(\mathbf{r})$'s being orthonormal in the ASA, the LMTO overlap matrix in (7) is seen to factorize to 1st order, and it is therefore simple to transform to a set of *nearly orthonormal* LMTOs:

$$\hat{\chi}(\mathbf{r}) = \chi(\mathbf{r})\left[1 + \left\langle \phi \mid \dot{\phi} \right\rangle (H - \varepsilon_\nu) \right]^{-1} \qquad (8)$$

$$\langle \hat{\chi} | \mathcal{H} - \varepsilon_\nu | \hat{\chi} \rangle \equiv \hat{H} - \varepsilon_\nu = \left[1 + (H - \varepsilon_\nu)\left\langle \dot{\phi} \mid \phi \right\rangle \right]^{-1}(H - \varepsilon_\nu)$$
$$= H - \varepsilon_\nu - (H - \varepsilon_\nu)\left\langle \dot{\phi} \mid \phi \right\rangle (H - \varepsilon_\nu) + \dots$$

$$\langle \hat{\chi} \mid \hat{\chi} \rangle = 1 + \left(\hat{H} - \varepsilon_\nu \right)\left\langle \dot{\hat{\phi}} \mid \dot{\hat{\phi}} \right\rangle\left(\hat{H} - \varepsilon_\nu \right)$$

$$\dot{\hat{\phi}}(\mathbf{r}) \equiv \dot{\phi}(\mathbf{r}) - \phi(\mathbf{r})\left\langle \phi \mid \dot{\phi} \right\rangle.$$

Here, the energy-derivative function, $\dot{\hat{\phi}}(\mathbf{r})$, equals $\dot{\phi}(\mathbf{r})$, orthogonalized to $\phi(\mathbf{r})$. Finally, we may transform to a set of *orthonormal* LMTOs:

$$\check{\chi}(\mathbf{r}) = \hat{\chi}(\mathbf{r}) \left[1 + \left(\hat{H} - \varepsilon_\nu\right) \left\langle \dot{\hat{\phi}} \mid \dot{\hat{\phi}} \right\rangle \left(\hat{H} - \varepsilon_\nu\right)\right]^{-1/2} = \tag{9}$$

$$\hat{\chi}(\mathbf{r}) \left[1 - \frac{1}{2}\left(\hat{H} - \varepsilon_\nu\right) \left\langle \dot{\hat{\phi}} \mid \dot{\hat{\phi}} \right\rangle \left(\hat{H} - \varepsilon_\nu\right) + ..\right]$$

$$\left\langle \check{\chi} \mid \mathcal{H} - \varepsilon_\nu \mid \check{\chi} \right\rangle \equiv \check{H} - \varepsilon_\nu = \hat{H} - \varepsilon_\nu -$$

$$\frac{1}{2}\left(\hat{H} - \varepsilon_\nu\right) \left\langle \dot{\hat{\phi}} \mid \dot{\hat{\phi}} \right\rangle \left(\hat{H} - \varepsilon_\nu\right)^2 - \frac{1}{2}\left(\hat{H} - \varepsilon_\nu\right)^2 \left\langle \dot{\hat{\phi}} \mid \dot{\hat{\phi}} \right\rangle \left(\hat{H} - \varepsilon_\nu\right) + ..$$

We thus realize that of the Hamiltonians considered, H is of 1st, $\hat{H}$ is of 2nd, and $\check{H}$ is of 3rd order. As the order increases, and the energy window – inside which the eigenvalues of the Hamiltonian are useful as single-electron energies – widens, the real-space *range* of the Hamiltonian increases. For real-space calculations [3–7], it is therefore important to be able to express a higher-order Hamiltonian as a power series in a lower-order Hamiltonian like in (8) and (9), because such a series may be truncated when the energy window is sufficiently wide.

The energy-derivative of the radial function $\varphi(\varepsilon, r)$ depends on the *energy derivative* of its *normalization*. If we choose to normalize according to: $\int_0^s \hat{\varphi}(\varepsilon, r)^2 r^2 dr = 1$, then it follows that $\int_0^s \hat{\varphi}(r)\dot{\hat{\varphi}}(r) r^2 dr = 0$. Choosing another energy-dependent normalization: $\varphi(\varepsilon, r) \equiv \hat{\varphi}(\varepsilon, r)[1 + (\varepsilon - \varepsilon_\nu)o]$, specified by a constant o, then we see that: $\dot{\varphi}(r) = \dot{\hat{\varphi}}(r) + \varphi(r)o$. Changing the energy derivative of the normalization thus adds some $\varphi(r)$ to $\dot{\hat{\varphi}}(r)$ and thereby changes the shape of the 'tail function' $\dot{\varphi}(r)$. Since all LMTOs (1) should remain smooth upon this change, also H must change, and so must all LMTOs in the set. The diagonal matrix $\left\langle \dot{\phi} \mid \dot{\phi} \right\rangle$, whose elements are the radial overlap integrals: $o = \int_0^s \varphi(r)\dot{\varphi}(r) r^2 dr$, thus determines the LMTO *representation*, and the first and the last equations (8) specify the linear transformation between representations. Values of the diagonal matrix $\left\langle \dot{\phi} \mid \dot{\phi} \right\rangle$ exist, which yield *short range* for the 1st-order Hamiltonian H and, hence, for the LMTO set (1). Such an H is therefore a *two-center TB Hamiltonian* and such an LMTO set is a *first-principles TB basis*.

In order to obtain an explicit expression for H, one needs to find the spherical-harmonics expansions about the various sites for a set of *smooth MTO envelope functions*. For a MT-potential, which is flat in the interstitial, the envelope functions are wave-equation solutions with pure spherical-harmonics character near the sites. Consistent with the idea behind the ASA – to use 'space-filling spheres' – is the use of envelope functions with *fixed* energy, specifically *zero*, which is a reasonable approximation for the kinetic energy between the atoms for a valence state. The envelope functions in the ASA are thus *screened multipole potentials*, with the screening specified by a *diagonal matrix* of *screening constants*, α_{Rl}, related to the radial overlaps o_{Rl}. The expansion of a *bare* multipole potential at site R about a different site R' is well known:

$$\frac{Y_L\left(\hat{\mathbf{r}}_R\right)}{r_R^{l+1}} \sim \sum_{R'L'} r_{R'}^{l'} Y_{L'}\left(\hat{\mathbf{r}}_{R'}\right) \frac{Y_{l''m''}\left(\widehat{\mathbf{R}'-\mathbf{R}}\right)}{\left|\mathbf{R}'-\mathbf{R}\right|^{l''+1}} \sim \sum_{R'L'} r_{R'}^{l'} Y_{L'}\left(\hat{\mathbf{r}}_{R'}\right) S^0_{R'L',RL}.$$

Here, $l'' \equiv l' + l$ and $m'' \equiv m' - m$. With suitable normalizations, the *bare structure matrix*, S^0, can be made Hermitian. The *screened* structure matrix is now related to the bare one through a Dyson equation:

$$\left(S^\alpha\right)^{-1} = \left(S^0\right)^{-1} - \alpha, \tag{10}$$

which may be solved by inversion of the matrix $S^0 - \alpha^{-1}$. This inversion may be performed in *real space*, that is in $\mathbf{R}$- rather than in $\mathbf{k}$-representation, provided that the screening constants take values known from experience to give a short-ranged S^α.

In the end, it turns out that *all* ingredients to the LMTO Hamiltonian and overlap integrals, H, $\left\langle \phi \mid \dot{\phi} \right\rangle$, and $\left\langle \dot{\phi} \mid \dot{\phi} \right\rangle$, may be obtained from the screened Korringa-Kohn-Rostoker (KKR) matrix in the ASA:

$$K^\alpha_{R'L',RL}\left(\varepsilon\right) \equiv p^\alpha_{Rl}\left(\varepsilon\right) \delta_{R'R}\delta_{L'L} - S^\alpha_{R'L',RL}. \tag{11}$$

Here, $p^0\left(\varepsilon\right)$ is a diagonal matrix of potential functions obtained from the radial logarithmic derivative functions, $\partial\left\{\varphi\left(\varepsilon,s\right)\right\} \equiv \partial\ln\left|\varphi\left(\varepsilon,r\right)\right|/\partial\ln r|_s$, evaluated at the MT-radius, and $p^\alpha\left(\varepsilon\right)$ is related to $p^0\left(\varepsilon\right)$ via the diagonal version of Equation (10). The results are:

$$H = \varepsilon_\nu - K = \varepsilon_\nu - p\dot{p}^{-1} + \dot{p}^{-\frac{1}{2}} S \dot{p}^{-\frac{1}{2}} \equiv c + d^{\frac{1}{2}} S d^{\frac{1}{2}},$$

$$\left\langle \phi \mid \dot{\phi} \right\rangle = \frac{\ddot{K}}{2!} = \frac{1}{2!}\frac{\ddot{p}}{\dot{p}}, \qquad \left\langle \dot{\phi} \mid \dot{\phi} \right\rangle = \frac{\dddot{K}}{3!} = \frac{1}{3!}\frac{\dddot{p}}{\dot{p}}, \tag{12}$$

expressed in terms of the KKR matrix, renormalized to have $\dot{K} = 1$:

$$K\left(\varepsilon\right) \equiv \dot{K}^{-\frac{1}{2}} K\left(\varepsilon\right) \dot{K}^{-\frac{1}{2}} = p\left(\varepsilon\right)\dot{p}^{-1} - \dot{p}^{-\frac{1}{2}} S \dot{p}^{-\frac{1}{2}}. \tag{13}$$

This corresponds to the partial-wave normalization: $\int_0^s \varphi\left(r\right)^2 r^2 dr = 1$, and $K\left(\varepsilon\right)$ is what in the 2nd-generation method [1,2] is denoted $-h\left(\varepsilon\right)$, but since the current notation identifies matrices by capitals, we cannot use h. The LMTO Hamiltonian and overlap matrices are thus expressed solely in terms of the structure matrix S and the potential functions $p\left(\varepsilon\right)$, specifically the diagonal matrices p, $\dot{p}$, $\ddot{p}$, and $\dddot{p}$. It may be realized that the nearly-orthonormal representation is generated if the diagonal screening matrix in (10) is set to the value γ, which makes $\dot{p}^\gamma$ vanish.

For calculations [8–10] which employ the *coherent-potential approximation* (CPA) to treat substitutional disorder, it is important to be able to perform screening transformations of the *Green matrix*:

$$G^\alpha\left(z\right) \equiv K^\alpha\left(z\right)^{-1} = \left[p^\alpha\left(z\right) - S^\alpha\right]^{-1}, \tag{14}$$

also called the resolvent, or the scattering path operator in multiple scattering theory [11]. In the 2nd generation MTO formalism, $\mathsf{G}^a(\varepsilon)$ was denoted $g^\alpha(\varepsilon)$. This screening transformation is:

$$\mathsf{G}^\beta(z) = (\beta - \alpha)\frac{\mathsf{p}^\alpha(z)}{\mathsf{p}^\beta(z)} + \frac{\mathsf{p}^\alpha(z)}{\mathsf{p}^\beta(z)}\mathsf{G}^\alpha(z)\frac{\mathsf{p}^\alpha(z)}{\mathsf{p}^\beta(z)}, \tag{15}$$

and is seen to involve no matrix multiplications, but merely energy-dependent rescaling of matrix elements. As a transformation between the nearly orthonormal, $\beta=\gamma$, and the short-ranged TB-representation, Eq. (15) has been useful also in *Green-function* calculations for extended defects, surfaces, and interfaces [8,10,12–14]. However, calculations which start out from the unperturbed Green matrices most natural for the problem – namely those obtained from LMTO band-structure calculations in the nearly orthonormal representation for the *bulk* systems – have usually been limited to *2nd-order* in $z - \varepsilon_\nu$, because $\mathsf{p}^\gamma(z)$ is linear to this order, and because attempts to use 3rd-order expressions for $\mathsf{p}^\gamma(z)$ employing the potential parameter $\ddot{\mathsf{p}}^\gamma = 3!\dot{\mathsf{p}}^\gamma\left\langle\dot{\phi}\mid\dot{\phi}\right\rangle$, induced false poles in the Green matrix.

What is *not* intelligible in the TB-LMTO-ASA method is that the LMTO *expansion* (1) must include *all* L'''s until convergence is reached throughout each sphere, and *all* R'''s until space is covered with spheres. This means that the LMTO-ASA basis is *minimal* – at most – for elemental, closely packed transition metals, the case for which it was in fact invented [15]. The supreme computational efficiency of the method soon made *self-consistent* density-functional [16] calculations possible, and not only for elemental transition metals, but also for compounds. In order to treat open structures such as diamond, *empty* spheres were introduced as a device for describing the repulsive potentials in the interstices [17]. All of this then, led to misinterpretations of the wave-function related output of such calculations in terms of the components of the one-center expansions (1), typically the numbers of s, p, and d electrons on the various atoms (including in the empty spheres!) and the charge transfers between them. Absurd statements to the effect that CsCl is basically a neutral compound with the Cs electron having a bit of s-, more p-, quite some d-, and a bit of f-character were not uncommon. Many practitioners of the ASA method did not realize that the role of the MT-spheres is to describe the *input potential,* rather than the output wave-functions. For the latter, the one-center expansions truncated outside the spheres constitute merely a decomposition which is used in the code for self-consistent calculations. The strange Cs electron is therefore little more than the expansion about the Cs site of the tails of the neighboring Cl p electrons spilling into the Cs sphere. That latter MT-sphere must of course be chosen to have about the same size as that of Cl, because only then is the shape of the Cs^+Cl^- *potential* in the bi-partitioned structure well described.

Now, the so-called *high* partial waves – they are those which are shaped like r^l in the outer part of the sphere where the potential flattens out – do enter the LMTO expansion (1), but *not* the eigenvalue problem (5) or the equivalent

KKR equation:

$$K\left(\varepsilon_i\right)c_i = 0, \qquad (16)$$

because they are part of the MTO envelope functions. This property of having the high-l limit correct is a strength of the MTO method, not shared by for instance Gaussian orbitals, which are solutions of (2) for a *parabolic* potential. There are, however, also other partial waves – like the Cs s-waves, d-waves in non-transition metal atoms, f-waves in transition-metal atoms, s-waves in oxygen and fluorine, and in positive alkaline ions, and all partial waves in empty spheres – which for the problem at hand are judged to be *inactive* and should therefore not have corresponding LMTOs in the basis. In order to get rid of such inactive LMTOs, one must first – by means of (10) or (15) – transform to a representation in which the inactive partial waves appear only in the 'tails' (second term of (1)) of the remaining LMTOs; only thereafter, the inactive LMTOs can be deleted. This *down-folding* procedure works for the LMTO-ASA method, but it messes up the connection between the LMTO Hamiltonian (7)-(13) and the KKR Green-function formalisms (12)-(16), and it is not as efficient as one would have liked it to be [2]. E.g., the Si valence band cannot be described with an sp LMTO basis set derived by down-folding of the Si d- as well as all empty-sphere partial waves [18].

The basic reason for these failures is that the ASA envelopes are chosen to be independent of energy – in order to avoid energy dependence of the structure matrix – because this is what forces us to carry out *explicitly* the integrals involving all partial waves in all spheres throughout space. What should be done is to include all inactive waves, $\varphi_I\left(\varepsilon, r\right)$, in energy-*dependent* MTO-*envelopes*, and *then* to linearize these MTOs to form LMTOs. This has been achieved with the development of the LMTO method of the *3rd-generation* [19,20], and will be dealt with in the present paper. The reason why energy linearization still works in a window of useful width, now that the energy dependence is kept throughout space, is due to the *screening* of the wave-equation solutions used as envelope functions [21].

As an extreme example, it was demonstrated in Fig. 7 of Ref. [20] – and we shall present further results in Fig. 11 below – how with this method one may pick the orbital of *one* band, with a particular local symmetry and energy range, out of a complex of overlapping bands. This goes beyond the construction of a Wannier function and has relevance for the treatment of correlated electrons in narrow bands [22,23]. Another example to be treated in the present paper is the valence and low-lying conduction-band structure of GaAs calculated with the minimal Ga spd As sp basis [24]. Other examples, not treated in this paper, concern the calculation of *chemical indicators,* such as the crystal-orbital-overlap-projected densities of states (COOPs) [25] for describing chemical pair bonding. These indicators were originally developed for the empirical Hückel method where all parameters have been standardized. When one tries to take this over to an *ab initio* method, one immediately gets confronted with the problems of *representation.* For instance, COOPs will vanish in a basis of orthonormal orbitals. Therefore, the COOPs first had to be substituted by COHPs, which are Hamiltonian- rather than overlap projections, but still, the LMTO-ASA method often gave strange results – for the above mentioned reasons [26]. What one has to do is – through downfolding – to chose the *chemically-correct* LMTO Hilbert space and – through screening – choose the chemically correct axes (orbitals) in this space. Only with such orbitals, does it make sense to compute indicators [27,28].

A current criterion for an electronic-structure method to be *accurate* and *robust* is that it can be used in *ab initio* density-functional molecular-dynamics (DF-MD) calculations [29]. According to this criterion, *hardly* any existing LMTO method – and the LMTO-ASA least of all – is accurate and robust.

Most LMTO calculations include non-ASA corrections to the Hamiltonian and overlap matrices, such as the *combined correction* for the neglected integrals over the interstitial region and the neglected high partial waves. This brings in the first energy derivative of the structure matrix, $\dot{S}$, in a way which makes the formalism clumsy [2]. The code [30] for the 2nd-generation LMTO method is useful [31] and quite accurate for calculating energy bands, because it includes downfolding in addition to the combined correction, as well as an automatic way of dividing space into MT-spheres, but the underlying formalism is complicated.

There certainly *are* LMTO methods sufficiently accurate to provide structural energies and forces within density-functional theory [8,9,34–36,7,38–40], but their basis functions are defined with respect to MT-potentials which do not overlap. As a consequence, in order to describe adequately the correspondingly large interstitial region, these LMTO sets must include *extra* degrees of freedom, such as LMTOs centered at interstitial sites and LMTOs with more than one radial quantum number. The latter include LMTOs with tails of different kinetic energies (multiple kappa-sets) and LMTOs for semi-core states. Moreover, these methods usually do not employ short-ranged representations. Finally, since a non-overlapping MT potential is a poor approximation to the self-consistent potential, these methods are forced to include the matrix elements of the *full potential*. Existing full-potential methods are thus set up to provide final, numerical results at relatively low cost, but since they are complicated, they have sofar lacked the robustness needed for DF-MD, and their *formalisms* provide little insight to the physics and chemistry of the problem.

One of the early full-potential MTO methods did fold down extra orbitals and furthermore contained a scheme by which the matrix elements of the full potential could be efficiently approximated by integrals in *overlapping* spheres [38]. The formalism however remained complicated, and the method apparently never took off. A decade later, it was shown [21,20] that the MT-potential, which defines the MTOs – and to which the Hamiltonian (4) refers – *may* in fact have some overlap: If one solves the *exact* KKR equations [41] with phase shifts calculated for MT-wells which overlap, then the resulting wave function is the one for the *superposition* of these MT-wells, plus an error of *2nd order* in the potential-overlap. This proof will be repeated in Eq. (28) of the present paper, and in Figs. 14 and 13 we shall supplement the demonstration in Ref. [20] that this may be exploited to make the kind of extra LMTOs mentioned above super-fluous, provided that the MTO-envelopes have the proper energy dependence, that is, provided that 3rd generation LMTOs are used. Presently we can han-dle MT-potentials with up to ~60% radial overlap $\left(s_R + s_{R'} < 1.6\left|\mathbf{R} - \mathbf{R'}\right|\right)$, and it seems as if such potentials, with the MT-wells centered exclusively on the atoms, are sufficiently realistic that we only need the minimal LMTO set defined therefrom [20,42]. It may even be that such fat MT-potentials, without full-potential corrections to the Hamiltonian matrix, will yield output charge densities which, when used in connection with the Hohenberg-Kohn variational principle for the total energy [16], will yield good structural energies [43]. Hence, we are getting rid of one of the major obstacles to LMTO DF-MD calculations, the empty spheres.

Soon after the development of the TB-LMTO-ASA method, it was realized [44] that the *full charge density* produced with this method – for cases where atomic and interstitial MT-spheres fill space well – is so accurate, that it should suffice for the calculation of total energies, provided that this charge density is used in connection with a variational principle. However, it took ten years before the first successful implementation was published [45]. The problem is as follows: The charge density, $\rho\left(\mathbf{r}\right) = \sum_i^{occ} \left|\Psi_i\left(\mathbf{r}\right)\right|$, is most simply obtained in the form of one-center expansions:

$$\rho\left(\mathbf{r}\right) = \sum_R \sum_{LL'} \int_{occ} \phi_{RL}\left(z, \mathbf{r}_R\right) \operatorname{Im} G_{RL,RL'}\left(z\right) \phi_{RL'}\left(z, \mathbf{r}_R\right)^* \frac{dz}{\pi}, \qquad (17)$$

where $G\left(z\right) \equiv K\left(z\right)^{-1}$, as can be seen from (1) and (3), but these expansions have terribly bad L-convergence in the region between the atoms and cannot even be used to plot the charge-density in that region. That was made possible by the transformation to a short-ranged representation, because one could now use:

$$\rho\left(\mathbf{r}\right) = \sum_{RL} \sum_{R'L'} \chi_{RL}\left(\mathbf{r}_R\right) \left[\int_{occ} \operatorname{Im} G_{RL,R'L'}\left(z\right) \frac{dz}{\pi}\right] \chi_{R'L'}\left(\mathbf{r}_{R'}\right)^*, \qquad (18)$$

where the L-sums only run over active values, and where the double-sum over sites converges fast. Nevertheless, to compute a value of $\chi_{RL}\left(\mathbf{r}\right)$ with $\mathbf{r}$ far away from a site, one must evaluate the LMTO envelope function, which is a superpo-sition of the bare ones, $Y_L\left(\hat{\mathbf{r}}_R\right)/r_R^{l+1}$, and this means that (18) actually contains a 4-double summation over sites. At that time, this appeared to make the eval-uation of $\rho\left(\mathbf{r}\right)$ at a sufficient number of interstitial points too time-consuming for DF-MD, although the full charge density from (18) was used routinely for plotting the charge-density, the electron-localization function [46], a.s.o. In order to evaluate the total energy, the full charge density must also be expressed in a form practical for solving the Poisson equation. If one insists on a real-space method, then fast Fourier transformation is not an option. In Fig. 12 of the present paper, we shall present results of a real-space scheme [47,48] used in connection with 3rd-generation LMTOs for the phase diagram of Si [49]. This scheme is presently not a full-potential, but a full charge-density scheme, and the calculation of inter-atomic forces has still not been implemented.

With 3rd generation LMTOs [19,20], the simple ASA expressions (1)-(18) still hold, provided that $\phi(\varepsilon, \mathbf{r})$ is suitably redefined, and that $K(\varepsilon)$ is substituted by the proper screened KKR matrix whose *structure matrix depends on energy*. The LMTO Hamiltonian and overlap matrices are given in terms of K, and its first three energy derivatives, $\dot{K}$, $\ddot{K}$, and $\dddot{K}$, which are not diagonal. Downfolding, the interstitial region, and potential-overlap to first order are now all included in this simple ASA-like formalism [1]. In due course, we thus hope to be able to perform DF-MD calculations with an electronic Hamiltonian which is little more complicated than (7), (8), or (9).

A final problem with the LMTO basis is that even with the conventional *spd*-basis and space-filling spheres, the LMTO set is insufficient for cases where semi-core states and excited states must be described by *one* minimal basis set, and in *one* energy panel. This problem becomes even more acute in the 3rd-generation method where, due to the proper treatment of the interstitial region, the expansion energy ε_ν must be *global*, that is, ε_ν is now the *unit* matrix times ε_ν, rather than a *diagonal* matrix with elements $\varepsilon_{\nu Rl}\delta_{RR'}\delta_{LL'}$. The same problem was met when attempting to apply the formally elegant relativistic, spin-polarized LMTO method of Ref. [50] to narrow, spin-orbit split f-bands. Finally, as MT-spheres get larger, and as more partial waves are being folded into the MTO envelopes, the energy window inside which the LMTO basis gives accurate results shrinks. This means, that the 3rd-generation LMTO method described in [20] may not be sufficiently *robust*.

The idea emerging from the LMTO construction (1) seems to be: Divide space into local regions inside which Schrödinger's equation separates due to spherical symmetry and which are so small that the energy dependence of the radial functions is weak over the energy range of interest. Then expand this energy dependence in a Taylor series to first order around the energy ε_ν at the center of interest: $\phi(\varepsilon, \mathbf{r}) \approx \phi(\mathbf{r}) + (\varepsilon - \varepsilon_\nu)\dot{\phi}(\mathbf{r})$. Finally, substitute the energy by a Hamiltonian to obtain the energy-*in*dependent LMTO. The question therefore arises (Fig. 1): Can we develop a more general, *polynomial* MTO scheme of degree N, which allows us to use an Nth-order Taylor series or – more generally – allows us to use a mesh of $N + 1$ *discrete* energy points, and thereby obtain good results over a wider energy range, *without increasing the size of the basis set*? Such an NMTO scheme has recently been developed [51] and shown to be very powerful [24]. We shall preview it in the present paper.

Most aspects of the 3rd-generation LMTO method have been dealt with in a set of lecture notes [19] and a recent review [20]. Here, we shall try to avoid repetition but, nevertheless, give a self-contained description of two selected aspects of the new method: the *basic concepts* and the new *polynomial NMTO scheme*, to be presented here for the first time.

We first explain (Sect. 2) what the functions $\phi(\varepsilon, \mathbf{r})$ actually are in the 3rd-generation formalism. This we do using conventional notation in terms of spherical Bessel functions and phase shifts – like in Ref. [21] – and only later, we

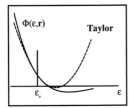 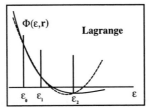

Fig. 1. Quadratic approximation to the energy dependence of a partial wave for a condensed (Taylor) and a discrete (Lagrange) mesh.

renormalize to the notation used in Refs. [19] and [20]. It turns out that the *bare* ϕ's are the energy-*de*pendent MTOs of the 1st generation [52]. The *screened* ϕ's are the screened, energy-dependent MTOs of the 2nd generation [21], with the proviso that $\kappa^2 \equiv \varepsilon$. This proviso – together with truncations of the screening divergencies at the sites, inside the so-called *screening spheres* – is what makes the screened ϕ's equal to the so-called *unitary* [19] or *kinked* [20] *partial waves* in the formalism of the 3rd generation. We then derive the screened KKR equations and repeat the proof from Refs. [21] and [20] that overlapping MT-potentials are treated correctly to leading (1st) order in the potential overlap. Towards the end of this first section, we introduce the so-called *contracted* Green function $\phi(\varepsilon, \mathbf{r}) G(\varepsilon)$, which will play a crucial role in the development of the polynomial NMTO scheme, and we derive the 3rd-generation version of the scaling relation (15) for screening the Green function.

In Sect. 3 we show how to get rid of the energy dependence of the kinked-partial wave set: We first introduce a set of energy-dependent NMTOs, $\chi^{(N)}(\varepsilon, \mathbf{r})$, which – like the $\phi(\varepsilon, \mathbf{r})$ set – spans the solutions of Schrödinger's equation for the chosen MT-potential, and whose contracted Green function, $\chi^{(N)}(\varepsilon, \mathbf{r}) G(\varepsilon)$, differs from $\phi(\varepsilon, \mathbf{r}) G(\varepsilon)$ by a function which is *analytical in energy*. Like in classical polynomial approximations, we choose a mesh of arbitrarily spaced energies, $\varepsilon_0, ..., \varepsilon_N$, and subsequently adjust the analytical function in such a way that, $\chi^{(N)}(\varepsilon_0, \mathbf{r}) = ... = \chi^{(N)}(\varepsilon_N, \mathbf{r})$. The latter then, constitutes the set of energy-*in*dependent NMTOs. The 0th-order set, $\chi^{(0)}(\mathbf{r})$, is seen to be the set of kinked partial waves, $\phi(\varepsilon_0, \mathbf{r})$, at the energy ε_0, and the 1st-order set, $\chi^{(1)}(\mathbf{r})$, to be the set of tangent or chord-LMTOs – depending on whether the mesh is condensed or discrete. For the case of a condensed mesh – which is the simplest – the matrices, which substitute for the energies in the Taylor series (1) – generalized to Nth order – turn out to be:

$$ E^{(M)} - \varepsilon_\nu = \frac{\overset{(M-1)}{G}}{(M-1)!} \left(\frac{\overset{(M)}{G}}{M!} \right)^{-1}, \qquad \text{for } 1 \leq M \leq N, \qquad (19) $$

in terms of the Mth and the $(M-1)$st energy derivatives of the Green matrix. Moreover, the expressions for the Hamiltonian and overlap matrices are:

$$
\left\langle \chi^{(N)} \, |\mathcal{H} - \varepsilon_\nu| \, \chi^{(N)} \right\rangle = -\left(\frac{\overset{(N)}{G}}{N!} \right)^{-1} \frac{\overset{(2N)}{G}}{(2N)!} \left(\frac{\overset{(N)}{G}}{N!} \right)^{-1}, \tag{20}
$$

$$
\left\langle \chi^{(N)} \, | \, \chi^{(N)} \right\rangle = -\left(\frac{\overset{(N)}{G}}{N!} \right)^{-1} \frac{\overset{(2N+1)}{G}}{(2N+1)!} \left(\frac{\overset{(N)}{G}}{N!} \right)^{-1},
$$

which, for $N = 1$, are easily seen to reduce to (7) upon insertion of (12). In retrospect, it is convenient that these basic NMTO results are expressed in terms of energy derivatives of the Green matrix $G(\varepsilon)$ – rather than in terms of those of its inverse, $K(\varepsilon)$, as we are used to from the LMTO-ASA method (12) – because if we imagine generalizing (1) to Nth order and using it to form the Hamiltonian and overlap matrices like in (7), then each matrix will consist of N^2 terms, among which a number of relations can be shown to exist. We also realize, that the problem mentioned above about using Green matrices beyond 2nd order in $z - \varepsilon_\nu$, is solved by using – instead of $G(z)$ – the NMTO Green function:

$$
\left\langle \chi^{(N)} \, |z - \mathcal{H}| \, \chi^{(N)} \right\rangle^{-1} = \frac{\overset{(N)}{G}}{N!} \left[\frac{\overset{(2N)}{G}}{(2N)!} - (z - \varepsilon_\nu) \frac{\overset{(2N+1)}{G}}{(2N+1)} \right]^{-1} \frac{\overset{(N)}{G}}{N!}, \tag{21}
$$

which equals $G(z)$ to $(2N+1)$st order. This Green function has the additional advantage of allowing for a simple treatment of non-MT perturbations. We admit that this route to energy-independent MTO basis sets has little in common with the twisted path we cut the first time, but once found, it is easy to accept and understand the results – which are simple.

In practice, it is cumbersome to differentiate a KKR matrix – not to speak of a Green matrix – many times with respect to energy. Hence, one uses a discrete energy mesh. With that, the derivatives in (19) and the pre- and post factors in (20) and (21) turn out to be *divided differences,* while those at the centers of (20) turn out to be the highest derivative of that approximating polynomial which is fitted not only to the values of $G(\varepsilon)$ at the mesh points, but also to its slopes. Hence, they are related to classical *Hermite interpolation* [53].

In both Sections 2 and 3, special attention is paid to the so-called *triple-valuedness,* because this was not previously explained in any detail, but has turned out to be crucial for the further developments and will be even more so when we come to evaluate the inter-atomic forces. A related aspect is the fact that a *screening transformation* in the formalism of the 3rd-generation is *linear* as regards the envelope functions, but *non-linear* as regards the NMTOs. This means, that changing the screening, changes the NMTO Hilbert space. This was not the case for 2nd-generation LMTOs. This is the reason why we took care to denote the nearly-orthonormal and orthonormal LMTO sets arrived at by the *linear* transformations (8) and (9) by respectively $\hat{\chi}$ and $\check{\chi}$, rather than by χ^γ and $\chi^\perp$, as in the 2nd-generation LMTO scheme, where screening transformations were linear and denoted by superscripts. Screening transformations like (10) and (15) still hold for the 3rd-generation structure- and Green-matrices, but the *partial waves* providing the spatial factors of the Green function (see(17)) are *different*: they have tails extending into the interstitial region. A tail is attached continuously, but with a kink, at the screening sphere, which is concentric with, but smaller than, its own MT-sphere, and the resulting kinked partial wave, or 0th-order energy-dependent MTO, is – for the purpose of evaluating its properties in a simple, approximate way – triple-valued in the shell between these two spheres. The radii, a_{RL}, define the screening and determine the shape of the MTO envelopes. Now, for a superposition of kinked partial waves given by a solution of the KKR equations (16), the kinks and the triple-valuedness cancel, but for a *single* NMTO, a triple-valuedness of order $(r-a)^{2N+1}\,(\varepsilon_i - \varepsilon_0)\,...\,(\varepsilon_i - \varepsilon_N)$ – which is the same as the error caused by the energy interpolation – remains. For this reason: The smaller the screening radii – i.e. the weaker the screening – the smaller the energy window inside which an energy-*in*dependent NMTO set gives good results. The extreme case is the *bare* $(a \to 0)$ $N = 0$ set, which is the set of 1st-generation MTOs [52], but defined *without* freezing the energy dependence outside the central MT-sphere. The tail-cancellation condition for this set leads to the original KKR equations [41], which – we know – must be solved energy-by-energy, that is, the energy window can be very narrow, depending on the application. Specifically, for free electrons the width is zero.

At the end of Sect. 3, we demonstrate the power of the new NMTO methods by applying the differential and discrete LMTO, QMTO, and CMTO variational methods to the valence and conduction-band structure of GaAs using a minimal Ga *spd* As *sp* basis, and to the conduction band of $CaCuO_2$ using only *one* orbital, all others being removed by massive downfolding [24]. We also give simple expressions for the charge density and show the total energy as a function of volume for the various crystalline phases of Si calculated with the full-charge, differential LMTO method [47–49]. Finally, numerical results are presented for the error of the valence-band energy of diamond-structured Si – as a function of the potential overlap – obtained from LMTOs constructed for a potential whose MT-wells are centered exclusively on the atoms. In addition, results of a scheme which corrects for the error of 2nd order in the overlap will be presented [42].

In Sect. 4 we show that energy-dependent, linear transformations of the set of kinked partial waves – such as a normalization – merely leads to similarity transformations among the NMTO basis functions and, hence, does not change the Hilbert space spanned by the NMTO set.

This is exploited in Sect. 5 to generate nearly orthonormal basis sets, $\hat{\chi}^{(N)}\,(\mathbf{r})$, for which the energy matrices defined in (19) become Hermitian, Hamiltonian matrices, $\hat{H}^{(M)}$. We also show how to generate *orthonormal* sets, $\check{\chi}^{(N)}\,(\mathbf{r})$, of general order, and we demonstrate by the example of the minimal MTO set for GaAs that this technique works numerically efficiently – at least up to and including $N = 3$. This development of orthonormal basis sets should be important e.g. for the construction of correlated, multi-orbital Hamiltonians for real materials [23,54].

In the last Sect. 6 we show explicitly how – for $N = 1$ and a condensed mesh – the general, nearly-orthonormal NMTO formalism reduces to the simple ASA formalism of the present Overview.

In the Appendix we have derived those parts of the classical formalism for polynomial approximation – Lagrange, Newton, and Hermite interpolation – needed for the development of the NMTO method for discrete meshes [53].

2   Kinked Partial Waves

In this section we shall define 0th-order energy-dependent MTOs and show that linear combinations can be formed which solve Schrödinger's equation for the MT-potential used to construct the MTOs. The coefficients of these linear combinations are the solutions of the (screened) KKR equations. By renormalization and truncation of the irregular parts of the screened MTOs inside appropriately defined screening spheres, these 0th-order energy-dependent MTOs become the kinked partial waves of the 3rd generation.

If we continue the regular solution $\varphi_{Rl}(\varepsilon, r)$ of the radial Schrödinger equation (2) for the single potential well, $v_R(r)$, smoothly outside that well, it becomes:

$$\varphi_{Rl}(\varepsilon, r) = n_l(\kappa r) - j_l(\kappa r)\cot\eta_{Rl}(\varepsilon) \equiv \varphi_{Rl}^\circ(\varepsilon, r), \quad \text{for } r > s_R, \tag{22}$$

in terms of the spherical Bessel and Neumann functions, $j_l(\kappa r)$ and $n_l(\kappa r)$, which are regular respectively at the origin and at infinity, and a phase shift defined by:

$$\cot\eta(\varepsilon) = \frac{n(\kappa s)}{j(\kappa s)}\frac{\partial\ln|\varphi(\varepsilon, r)|/\partial\ln r|_s - \partial\ln|n(\kappa r)|/\partial\ln r|_s}{\partial\ln|\varphi(\varepsilon, r)|/\partial\ln r|_s - \partial\ln|j(\kappa r)|/\partial\ln r|_s}.$$

In the latter expression, we have dropped the subscripts. Note that we no longer distinguish between 'inside' and 'outside' kinetic energies, $\varepsilon - v(r)$ and $\kappa^2 \equiv \varepsilon - V_{mtz}$, and that we have returned to the common practice of setting $V_{mtz} \equiv 0$. If the energy is negative, $n_l(\kappa r)$ denotes a spherical, exponentially decreasing Hankel function. Note also that – unlike in the ASA – the radial function is not truncated outside its MT-sphere, and is not normalized to unity inside. In fact, we shall meet three different normalizations throughout the bulk of this paper, and (22) is the first.

Si p unscreened kpw

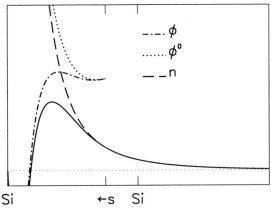

Fig. 2. Bare Si p MTO according to Eq.(23)

2.1   Bare MTOs

The bare, energy-dependent muffin-tin orbital (MTO) remains the one of the 1st generation [52]:

$$\phi_{RL}(\varepsilon, \mathbf{r}) \equiv Y_L(\hat{\mathbf{r}}) \left[\varphi_{Rl}(\varepsilon, r) + j_l(\kappa r) \cot \eta_{Rl}(\varepsilon) \right]$$

$$= Y_L(\hat{\mathbf{r}}) \begin{cases} \varphi_{Rl}(\varepsilon, r) + j_l(\kappa r) \cot \eta_{Rl}(\varepsilon) & \text{for} \quad r \leq s_R \\ n_l(\kappa r) & \text{for} \quad r > s_R \end{cases}$$

$$= Y_L(\hat{\mathbf{r}}) \left[\varphi_{Rl}(\varepsilon, r) - \varphi^{\circ}_{Rl}(\varepsilon, r) + n_l(\kappa r) \right], \tag{23}$$

and is seen to have pure angular momentum and to be regular in all space. The reason for denoting this 0th-order MTO $\phi(\varepsilon, \mathbf{r})$, rather than $\chi^{(N=0)}(\varepsilon, \mathbf{r})$, should become clear later.

In Fig. 2 we show the radial part of this MTO for a Si p-orbital, a MT-sphere which is so large that it reaches 3/4 the distance to the next site in the diamond lattice, and an energy in the valence-band, which – in this case of a large MT-sphere – is slightly negative (see Fig. 11 in Ref. [20]). The full line shows the MTO as defined in (23), while the various broken lines show it 'the 3-fold way': The radial Schrödinger equation for the potential $v(r)$ is integrated outwards, from the origin to the MT radius, s, yielding the regular solution, $\varphi(\varepsilon, r)$, shown by the dot-dashed curve. At s, the integration is continued with reversed direction and with the potential substituted by the flat potential, whose value is defined as the zero of energy. This inwards integration results in the radial function 'seen from the outside of the atom', $\varphi^{\circ}(\varepsilon, r)$, shown by the dotted curve. The inwards integration is continued to the origin, where $\varphi^{\circ}(\varepsilon, r)$ joins the 'outgoing' solution for the flat potential, that is the one which is regular at infinity: $n(\kappa r)$. The latter is the envelope function for the bare MTO.

As usual, the envelope-function for the MTO centered at $\mathbf{R}$ may be expanded in spherical-harmonics about another site $\mathbf{R}'$ ($\neq \mathbf{R}$):

$$\kappa n_l \left(\kappa r_R\right) Y_L \left(\hat{\mathbf{r}}_R\right) = \sum_{L'} j_{l'} \left(\kappa r_{R'}\right) Y_{L'} \left(\hat{\mathbf{r}}_{R'}\right) B_{R'L',RL} \left(\varepsilon\right),$$

where the expansion coefficients form the Hermitian KKR structure matrix:

$$B_{R'L',RL} \left(\varepsilon\right) \equiv \sum_{l''} 4\pi \, i^{-l+l'-l''} C_{LL'l''} \, \kappa n_{l''} \left(\kappa \left|\mathbf{R} - \mathbf{R}'\right|\right) Y_{l'',m''}^* \left(\widehat{\mathbf{R} - \mathbf{R}'}\right) \quad (24)$$

as conventionally [41] defined, albeit in R-space. The spherical harmonics are as defined by Condon and Shortley, $m'' \equiv m' - m$, the summation runs over $l'' = |l' - l|, \; |l' - l| + 2, \; ..., \; l' + l$, and $i^{-l+l'-l''}$ is real, because $C_{LL'L''} \equiv \int Y_L(\hat{r}) Y_{L'}^*(\hat{r}) Y_{L''}(\hat{r}) d\hat{r}$.

If for the on-site elements of $B\left(\varepsilon\right)$, we define: $B_{RL,RL'} \left(\varepsilon\right) \equiv 0$, and use the notation: $f_L \left(\varepsilon, \mathbf{r}_R\right) \equiv f_l \left(\kappa r_R\right) Y_L \left(\hat{\mathbf{r}}_R\right)$, as well as the vector-matrix notation introduced in connection with (7), we may express the spherical-harmonics expansion of the bare envelope about any site symbolically as:

$$\kappa n \left(\varepsilon, \mathbf{r}\right) = j \left(\varepsilon, \mathbf{r}\right) B \left(\varepsilon\right) + \kappa n \left(\varepsilon, \mathbf{r}\right). \quad (25)$$

If we now form a linear combination, $\sum_{RL} \phi_{RL} \left(\varepsilon, \mathbf{r}_R\right) c_{RL}$, of energy-dependent MTOs (23), and require that it be a solution of Schrödinger's equation, then the condition is that, inside any MT-sphere (R') and for any angular momentum (L'), the contributions from the tails should cancel the $j_{l'} \left(\kappa r\right) \cot \eta_{R'l'} \left(\varepsilon\right)$-term from their own MTO, $\phi_{R'l'} \left(\varepsilon, \mathbf{r}_{R'}\right)$, thus leaving behind the term $\varphi_{R'l'} \left(\varepsilon, r\right)$, which is a solution by construction. This gives rise to the original KKR equations [41]:

$$\sum_{RL} \left[B_{R'L',RL} \left(\varepsilon_i\right) + \kappa \cot \eta_{Rl} \left(\varepsilon_i\right) \delta_{R'R} \delta_{L'L}\right] c_{RL,i}$$
$$\equiv \sum_{RL} K_{R'L',RL} \left(\varepsilon_i\right) c_{RL,i} = 0, \quad (26)$$

which have non-zero solutions, $c_{RL,i}$, for those energies, ε_i, where the determinant of the KKR matrix vanishes.

With those equations satisfied, the wave function is

$$\sum_{RL} \phi_{RL} \left(\varepsilon_i, \mathbf{r}_R\right) c_{RL,i} = \sum_{l'=0}^{\infty} \sum_{m'=-l'}^{l'} \varphi_{R'l'} \left(\varepsilon_i, r_{R'}\right) Y_{L'} \left(\hat{\mathbf{r}}_{R'}\right) c_{R'L',i} + \quad (27)$$
$$\sum_{R \neq R'} \sum_{L} \left[\varphi_{Rl} \left(\varepsilon_i, r_R\right) - \varphi_{Rl}^\circ \left(\varepsilon_i, r_R\right)\right] Y_L \left(\hat{\mathbf{r}}_R\right) c_{RL,i}$$

near site R'. Since according to (22) the function $\varphi - \varphi^\circ$ vanishes outside its own MT-sphere, the terms in the second line vanish for a non-overlapping MT-potential so that, in this case, (27) solves Schrödinger's equation exactly. If the potential from a neighboring site (R) *overlaps* the central site (R'), then $\varphi_{RL} - \varphi_{RL}^\circ$ *tongues* stick into the MT-sphere at R'. The radial part of such a tongue is $\frac{1}{2}(s_R - r_R)^2 v_R(s_R) \varphi_{RL}(s_R)$, to lowest order in $s_R - r_R$, as may be seen from the radial Schrödinger equation (2). Let us now operate on the smooth function $\Psi_i(\mathbf{r}) \equiv \sum_{RL} \phi_{RL}(\varepsilon_i, \mathbf{r}_R) c_{RL,i}$, of which (27) is the expansion around site R', with $\mathcal{H} - \varepsilon_i$ as given by (4) to find the error:

$$(\mathcal{H} - \varepsilon_i)\Psi_i(\mathbf{r}) =$$

$$\sum_{R'} v_{R'}(r_{R'}) \sum_{R \neq R'} \sum_L [\varphi_{Rl}(\varepsilon_i, r_R) - \varphi_{Rl}^\circ(\varepsilon_i, r_R)] Y_L(\hat{\mathbf{r}}_R) c_{RL,i} \qquad (28)$$

$$\sim \frac{1}{2} \sum_{RR'}^{pairs} v_{R'}(s_{R'}) \left[(s_{R'} - r_{R'})^2 + (s_R - r_R)^2\right] v_R(s_R) \Psi_i(\mathbf{r}).$$

This shows that the wave function (27) solves Schrödinger's equation for the *superposition* of MT-wells to within an error, which is of *second* order in the potential overlap [21,20].

2.2   Screened MTOs

Screening is the characteristic of 2nd-generation MTOs and was first discovered as the transformation (8) to a *nearly-orthonormal* representation, in which the Hamiltonian is of *second* order [55,56]. Shortly thereafter it was realized that there exists a whole set of screening transformations which may be used to make the orbitals *short ranged*, so that the structure matrix may be generated in real space. It was also realized that the screening transformation could be used to downfold inactive channels and, hence, to produce *minimal* basis sets [1,18,44]. These applications were all for the ASA with $\kappa^2{=}0$. Only long time after [21], did it become clear that screening would work for *positive* energies as well, and at that time a fourth virtue of screening became clear, namely, that sceening the range of the orbitals, simultaneously reduces their energy dependence *to the extent* that the full energy dependence may be kept in the interstitial region, thus making the $\kappa^2{=}0$-part of the ASA superfluous. Most of this was shown in the last paper on the 2nd-generation formalism [21]. Nevertheless, this paper was unable to devise a generally useful recipe for choosing the energy-dependent screening constants, it failed to realize that screening allows the return to: $\kappa^2{=}\varepsilon$, and for those reasons it missed the elegant energy-linearization of the MTOs achieved by the 3rd generation.

The *screened envelopes* of the 2nd-generation method are linear superpositions,

$$n^\alpha(\varepsilon, \mathbf{r}) \equiv n(\varepsilon, \mathbf{r}) S^\alpha(\varepsilon), \qquad (29)$$

of the envelope functions, $n(\varepsilon, \mathbf{r})$, with the property that the spherical-harmonics expansions of the set of screened envelopes be:

$$\kappa n(\varepsilon, \mathbf{r}) S^\alpha(\varepsilon) \equiv \kappa n^\alpha(\varepsilon, \mathbf{r}) = j^\alpha(\varepsilon, \mathbf{r}) B^\alpha(\varepsilon) + \kappa n(\varepsilon, \mathbf{r}), \qquad (30)$$

which are (25) with the substitutions:

$$j_l(\kappa r) \quad \rightarrow \quad j^\alpha_{Rlm}(\varepsilon, r) \equiv j_l(\kappa r) - n_l(\kappa r)\tan\alpha_{Rlm}(\varepsilon), \tag{31}$$

and: $B(\varepsilon) \rightarrow B^\alpha(\varepsilon)$, which will be determined below. In contrast to its bare counterpart, a screened envelope does *not* have pure angular momentum, i.e., cannot be factorized as a radial function times a spherical harmonics, and it depends *explicitly* on its surroundings. The *background phase shifts* $\alpha(\varepsilon)$ – which may even depend on m (see for instance Fig. 11) – specify the *shapes* of the screened *envelopes*. Whereas the bare envelopes are regular in all space – except at their own site where they diverge like $Y_{lm}(\hat{\mathbf{r}})/r^{l+1}$ – the screened envelopes *diverge* at any site where there is a finite background phase shift in at least one L-channel.

Note that only in the Overview did we use ASA $\kappa^2=0$-notation with Greek letters denoting screening constants and S^α the structure matrix. In the bulk of the present paper, we use Greek letters to denote background phase shifts, and B^α and S^α to denote respectively the structure matrix and the screening transformation.

We now find the screened structure matrix and the transformation matrix by expanding also the bare envelope on the left hand side of (30) by means of (25). Comparisons of the coefficients to $\kappa n_{L'}(\varepsilon, \mathbf{r}_{R'})$ and $j_{L'}(\varepsilon, \mathbf{r}_{R'})$ yield respectively:

$$S^\alpha(\varepsilon) = 1 - \frac{\tan\alpha(\varepsilon)}{\kappa}B^\alpha(\varepsilon), \quad \text{and}: \quad B^\alpha(\varepsilon) = B(\varepsilon)S^\alpha(\varepsilon) \tag{32}$$

with the quantities regarded as matrices, e.g. $\kappa^{-1}\tan\alpha$ is considered a diagonal matrix with elements $\kappa^{-1}\tan\alpha_{RL}\,\delta_{RR'}\delta_{LL'}$. As a result of (32):

$$B^\alpha(\varepsilon)^{-1} = B(\varepsilon)^{-1} + \frac{\tan\alpha(\varepsilon)}{\kappa}, \tag{33}$$

which shows that, like the bare structure matrix, also the screened one is Hermitian. In contrast to the bare structure matrix, the screened one has non-vanishing on-site elements. For background phase shifts known to give a short-ranged $B^\alpha(\varepsilon)$, the inversion of the matrix $B(\varepsilon) + \kappa\cot\alpha(\varepsilon)$, implied by (33), may be performed in real space, although the *bare* structure matrix is long-ranged. Eq. (33) is the $\kappa^2=\varepsilon$ equivalent of the ASA 'Dyson equation' (10).

For the *inactive* channels $(RL \equiv I)$, we choose the background phase shifts to be equal to the *real* phase shifts:

$$\alpha_I(\varepsilon) \equiv \eta_I(\varepsilon) \tag{34}$$

so that for these channels,

$$j^\alpha_I(\varepsilon, r) \;=\; j_I(\kappa r) - n_I(\kappa r)\tan\eta_I(\varepsilon) \;=\; -\varphi^\circ_I(\varepsilon, r)\tan\eta_I(\varepsilon).$$

That is, we shape the set of screened envelope functions in such a way that, for the inactive channels, the radial functions, $\varphi^\circ_I(\varepsilon, r)$, may be *substituted smoothly* by the regular solutions, $\varphi_I(\varepsilon, r)$, of the radial *Schrödinger* equation. This is what we call *downfolding*. This substitution makes the screened envelopes become the so-called *screened spherical waves*, ψ, of the 3rd-generation method. Only the screened spherical waves corresponding to the remaining, so-called *active* channels ($RL = A$) will be used to construct the MTO; they are:

$$\psi_{RL}^\alpha(\varepsilon, \mathbf{r}_R) \equiv n_{RL}^\alpha(\varepsilon, \mathbf{r}_R) + \tag{35}$$

$$\sum_I [\varphi_I^\circ(\varepsilon, r_{R'}) - \varphi_I(\varepsilon, r_{R'})] \frac{\tan \eta_I(\varepsilon)}{\kappa} Y_I(\hat{\mathbf{r}}_{R'}) B_{I,RL}^\alpha(\varepsilon),$$

which – in contrast to $n_{RL}^\alpha(\varepsilon, \mathbf{r}_R)$ – are *regular* in all inactive channels, albeit *irregular* in the active channels. In (35), $I \equiv R'L'$. Below, we shall choose to truncate the active channels inside their screening spheres. Due to the augmentation (substitution), the screened spherical waves do *not* transform linearly like (29).

For the partial waves of high l, the phase shifts vanish due to the dominance of the centrifugal term over the potential term in the radial Schrödinger equation (2). As a consequence, the matrices involved in the Dyson equation (33) – whose indices run over all active as well as inactive channels – truncate above a certain l of about $3 - 4$.

Before specifying our choice of background phase shifts for the *active* channels, let us define the energy-dependent, *screened* MTO analogous to the third equation (23) as the (augmented) envelope function, plus a term proportional to the function $\varphi - \varphi^\circ$, which vanishes (quadratically) outside the central MT-sphere and has pure angular-momentum character. That is:

$$\phi_{RL}^\alpha(\varepsilon, \mathbf{r}_R) \equiv Y_L(\hat{\mathbf{r}}_R) [\varphi_{Rl}(\varepsilon, r_R) - \varphi_{Rl}^\circ(\varepsilon, r_R)] \frac{\tan \eta_{Rl}(\varepsilon)}{\tan \eta_{RL}^\alpha(\varepsilon)} + \psi_{RL}^\alpha(\varepsilon, \mathbf{r}_R)$$

$$\equiv Y_L(\hat{\mathbf{r}}_R) [\varphi_{Rl}^\alpha(\varepsilon, r_R) - \varphi_{Rl}^{\circ\,\alpha}(\varepsilon, r_R)] + \psi_{RL}^\alpha(\varepsilon, \mathbf{r}_R) \tag{36}$$

and $RL \in A$. Here, the coefficient to $\varphi - \varphi^\circ$ has been chosen in such a way that, in its own channel and outside any other MT-sphere, the screened MTO is $\varphi^\alpha + j^\alpha \cot \eta^\alpha$ plus a term from the diagonal element of the screened structure matrix.

To check this, we project onto the 'eigen-channel,' making use of (35), (30), (22), and (31), and neglecting any contribution from $\varphi_I(\varepsilon, r_{R'})$'s from overlapping neighboring MT-spheres:

$$\mathcal{P}_{RL} \phi_{RL}^\alpha(\varepsilon, \mathbf{r}_R) = \varphi_{Rl}^\alpha(\varepsilon, r_R) - \varphi_{Rl}^{\circ\,\alpha}(\varepsilon, r_R) + \mathcal{P}_{RL} \psi_{RL}^\alpha(\varepsilon, \mathbf{r}_R)$$

$$= [\varphi - n + (j^\alpha + n \tan \alpha) \cot \eta] \frac{\tan \eta}{\tan \eta^\alpha} + n + j^\alpha \frac{B^\alpha}{\kappa}$$

$$= \varphi^\alpha + j^\alpha \cot \eta^\alpha - n \frac{\tan \eta - \tan \alpha}{\tan \eta^\alpha} + n + j^\alpha \frac{B^\alpha}{\kappa}$$

$$= \varphi^\alpha + j^\alpha \cot \eta^\alpha + j^\alpha \frac{B^\alpha}{\kappa} \tag{37}$$

For simplicity, we have dropped all arguments and indices in the last three lines. We see that the new phase shift, η^α, is given by:

$$\tan \eta^\alpha_{RL} (\varepsilon) \equiv \tan \eta_{Rl} (\varepsilon) - \tan \alpha_{RL} (\varepsilon), \tag{38}$$

as expected for the phase shift on the background of α. This is the same transformation as the one obtained from (33) for $-B^\alpha (\varepsilon)^{-1}$. The definition of the renormalized free radial solution given in (36) may be written as:

$$\varphi^{\circ\,\alpha}_{RL} (\varepsilon, r) \equiv n_l (\kappa r) - j^\alpha_{RL} (\varepsilon, r) \cot \eta^\alpha_{RL} (\varepsilon) \tag{39}$$
$$= [n_l (\kappa r) \tan \eta_{Rl} (\varepsilon) - j_l (\kappa r)] \cot \eta^\alpha_{RL} (\varepsilon),$$

and $\varphi^\alpha_{Rl} (\varepsilon, r_R)$ is the solution of the radial Schrödinger equation, normalized in such a way that it matches onto $\varphi^{\circ\,\alpha}_{RL} (\varepsilon, r)$ at the MT radius, s_R. The definition (39) reduces to (22) when $\alpha = 0$.

The *set* of screened MTOs now consists of the screened MTOs (36) of all active channels. Since the $\varphi - \varphi^\circ$ function has pure angular-momentum character, the mixed character of the screened MTO stems solely from the ψ-function. The result of projecting the screened MTO onto an active channel $R'L'$ different from its own is seen from (30) to be:

$$\mathcal{P}_{R'L'} \phi^\alpha_{RL} (\varepsilon, \mathbf{r}_R) = \mathcal{P}_{R'L'} \psi^\alpha_{RL} (\varepsilon, \mathbf{r}_R) = j^\alpha_{R'L'} (\varepsilon, r_{R'}) \frac{B^\alpha_{R'L',RL} (\varepsilon)}{\kappa}, \tag{40}$$

when $r_{R'}$ is so small that $\mathbf{r}$ lies inside only *one* MT-sphere, the one centered at R'. From (40) and (37) it is then obvious that, in order to get a *smooth* linear combination $\sum_A \phi^\alpha_A (\varepsilon, \mathbf{r}_A) c^\alpha_A$ of screened MTOs, *all j^α-functions must cancel*. This leads to the condition that the energy must be such that the coefficients can satisfy

$$\sum_A [B^\alpha_{A'A} (\varepsilon_i) + \kappa \cot \eta^\alpha_A (\varepsilon_i) \delta_{A'A}] c^\alpha_{A,i} \equiv \sum_A K^\alpha_{A'A} (\varepsilon_i) c^\alpha_{A,i} = 0, \tag{41}$$

for all active $R'L' \equiv A'$. These are the *screened KKR equations*, and $K^\alpha (\varepsilon)$ is the screened KKR matrix. If these equations are satisfied, the linear combination of screened MTOs is:

$$\sum_A \phi^\alpha_A (\varepsilon_i, \mathbf{r}_R) c^\alpha_{A,i} = \sum_{l'=0}^{\infty} \sum_{m'} \varphi^\alpha_{R'L'} (\varepsilon_i, r_{R'}) Y_{L'} (\hat{\mathbf{r}}_{R'}) c^\alpha_{R'L',i} + \tag{42}$$
$$\sum_{R \neq R'} \sum_L [\varphi^\alpha_{RL} (\varepsilon_i, r_R) - \varphi^{\circ\,\alpha}_{RL} (\varepsilon_i, r_R)] Y_L (\hat{\mathbf{r}}_R) c^\alpha_{RL,i}$$

near site R'. As long as the MT-spheres do not overlap, this is a solution of Schrödinger's equation for the MT-potential and, if the potentials overlap, then the $\varphi - \varphi^\circ$ tongues from the neighboring sites in the second line of (42) make the wave function correct to first order in the overlap [20]. This is exactly as in (27). The summation over spherical-harmonics around the central site includes the contributions $-\varphi_I (\varepsilon, \mathbf{r}_{R'}) \kappa^{-1} \tan \eta_I (\varepsilon) \sum_A B^\alpha_{I,A} (\varepsilon) c^\alpha_{A,i}$ provided by the screened-spherical-wave part of the MTO (see (36) and (35)).

Although energy-dependent MTO sets with different screenings are not linearly related, they all solve Schrödinger's equation for the MT-potential used for their construction via the corresponding KKR equation. E.g. had one chosen a representation in which a channel making a significant contribution to a wave function $\Psi_i(\mathbf{r})$ with energy $\varepsilon_i = \varepsilon$ were downfolded, then the corresponding solution of the KKR equation (41) would arise from $B^\alpha(\varepsilon)$ being long ranged and, as a function of ε, going through a zero-pole pair near ε_i. If the energy were now *fixed* at some energy ε_ν, and the energy-independent set $\phi^\alpha(\varepsilon_\nu, \mathbf{r})$ were used as the 0th-order MTO basis in a variational calculation, then a useful result could in principle be obtained, but only if ε_ν were chosen very close to ε_i.

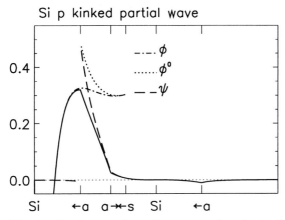

Fig. 3. Si p_{111} member of a screened *spd*-set of 0th-order MTOs (see text and Eqs.(36),(44)-(47)).

2.3   Hard-Sphere Interpretation and Redefinitions

We now wish to choose the background phase shifts for the active channels in a way which reduces the spatial range and the energy dependence of the MTO envelopes. It is obvious, that for the orbitals to be localized, they must have energies *below the bottom of the continuum of the background* – defined as the system which has the same structure as the real system, but has all phase shifts equal to those of the background. Hence, the active $\alpha(\varepsilon)$'s should be defined in such a way that the energy band defined by: $\left| B^0(\varepsilon) + \kappa \cot \alpha(\varepsilon) \right| = 0$, lie as high as possible.

The discovery of a useful way of determining this background, turned out to be the unplanned birth of the 3rd MTO generation [19,20]. Realizing that the weakest point of the ASA was its solution of Poisson's – and not Schrödinger's – equation, and unhappy with the complexities of existing full-potential schemes, we [57] were looking for those linear combinations of Hankel functions – like (29) – which would fit the charge density continuously at spheres. With Methfessel's formulation [35]: What we wanted was those solutions of the wave equation which are $Y_L(\hat{\mathbf{r}}_R)$ at their own sphere and for their own angular momentum, and zero at all other spheres and for all other angular momenta. This set was therefore named *unitary* spherical waves. The solution to this boundary-value problem is of course a particular screening transformation (33).

Our way of defining the background was thus in terms of hard *screening-spheres* for the active channels; the larger the screening spheres, the larger the excluded volume and the higher the bottom of the continuum. The screening spheres are not allowed to overlap – at least not if all l-channels were active, because then a unitary spherical wave would be asked to take both values, 1 *and* 0, on the circle common to the central and an overlapping sphere. As a consequence, in order to reduce the range and the energy dependence of the MTO envelope functions, the screening spheres should in general be *nearly touching*. Now, since the screening radii, , control the shapes of the envelopes, the *relative* sizes of the screening spheres should be determined by *chemical* considerations, i.e. *the a's may be covalent- or ionic radii* in order that results obtained from an electronic-structure calculation be interpretable in terms of covalency, ionicity etc. Referring to the discussion in the Overview, one could say: The MT-spheres (s) are potential-spheres and the screening-spheres (a) are charge-spheres.

Inspired by Ref. [21], practitioners of multiple-scattering theory – who traditionally take the Kohn-Rostoker [41] Green-function point of view – found another useful way of determining the background phase shifts, namely in terms of *repulsive potentials* [58].

For a given active channel ($RL = A$), the radial positions, $r = a_A(\varepsilon)$, of the *nodes* of the background functions j^α given by (31) are the solutions of the equation:

$$0 = j_A^\alpha(\varepsilon, a_A(\varepsilon)) = j_l(\kappa a_A(\varepsilon)) - n_l(\kappa a_A(\varepsilon)) \tan \alpha_A(\varepsilon).$$

Whereas *attractive* potentials usually do not give positive radii – for an example, see the dotted curve in Fig. 2 – repulsive potentials do, as may be seen from the radial Schrödinger equation (2). For a *hard* repulsive potential, the position of the node is *independent* of energy and of l. What we shall use for the active channels are therefore screening-sphere radii, a_A, which are independent of energy and which usually depend little on L among the active channels. In terms of such a screening radius, the corresponding background phase shift is given by:

$$\tan \alpha_A(\varepsilon) = j_l(\kappa a_A) / n_l(\kappa a_A). \tag{43}$$

Now, instead of having screened spherical waves (35) and MTOs (36) whose active channels are irregular at the origin – the irregularities of the *in*active channels were already gotten rid of by downfolding, followed by $\varphi_I^\circ(\varepsilon, r) \rightarrow \varphi_I(\varepsilon, r)$ substitutions – we prefer that the active channels have merely *kinks*. This is achieved by *truncating* all *active* j^α-functions *inside* their *screening spheres*, that is, we perform the substitution:

$$j_A^\alpha(\varepsilon, r) \rightarrow \begin{cases} 0 & \text{for } r < a_A, \\ j_l(\kappa r) - n_l(\kappa r) j_l(\kappa a_A) / n_l(\kappa a_A) & \text{for } r \geq a_A \end{cases}, \tag{44}$$

which is continuous but not differentiable, for the screened spherical waves and for its own j^α-function of the MTO – that is the second term on the last two lines of (37). With that substitution, a screened spherical wave, $\psi_{RL}^\alpha(\varepsilon, \mathbf{r}_R)$, vanishes inside all screening spheres of the active channels – except inside its own, where it equals $n_l(\kappa r_R) Y_L(\hat{\mathbf{r}}_R)$. This may be seen from (40) and the two first lines of (37). Finally, if we renormalize according to:

$$\psi_{RL}^a(\varepsilon, \mathbf{r}_R) \equiv \psi_{RL}^\alpha(\varepsilon, \mathbf{r}_R) / n_l(\kappa a_{RL}) \qquad (45)$$

– note the difference between the superscripts a and α – we finally arrive at the screened (unitary) spherical wave as defined in Refs. [19,20].

$\psi_{RL}^a(\varepsilon, \mathbf{r}_R)$ is that solution of the wave equation which is $Y_L(\hat{\mathbf{r}}_R)$ on its own screening sphere, has vanishing $Y_{L'}(\hat{\mathbf{r}}_{R'})$-average on the screening spheres of the other active channels, and joins smoothly onto the regular solutions of the radial Schrödinger equations of the inactive channels. In those, the regular Schrödinger solutions are, in fact, substituted for the wave-equation solutions.

It is now obvious, that overlap of screening spheres will cause complicated, and hence long-ranged spatial behavior of the screened spherical waves, and the worse, the more spherical harmonics are active.

With the normalization (45), there is apparently no need for functions, like spherical Bessel and Neumann or Hankel functions, which have a branch-cut at zero energy, and this was the point of view taken in the first accounts [19,20] of the 3rd-generation method. However, the normalization (45) is not appropriate for $a=0$, and expressing the screened structure matrix in terms of the bare one (24) – which is the only one computable in terms of elementary functions – was slightly painful in Ref. [19]; moreover, in that paper downfolding was not presented in its full generality. In these respects, the present, conventional derivation is simpler, but it takes more equations.

With the $\alpha \rightarrow a$ redefinitions (44)-(45), the MTO remains as defined by (36), but with the screened spherical waves and its own j^α-function truncated as described above. We may also renormalize the MTO like in (45):

$$\phi_{RL}^a(\varepsilon, \mathbf{r}_R) \equiv \phi_{RL}^\alpha(\varepsilon, \mathbf{r}_R) / n_l(\kappa a_{RL}), \qquad (46)$$

whereby these energy-dependent 0th-order MTOs become identical with the *kinked partial waves* of Refs. [19,20]. This normalization corresponds to:

$$\varphi_{Rl}^{\circ\, a}(\varepsilon, a_{RL}) \equiv 1. \qquad (47)$$

Note that this will cause the normalization of the radial Schrödinger-equation solution, $\varphi^a(\varepsilon, r)$, to depend on m in case the corresponding screening radius is chosen to do so.

In Fig. 3 we show the screened counterpart of the bare Si p orbital in Fig. 2. Since only the two first terms of (36) – but not the screened spherical wave – has pure angular momentum, we cannot plot just the radial wave function like in Fig. 2. Rather, we show the MTO together with its three parts along the [111]-line between the central atom and one of its four nearest neighbors in the diamond structure. The positions of the central and the nearest-neighbor atoms are indicated on the axis (Si), and so is the intersection with the *central* MT-sphere (s). The p orbital chosen is the one pointing along this [111] direction. The Si *spd* channels were taken as active, and to have one and the same screening radius, $a = 0.75t$, where t is half the nearest-neighbor distance, i.e., the touching-sphere radius. The places where the central and the nearest-neighbor screening spheres intersect the [111]-line are indicated by '$\leftarrow a$' and '$a \rightarrow$' with the arrow pointing towards the respective center. We see that the central MT-sphere is so large, that it overlaps the screening sphere of the neighboring atom. Like in Fig. 2, the full curve shows the MTO (ϕ^a), and the dot-dashed ($\varphi^a Y$), the dotted ($\varphi^{\circ\, a} Y$), and the dashed (ψ^a) curves show the three terms in the renormalized version of equation (36). The dot-dashed and the dotted curves are identical with those in Fig. 2, except for the normalization; they are the outwards-integrated solution ($\varphi^a Y$) of the radial Schrödinger equation, continued by the inwards-integrated solution ($\varphi^{\circ\, a} Y$) for the flat potential. These two curves have been deleted outside the central MT-sphere where their contribution to the MTO (36) cancels. The inwards integration ends at the screening sphere, inside which $\varphi^{\circ\, a}$ – with j^a truncated – cancels its own-part, $n_l(\kappa r)/n_l(\kappa a)$, of the screened spherical wave, ψ, shown by the dashed curve (see Eqs. (37) and (44)). Neither of these cancelling parts are shown in the figure, and the dashed curve inside the central screening sphere therefore merely shows the contribution to the screened spherical wave from the inactive channels ($l \geq 3$). Due to the j^a-truncations, the screened spherical wave has kinks at *all* screening spheres and, inside these spheres, only the contribution from the inactive partial waves – which are regular solutions of the radial Schrödinger equations – remain. The full curve is the MTO, which is identical with the screened spherical wave outside its own MT-sphere. At its own screening sphere, its kink differs from that of the screened spherical wave due to the truncation of the j^a-contribution to $\varphi^{\circ\, a}$. Compared with the bare MTO in Fig. 2, the screened MTO in Fig. 3 is considerably more localized, even though a negative energy was chosen.

If one demands that the valence band – as well as the lower part of the conduction band – of Si be described from first principles using merely the minimal 4 orbitals per atom, one cannot use a set with p orbitals such as those shown in Figs. 2 and 3; the d-MTOs must be folded into the envelopes of the remaining *sp* set by use of the appropriate structure matrix obtained from Eq. (33) with the choice (34) for the Si d-channels. The corresponding Si p_{111}-MTO is shown in Fig. 4. Little is changed inside the central screening sphere, but the tail extending into the nearest-neighbor atom has attained a lot of d-character around that site, and the MTO is correspondingly more delocalized.

The Si p_{111}-MTO for use in an *sp* MTO basis constructed from the conventional Si+E potential – for which the diamond structure is packed bcc with equally large space-filling spheres – is obtained by down-folding of the Si d and all empty-sphere channels. It turns out to be so similar to the one obtained from the fat Si-centered potential shown in Fig. 4, that we will not take the space to show it.

Whereas the bare MTO in Fig. 2 is what has always been called a bare MTO, the screened ones in Figs. 3 and 4 look more like a partial wave, φY, with a tail attached at its own screening sphere – and with kinks at all screening spheres. Hence the name 'kinked partial wave' given in Ref. [19]. In this original derivation, kinked partial waves with $a = s \leq t$ were considered first, and only later, the limiting case $a \to 0$ gave rise to a painful exercise. The kinked partial waves have in common with Slater's original Augmented Plane Waves (APWs) [59], that they are partial waves, $\varphi(\varepsilon, r) Y$, of the proper energy inside non-overlapping spheres, which are joined continuously – but with kinks – to wave-equation solutions in the interstitial. In that region, the APW is a wave-equation solution with a given *wave-vector*, whereas the MTO is a solution with the same *energy*. Moreover, whereas the APW method uses identical potential and augmentation spheres, this is not the case for MTOs.

If – for the third time in this section – we make a linear combination of MTOs – this time defined with kinks – and demand that it solves Schrödinger's equation, then the condition is, that the kinks – rather than the j^α-functions – from the tails should cancel the ones in the head. This condition is of course equivalent with the one for j^α-cancellation. Nevertheless, let us express the KKR equations in this language because it will turn out to have three further advantages: The artificial dependence on $\kappa \equiv \sqrt{\varepsilon}$ and the associated change between Neumann and decaying Hankel functions will disappear, there will be a simple expression for the integral of the product of two MTOs, and we will be led to a contracted Green function of great use in the following section.

Since the kinks arise because the j^α-functions are truncated inside their screening spheres, the kink in a certain active channel of an MTO is proportional to the slope of the corresponding j^α-function at a_+. An expression for this slope is most easily found from the Wronskian, which in general is defined as: $r^2 [f(r) g'(r) - g(r) f'(r)] \equiv \{f, g\}_r$, and is independent of r when the two functions considered are solutions of the same linear, second-order differential

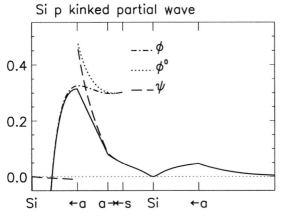

Fig. 4. Si p_{111} member of a screened minimal sp-set of 0th-order MTOs (see text).

equation. As a consequence, $\{n, j^\alpha\} = \{n, j - n \tan \alpha\} = \{n, j\} = -\kappa^{-1}$, and therefore:

$$\partial j^\alpha \left(\varepsilon, r\right) / \partial r |_{a_+} = - \left[a^2 \kappa n \left(\kappa a\right)\right]^{-1}. \tag{48}$$

We now define the elements $K^a_{R'L',RL}\left(\varepsilon\right)$ – where $R'L'$ and RL both refer to active channels – of a *kink matrix* [19,20] as $a^2_{R'L'}$ times the kink in the $R'L'$-channel of $\phi^a_{RL}\left(\varepsilon, \mathbf{r}_R\right)$. From the expression for $\partial j^\alpha / \partial r |_{a_+}$, the last forms of the spherical-harmonics expansions (37) and (40), the definition (41) of the screened KKR matrix, and the renormalization (46), this is seen to be:

$$K^a_{R'L',RL}\left(\varepsilon\right) = \frac{-K^\alpha_{R'L',RL}\left(\varepsilon\right)}{\kappa n_{l'}\left(\kappa a_{R'L'}\right) \kappa n_l\left(\kappa a_{RL}\right)}. \tag{49}$$

Note that this is the kink matrix as defined in Ref. [20], whereas the one defined in Ref. [19] has the opposite sign. As presently defined, the energy derivative of the kink matrix is positive definite, as we shall se in the next section.

Screening and the definition (49) have removed the spurious energy dependencies of $K^{\alpha=0}\left(\varepsilon\right)$. To see this more clearly, let us use the first – rather than the last – forms of the spherical-harmonics expansions (37) and (40), which are also more closely related to the definition (36) of the MTO, and to Figs. 3 and 4: The kink matrix for $\psi^a_A\left(\varepsilon, \mathbf{r}_R\right)$ is $- \left[\kappa n_l\left(\kappa a_{A'}\right)\right]^{-1} B^\alpha_{A'A}\left(\varepsilon\right) \left[\kappa n_l\left(\kappa a_A\right)\right]^{-1}$. Moreover, $\psi^a_A\left(\varepsilon, \mathbf{r}_R\right)$ contains the diverging term $n\left(\kappa r\right) / n\left(\kappa a\right)$ in its own channel, which in the MTO is being cancelled by a term from $\varphi^{\circ a}$ (see the third equation (37) and (38)). The kink matrix for the MTO set is now seen to equal the one for the set of screened spherical waves, plus – in the diagonal – the kink in the function $\varphi^a - \varphi^{\circ a} + n\left(\kappa r\right) / n\left(\kappa a\right)$. Since $\varphi - \varphi^\circ$ is smooth, this kink is the one between the radial functions $\varphi^{\circ a}\left(\varepsilon, r\right)$ and $n\left(\kappa r\right) / n\left(\kappa a\right)$. We thus arrive at the expression:

$$K^a_{R'L',RL}\left(\varepsilon\right) = - \frac{B^\alpha_{R'L',RL}\left(\varepsilon\right)}{\kappa n_{l'}\left(\kappa a_{R'L'}\right) \kappa n_l\left(\kappa a_{RL}\right)} \tag{50}$$

$$+ a_{RL} \left[\partial \left\{n_l\left(\varepsilon, a\right)\right\} - \partial \left\{\varphi^\circ_l\left(\varepsilon, a\right)\right\}\right] \delta_{R'R} \delta_{L'L}$$

$$= a^2_{R'L'} \frac{\partial}{\partial r} \mathcal{P}_{R'L'} \psi^a_{RL}\left(\varepsilon, \mathbf{r}_R\right) \Big|_a - a_A \partial \left\{\varphi^\circ_l\left(\varepsilon, a\right)\right\} \delta_{R'R} \delta_{L'L}$$

$$\equiv B^a_{R'L',RL}\left(\varepsilon\right) - a_{RL} \partial \left\{\varphi^\circ_l\left(\varepsilon, a\right)\right\} \delta_{R'R} \delta_{L'L}, \tag{51}$$

in terms of the *logarithmic-derivative function* at the screening sphere of the inwards-integrated radial function, $\partial \left\{\varphi^\circ_l\left(\varepsilon, a\right)\right\} \equiv \partial \ln |\varphi^\circ_l\left(\varepsilon, r\right)| / \partial \ln r |_a$. Remember that RL and $R'L'$ refer to active channels.

In the third line of (50) we have pointed to the fact that the first, *potential-independent* part of the kink matrix is $a^2_{A'}$ times the outwards slope of the screened spherical wave and in (51) we have denoted this *slope matrix* $B^a_{R'L',RL}\left(\varepsilon\right)$. Note that, as presently defined, this slope matrix is Hermitian and equals $a_{R'L'}$ times the non-Hermitian slope matrix defined in Refs. [19,20]; moreover, the transformation from B^α to B^a is not quite (49), but differs from it by the term $a\partial \ln |n_l (\kappa r)| /\partial \ln r|_a$. We may switch from Neumann to Bessel functions, using again that $j_l (\kappa a) = n_l (\kappa a) \tan \alpha$, and that $\{j, n\} = 1/\kappa$. We get:

$$B^a (\varepsilon) = -\frac{\tan \alpha (\varepsilon)}{\kappa j (\kappa a)} [B^\alpha (\varepsilon) - \kappa \cot \alpha (\varepsilon)] \frac{\tan \alpha (\varepsilon)}{\kappa j (\kappa a)} + a\partial \{j (\kappa a)\}$$

$$= \frac{1}{j (\kappa a)} [B (\varepsilon) + \kappa \cot \alpha (\varepsilon)]^{-1} \frac{1}{j (\kappa a)} + a\partial \{j (\kappa a)\}, \qquad (52)$$

where the last equation has been obtained with the help of (33), and where $B (\varepsilon) \equiv B^{\alpha=0} (\varepsilon)$ is the bare KKR structure matrix (24). The matrix $B (\varepsilon) + \kappa \cot \alpha (\varepsilon)$ is the bare KKR matrix for the background-potential and has dimension $(A + I)^2$; it only truncates when $\alpha_l (\varepsilon) \equiv \eta_l (\varepsilon) = 0$, as it happens for high l.

Computational Procedure. The recipe for a computation could be: Solve the radial Schrödinger equations outwards, and then inwards to $a \sim 0.8t$, for all channels up $l \lesssim 3$. Then, compute the Green matrix of the background, $G^{\alpha=0} (\varepsilon) \equiv [B^{\alpha=0} (\varepsilon) + \kappa \cot \alpha (\varepsilon)]^{-1}$, by inversion in real space, choosing the strong screening just mentioned, i.e. nearly touching screening spheres for all $spd (f)$ channels. This gives the strongly screened structure matrix, $B^\alpha (\varepsilon)$ or $B^a (\varepsilon)$, according to (52), and the KKR matrix, $K^\alpha (\varepsilon)$ or $K^a (\varepsilon)$, for the real potential in the strongly screened representation according to (41) or (51). For a crystal, Bloch-sum the KKR matrix. Now, invert this matrix in real space to obtain the Green matrix, $G^\alpha (\varepsilon) \equiv K^\alpha (\varepsilon)^{-1}$ or $G^a (\varepsilon) \equiv K^a (\varepsilon)^{-1}$. Next, choose the physically and chemically motivated screening (β) and rescreen the Green matrix to the downfolded representation, $G^\beta (\varepsilon)$ or $G^b (\varepsilon)$, using the scaling relations (53) or (55) derived below. As will be explained in the following Sect. 3, this should be done for a number of energies. In addition, one will need the first energy derivatives $\dot{G}^b (\varepsilon)$. The latter may be obtained from $\dot{K}^a (\varepsilon)$ via numerical differentiation of the weakly energy dependent structure matrix, $B^a (\varepsilon)$, and calculation of $\int_0^s \varphi^a (\varepsilon, r)^2 r^2 dr - \int_a^s \varphi_{RL}^{oa} (\varepsilon, r)^2 r^2 dr$ for the energy derivative of the logarithmic derivative function in (51), as will be shown in (61)-(63) below. With this $\dot{K}^a (\varepsilon)$, compute $\dot{G}^a (\varepsilon)$ from (63) and, finally, rescreen to $\dot{G}^b (\varepsilon)$ using the energy derivative of (55) given below.

In order to evaluate the wave function (42), one needs in addition to $B_{A'A}^b (\varepsilon)$, the block $B_{IA}^b (\varepsilon)$, and this may be obtained from (52).

The relation of the screening constants, the structure matrix, and the KKR matrix to those – see (10) and (11) – of the conventional ASA is simple, but not as straightforward as the α-to-a transformations of the present section, so for this topic we refer to Refs. [19,20].

This completes our *exact* transformation of the original KKR matrix (26) which has long range and strong energy dependence $[B^0 (\varepsilon, \mathbf{k})$ has poles at the free-electron parabola: $\sum_G |\mathbf{k} + \mathbf{G}|^2 =\varepsilon$] to a screened and renormalized KKR matrix which – depending on the screening – may be short ranged and weakly energy dependent. The kink matrix is expressed in terms of a slope matrix, which only depends on the energy and the structure of the background, and the logarithmic derivatives of the active radial functions extrapolated inwards to the appropriate screening radius.

2.4   Re-screening the Green Matrix

In the ASA, it is simpler to re-screen the Green matrix (15) than the structure matrix (10), because the former involves additions to the diagonal and energy-dependent rescaling of rows and columns, but no matrix inversions. The same holds for the fully energy-dependent matrices of the 3rd-generation, as may be seen from (33) or (52) for the structure matrix. For the Green matrix (41), we get with the help of (52) and a bit of algebra:

$$G^{\alpha}\left(\varepsilon\right) \equiv K^{\alpha}\left(\varepsilon\right)^{-1} = \kappa^{-1}\tan\alpha\left(\varepsilon\right)\left[1 - \tan\alpha\left(\varepsilon\right)\cot\eta\left(\varepsilon\right)\right]$$
$$+ \left[1 - \tan\alpha\left(\varepsilon\right)\cot\eta\left(\varepsilon\right)\right]G^{\alpha=0}\left(\varepsilon\right)\left[1 - \tan\alpha\left(\varepsilon\right)\cot\eta\left(\varepsilon\right)\right],$$

which has the form (15). Solving for $G^{\alpha=0}\left(\varepsilon\right)$ and setting the result equal to $G^{\beta}\left(\varepsilon\right)$ yields the following relation for re-screening of the Green matrix:

$$G^{\beta}\left(\varepsilon\right) = \frac{\tan\eta^{\beta}\left(\varepsilon\right)}{\tan\eta^{\alpha}\left(\varepsilon\right)}G^{\alpha}\left(\varepsilon\right)\frac{\tan\eta^{\beta}\left(\varepsilon\right)}{\tan\eta^{\alpha}\left(\varepsilon\right)} - \frac{\tan\alpha\left(\varepsilon\right) - \tan\beta\left(\varepsilon\right)}{\kappa}\frac{\tan\eta^{\beta}\left(\varepsilon\right)}{\tan\eta^{\alpha}\left(\varepsilon\right)}. \tag{53}$$

In a-language, where according to (49): $G^{a}\left(\varepsilon\right) = -\kappa n\left(\kappa a\right)G^{\alpha}\left(\varepsilon\right)\kappa n\left(\kappa a\right)$, the diagonal matrices in (53) become $\left[n\left(\kappa b\right)/n\left(\kappa a\right)\right]\left[\tan\eta^{\beta}\left(\varepsilon\right)/\tan\eta^{\alpha}\left(\varepsilon\right)\right]$ and $\kappa n\left(\kappa a\right)n\left(\kappa b\right)\left[\tan\alpha\left(\varepsilon\right) - \tan\beta\left(\varepsilon\right)\right]$ and may, in fact, be expressed more simply in terms of the inwards-integrated radial wave function, renormalized according to (47). In order to see this, we first use the form (39):

$$\varphi^{\circ\,a}\left(\varepsilon,r\right) = \frac{n\left(\kappa r\right)\tan\eta^{\alpha}\left(\varepsilon\right) - j^{\alpha}\left(\varepsilon,r\right)}{n\left(\kappa a\right)\tan\eta^{\alpha}\left(\varepsilon\right)},$$

and then evaluate this at the screening-radius b :

$$\varphi^{\circ\,a}\left(\varepsilon,b\right) = \frac{n\left(\kappa b\right)\tan\eta^{\alpha}\left(\varepsilon\right) - j^{\alpha}\left(\varepsilon,b\right)}{n\left(\kappa a\right)\tan\eta^{\alpha}\left(\varepsilon\right)} = \frac{n\left(\kappa b\right)\tan\eta^{\beta}\left(\varepsilon\right)}{n\left(\kappa a\right)\tan\eta^{\alpha}\left(\varepsilon\right)}.$$

To obtain this result, we have also used:

$$j^{\alpha}\left(\varepsilon,b\right) = j\left(\kappa b\right) - n\left(\kappa b\right)\tan\alpha\left(\varepsilon\right) = n\left(\kappa b\right)\left[\tan\beta\left(\varepsilon\right) - \tan\alpha\left(\varepsilon\right)\right],$$

from (31) and (43). The second, readily computable function is that solution of the radial wave equation which vanishes at a with slope $1/a^{2}$:

$$j^{a}\left(\varepsilon,r\right) \equiv \frac{j^{\alpha}\left(\varepsilon,r\right)}{a^{2}\partial j^{\alpha}\left(\varepsilon,r\right)/\partial r|_{a}} = -\kappa n\left(\kappa a\right)j^{\alpha}\left(\varepsilon,r\right). \tag{54}$$

Evaluation at $r = b$ yields:

$$j^a (\varepsilon, b) = -\kappa n (\kappa a) \, j^\alpha (\varepsilon, b) = \kappa n (\kappa a) \, n (\kappa b) \left[\tan \alpha (\varepsilon) - \tan \beta (\varepsilon) \right],$$

which is the second function needed. Hence, we have found the following simple and practical scaling relation for re-screening of the Green matrix:

$$G^b (\varepsilon) = \varphi^{\circ a} (\varepsilon, b) \, G^a (\varepsilon) \, \varphi^{\circ a} (\varepsilon, b) + j^a (\varepsilon, b) \, \varphi^{\circ a} (\varepsilon, b). \tag{55}$$

2.5   Green Functions, Matrix Elements, and Charge Density

The kinked partial wave is the solution of the inhomogeneous Schrödinger equation:

$$(\mathcal{H} - \varepsilon) \, \phi^a_{R'L'} (\varepsilon, \mathbf{r}) = - \sum_{RL} \delta \left(r_R - a_{RL} \right) Y_L (\hat{\mathbf{r}}_R) \, K^a_{RL, R'L'} (\varepsilon), \tag{56}$$

provided that we define the MTO (36) the 3-fold way indicated in Figs. 2 – 4, and therefore – for the MT-Hamiltonian $\mathcal{H}$ (4) – use the radial Schrödinger equation (2) channel-wise.

The *kinks* of the MTO are given correctly by (56), but the proper MTO does not solve Schrödinger's differential equation in the shells between the screening and the MT-spheres; here we need the 3-fold way. This way *must* not be an approximation: For instance, when applied to those *linear combinations* of MTOs which solve the KKR equations – and hence Schrödinger's equation – equation (56) is correct (and yields zero), because for each active channel, A', the two solutions, $\mathcal{P}_{A'} \sum_A \psi^a_A (\varepsilon, \mathbf{r}_R) c^a_A$ and $\varphi^{\circ a}_{A'} (\varepsilon, r_{R'}) c^a_{A'}$, of the radial wave equation match in value and slope at $a_{R'L'}$, and therefore cancel *throughout* the shell $s_{R'} - a_{R'L'}$. Expressed in another way: For energy-dependent MTOs, kink-cancellation leads to cancellation of the triple-valuedness. For the energy-independent NMTOs to be derived in the next section, special considerations will be necessary.

Solving (56) for $\delta \left(r_R - a_{RL} \right) Y_L (\hat{\mathbf{r}}_R)$, leads to:

$$(\mathcal{H} - \varepsilon) \sum_{R'L'} \phi^a_{R'L'} (\varepsilon, \mathbf{r}) \, G^a_{R'L', RL} (\varepsilon) = -\delta \left(r_R - a_{RL} \right) Y_L (\hat{\mathbf{r}}_R) \tag{57}$$

which shows that the linear combinations

$$\gamma^a_{RL} (\varepsilon, \mathbf{r}) = \sum_{R'L'} \phi^a_{R'L'} (\varepsilon, \mathbf{r}) \, G^a_{R'L', RL} (\varepsilon), \tag{58}$$

of MTOs – all with the same energy and screening – is a *contraction* of $\mathbf{r}'$ onto the screening spheres $(\mathbf{r}' \to a_{RL}, RL)$ *of the Green function* defined by:

$$(\mathcal{H}_\mathbf{r} - \varepsilon) \, G (\varepsilon; \mathbf{r}, \mathbf{r}') = -\delta (\mathbf{r} - \mathbf{r}').$$

The *contracted Green function* $\gamma_{RL}^a (\varepsilon, \mathbf{r})$ has kink 1 in its own channel and kink 0 in all other active channels ($\neq RL$). This function is therefore a solution of the Schrödinger equation (defined the 3-fold way) which is smooth everywhere except at its own screening sphere. $\gamma_{RL}^a (\varepsilon, \mathbf{r})$ is usually *de*localized, and when the energy, ε, coincides with a pole, ε_j, of the Green matrix, $\gamma_{RL}^a (\varepsilon, \mathbf{r})$ diverges everywhere in space. This means, that when $\varepsilon = \varepsilon_j$, then the *renormalized* function is smooth also at its own sphere, and it therefore solves Schrödinger's equation. In vector-matrix notation, equations (56) and (57) become:

$$(\mathcal{H} - \varepsilon) \phi^a (\varepsilon, \mathbf{r}) \qquad = -\delta^a (\mathbf{r}) K^a (\varepsilon),$$
$$(\mathcal{H} - \varepsilon) \phi^a (\varepsilon, \mathbf{r}) G^a (\varepsilon) \equiv (\mathcal{H} - \varepsilon) \gamma^a (\varepsilon, \mathbf{r}) = -\delta^a (\mathbf{r}),$$

where we have defined a *set* of spherical harmonics on the a-shells with the following members:

$$\delta_{RL}^a (\mathbf{r}_R) \equiv \delta (r_R - a_{RL}) Y_L (\hat{\mathbf{r}}_R). \tag{59}$$

If expressed in real space, our Green matrix, $G^a (\varepsilon)$, is what in multiple-scattering theory [11] is usually called the scattering path operator and denoted $\tau (\varepsilon)$. In the 2nd-generation LMTO formalism, it was denoted $g (\varepsilon)$, but in the present paper we denote matrices by capitals.

Since in the 3-fold way, an MTO takes the value one at its own screening sphere and zero at all other screening spheres, expression (56) yields for the matrix element of $\mathcal{H} - \varepsilon$ with another, or the same, MTO in the set:

$$\langle \phi_{R'L'}^a (\varepsilon) | \mathcal{H} - \varepsilon | \phi_{RL}^a (\varepsilon) \rangle = -K_{R'L',RL}^a (\varepsilon) \equiv -G_{R'L',RL}^a (\varepsilon)^{-1}, \tag{60}$$

which says that the negative of the kink matrix is the *Hamiltonian* matrix, minus the energy, in the basis of energy-dependent 0th-order MTOs.

For the *overlap integral* between screened spherical waves, with possibly different energies and in the interstitial between the screening spheres, defined channel-by-channel, we obtain the simple expression [19]:

$$\langle \psi_{R'L'}^a (\varepsilon') | \psi_{RL}^a (\varepsilon) \rangle = \frac{B_{R'L',RL}^a (\varepsilon') - B_{R'L',RL}^a (\varepsilon)}{\varepsilon' - \varepsilon} \tag{61}$$

$$\longrightarrow \dot{B}_{R'L',RL}^a (\varepsilon) \quad \text{if} \quad \varepsilon' \to \varepsilon$$

by use of Green's second theorem, together with expression (51) for the surface integrals. Note that, *neither* active channels different from the eigen-channels, $R'L'$ and RL, *nor* the inactive channels contribute to the surface integrals. The reasons are that $\psi_{R'L'}^a (\varepsilon', \mathbf{r})$ and $\psi_{RL}^a (\varepsilon, \mathbf{r})$ vanish on all 'other' screening spheres, and that they are regular in the inactive channels. The latter means that, in the inactive channels, the 'screening-sphere interstitial' extends all the way to the sites ($a_I \to 0$). For the overlap integral between kinked partial waves, the 3-fold way yields:

$$\langle \phi^a_{R'L'}(\varepsilon') \mid \phi^a_{RL}(\varepsilon) \rangle \equiv \langle \psi^a_{R'L'}(\varepsilon') \mid \psi^a_{RL}(\varepsilon) \rangle + \delta_{R'R}\delta_{L'L} \times$$

$$\left(\int_0^{s_R} \varphi^a_{RL}(\varepsilon',r)\,\varphi^a_{RL}(\varepsilon,r)\,r^2 dr - \int_{a_{RL}}^{s_R} \varphi^{\circ a}_{RL}(\varepsilon',r)\,\varphi^{\circ a}_{RL}(\varepsilon,r)\,r^2 dr \right)$$

$$= \frac{K^a_{R'L',RL}(\varepsilon') - K^a_{R'L',RL}(\varepsilon)}{\varepsilon' - \varepsilon} \quad \longrightarrow \quad \dot{K}^a_{R'L',RL}(\varepsilon) \text{ if } \varepsilon' \to \varepsilon. \tag{62}$$

For the overlap matrix for the set of contracted Green functions, this gives:

$$\langle \gamma^a(\varepsilon') \mid \gamma^a(\varepsilon) \rangle = -\frac{G^a(\varepsilon') - G^a(\varepsilon)}{\varepsilon' - \varepsilon} \tag{63}$$

$$\longrightarrow -\dot{G}^a(\varepsilon) = G^a(\varepsilon)\,\dot{K}^a(\varepsilon)\,G^a(\varepsilon) \quad \text{if } \varepsilon' \to \varepsilon.$$

We see that $\dot{B}^a(\varepsilon)$, $\dot{K}^a(\varepsilon)$, and $\dot{G}^a(\varepsilon)$ are Hermitian, just like $B^a(\varepsilon)$, $K^a(\varepsilon)$, and $G^a(\varepsilon)$. Whereas $\dot{B}^a(\varepsilon)$ and $\dot{K}^a(\varepsilon)$ are positive definite matrices, that is, their eigenvalues are positive or zero, $\dot{G}^a(\varepsilon)$ is negative definite. For well-screened MTOs, the logarithmic derivative functions in the diagonal of the kink matrix (51) depend more strongly on energy than the slope matrix. The way to compute the energy derivative $\dot{K}^a(\varepsilon)$ is therefore to compute $\dot{B}^a(\varepsilon)$ by numerical differentiation, and the remaining terms by integration as in (62).

In the following we shall stay with the normalization (45)-(47) denoted by Latin – rather than Greek – superscripts and shall rarely change the screening. We therefore usually drop the superscript a altogether. Some well-screened representation is usually what we have in mind, but also heavily down-folded – and therefore long-ranged – representations will be considered. In those cases, some parts of the computation must of course be performed in the Bloch – or **k**-space – representation.

The wave function is $\Psi_i(\mathbf{r}) = \phi(\varepsilon_i, \mathbf{r})\,c_i$, where the eigen(column)vector c_i solves the KKR equations, $K(\varepsilon_i)\,c_i = 0$, and is normalized according to: $1 = c_i^\dagger \dot{K}(\varepsilon_i)\,c_i$, in order that $\langle \Psi_i \mid \Psi_i \rangle = 1$. From the definition (36) of the MTO, we see that an accurate approximation for the *charge density*, which is consistent with the 3-fold way and, hence, with the normalization, has the simple form:

$$\rho(\mathbf{r}) = \rho^\psi(\mathbf{r}) + \sum_R \left[\rho^\varphi_R(\mathbf{r}_R) - \rho^{\varphi^\circ}_R(\mathbf{r}_R) \right] \tag{64}$$

where the global contribution is:

$$\rho^\psi(\mathbf{r}) \equiv \sum_{RR'} \sum_{LL'} \int^{\varepsilon_F} \psi_{RL}(\varepsilon, \mathbf{r}_R)\,\Gamma_{RL,R'L'}(\varepsilon)\,\psi_{R'L'}(\varepsilon, \mathbf{r}_{R'})^*\, d\varepsilon \tag{65}$$

and the local contributions, $\rho_R^\varphi(\mathbf{r}_R) - \rho_R^{\varphi^\circ}(\mathbf{r}_R)$, which vanish smoothly at their respective MT-sphere, are given by:

$$\rho_R^\varphi(\mathbf{r}) = \sum_{LL'} Y_L(\hat{\mathbf{r}}) Y_{L'}^*(\hat{\mathbf{r}}) \int^{\varepsilon_F} \varphi_{Rl}(\varepsilon, r)\, \Gamma_{RL,RL'}(\varepsilon)\, \varphi_{Rl'}(\varepsilon, r)\, d\varepsilon$$

$$\rho_R^{\varphi^\circ}(\mathbf{r}) = \sum_{LL'} Y_L(\hat{\mathbf{r}}) Y_{L'}^*(\hat{\mathbf{r}}) \int^{\varepsilon_F} \varphi_{Rl}^\circ(\varepsilon, r)\, \Gamma_{RL,RL'}(\varepsilon)\, \varphi_{Rl'}^\circ(\varepsilon, r)\, d\varepsilon. \quad (66)$$

The common density-of-states matrix in these equations is:

$$\Gamma_{RL,R'L'}(\varepsilon) = \sum_i^{occ} c_{RL,i}\, \delta(\varepsilon - \varepsilon_i)\, c_{R'L',i}^* = \frac{1}{\pi} \mathrm{Im}\, G_{RL,R'L'}(\varepsilon + i\delta). \quad (67)$$

The approximations inherent in (64) are that all cross-terms between products of ψ-, φ-, and φ°-functions, and between φ- or φ°-functions on different sites are neglected.

3   Polynomial MTO Approximations

In this section we shall show how energy-*in*dependent basis sets may be derived from the kinked partial waves, that is, how we get rid of the energy dependence of the MTOs. Specifically, we shall preview the generalization [51,24] of the 3rd-generation LMTO method [19,20] mentioned in connection with Fig. 1. This generalization is to an 'N'MTO method in which the basis set consists of energy-*in*dependent NMTOs,

$$\chi_{RL}^{(N)}(\mathbf{r}) = \sum_{n=0}^{N} \sum_{R'L'} \phi_{R'L'}(\varepsilon_n, \mathbf{r})\, L_{R'L',RL;n}^{(N)}, \quad (68)$$

$$\text{where} \quad \sum_{n=0}^{N} L_{R'L',RL;n}^{(N)} = \delta_{R'R}\delta_{L'L},$$

constructed as linear combinations of the kinked partial waves at a mesh of $N+1$ energies, in such a way that the NMTO basis can describe the solutions, $\Psi_i(\mathbf{r})$, of Schrödinger's equation correctly to within an error proportional to $(\varepsilon_i - \varepsilon_0)(\varepsilon_i - \varepsilon_1)\dots(\varepsilon_i - \varepsilon_N)$. Note the difference between one-electron energies denoted ε_i and ε_j, and mesh points denoted ε_n and ε_m, with n and m taking integer values. The set, $\chi^{(N=0)}(\mathbf{r})$, is therefore simply $\phi(\varepsilon_0, \mathbf{r})$, and this is the reason why, right at the beginning of the previous section, $\phi(\varepsilon, \mathbf{r})$ was named the set of 0th-order energy-*de*pendent MTOs. For $N > 0$, the NMTOs are smooth and their triple-valuedness decreases with increasing N. For the mesh condensing to one energy, ε_ν, the NMTO basis is of course constructed as linear combinations of $\phi(\varepsilon_\nu, \mathbf{r})$ and its first N energy derivatives at ε_ν. For $N=1$, this is the well-known LMTO set.

The immediate practical use of this new development is to widen and sharpen the energy window inside which the method gives good wave functions, *without increasing the size of the basis set*. One may even *decrease* the size of the basis through downfolding, and still maintain an acceptable energy window by increasing the *order* of the basis set. The prize for increasing N is: More computation and increased range of the basis functions.

3.1 Energy-Independent NMTOs

What we have done in the previous sections – one might say – is to factorize out of the contracted Green function, $\gamma(\varepsilon, \mathbf{r})$, some spatial functions, $\phi_{RL}(\varepsilon, \mathbf{r})$, which are so localized that, for two energies inside the energy-window of interest, the corresponding functions, $\phi_{RL}(\varepsilon, \mathbf{r})$ and $\phi_{RL}(\varepsilon', \mathbf{r})$, *cannot be orthogonal*. In other words: The kinked partial waves are so well separated through localization and angular symmetry that we need only *one radial quantum number* for each function.

Now, we want to get rid of the kinks and to *reduce the triple-valuedness and the energy dependence* of *each* kinked partial wave – retaining its RL-character – to a point where the triple-valuedness and the energy-dependence may both be neglected. This we do, first by passing from the set $\phi(\varepsilon, \mathbf{r})$ to a set of so-called Nth-order energy-dependent MTOs, $\chi^{(N)}(\varepsilon, \mathbf{r})$, whose contracted Green function,

$$\chi^{(N)}(\varepsilon, \mathbf{r})\, G(\varepsilon) \equiv \phi(\varepsilon, \mathbf{r})\, G(\varepsilon) - \sum_{n=0}^{N} \phi(\varepsilon_n, \mathbf{r})\, G(\varepsilon_n)\, A_n^{(N)}(\varepsilon), \qquad (69)$$

differs from $\phi(\varepsilon, \mathbf{r})\, G(\varepsilon)$ by a function which remains in the Hilbert space spanned by the set $\phi(\varepsilon, \mathbf{r})$ with energies inside the window of interest, and which is *analytical in energy*. The two contracted Green functions thus have the same poles, and both energy-dependent basis sets, $\phi(\varepsilon, \mathbf{r})$ and $\chi^{(N)}(\varepsilon, \mathbf{r})$, can therefore yield the exact Schrödinger-equation solutions. The analytical functions of energy we wish to determine in such a way that $\chi^{(N)}(\varepsilon, \mathbf{r})$ takes the *same* value, $\chi^{(N)}(\mathbf{r})$, at the $N+1$ points, $\varepsilon_0, ..., \varepsilon_N$. With the set $\chi^{(N)}(\varepsilon, \mathbf{r})$ defined that way, we can finally *neglect* its energy dependence, and the resulting $\chi^{(N)}(\mathbf{r})$ is then the set of Nth-order energy-*in*dependent MTOs.

Other choices for the analytical functions of energy, involving for instance complex energies or Chebyshev polynomials, await their exploration.

One solution with the property that $\chi_{RL}^{(N)}(\varepsilon, \mathbf{r})$ takes the same value for ε at any of the $N+1$ mesh points, is of course given by the *polynomial*:

$$A_{n; R'L', RL}^{(N)}(\varepsilon) = \delta_{R'R}\delta_{L'L} \prod_{m=0, \neq n}^{N} \frac{\varepsilon - \varepsilon_m}{\varepsilon_n - \varepsilon_m},$$

of Nth degree. But this solution is useless, because it yields: $\chi^{(N)}(\mathbf{r}) = 0$. If, instead, we try a polynomial of $(N-1)$st degree for the analytical function, then we can write down the corresponding expression for the set $\chi^{(N)}(\mathbf{r})$ without explicitly solving for the $(N+1)^2$ matrices $A_n^{(N)}(\varepsilon_m)$, and then prove afterwards that each basis function has its triple-valuedness reduced *consistently* with the remaining error $\propto (\varepsilon_i - \varepsilon_0)(\varepsilon_i - \varepsilon_1)\dots(\varepsilon_i - \varepsilon_N)$ of the set.

Since we want $\chi^{(N)}(\varepsilon_n, \mathbf{r})$ to be independent of n for $0 \leq n \leq N$, *all* its *divided differences* on the mesh – up to and including the divided difference of order N – *vanish*, with the exception of the 0th divided difference, which is $\chi^{(N)}(\mathbf{r})$. As a consequence, the Nth divided difference of $\chi^{(N)}(\varepsilon, \mathbf{r}) G(\varepsilon)$ on the left-hand side of (69) is $\chi^{(N)}(\mathbf{r})$ times the Nth divided difference of the Green matrix. Now, the Nth divided difference of the last term on the right-hand side vanishes, because it is a polynomial of order $N-1$, and as a consequence,

$$\chi^{(N)}(\mathbf{r}) = \frac{\Delta^N \phi(\mathbf{r}) G}{\Delta[0\dots N]} \left(\frac{\Delta^N G}{\Delta[0\dots N]}\right)^{-1}. \tag{70}$$

This basically solves the problem of finding the energy-independent NMTOs! What remains, is to factorize the divided difference of the product $\phi(\varepsilon, \mathbf{r}) G(\varepsilon)$ into spatial functions, $\phi(\varepsilon_n, \mathbf{r})$, which are vectors in RL, and matrices, $G(\varepsilon_n)$, with $n = 0, \dots, N$. Equivalently, we could use a binomial divided-difference series in terms of $\phi(\varepsilon_0, \mathbf{r})$ and its first N divided differences on the mesh together with $G(\varepsilon_N)$ and its corresponding divided differences.

For a *condensed* energy mesh, defined by: $\varepsilon_n \to \varepsilon_\nu$ for $0 \leq n \leq N$, the Nth divided difference becomes $\frac{1}{N!}$ times the Nth derivative:

$$\frac{\Delta^N f}{\Delta[0\dots N]} \equiv f[0\dots N] \quad \to \quad \frac{1}{N!} \left.\frac{d^N f(\varepsilon)}{d\varepsilon^N}\right|_{\varepsilon_\nu}, \tag{71}$$

but since a discrete mesh with arbitrarily spaced points is much more powerful in the present case where the time-consuming part of the computation is the evaluation of the Green matrix (and its first energy derivative for use in Eq. (63)) at the energy points, we shall proceed using the language appropriate for a discrete mesh. In (71) we have introduced the form $f[0\dots N]$ because it may – more easily than $\Delta^N f/\Delta[0\dots N]$ – be modified to include another kind of divided differences, the so-called Hermite divided differences, which we shall meet later.

Readers interested in the details of the discrete formalism are referred to the Appendix where we review relevant parts of the classical theory of polynomial approximation, and derive formulae indispensable for the NMTO formalism for discrete meshes. Readers merely interested in an overview, may be satisfied with the formalism as applied to a *condensed* mesh and for this, they merely need the translation (71) together with the divided-difference form of the NMTO to be described in the following. Details about the Lagrange form may be ignored.

Lagrange form. We first use the Lagrange form (149) of the divided difference to factorize the energy-independent NMTO (70) and obtain:

$$\chi^{(N)}(\mathbf{r}) = \sum_{n=0}^{N} \frac{\phi_n(\mathbf{r}) G_n}{\prod_{m=0,\neq n}^{N}(\varepsilon_n - \varepsilon_m)} G[0..N]^{-1}, \tag{72}$$

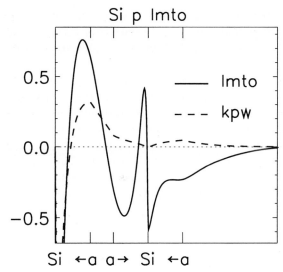

Fig. 5. Si p_{111} member of the *spd*-set of 0th (dottet) and 1th-order MTOs (see text and Eq.(74)).

Here and in the following, $\phi_n(\mathbf{r}) \equiv \phi(\varepsilon_n, \mathbf{r})$ and $G_n \equiv G(\varepsilon_n)$. Eq. (72) has the form (68) and we see, that the weight with which the MTO set at ε_n enters the NMTO set, is:

$$L_n^{(N)} = \frac{G_n}{\prod_{m=0,\neq n}^{N} (\varepsilon_n - \varepsilon_m)} G[0..N]^{-1}. \tag{73}$$

By application of (149) to the Green matrix, we may verify that these Lagrange weights sum up to the unit matrix. For this reason, the RL characters of the NMTO basis functions will correspond to those of the kinked partial waves.

As an example, for $N=1$ we get the so-called chord-LMTO:

$$\begin{aligned}
\chi^{(1)}(\mathbf{r}) &= \phi_0(\mathbf{r}) G_0 (G_0 - G_1)^{-1} + \phi_1(\mathbf{r}) G_1 (G_1 - G_0)^{-1} \\
&= \phi_0(\mathbf{r}) (K_1 - K_0)^{-1} K_1 + \phi_1(\mathbf{r}) (K_0 - K_1)^{-1} K_0 \quad (74) \\
&= \phi_0(\mathbf{r}) - \phi([01], \mathbf{r}) K[01]^{-1} K_0 \\
&\to \phi(\mathbf{r}) - \dot{\phi}(\mathbf{r}) \dot{K}^{-1} K.
\end{aligned}$$

In this case, there is only one energy difference, $\varepsilon_0 - \varepsilon_1$, so it cancels out. In the 3rd line, we have reordered the terms in such a way that the *Newton form,* to be derived for general N in (88) and (90) below, is obtained. In the 4th line, we have condensed the mesh onto ε_ν, whereby the well-known tangent-LMTO [19,20] is obtained. The latter is shown by the full curve in Fig.5 for the case of the Si p_{111}-orbital belonging to an sp set. The dashed curve is the corresponding kinked partial wave, $\phi(\mathbf{r})$, shown by the full curve in Fig. 4. Compared to the latter, $\chi^{(1)}(\mathbf{r})$ is smooth, but has longer range. The strong contributions to the tail of the LMTO from $\dot{\phi}\,(\mathbf{r})$'s on the nearest neighbor are evident. It is also clear, that for computations involving wave functions – e.g. of the charge density – the building blocks will rarely be the NMTOs, but the kinked partial waves, $\phi_n\,(\mathbf{r})$, which are more compact.

One might fear that the discrete NMTO scheme would fail when one of the mesh points is close to a one-electron energy, that is, to a pole of the Green matrix, but that does not happen: If one of the G_n's diverges, this just means that the corresponding Lagrange weight is 1, and the others 0. Hence, in this case the NMTO is just $\phi_n\,(\mathbf{r})$, and this is the correct result. Moreover, the kink of this single $\phi_n\,(\mathbf{r})$ does not matter, because in this case where $G\,(\varepsilon)$ is at a pole, the determinant of its inverse vanishes, so that the kink-cancellation equations, $K_n c_n = 0$, have a non-zero solution, c_n, which yields a smooth linear combination, $\phi_n\,(\mathbf{r})\,c_n$, of NMTOs.

Kinks and triple-valuedness. The energy-independent NMTOs have been defined through (69) and (70) in such a way that $\chi^{(N)}\,(\varepsilon,\mathbf{r}) - \chi^{(N)}\,(\mathbf{r}) \propto (\varepsilon - \varepsilon_0)\ldots$ $(\varepsilon - \varepsilon_N)$. We now show, that also the kink-and-triple-valuedness of $\chi^{(N)}\,(\mathbf{r})$ is of that order, and therefore negligible.

The result of projecting the energy-dependent MTO onto $Y_{L'}\,(\hat{\mathbf{r}}_{R'})$ for an active channel was given in (37) for its own channel, and in (40) for any other active channel. Together, these results may be expressed as:

$$\mathcal{P}_{R'L'}\phi^\alpha_{RL}\,(\varepsilon,\mathbf{r}_R) = \varphi^\alpha_{Rl}\,(\varepsilon,r_R)\,\delta_{R'R}\delta_{L'L} + j^\alpha_{R'L'}\,(\varepsilon,r_{R'})\,\kappa^{-1}\times$$
$$\left[\kappa\cot\eta^\alpha_{RL}\,(\varepsilon)\,\delta_{R'R}\delta_{L'L} + B^\alpha_{R'L',RL}\,(\varepsilon)\right]$$

or, in terms of the renormalized functions (44), (46), (47), and (54), as well as the kink matrix defined in (49), as:

$$\mathcal{P}_{R'L'}\phi^a_{RL}\,(\varepsilon,\mathbf{r}_R) = \varphi^a_{Rl}\,(\varepsilon,r_R)\,\delta_{R'R}\delta_{L'L} + j^a_{R'L'}\,(\varepsilon,r_{R'})\,K^a_{R'L',RL}\,(\varepsilon).$$

Here, like in (37) and (40), contributions from MT-overlaps – which are irrelevant for the present discussion – have been neglected. Without kinks and triple-valuedness, $\mathcal{P}_{R'L'}\phi^a_{RL}\,(\varepsilon,\mathbf{r}_R)$ would be given by the first term, and the kinks and the triple-valuedness are therefore given by the second term:

$$\mathcal{T}_{R'L'}\phi^a_{RL}\,(\varepsilon,\mathbf{r}_R) = j^a_{R'L'}\,(\varepsilon,r_{R'})\,K^a_{R'L',RL}\,(\varepsilon). \qquad (75)$$

This vanishes for those *linear combinations* of MTOs which solve the kink-cancellation conditions.

What now happens for the energy-independent approximation, $\chi^{(0)}\,(\mathbf{r}) \equiv \phi_0\,(\mathbf{r})$, to the 0th-order energy-dependent MTO, $\chi^{(0)}\,(\varepsilon,\mathbf{r}) \equiv \phi\,(\varepsilon,\mathbf{r})$, is that the former has kinks and triple-valuedness, but both are proportional to $K\,(\varepsilon_0)$ which – according to (56) – is proportional to $\mathcal{H}-\varepsilon_0$ and, hence, to $\varepsilon_i-\varepsilon_0$. The kinks and triple-valuedness are thus of the same order as the error of $\chi^{(0)}\,(\mathbf{r})$. Similarly, for $N > 0$, the fact that the $A^{(N)}_n\,(\varepsilon)$'s are polynomials of $(N-1)$st degree, reduces the triple-valuedness of $\chi^{(N)}\,(\mathbf{r})$ to being proportional to $(\varepsilon - \varepsilon_0)\ldots(\varepsilon - \varepsilon_N)$, as we shall now see: Multiplication of (75) with $G^a(\varepsilon)$ from the right yields: $\mathcal{T}\phi^a(\varepsilon,\mathbf{r})G^a(\varepsilon) = j^a(\varepsilon,\mathbf{r})$, and for the kinks and the triple-valuedness of the contracted Green function (69) we therefore get:

$$\mathcal{T}\chi^{(N)}(\varepsilon,\mathbf{r})G(\varepsilon) = j^a(\varepsilon,r) - \sum_{n=0}^{N} j^a(\varepsilon_n,r)A_n^{(N)}(\varepsilon).$$

Taking again the Nth divided difference for the mesh on which $\chi^{(N)}(\varepsilon,\mathbf{r})$ is constant yields:

$$\mathcal{T}\chi^{(N)}(\mathbf{r}) = j^a([0...N],r)G^a[0...N]^{-1} \tag{76}$$

$$= -j^a([0...N],r)\left(E^{(0)} - \varepsilon_0\right)\left(E^{(1)} - \varepsilon_1\right)...\left(E^{(N)} - \varepsilon_N\right),$$

for the kinks and the triple-valuedness of the energy-independent NMTO. In the last line, we have used an expression – which will be proved in (83) – for the inverse of the Nth divided difference of the Green matrix in terms of the product of energy matrices to be defined in (81). At present, it suffices to note that differentiation of the Green function,

$$\check{G}(\varepsilon) \equiv \sum_j \frac{1}{\varepsilon - \varepsilon_j}, \tag{77}$$

for a model with *one, normalized orbital* yields:

$$\left[\frac{1}{N!}\frac{d^N\check{G}(\varepsilon)}{d\varepsilon^N}\bigg|_{\varepsilon_\nu}\right]^{-1} = -\left[\sum_j \frac{1}{(\varepsilon_j - \varepsilon_\nu)^{N+1}}\right]^{-1} \approx -(\varepsilon_i - \varepsilon_\nu)^{N+1},$$

where the last approximation holds when the mesh is closer to the one-electron energy of interest, ε_i, than to any other one-electron energy, $\varepsilon_j \neq \varepsilon_i$. Note that j – and not n – denotes the radial quantum number. Similarly, this model Green function has a divided difference on a discrete mesh of $N+1$ points, whose inverse is:

$$\check{G}[0..N]^{-1} = -\left[\sum_j \frac{1}{\prod_{n=0}^{N}(\varepsilon_j - \varepsilon_n)}\right]^{-1} \approx -\prod_{n=0}^{N}(\varepsilon_i - \varepsilon_n), \tag{78}$$

as proved in Eq. (159) of the Appendix. We have thus seen that the triple-valuedness is of the *same* order as the error present in $\chi^{(N)}(\mathbf{r})$ due to the neglect of the energy-dependence of $\chi^{(N)}(\varepsilon,\mathbf{r})$.

The radial function $j^a(\varepsilon,r)$ in (75) vanishes for $r \leq a$, where it has a kink of value $1/a^2$, and it solves the radial wave equation for $r \geq a$. As shown in [51], its expansion in powers of $r - a \geq 0$ is:

$$rj^a(\varepsilon,r) = \frac{r-a}{a} + \frac{1}{3!}[l(l+1) - \varepsilon a^2]\left(\frac{r-a}{a}\right)^3 - \frac{l(l+1)}{3!}\left(\frac{r-a}{a}\right)^4$$

$$+ \frac{1}{5!}\left[18l(l+1) + (l(l+1) - \varepsilon a^2)^2\right]\left(\frac{r-a}{a}\right)^5 + ...$$

This means the Nth divided-difference function entering (76) satisfies:

$$j^a\left([0...N],r\right) \propto (r-a)^{2N+1}.$$

The *kink and triple-valuedness* (76) in the $s-a$ shell of $\chi^{(N)}(r)$ is thus proportional to $(r-a)^{2N+1}\prod_{n=0}^{N}(\varepsilon_i-\varepsilon_n)$, and for this reason the energy-window *widens* as $s-a$ decreases, that is, as the screening *increases*.

Transfer matrices and correspondence with Lagrange interpolation.
We need to work out the effect of the Hamiltonian on the NMTO set. Since the NMTOs with $N>0$ are smooth, the contributions from the delta-function on the right-hand side of (57) for the contracted Green function will cancel in the end. Operation on (69) therefore yields:

$$\mathcal{H}\left[\phi\left(\varepsilon,\mathbf{r}\right)-\chi^{(N)}\left(\varepsilon,\mathbf{r}\right)\right]G\left(\varepsilon\right)=\phi\left(\varepsilon,\mathbf{r}\right)\varepsilon G\left(\varepsilon\right)-\mathcal{H}\chi^{(N)}\left(\varepsilon,\mathbf{r}\right)G\left(\varepsilon\right)$$
$$=\sum_{n=0}^{N}\phi_n\left(\mathbf{r}\right)\varepsilon_n G_n A_n^{(N)}\left(\varepsilon\right)$$

and by taking the Nth divided difference for the mesh on which $\chi^{(N)}\left(\varepsilon,\mathbf{r}\right)$ is constant, we obtain:

$$\mathcal{H}\gamma\left([0...N],\mathbf{r}\right)=\mathcal{H}\chi^{(N)}\left(\mathbf{r}\right)G\left[0...N\right]=\left(\phi\varepsilon G\right)\left([0...N],\mathbf{r}\right)$$
$$=\gamma\left([0..N-1],\mathbf{r}\right)+\varepsilon_N\gamma\left([0...N],\mathbf{r}\right), \tag{79}$$

using (151) with the choice of the last point on the mesh. Solving for the NMTOs yields:

$$\left(\mathcal{H}-\varepsilon_N\right)\chi^{(N)}\left(\mathbf{r}\right)=\chi^{(N-1)}\left(\mathbf{r}\right)\left(E^{(N)}-\varepsilon_N\right) \tag{80}$$

where $\chi^{(N-1)}\left(\mathbf{r}\right)\equiv\gamma\left([0..N-1],\mathbf{r}\right)G\left[0..N-1\right]^{-1}$ is the energy-independent MTO of order $N-1$, obtained by *not* using the last point. Moreover,

$$E^{(N)}\equiv\varepsilon_N+G\left[0..N-1\right]G\left[0...N\right]^{-1}=\left(\varepsilon G\right)\left[0...N\right]G\left[0...N\right]^{-1}$$
$$=\sum_{n=0}^{N}\frac{\varepsilon_n G_n}{\prod_{m=0,\neq n}\left(\varepsilon_n-\varepsilon_m\right)}G\left[0...N\right]^{-1}=\sum_{n=0}^{N}\varepsilon_n L_n^{(N)}, \tag{81}$$

is the *energy matrix* which – in contrast to $\chi^{(N-1)}\left(\mathbf{r}\right)$ – is independent of which point on the mesh is omitted. The first equation (81) shows how to compute $E^{(N)}$ and the last equation shows that $E^{(N)}$ is the energy *weighted* on the $0...N$-mesh by the Lagrange matrices (73). For a condensed mesh, the results is the simple one (19) quoted in the Overview.

We now consider a sequence of energy meshes, starting with the single-point mesh, ε_0, then adding ε_1 in order to obtain the two-point mesh $\varepsilon_0, \varepsilon_1$, then adding ε_2 obtaining the three-point mesh $\varepsilon_0, \varepsilon_1, \varepsilon_2$, a.s.o. Associated with these meshes we obtain a sequence of NMTO sets: the kinked-partial wave set, $\chi^{(0)}(\mathbf{r})$, the LMTO set, $\chi^{(1)}(\mathbf{r})$, the QMTO set, $\chi^{(2)}(\mathbf{r})$, a.s.o. Working *downwards*, we thus always *delete* the point with the *highest* index. Equation (80) now shows that $\mathcal{H} - \varepsilon_N$ may be viewed as the *step-down* operator and $E^{(N)} - \varepsilon_N$ as the corresponding *transfer matrix* with respect to the *order* of the NMTO set.

In this sequence we may include the case $N=0$, provided that we define:

$$E^{(0)} - \varepsilon_0 \equiv -K(\varepsilon_0) \quad \text{and} \quad \chi^{(-1)}(\mathbf{r}) \equiv \delta(\mathbf{r}). \tag{82}$$

$N + 1$ successive step-down operations on the NMTO set thus yield:

$$(\mathcal{H} - \varepsilon_0) \dots (\mathcal{H} - \varepsilon_N) \, \chi^{(N)}(\mathbf{r}) = \delta(\mathbf{r}) \left(E^{(0)} - \varepsilon_0 \right) \dots \left(E^{(N)} - \varepsilon_N \right)$$

which, first of all, tells us that one has to operate N times with ∇^2 – that is, with ∇^{2N} – before getting to the non-smoothness of an NMTO. This is consistent with the conclusion about kinks and triple-valuedness reached in the preceding sub-section. Secondly, it tells us that the higher the N, the more spread out the NMTOs; if we let $r(M)$ denote the range of the $E^{(M)}$-matrix, then the range of the NMTO is roughly $\sum_{M=0}^{N} r(M)$.

The product of $E^{(0)} - \varepsilon_0$ and all the transfer matrices on the right-hand side of the above equation is seen from (81) and (82) to be simply: $-G[0...N]^{-1}$. Hence, we have found the *matrix equivalent* of the elementary relation (78):

$$-G[0...N]^{-1} = \left(E^{(0)} - \varepsilon_0 \right) \left(E^{(1)} - \varepsilon_1 \right) \dots \left(E^{(N)} - \varepsilon_N \right). \tag{83}$$

The other way around: Recursive use of (83) with increasing N, will generate the transfer matrices and will lead to the first equation (81). Note that although the order of the arguments in the divided difference on the left-hand side is irrelevant, the order of the factors on the right-hand side is *not*, since the transfer matrices *do not commute*. That $G[0...N]$ is Hermitian, is not so obvious from (83) either. Finally, we may note that $G[0..n-1, n+1..N]$ is *not* defined by (83) but by (148):

$$G[0..n-1, n+1..N] \equiv G[0...N-1] + (\varepsilon_N - \varepsilon_n) G[0....N].$$

Relation (83) now gives the following form for the Lagrange weights (73):

$$L_n^{(N)} = \left(E^{(n)} - \varepsilon_n \right)^{-1} \frac{\left(E^{(0)} - \varepsilon_0 \right) \dots \left(E^{(n)} - \varepsilon_n \right) \dots \left(E^{(N)} - \varepsilon_N \right)}{(\varepsilon_n - \varepsilon_0) \dots (\varepsilon_n - \varepsilon_{n-1})(\varepsilon_n - \varepsilon_{n+1}) \dots (\varepsilon_n - \varepsilon_N)}, \tag{84}$$

and this is seen to pass over to the classical expression (146) for the Lagrange coefficients if we substitute all energy matrices by the energy: $E^{(M)} \to \varepsilon$. This correspondence between – on the one side – the set $\phi(\varepsilon, \mathbf{r})$ and the Lagrange polynomial approximation (146) to its energy dependence (Fig. 1) and – on the other side – the set $\chi^{(N)}(\mathbf{r})$ expressed by (68) with the matrix form (84), is conceptually very pleasing. What is not so obvious – but comforting – is that the Hilbert space spanned by the NMTO set is invariant under energy-dependent linear transformations, $\hat{\phi}(\varepsilon, \mathbf{r}) \equiv \phi(\varepsilon, \mathbf{r}) T(\varepsilon)$, of the kinked partial waves. This will be shown in a later section.

By taking matrix elements of (80), the transfer matrix may be expressed as:

$$E^{(N)} - \varepsilon_N = \left\langle \chi^{(N)} \mid \chi^{(N-1)} \right\rangle^{-1} \left\langle \chi^{(N)} \mid \mathcal{H} - \varepsilon_N \mid \chi^{(N)} \right\rangle. \tag{85}$$

This holds also for $N=0$, provided that we take the value of $\chi^{(0)}(\mathbf{r})$ at its screening sphere to be $\varphi^{\circ a}(\varepsilon, a) = 1$ – as dictated by the 3-fold way – so that $\left\langle \chi^{(0)} \mid \chi^{(-1)} \right\rangle = 1$. The form (85) shows that the transfer matrices with $N \geq 1$ are *not* Hermitian, but short ranged, as one may realize by recursion starting from $N=0$. Finally, it should be remembered that the NMTOs considered sofar have particular normalizations, which are *not:* $\left\langle \chi^{(N)} \mid \chi^{(N)} \right\rangle = 1$, and so do the transfer matrices. We shall return to this point.

Newton form. Instead of using the Lagrange form (149) to factorize the NMTO (70), we may use the divided-difference expression (150). With the substitutions: $f(\varepsilon) \to G(\varepsilon)$ and $g(\varepsilon) \to \phi(\varepsilon, \mathbf{r})$, we obtain the Newton form for the NMTO which most clearly exhibits the step-down property (80):

$$\chi^{(N)}(\mathbf{r}) = \sum_{M=N}^{0} \phi([M..N], \mathbf{r}) G[0..M] G[0...N]^{-1}$$

$$= \phi_N(\mathbf{r}) + \phi([N-1, N], \mathbf{r})\left(E^{(N)} - \varepsilon_N\right) + .. \tag{86}$$

$$.. + \phi([0...N], \mathbf{r})\left(E^{(1)} - \varepsilon_1\right) .. \left(E^{(N)} - \varepsilon_N\right),$$

since, from (56) and (79),

$$(\mathcal{H} - \varepsilon_N)\,\phi_N(\mathbf{r}) \qquad = -\delta_{N,0}\delta(\mathbf{r})\,K_0,$$

$$(\mathcal{H} - \varepsilon_N)\,\phi([M...N], \mathbf{r}) = \phi([M..N-1], \mathbf{r}). \tag{87}$$

We thus realize that the energy matrices in the Newton series for the NMTO set are the matrices for stepping down to the sets of lower order. For some purposes, the 'reversed' series, obtained from (150) with $f(\varepsilon) \to \phi(\varepsilon, \mathbf{r}) G(\varepsilon)$ and $g(\varepsilon) \to G(\varepsilon)$:

$$\chi^{(N)}(\mathbf{r}) = \sum_{M=0}^{N} \phi([0..M], \mathbf{r}) G[M..N] G[0...N]^{-1}$$

$$= \phi_0(\mathbf{r}) + \phi([01], \mathbf{r})\left(E^{(N)} - \varepsilon_0\right) + .. \tag{88}$$

$$.. + \phi([0...N], \mathbf{r})\left(E^{(1)} - \varepsilon_{N-1}\right) .. \left(E^{(N)} - \varepsilon_0\right),$$

is more convenient. This expression clearly exhibits the correspondence with the Newton polynomial approximation (147) to the energy dependence of $\phi(\varepsilon, \mathbf{r})$. Conceptually, a divided-difference series is more desirable than the Lagrange series, because the Lagrange weights (84) 'fluctuate wildly' as a function of n, taken in the order of monotonically increasing energies.

For a condensed mesh, (86) and (88) obviously reduce to one-and-the-same matrix-equivalent of the Taylor series for $\phi(\varepsilon, \mathbf{r})$:

$$\chi^{(N)}(\mathbf{r}) \rightarrow \phi(\mathbf{r}) + \dot{\phi}(\mathbf{r}) \left(E^{(N)} - \varepsilon_\nu \right) + ..$$

$$.. + \frac{1}{N!} \overset{(N)}{\phi}(\mathbf{r}) \left(E^{(1)} - \varepsilon_\nu \right) .. \left(E^{(N)} - \varepsilon_\nu \right),$$

and (87) becomes:

$$(\mathcal{H} - \varepsilon_\nu) \phi(\mathbf{r}) = -\delta_{N,0}\delta(\mathbf{r})K, \quad (\mathcal{H} - \varepsilon_\nu) \frac{\overset{(N-M)}{\phi}(\mathbf{r})}{(N-M)!} = \frac{\overset{(N-M-1)}{\phi}(\mathbf{r})}{(N-M-1)!}.$$

Readers used to the LMTO-ASA method, where – according to (12) – the KKR matrix is basically the two-center TB Hamiltonian, may not like the thought of having to differentiate its inverse, the Green matrix, with respect to energy. (The computer seems to work well with the formalism based on the Green matrix). Such readers might therefore prefer an NMTO formalism in terms of kink matrices. For a discrete mesh many ugly relations exist, but the one relation which is conceptually pleasing is the following:

$$0 = \qquad\qquad\qquad\qquad\qquad\qquad\qquad\qquad\qquad (89)$$

$$K_0 + K[01] \left(E^{(N)} - \varepsilon_0 \right) + .. + K[0..N] \left(E^{(1)} - \varepsilon_{N-1} \right) .. \left(E^{(N)} - \varepsilon_0 \right),$$

because it looks like the matrix form of the secular KKR equation: $|K(\varepsilon)| = 0$. This relation may be obtained by taking the Nth divided difference of the equation: $K(\varepsilon)\,G(\varepsilon) \equiv 1$, using the binomial expression (150) for a product like in (88), but with $K(\varepsilon)$ substituted for $\phi(\varepsilon, \mathbf{r})$, and multiplying the result from the right by $G[0...N]^{-1}$. To find the transfer matrices from (89), we may solve for $E^{(N)} - \varepsilon_0$ and do recursion starting from $N=1$. The results are:

$$E^{(1)} - \varepsilon_0 = -K[01]^{-1} K_0 \qquad \rightarrow \qquad -\dot{K}^{-1}K,$$

$$E^{(2)} - \varepsilon_0 = -\left(K[01] + K[012] \left(E^{(1)} - \varepsilon_1 \right) \right)^{-1} K_0 \qquad (90)$$

$$\rightarrow -\left(\dot{K} - \ddot{K}\dot{K}^{-1}K/2 \right)^{-1} K,$$

a.s.o. These low-N expressions are reasonably simple. For $N=1$, the discrete form is seen to be identical with (74) and, for a condensed mesh, it reduces to the well-known expression for the 3rd-generation LMTO. We conclude that the energy matrices, $E^{(M)}$, are well-behaved functions of the kink matrix and its divided differences, up to and including Mth order. With M increasing, the corresponding expressions for $E^{(M)}$ however become more and more complicated. The simplest expression for $E^{(M)}$ is therefore (81), the one which uses G-language.

3.2   Variational NMTO Method

The NMTO set has been defined through (69) and (70) in such a way that its leading errors are proportional to $(\varepsilon - \varepsilon_0) .. (\varepsilon - \varepsilon_N)$. By virtue of the variational principle, solution of the generalized eigenvalue problem (5) with this basis set will therefore provide *one-electron energies*, ε_i, with a leading error $\propto (\varepsilon_i - \varepsilon_0)^2 .. (\varepsilon_i - \varepsilon_N)^2$. The error of the *wave function* will of course still be of order $(\varepsilon_i - \varepsilon_0) .. (\varepsilon_i - \varepsilon_N)$, but that is usually all right because, as mentioned at the beginning of the present section, the MTO scheme is based on the factorization: $\gamma (\varepsilon, \mathbf{r}) = \phi (\varepsilon, \mathbf{r}) G (\varepsilon)$, where $\phi (\varepsilon, \mathbf{r})$ has a *smooth* energy dependence and $G (\varepsilon)$ provides the poles at the one-electron energies.

Hamiltonian and overlap matrices. For a variational calculation, we need expressions for the NMTO overlap and Hamiltonian matrices, $\langle \chi^{(N)} \mid \chi^{(N)} \rangle$ and $\langle \chi^{(N)} \mid \mathcal{H} \mid \chi^{(N)} \rangle$. From (69), the Nth divided difference of the contracted Green function (58) is:

$$\gamma^{(N)} ([0..N], \mathbf{r}) = \chi^{(N)} (\mathbf{r}) G [0..N] = \sum_{n=0}^{N} \frac{\phi_n (\mathbf{r}) G_n}{\prod_{m=0,\neq n}^{N} (\varepsilon_n - \varepsilon_m)} \qquad (91)$$

and using now (63), we obtain for the integral over the product of the Mth and Nth divided differences of contracted Green functions:

$$\langle \gamma [0...M] \mid \gamma [0....N] \rangle = \sum_{n=0}^{N} \sum_{n'=0}^{M} \frac{-G [n, n']}{\prod_{m=0,\neq n}^{N} (\varepsilon_n - \varepsilon_m) \prod_{m'=0,\neq n'}^{M} (\varepsilon_{n'} - \varepsilon_{m'})}$$

$$= -G [[0...M]..N] \; \rightarrow \; -\frac{\overset{(M+N+1)}{G}}{(M + N + 1)!}. \qquad (92)$$

This is simply the negative of the $(M + N + 1)$st *Hermite divided difference* (152) of the Green matrix, as proved in Eq. (160) in the Appendix!

Note that the meaning of a matrix equation like (63) is:

$$\langle \gamma_{RL} (\varepsilon_n) \mid \gamma_{R'L'} (\varepsilon_{n'}) \rangle = -G_{RL,R'L'} [n, n']$$
$$= -G_{RL,R'L'} [n', n] = \langle \gamma_{RL} (\varepsilon_{n'}) \mid \gamma_{R'L'} (\varepsilon_n) \rangle .$$

In *matrix* notation, that is: $\langle \gamma_n \mid \gamma_{n'} \rangle = \langle \gamma_{n'} \mid \gamma_n \rangle$, and not: $\langle \gamma_n \mid \gamma_{n'} \rangle = \langle \gamma_{n'} \mid \gamma_n \rangle^*$. Even without the symmetry of the matrix $G [n, n']$ with respect to the exchange of n and n', it is of course always true that

$$\langle \gamma_{RL} (\varepsilon_n) \mid \gamma_{R'L'} (\varepsilon_{n'}) \rangle = \langle \gamma_{R'L'} (\varepsilon_{n'}) \mid \gamma_{RL} (\varepsilon_n) \rangle^* ,$$

i.e. that a matrix like $\langle \gamma_n \mid \gamma_{n'} \rangle$ is Hermitian: $\langle \gamma_n \mid \gamma_{n'} \rangle = \langle \gamma_{n'} \mid \gamma_n \rangle^\dagger$. The point is, that n is an argument – not an index – of a matrix. Similarly, N and M

are not matrix indices in (92). Since the first expression (92) is symmetric under exchange of N and M, because $G[n, n']$ is symmetric, we may choose $M \leq N$, and this has in fact been done in the second expression.

From (79) and (92), we now see that the Hamiltonian matrix between the Nth divided differences of contracted Green functions becomes:

$$\langle \gamma[0...N] | \mathcal{H} - \varepsilon_N | \gamma[0...N] \rangle = \langle \gamma[0...N] | \gamma[0..N-1] \rangle$$

$$= -G[[0..N-1]N] \rightarrow -\frac{\overset{(2N)}{G}}{(2N)!}. \qquad (93)$$

Hence, we have arrived at the important results: The NMTO overlap matrix may be expressed in terms of the Nth-order divided difference and the $(2N+1)$st Hermite divided difference of the Green matrix as:

$$\left\langle \chi^{(N)} | \chi^{(N)} \right\rangle = -G[0...N]^{-1} G[[0...N]] G[0...N]^{-1}, \qquad (94)$$

where the – even simpler – result for a condensed mesh was quoted in the Overview (20). The Hermite derivative $G[[0, ..., N]]$ is thus negative definite. The NMTO Hamiltonian matrix may be expressed analogously, in terms of a $2N$th-order Hermite divided difference:

$$\left\langle \chi^{(N)} | \mathcal{H} - \varepsilon_N | \chi^{(N)} \right\rangle = -G[0...N]^{-1} G[[0..N-1]N] G[0...N]^{-1}. \qquad (95)$$

Here again, the result given in (20) for a condensed mesh is even simpler. The NMTO Green function is

$$\left\langle \chi^{(N)} | z - \mathcal{H} | \chi^{(N)} \right\rangle^{-1} =$$

$$G[0...N] \{G[[0..N-1]N] - (z - \varepsilon_N) G[[0...N]]\}^{-1} G[0...N]$$

Expressions (94) and (95) for the NMTO overlap and Hamiltonian matrices are not only simple and beautiful, but they also offer sweet coding and speedy computation. For a crystal, and transforming to **k**-representation, one may even use the representation of contracted Green functions where the overlap and Hamiltonian matrices – according to (92) and (93) – are merely $-G[[0...N]]$ and $-G[[0..N-1]N]$. In Section 4 we shall see that an energy-dependent linear transformation of the kinked partial waves does *not* change the Hilbert space spanned by an energy-independent NMTO set – but only the individual basis functions. Therefore, we might also use kinked partial waves $\phi^\alpha(\varepsilon, \mathbf{r})$ and the Green matrix $G^\alpha(\varepsilon)$ with phase-shift normalization.

In summary: The variational NMTO scheme requires computation of the kink matrix and its first energy derivative at the $N+1$ mesh points. It delivers energies and wave functions which are correct to order $2N + 1$ and N, respectively. This lower accuracy of the wave functions is appropriate because the kinked partial waves are rather smooth functions of energy. For the computation of the $\dot{\partial}_n$'s entering $\dot{K}_n \equiv a \left(\dot{B}_n - \dot{\partial}_n \right)$, radial normalization-integrals should be used.

As an example, for the LMTO method, the Hamiltonian and overlap matrices are respectively:

$$\left\langle \chi^{(1)} \middle| \mathcal{H} - \varepsilon_1 \middle| \chi^{(1)} \right\rangle = -G\,[01]^{-1}\,G\,[[0]\,1]\,G\,[01]^{-1}$$

$$= (\varepsilon_0 - \varepsilon_1)\,(G_0 - G_1)^{-1}\left(-\dot{G}_0 + G\,[01]\right)(G_0 - G_1)^{-1} \tag{96}$$

$$\rightarrow -\dot{G}^{-1}\frac{\ddot{G}}{2!}\dot{G}^{-1} = -K + K\dot{K}^{-1}\frac{\ddot{K}}{2!}\dot{K}^{-1}K,$$

and

$$\left\langle \chi^{(1)} \middle| \chi^{(1)} \right\rangle = -G\,[01]^{-1}\,G\,[[01]]\,G\,[01]^{-1}$$

$$= (G_0 - G_1)^{-1}\left(-\ddot{G}_0 + 2G\,[01] - \ddot{G}_1\right)(G_0 - G_1)^{-1} \tag{97}$$

$$\rightarrow -\dot{G}^{-1}\frac{\dddot{G}}{3!}\dot{G}^{-1} = \dot{K} - K\dot{K}^{-1}\frac{\ddot{K}}{2!} - \frac{\ddot{K}}{2!}\dot{K}^{-1}K + K\dot{K}^{-1}\frac{\dddot{K}}{3!}\dot{K}^{-1}K.$$

The result for a condensed mesh in terms of the kink matrix and its first three energy derivatives is seen to be almost identical to the one (16), which in previous LMTO generations required the ASA. To get exactly to (16), one needs to transform to the LMTO set: $\hat{\chi}^{(1)}(\mathbf{r}) \equiv \chi^{(1)}(\mathbf{r})\,\dot{K}^{-1/2}$, which in fact corresponds to a Löwdin orthonormalization of the 0th-order set. We shall return to this matter in Sect. 6. From the above relations we realize that – even for a condensed mesh and N as low as 1 – G-language is far simpler than K-language.

Orthonormal NMTOs. In many cases one would like to work with a representation of *orthonormal* NMTOs, which preserves the RL-character of each NMTO. In order to arrive at this, we should – in the language of Löwdin – perform a *symmetrical orthonormalization* of the NMTO set. According to (94) such a representation is obtained by the following transformation:

$$\check{\chi}^{(N)}(\mathbf{r}) = \chi^{(N)}(\mathbf{r})\,G\,[0...N]\,\sqrt{-G\,[[0...N]]}^{\,-1}, \tag{98}$$

because it yields:

$$\left\langle \check{\chi}^{(N)} \middle| \check{\chi}^{(N)} \right\rangle = -\sqrt{-G\,[[0...N]]}^{\,-1\dagger}\,G\,[[0...N]]\,\sqrt{-G\,[[0...N]]}^{\,-1} = 1.$$

Note that this means: $-G\,[[0..N]] = \sqrt{-G\,[[0..N]]}^{\,\dagger}\sqrt{-G\,[[0..N]]}$. In this orthonormal representation, the Hamiltonian matrix becomes

$$\left\langle \check{\chi}^{(N)} \middle| \mathcal{H} - \varepsilon_N \middle| \check{\chi}^{(N)} \right\rangle = -\sqrt{-G\,[[0...N]]}^{\,-1\dagger} \times \tag{99}$$

$$G\,[[0..N-1]\,N]\,\sqrt{-G\,[[0...N]]}^{\,-1}.$$

To find an efficient way to *compute* the *square root* of the Hermitian, positive definite matrix $-G\,[[0...N]]$ may be a problem. Of course one may diagonalize the matrix, take the square root of the eigenvalues, and then back-transform, but this is time consuming. Cholesky decomposition is a better alternative, but that usually amounts to staying in the original representation. Löwdin orthogonalization works if the set is nearly orthogonal, because then the overlap matrix is nearly diagonal, and Löwdin's solution was to normalize the matrix such that it becomes 1 along the diagonal and then expand in the off-diagonal part, O :

$$\sqrt{1+O}^{-1} = 1 - \frac{1}{2}O + \frac{3}{8}O^2 - \dots \qquad (100)$$

This should work for the NMTO overlap matrix (94) when the NMTOs are nearly orthogonal, but it hardly works for $-G\left[\left[0...N\right]\right]$. There is therefore no advantage in pulling out the factor $G\left[0...N\right]$, on the contrary. The other way around: In order to take the square root of $-G\left[\left[0...N\right]\right]$, we should find a transformation, T, such that $T^\dagger G\left[\left[0...N\right]\right]T$ is nearly diagonal, and then perform the Löwdin orthonormalization on the latter matrix. We shall return to this problem in Sect. 5.

One-orbital model: switching behavior of $H^{(N)}$, $L_n^{(N)}$, and the variational energy. Our development of the NMTO formalism has been focused on its matrix aspects and, through the introduction of energy *matrices* and by pointing to the correspondence with classical Lagrange and Newton interpolation of the energy-dependent kinked partial waves, we have tried to make the reader accept the seemingly uncomfortable fact, that the quantities of interest do arise by energy differentiations of a Green matrix.

Let us now illustrate the Green-function aspects by considering the 1×1 Green matrix (77) for *one, normalized orbital:* $\check{\chi}^{(N)}(\mathbf{r}) = \Psi_j(\mathbf{r})$ with $\left\langle \left| \check{\chi}^{(N)} \right|^2 \right\rangle = 1$. Note that in this model, j runs over the one-electron energies, which is a different set – with much larger spacing – than the energy mesh whose points are denoted n and m. For a crystal, and using Bloch-symmetrized NMTOs and Green matrices, $\check{\chi}^{(N)}(\mathbf{k},\mathbf{r})$ and $\check{G}(\varepsilon,\mathbf{k})$, this would be an s-band model with j being the radial quantum number. We want the NMTO to describe the i-band and therefore choose the mesh between $\varepsilon_{i-1}(\mathbf{k})$ and $\varepsilon_{i+1}(\mathbf{k})$. In the following we shall drop the Bloch vector and not necessarily consider a crystal.

We first demonstrate how $\check{E}^{(N)} \equiv H^{(N)}$ – in this case a 1×1 Hamiltonian (see Sect.5) – expressed in terms of ratios of energy derivatives of a Green function, with its singular behavior, produces correct results for the one-electron energy and how, when the mesh is swept over a large energy interval, $H^{(N)}$ switches between bands with different radial quantum numbers. From (81) and (78) we get:

$$H^{(N)} - \varepsilon_N = \frac{\check{G}\left[0..N-1\right]}{\check{G}\left[0...N\right]} = (\varepsilon_i - \varepsilon_N) \frac{1 + \sum\limits_{\substack{j \neq i}} \prod\limits_{m=0}^{N-1} \frac{\varepsilon_i - \varepsilon_m}{\varepsilon_j - \varepsilon_m}}{1 + \sum\limits_{\substack{j \neq i}} \prod\limits_{m=0}^{N} \frac{\varepsilon_i - \varepsilon_m}{\varepsilon_j - \varepsilon_m}}.$$

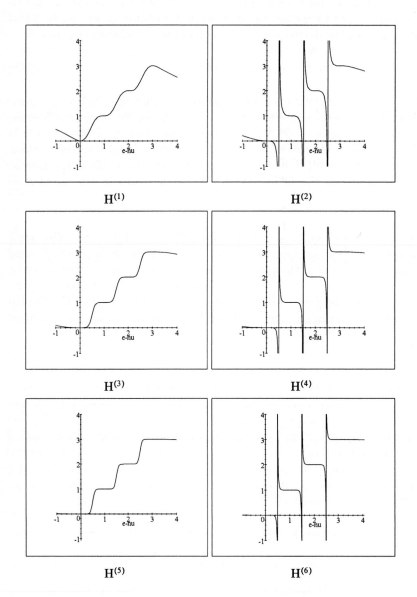

Fig. 6. Switching behavior of $E^{(N)}\left(\varepsilon_\nu\right) \equiv H^{(N)}\left(\varepsilon_\nu\right)$ for the orthonormal one-orbital model defined by Eq. (77) with 4 radial levels: $\varepsilon_j = 0, 1, 2, 3$.

Hence, for the model and an energy mesh with $N+1$ points, $H^{(N)}$ equals ε_i to order N, with an error proportional to $(\varepsilon_i - \varepsilon_0) .. (\varepsilon_i - \varepsilon_N)$, which for a condensed mesh becomes $(\varepsilon_i - \varepsilon_\nu)^{N+1}$. In Fig. 6 we show $H^{(N)}(\varepsilon_\nu)$ for $N = 1$ to 6, computed from the above expression for a four-level model with $\varepsilon_j = 0, 1, 2,$ and 3, and a condensed mesh. We see that $H^{(N)}(\varepsilon_\nu)$ behaves as it should: It switches from one level to the next, with the plateau around each level flattening out as N increases. For N odd, the switching-curve is step-like and, for N even, the switching is via $-\infty \rightarrow +\infty$. This comes from the ability of the denominator in the expression for $H^{(N)}$ to be zero when $N+1$ is odd. An energy-*independent* orbital, as considered in the present model, can of course only describe *one* band. With the NMTO defined for a mesh condensed onto a chosen energy ε_ν, we want to describe the band near ε_ν as well as possible – also if the distance to the next band is small – and with a result which over a large region is insensitive to the choice of ε_ν. In a *multi-orbital* calculation, we should fold down those channels which are switching in the energy range of interest into the screened spherical waves. This will remove schizophrenic members of the NMTO set and prevent the possible occurrence of ghost bands.

In the one-orbital model, the estimate of a true, normalized wave function, $\check{\phi}(\varepsilon_i, \mathbf{r})$, is the Nth-order muffin-tin orbital: $\check{\chi}^{(N)}(\mathbf{r}) = \sum_n^N \check{\phi}_n(\mathbf{r}) L_n^{(N)}$. If we now use (77) and (78) to evaluate expression (73) for the Lagrange weights, we find:

$$L_n^{(N)} = \frac{\sum_j \frac{1}{\varepsilon_j - \varepsilon_n}}{\sum_j \frac{1}{\varepsilon_j - \varepsilon_n} \prod_{m=0, \neq n}^N \frac{\varepsilon_n - \varepsilon_m}{\varepsilon_j - \varepsilon_m}} = l_n^{(N)}(\varepsilon_i) \frac{1 + \sum_{j \neq i} \frac{\varepsilon_i - \varepsilon_n}{\varepsilon_j - \varepsilon_n}}{1 + \sum_{j \neq i} \prod_{m=0}^N \frac{\varepsilon_i - \varepsilon_m}{\varepsilon_j - \varepsilon_m}},$$

where $l_n^{(N)}(\varepsilon)$ is the Lagrange polynomial (146) of degree N. We have therefore reached the conclusion that – in our orthonormal model, and to leading order – the wave function is the energy-dependent MTO, $\check{\phi}(\varepsilon, \mathbf{r})$, Lagrange interpolated over the $(N+1)$-point mesh.

Since the error of an NMTO set is of order $N+1$, use of the *variational* principle will reduce the error of the one-electron *energies*, ε_i, from that of the *highest* transfer matrix, $H^{(N)} - \varepsilon_N$, to order $2(N+1)$. The variational energies are thus correct to order $2N+1$. For a condensed mesh, this also follows trivially from (94)-(95), which show that the variational energy, with respect to ε_ν, is:

$$\frac{\langle \chi^{(N)} | \mathcal{H} - \varepsilon_\nu | \chi^{(N)} \rangle}{\langle \chi^{(N)} | \chi^{(N)} \rangle} = \frac{\overset{(2N)}{G}}{(2N)!} \bigg/ \frac{\overset{(2N+1)}{G}}{(2N+1)!} = H^{(2N+1)} - \varepsilon_\nu.$$

The odd-ordered switching curves $H^{(1)}(\varepsilon_\nu)$, $H^{(3)}(\varepsilon_\nu)$, and $H^{(5)}(\varepsilon_\nu)$ shown in the left-hand panel of Fig. 6 are thus the variational estimates resulting from the use of respectively the 0th, 1st, and 2nd-order NMTO, that is, the MTO, the LMTO, and the QMTO. These curves are well behaved.

The expression for the variational energy in the one-band model can be evaluated exactly, also for a discrete mesh, and yields a transparent result. We use the double-mesh procedure explained in the Appendix after (152), and let the differences $\epsilon_n \equiv \varepsilon_{n+N+1} - \varepsilon_n$ shrink to zero. From (78) we then get:

$$\check{G}\left[[0...N]\right] \qquad = -\sum_j \frac{1}{\prod_{m=0}^{N}\left(\varepsilon_j - \varepsilon_m\right)^2}, \qquad (101)$$

$$\check{G}\left[[0..N-1]\,N\right] = -\sum_j \frac{1}{\left(\varepsilon_j - \varepsilon_N\right)\prod_{m=0}^{N-1}\left(\varepsilon_j - \varepsilon_m\right)^2},$$

and for the variational energy (99):

$$\left\langle \check{\chi}^{(N)}\left|\mathcal{H}-\varepsilon_N\right|\check{\chi}^{(N)}\right\rangle = \left(\varepsilon_i-\varepsilon_N\right)\frac{1+\sum\limits_{j\neq i}\frac{\varepsilon_i-\varepsilon_N}{\varepsilon_j-\varepsilon_N}\prod\limits_{m=0}^{N-1}\left(\frac{\varepsilon_i-\varepsilon_m}{\varepsilon_j-\varepsilon_m}\right)^2}{1+\sum\limits_{j\neq i}\prod\limits_{m=0}^{N}\left(\frac{\varepsilon_i-\varepsilon_m}{\varepsilon_j-\varepsilon_m}\right)^2},$$

which of course agrees with the variational principle.

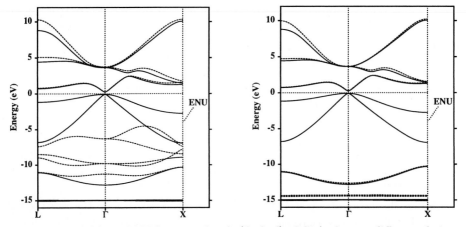

Fig. 7. Minimal-basis LMTO energy bands (dashed) of GaAs for two different choices of the screening-radii compared to the exact KKR band structure (solid). In the left-hand panel all screening radii were $\sim 0.8t$, while in the right-hand panel the Ga d radius was reduced to the radius of the Ga $3d$ core [24]. See text.

Treating semi-core and excited states: GaAs. An accurate description of the cohesive properties of GaAs requires a good band-structure calculation of the five Ga $3d^{10}$ semi-core, the As $4s^2$-band, and the three As $4p^2$ Ga $4sp^3$ valence bands. If also the four lowest conduction bands must be described, one is faced with the problem of computing a band structure containing extremely narrow as well as wide bands over a 20 eV-region. To do this *ab initio* with a *minimal* Ga spd As sp basis set (13 orbitals per GaAs), has hitherto not been possible.

With 1st and 2nd-generation LMTO-ASA methods one would normally use Rl-dependent ε_ν's and employ a 36-orbital-per-GaAs basis, consisting of the *spd* LMTOs centered on the Ga, the As, and the interstitial sites in the zincblende structure. The conduction-band errors arising from the choice $\kappa^2=0$ are so large that the combined correction is needed. Downfolding works for the *p* and *d* orbitals on the two interstitial spheres, but not for the interstitial *s* and the As *d* orbitals. With the 3rd-generation LMTO method, downfolding works much better, but the energy window is now screening dependent, and the use of Rl-dependent ε_ν's is avoided because it messes up the formalism.

In Fig. 7 we show – in full lines – the exact (up to 7eV) LDA band structure calculated by the screened KKR method, i.e. by the 3rd-generation LMTO method using *many* energy panels and the Ga *spd* As *sp* basis. The five Ga $3d^{10}$ semi-core bands are at -15 eV, the As $4s^2$-band is around -12 eV, and the three As $4p^2$ Ga $4sp^3$ valence bands extend from -7 to 0 eV. Above the gap, there are the four As $4p^4$ Ga $4sp^3$ conduction bands. The dotted lines give results of 3rd-generation LMTO variational calculations with a condensed mesh and an ε_ν in the middle of the three valence bands. In the left-hand

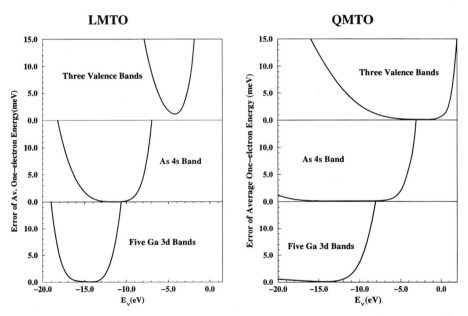

Fig. 8. Mean error in each of the three types of occupied valence bands in GaAs calculated with the LMTO and QMTO methods as a function of the expansion energy ε_ν for a condensed mesh [24]. See Fig. 7 and text.

figure, the screening-sphere radii for the active Ga *spd* and As *sp* channels were chosen at the Ga and As default values, respectively $0.82t$ and $0.78t$, where t is half the nearest-neighbor distance. We see that the entire valence-band structure is distorted by hybridization with Ga *d* ghost bands. The dotted bands in the right-hand figure result after changing the Ga d screening-sphere radius to $0.35t$, which is close to the actual radius of the Ga $3d$ core. Now, the band structure looks reasonable: The valence bands near ε_ν are perfect, but the Ga $3d$ bands are nearly 0.5 eV to high [24].

That the variational LMTO method with a minimal basis and a single ε_ν cannot describe all occupied states of GaAs with sufficient accuracy, becomes even more obvious from the left-hand side of Fig. 8, where we show – as functions of ε_ν – the average errors of the five Ga $3d$ bands, those of the As $4s$ band, and those of the three valence bands. The error $\propto (\varepsilon_i(\mathbf{k}) - \varepsilon_\nu)^4$ of the variational energy is clearly visible for the narrow Ga $3d$ and As $4s$ bands. With ε_ν's in a narrow range around -11 eV, the variational error in the sum of the one-electron energies gets down to about 250 meV per GaAs. On the right-hand side, we show the same quantities, but obtained with the QMTO method. Now the errors $\propto (\varepsilon_i(\mathbf{k}) - \varepsilon_\nu)^6$ are acceptable, and there is a comfortable range of ε_ν's around -10 eV where the error in the sum of the one-electron energies does not exceed 25 meV per GaAs. The screening-sphere radii chosen in these calculations [24] were: $0.93t$, $1.05t$, and $0.35t$ for respectively Ga s, p, and d, and $0.89t$ and $1.00t$ for respectively As s and p.

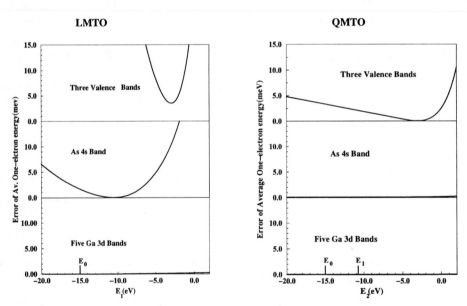

Fig. 9. Like Fig. 8, but calculated using discrete meshes and as functions of the position of the last energy point. The first energy points were fixed at the positions indicated on the abscissa [24]. See text.

In Fig. 9 we show the same kind of results, but this time obtained with the discrete (Lagrange) LMTO and QMTO methods. The size of the basis set, the screening-sphere radii, etc., were as in Fig. 8. For the LMTO method, ε_0 was fixed at the position of the Ga $3d$ bands and the figure shows the result of varying the position ε_1 of the other mesh point. The quadratic dependence on ε_1 of the variational energy-error $\propto \left(\varepsilon_i\left(\mathbf{k}\right) - \varepsilon_0\right)^2 \left(\varepsilon_i\left(\mathbf{k}\right) - \varepsilon_1\right)^2$ is clearly recognized. Compared with the results of the tangent LMTO method shown in the previous figure, those of the chord-LMTO are far superior: With ε_1's around -5 eV, the variational error in the sum of the one-electron energies gets down to about 30 meV per GaAs, and yet, for N given, the method employing a discrete mesh is computationally simpler than the one employing a condensed mesh. On the right-hand side of the figure, we show the QMTO results as functions of ε_2, with ε_0 fixed at the Ga $3d$ position, and ε_1 at the As $4s$ position. Here again, the quadratic dependence on ε_2 of the variational energy-error $\propto \left(\varepsilon_i\left(\mathbf{k}\right) - \varepsilon_0\right)^2 \left(\varepsilon_i\left(\mathbf{k}\right) - \varepsilon_1\right)^2 \left(\varepsilon_i\left(\mathbf{k}\right) - \varepsilon_2\right)^2$ may be seen. We realize, that with this discrete QMTO method, meV-accuracy for the sum of the one-electron energies can be reached.

Finally, in Fig 10 we show the GaAs band structure in a wide (40 eV) range around the gap. Further conduction bands now appear above 7 eV and we needed to employ a basis consisting of the Ga spd As $spdf$ 2E s QMTOs. ε_0 was chosen at the Ga $3d$ position, ε_1 near the gap, and ε_2 10 eV above the gap. The results of this discrete QMTO calculation shown by the dotted curves agree superbly with those of a multi-panel LMTO (=KKR) calculation shown in full line [24]. This proves the power of the 3rd-generation NMTO method.

Massive downfolding: $CaCuO_2$. An increasingly important field of research is the electronic structure of real materials with strongly correlated conduction electrons. Within a given class of materials, fine-tuning of the interesting properties will require detailed knowledge of the single-electron part – the orbitals, hopping integrals and basic on-site terms – of the correlated Hamiltonian. In the previous review [20] of the 3rd-generation 0th- and 1st-order differential MTO method, we demonstrated for the idealized high-temperature superconductor, $CaCuO_2$ with dimpled CuO_2 planes, how one could extract low-energy, few-band Hamiltonians by massive downfolding; in the extreme limit: Downfolding to *one* Cu $d_{x^2-y^2}$ orbital per Cu site [22,23]. Let us now reconsider this example in the light of the new NMTO methods.

In Fig. 11 the full lines in all four parts show the (same) full LDA band structure in a ± 3 eV region around the Fermi level, which for the doping levels of interests would be near the energy -0.8 eV of the so-called extended saddle-point at $\mathbf{X}$. The conduction band has mostly O-Cu anti-bonding $pd\sigma$-character (O p_x – Cu $d_{x^2-y^2}$) with the bonding partner lying 10 eV lower in energy. The bottom of the conduction band is seen to cross and hybridize with a multitude of O-Cu $pd\pi$-bands lying below -1.2 eV. The top of the conduction band hybridizes strongly with a broad O-Ca bonding $pd\pi$ (O p_x – Ca d_{xy}) band near $\mathbf{A}$. In this situation, one clearly does not want to use the rather ill-defined and very long-ranged Wannier orbital for describing the low-energy electronic structure. Rather, one wants an orbital which describes the band (including its dependence on other relevant low-energy excitations such as spin-fluctuations and phonons) in the ± 200meV range around ε_F, that is an NMTO with *all* channels, except

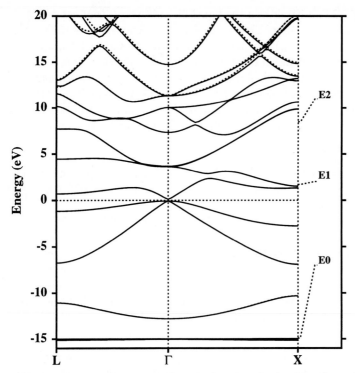

Fig. 10. Energy bands of GaAs calculated with the QMTO method and the energy mesh indicated on the right-hand side (dashed) as compared with the exact KKR result (solid) [24]. See text.

Cu $d_{x^2-y^2}$, downfolded and with as short a range as possible. The four dotted bands shown in each of the sub-figures result from such calculations [24]. In all cases, the screening-sphere radius of Cu $d_{x^2-y^2}$ was taken to be $0.62t$. The upper figures illustrate a problem with the 3rd-generation tangent LMTO method: If ε_ν is taken where we want it to be, at the -0.8 eV saddle-point deep down in the anti-bonding $pd\sigma$-band, then the method develops a *schizophrenia* near the top of the band, above 1 eV and near **M**, which is apparently sufficiently far away from ε_ν that the LMTO 'might consider' describing the bonding rather than the anti-bonding state.

The resulting orbital has very long range due to the high Fourier components caused by the schizophrenia and, as a result, we are forced to take ε_ν at a higher energy than we actually want. With $\varepsilon_\nu = -0.3$ eV, we still get long range as seen in the upper left-hand figure, and in order to cure that problem we need to go to $\varepsilon_\nu = +0.3$ eV, but then the description of the bottom of the anti-bonding band, the extended saddle-point in particular, has substantially deteriorated. In the lower left-hand figure we have now switched from the tangent to the chord LMTO, and that is seen to help considerably. Finally, the lower right-hand figure

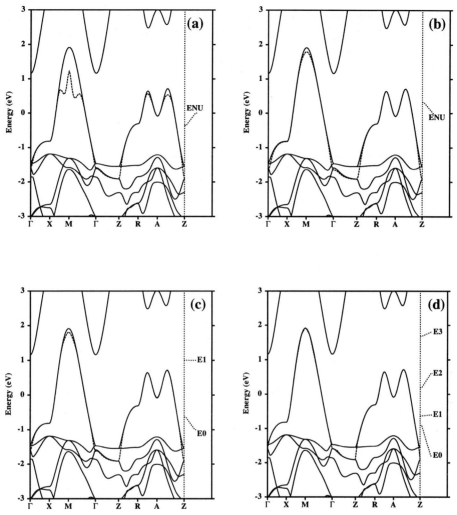

Fig. 11. Conduction band of $CaCuO_2$ calculated by massive downfolding to a single Cu $x^2 - y^2$ NMTO (dotted) compared with the full band structure (solid) [24]. See text.

presents what might be called an 'overkill': We have used the discrete CMTO (N=3) method, and the agreement with the exact result is superb.

Using integrals of divided differences of MTOs. In all previous derivations of the variational LMTO method, the LMTO was expressed as a matrix Taylor series (1) and the Hamiltonian and overlap matrices (7) were worked out using expressions (12) for $\left\langle \phi \mid \dot{\phi} \right\rangle$ and $\left\langle \dot{\phi} \mid \dot{\phi} \right\rangle$.

The same may be done for the general, discrete NMTO method, although the number of terms in the resulting series increases quadratically with N. For this, we first use a divided-difference form – such as (88) – for the NMTO and then need expressions for the overlap integrals, $\langle \phi \, [0..N] \mid \phi \, [0..M] \rangle$, and Hamiltonians, $\langle \phi \, [0..N] \mid \mathcal{H} \mid \phi \, [0..M] \rangle$, between divided differences of kinked partial waves. Since expressions (62) and (63) are formally equivalent, we find that, analogous to (92),

$$\langle \phi \, [0..M] \mid \phi \, [0...N] \rangle \;=\; \langle \phi \, [0...N] \mid \phi \, [0..M] \rangle \;=\; K \, [[0..M] \,.N] \quad (102)$$

$$\rightarrow \left\langle \frac{\overset{(M)}{\phi}}{M!} \,\Bigg|\, \frac{\overset{(N)}{\phi}}{N!} \right\rangle \;=\; \left\langle \frac{\overset{(N)}{\phi}}{N!} \,\Bigg|\, \frac{\overset{(M)}{\phi}}{M!} \right\rangle \;=\; \frac{\overset{(M+N+1)}{K}}{(M+N+1)!},$$

where we have assumed $M \leq N$. From this result for $M = N$, it follows that the odd-ordered Hermite divided differences of the kink matrix are positive definite. For a contracted mesh, this overlap matrix is seen to depend only on $M + N$.

For the matrix elements of the Hamiltonian we must use:

$$\langle \phi \, [0..M] \mid \mathcal{H} - \varepsilon_n \mid \phi \, [0...N] \rangle = \langle \phi \, [0..M] \mid \phi \, [0..n-1, n+1..N] \rangle$$

$$= \begin{cases} K \, [[0..n-1, n+1..\min(M,N)] \, n.. \max(M,N)] \\ K \, [[0.. \min(M,N)]..n-1, n+1.. \max(M,N)] \end{cases} \quad (103)$$

$$\rightarrow \left\langle \frac{\overset{(M)}{\phi}}{M!} \,\Big|\, \mathcal{H} - \varepsilon_\nu \,\Big|\, \frac{\overset{(N)}{\phi}}{N!} \right\rangle \;=\; \left\langle \frac{\overset{(M)}{\phi}}{M!} \,\Bigg|\, \frac{\overset{(N-1)}{\phi}}{(N-1)!} \right\rangle \;=\; \frac{\overset{(M+N)}{K}}{(M+N)!}.$$

The upper and lower results on the second line correspond to $n \lesseqgtr \min(M,N)$. Here again, for a condensed mesh the Hamiltonian matrix depends only on $M + N$.

The resulting expressions for $\langle \chi^{(N)} \mid \chi^{(N)} \rangle$ and $\langle \chi^{(N)} \mid \mathcal{H} - \varepsilon_n \mid \chi^{(N)} \rangle$ contain the above-mentioned integrals *times* products of $\left(E^{(N-M+1)} - \varepsilon_{M-1} \right)$-matrices. These expressions are by far not as explicit as equations (94) and (95), and they are more complicated for a discrete than for a condensed mesh. We shall now consider a more useful application of (102)-(103).

Charge density and total energy: Si phase diagram. The wave function obtained from a variational calculation is: $\Psi_i (\mathbf{r}) = \chi (\mathbf{r}) c_i$, where we have dropped the superscript (N) on the NMTO. The eigen(column)vector, c_i, of the generalized eigenvalue equation (5) should be normalized according to: $c_i^\dagger \langle \chi \mid \chi \rangle c_{i'} = \delta_{ii'}$, or – regarding $c_{RL,i}$ as a matrix – according to: $c^\dagger \langle \chi \mid \chi \rangle c = 1$. The charge density is now given by (18), which to a very good approximation is (64) with the energy-dependent wave functions in expressions (65)-(66) substituted by their matrix Lagrange or Newton series. The computer code would use the Lagrange form:

$$\rho (\mathbf{r}) = \chi (\mathbf{r}) \, cc^\dagger \chi (\mathbf{r})^\dagger = \sum_{nn'} \phi_n (\mathbf{r}) \, L_n cc^\dagger L_{n'}^\dagger \, \phi_{n'} (\mathbf{r})^\dagger ,$$

so that in this case, the density-of-states matrix $\Gamma(\varepsilon)$ in (67) should be substituted by:

$$\Gamma_{nn'} \equiv L_n \left(\sum_i^{occ} c_i c_i^\dagger \right) L_{n'}^\dagger. \qquad (104)$$

Equations (65)-(66) then become:

$$\rho^\psi(\mathbf{r}) \equiv \sum_{RR'} \sum_{LL'} \sum_{nn'} \psi_{RL,n}(\mathbf{r}_R) \, \Gamma_{RL,n;R'L',n'} \, \psi_{R'L',n'}(\mathbf{r}_{R'})^*, \qquad (105)$$

$$\rho_R^\varphi(\mathbf{r}) = \sum_{LL'} Y_L(\hat{\mathbf{r}}) Y_{L'}^*(\hat{\mathbf{r}}) \sum_{nn'} \varphi_{Rl,n}(r) \, \Gamma_{RL,n;RL',n'} \, \varphi_{Rl',n'}(r),$$

$$\rho_R^{\varphi^\circ}(\mathbf{r}) = \sum_{LL'} Y_L(\hat{\mathbf{r}}) Y_{L'}^*(\hat{\mathbf{r}}) \sum_{nn'} \varphi_{Rl,n}^\circ(r) \, \Gamma_{RL,n;RL',n'} \, \varphi_{Rl',n'}^\circ(r).$$

If one feels that, with the variational NMTO method, the KKR equations have been solved with sufficient accuracy, then one may even use (65)-(67) as they stand, and interpolate the energy dependences of the wave functions using the *classical* Lagrange or Newton methods (146) and (147).

In order to solve Poisson's equation and to compute the Coulomb- and exchange-correlation integrals for the total energy and forces, we need to fit the charge density by suitable functions. The properties of $\rho(\mathbf{r})$ to which we have most easy access are its spherical-harmonics expansions around the various sites. For the fitting we therefore choose atom-centered NMTO-like functions which have the following advantages: (1) they are the unitary functions for *continuous fitting* at non-touching a-spheres, (2) they are localized, (3) we know the result of operating on them with ∇^2, and (4) the integral of any product of two such functions is the energy derivative of a kink matrix (102)-(103).

Our fitting procedure [47] can be outlined as follows: We first place a set of screening spheres around each atomic site. This defines our screened Hankel functions (29) and divides space into non overlapping intra-sphere parts and an interstitial part. It is not necessary to place screening spheres at interstitial sites, even though the resulting interstitial can be very large. In the intra-sphere region we use a spherical-harmonics expansion of the charge density, with the components $\rho_{RL}(r)$ known on a radial mesh. As the screening spheres are relatively small this summation can be truncated at $l=3$ or 4. In the interstitial we expand in the screened Hankel functions, $n_{RL}^a(\varepsilon, \mathbf{r}_R)$, normalized as in (45) and with 3 different, negative energies, of which the lowest is about 4 times the work function, that is:

$$\rho(\mathbf{r}) \approx \sum_{n=0}^{2} \sum_{RL} n_{RL}^a(\varepsilon_n, \mathbf{r}_R) \, \lambda_{RL;n} = \sum_{RL} \breve{n}_{RL}^a(\mathbf{r}_R) \, \mu_{RL} + \qquad (106)$$

$$\sum_{RL} \left(n_{RL}^a(\mathbf{r}_R) \, \rho_{RL}(a) + n_{RL}^a([01], \mathbf{r}_R) \sum_{R'L'} X_{RL,R'L'} \, \rho_{R'L'} \right)$$

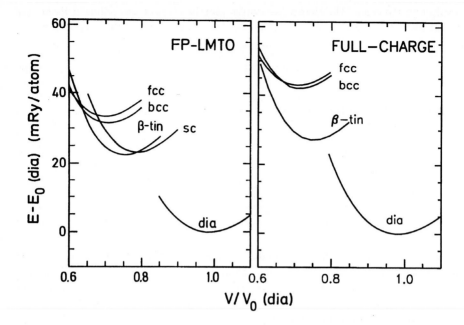

Fig. 12. Total energy of Si as a function of the atomic volume for different structures calculated with the full-potential LMTO method [34] and with the present full-charge scheme [49,47,48]. See text.

for all $r_R \geq a_{RL}$. With three energies, we can in principle fit continuously with continuous 1st and 2nd derivatives. However, in practice it is difficult to compute the 2nd radial derivatives of the high-l components of the charge density. We therefore determine the matrix X in such a way that the fitting is continuous and *once* differentiable, that is: $X = B^a [01]^{-1} (\partial \{\rho(a)\} - B_0^a)$. The functions $\breve{n}_{RL}^a (\mathbf{r}_R)$ in (106) are those linear combinations of the three $n^a (\varepsilon_n, \mathbf{r})$'s whose value and radial slope vanish in *all* channels at the screening spheres. These functions therefore peak in the interstitial region and their coefficients μ_{RL} are determined by a least squares fit in the region interstitial to the MT-spheres, by sampling the full charge (105), as well as the expansion (106) at a set of pseudo-random points. Once the expansion is obtained, it is very easy to solve Poisson's equation. In the intra-sphere part this is done numerically and in the interstitial analytically by virtue of the screened Hankel functions solving the wave equation. The same expansion procedure can be applied to the exchange-correlation energy density $\epsilon(\mathbf{r})$ and potential $\mu(\mathbf{r})$. This gives a full potential. The total energy E_{tot} is also easy to evaluate. The interstitial part of the integrals reduces simply to a summation over Hermite divided differences of the slope matrix.

We have applied this procedure to look at the total energy of various possible structures for silicon [49]. For each structure we perform a standard self consistent LMTO-ASA calculation. In the last iteration an expansion of the full charge density is made and E_{tot} evaluated correctly. The result is shown in Fig. 12 where, for comparison, we show the full-potential LMTO result from Ref. [34].

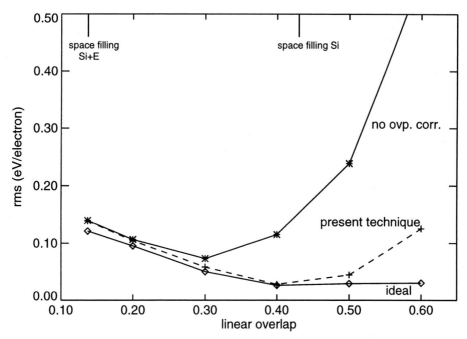

Fig. 13. Rms error of the valence-band energies in diamond-structured Si as a function of the overlap in the atom-centered MT-potential [42,43]. See text.

Overlapping MT-potential: Si without empty spheres. The phase diagram of Si just shown was calculated using LMTOs defined for MT-potentials with empty spheres. We now consider the possibility offered by Eq. (28) of allowing the atom-centered sphere a substantial overlap – like the 50% radial overlap shown in Figs. 2-5 – and, hence, of getting rid of the empty spheres.

The first question is: How to construct such a potential? Our answer is [42] that the potential should be constructed such as to minimize the mean squared deviation of the valence-band energies from the ones for the full potential. From this condition, it then follows that the overlapping MT-potential, $\sum_R v(r_R)$, should be the least-squares approximation to the full-potential, $V(\mathbf{r})$, weighted with the valence charge density. This yields a set of coupled equations for the *shape*, $f(r) \equiv v(r) - g$, and the *zero*, g, of the MT-potential. The equation which arises from requiring stationarity with respect to δg is of course: $\int (V - \sum v) \rho d^3 r = 0$, and it means that the error in the sum of the valence-band energies should vanish to leading order. The other equations, which arise by requiring stationarity with respect to $\delta f(r)$, are coupled integral equations, which are complicated due to the presence of the charge-density weighting. Taking the charge density to be constant in space, corresponds to minimizing the mean squared energy-deviation for the *entire* spectrum, rather than merely for the valence band. Now, in our present implementation, we only took the spatial behavior of the charge density into account in the δg-equation. The resulting potentials for diamond-structured Si were shown in Figs. 10 and 11 of Ref. [20]. We have recently succeeded in obtaining the overlapping MT potential from the full potential obtained from the charge density (106) [43], but in the present paper we shall only show results obtained by taking the full potential to be the Si+E ASA potential – like in Ref. [20].

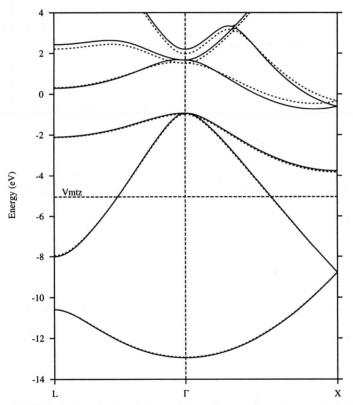

Fig. 14. Band structure of Si calculated with the 3rd-generation LMTO method for the self-consistent Si+E MT-potential (dashed) and for the Si-centered, 60%-overlapping MT-approximation to it (solid). The latter calculation included the correction for the kinetic-energy error Eq. (28) in the LMTO Hamiltonian, and the value of the MT-zero was adjusted in such a way that the average energy of the valence band was correct. Hence, the solid band structure corresponds to the last point on the curve marked 'ideal' in Fig. 13 [42,43]. See text.

Fig. 13 shows three different results for the rms error of the valence-band energy as a function of the linear overlap, $\omega \equiv (s/t)-1$. For the overlap increasing up to about 30%, the rms error falls in all cases, simply because the overlapping MT-potential becomes an increasingly better approximation to the full potential. Without any overlap correction, the kinetic-energy error (28), which is of second order in the potential overlap, initially rises proportional to $v(s)^2\,\omega^4$ [20], and this is seen to limit the maximum overlap to about 30%. We may, however, use the LMTO equivalent [43] of Eq. (28) to correct each band energy, $\varepsilon_i(\mathbf{k})$, and the results are shown by the two other curves. The dashed curve – marked 'present technique' – uses the δg-equation as given above, whereas the 'ideal' curve was obtained by adjusting g – iteratively, because g enters the $\delta f(r)$ equations – to have the mean error of the valence-band energy vanish exactly. It *is* possible to improve upon the 'present technique' without knowing the valence-band energy *a priori*, and we are currently including charge-density weighting in the $\delta f(r)$-equations. This makes the curve flatten out – like the one marked 'ideal' [43].

The solid curves in Fig. 14 show the Si band structure obtained with the 60% overlapping MT-potential, including the LMTO overlap correction, and determining g to yield vanishing mean error of the valence band. The dotted curve is the 'exact' result as obtained with a (3rd-generation) LMTO calculation for the Si+E potential. The errors seen in the valence band are certainly no larger than 30 meV, but those in the conduction band are larger.

4   Energy-Dependent Linear Transformations

If one considers Fig. 1, it might seem as if the energy-window over which an NMTO set yields good approximations to the wave functions will be wider if one starts out from energy-dependent linear combinations of kinked partial waves:

$$\hat{\phi}(\varepsilon, \mathbf{r}) \equiv \phi(\varepsilon, \mathbf{r})\,\hat{T}(\varepsilon), \qquad (107)$$

which have smoother energy dependencies. Normalized kinked partial waves and Löwdin orthonormalized kinked partial waves are examples of cases where the divergences of the kinked partial waves at the energies, ε_{RL}^a, where a node passes through the screening radius, are avoided. The transformation given by the – in general non-Hermitian – matrix $\hat{T}(\varepsilon)$ mixes kinked partial waves with the *same energy* and *different RL's linearly*. Although the Hilbert spaces spanned by the energy-dependent sets, $\phi(\varepsilon, \mathbf{r})$ and $\hat{\phi}(\varepsilon, \mathbf{r})$, are identical, it is not obvious that those spanned by the respective polynomial approximations, $\chi^{(N)}(\mathbf{r})$ and $\hat{\chi}^{(N)}(\mathbf{r})$, are also identical, particularly not if one bears only Fig. 1 in mind.

Depending on the transformation, the resulting $\hat{\phi}(\varepsilon, \mathbf{r})$ may completely have lost its original RL-character. Since the linear combination, $\hat{\phi}(\varepsilon, \mathbf{r})$, of kinked partial waves has *active* radial functions on *other* sites, as well as at its own site for other L's, it is *not* a kinked partial wave in the usual sense, that is, one which could have been obtained by a screening transformation. Remember, that for 3rd-generation kinked partial waves, a screening transformation is not linear.

In the following, we shall assume that the screening radii have been chosen at the previous step, in the screening calculation for the structure matrix, and perhaps by subsequent re-screening of the G_n's using (55).

A further motivation for considering transformed kinked partial waves is that they might provide the freedom to obtain energy matrices (81) which are *Hermitian*. This would simplify the finite-difference expressions (86) and (88) for the NMTO so that they take the simpler form (1) which then – like in (3) – could be diagonalized to leading order by the eigenvectors of $\hat{E}^{(N)}$. From expression (85) for the transfer matrix, we realize that the condition that a transformed $\hat{E}^{(M)}$ be a Hamiltonian matrix, is that we can find a transformation with the property that

$$\left\langle \hat{\chi}^{(M)} \mid \hat{\chi}^{(M-1)} \right\rangle = 1. \tag{108}$$

This formalism could therefore also be the basis for obtaining an *orthonormal* NMTO set.

Let us finally express the important equations (57)-(63) in terms of the transformed kinked partial waves:

$$(\mathcal{H} - \varepsilon)\,\hat{\phi}\,(\varepsilon, \mathbf{r}) = -\delta\,(\mathbf{r})\,\hat{K}\,(\varepsilon)\,, \quad (\mathcal{H} - \varepsilon)\,\hat{\phi}\,(\varepsilon, \mathbf{r})\,\hat{G}\,(\varepsilon) = -\delta\,(\mathbf{r})\,, \tag{109}$$

where we have defined the *non-Hermitian* matrices

$$\hat{K}\,(\varepsilon) \equiv K\,(\varepsilon)\,\hat{T}\,(\varepsilon)\,, \qquad \hat{G}\,(\varepsilon) \equiv \hat{K}\,(\varepsilon)^{-1} = \hat{T}\,(\varepsilon)^{-1}\,G\,(\varepsilon)\,. \tag{110}$$

Note that these definitions do *not* correspond to similarity transformations. The kink matrix, $K\,(\varepsilon)$, and thereby its inverse, $G\,(\varepsilon)$, were originally defined in such a way that they are Hermitian, but they are *inherently 'skew'*, because (109) tells us that it is the 'one-sided' contraction of the Green function,

$$\gamma\,(\varepsilon, \mathbf{r}) = \phi\,(\varepsilon, \mathbf{r})\,G\,(\varepsilon) = \hat{\phi}\,(\varepsilon, \mathbf{r})\,\hat{G}\,(\varepsilon)\,, \tag{111}$$

which is *invariant*. For the same reason, the integrals of the products of two contracted Green functions, with possibly different energies, form an overlap matrix,

$$\hat{G}\,(\varepsilon)^{\dagger} \left\langle \hat{\phi}\,(\varepsilon) \mid \hat{\phi}\,(\varepsilon') \right\rangle \hat{G}\,(\varepsilon') = -\frac{G\,(\varepsilon) - G\,(\varepsilon')}{\varepsilon - \varepsilon'}\,, \tag{112}$$

which is independent of $\hat{T}\,(\varepsilon)$.

Adding to the discussion following (92) about the meaning of the *matrix* equation $\langle \phi_n \mid \phi_{n'} \rangle = \langle \phi_{n'} \mid \phi_n \rangle$, note that this equation does not hold in a general representation: $\left\langle \hat{\phi}_n \mid \hat{\phi}_{n'} \right\rangle = \hat{T}_n^{\dagger} \hat{T}_{n'}^{\dagger -1} \left\langle \hat{\phi}_{n'} \mid \hat{\phi}_n \right\rangle \hat{T}_n^{-1} \hat{T}_{n'} \neq \left\langle \hat{\phi}_{n'} \mid \hat{\phi}_n \right\rangle$, unless $\hat{T}_n = \hat{T}_{n'}$. But it is of course always true that $\left\langle \hat{\phi}_n \mid \hat{\phi}_{n'} \right\rangle = \left\langle \hat{\phi}_{n'} \mid \hat{\phi}_n \right\rangle^{\dagger}$.

We now come to derive NMTOs from the transformed kinked partial waves (107). Since the arguments around expression (69) concerned the contracted

Green function, which according to (111) is invariant, (69) is unchanged but should be rewritten in the form:

$$\hat{\chi}^{(N)}\left(\varepsilon,\mathbf{r}\right)\hat{G}\left(\varepsilon\right) = \hat{\phi}\left(\varepsilon,\mathbf{r}\right)\hat{G}\left(\varepsilon\right) - \sum_{n=0}^{N}\hat{\phi}_n\left(\mathbf{r}\right)\hat{G}_n\,A_n^{(N)}\left(\varepsilon\right). \tag{113}$$

As a consequence, (70) should be substituted by:

$$\hat{\chi}^{(N)}\left(\mathbf{r}\right) = \frac{\Delta^N\hat{\phi}\left(\mathbf{r}\right)\hat{G}}{\Delta\left[0..N\right]}\left(\frac{\Delta^N\hat{G}}{\Delta\left[0..N\right]}\right)^{-1} = \frac{\Delta^N\phi\left(\mathbf{r}\right)G}{\Delta\left[0..N\right]}\left(\frac{\Delta^N\hat{G}}{\Delta\left[0..N\right]}\right)^{-1}. \tag{114}$$

The last equation (114) shows that the polynomial approximation to the transformed energy-dependent NMTO, $\hat{\chi}^{(N)}\left(\varepsilon,\mathbf{r}\right) = \chi^{(N)}\left(\varepsilon,\mathbf{r}\right)\hat{T}\left(\varepsilon\right)$, is

$$\hat{\chi}^{(N)}\left(\mathbf{r}\right) = \chi^{(N)}\left(\mathbf{r}\right)G\left[0...N\right]\hat{G}\left[0...N\right]^{-1}, \tag{115}$$

which is a *linear* transformation. Hence, regardless of the energy-dependent transformation $\hat{T}\left(\varepsilon\right)$ of the kinked partial waves, *all* NMTO sets span the *same* Hilbert space and all energy-windows are therefore identical. This disproves the above-mentioned naive conclusion drawn from Fig. 1. Since $G\left(\varepsilon\right) = \hat{T}\left(\varepsilon\right)\hat{G}\left(\varepsilon\right)$, we may express the NMTO transformation (115) as a Newton series (88) for $\hat{T}\left(\varepsilon\right)$:

$$G\left[0...N\right]\hat{G}\left[0...N\right]^{-1} = \left(\hat{T}\hat{G}\right)\left[0...N\right]\hat{G}\left[0...N\right]^{-1} \tag{116}$$

$$= \sum_{M=0}^{N}\hat{T}\left[0..M\right]\hat{G}\left[M..N\right]\hat{G}\left[0...N\right]^{-1}$$

$$= \hat{T}_0 + .. + \hat{T}\left[0...N\right]\left(\hat{E}^{(1)} - \varepsilon_{N-1}\right)..\left(\hat{E}^{(N)} - \varepsilon_0\right).$$

Since the contracted Green function is invariant, so are equations (92) and (93) which relate the overlap and Hamiltonian integrals of such functions to Hermite divided differences of $G\left(\varepsilon\right)$. For the NMTO overlap and Hamiltonian matrices, we therefore obtain (94) and (95), with the prefactor substituted by $\hat{G}\left[0..N\right]^{-1\dagger}$, the postfactor substituted by $\hat{G}\left[0..N\right]^{-1}$, and the Hermite divided differences of $G\left(\varepsilon\right)$ unaltered.

The first equation (114) shows that the expressions derived previously for the NMTOs, *excluding* those for *integrals* over NMTOs, may be taken over, after these expressions have been subject to the following substitutions:

$$\begin{array}{lll} \phi\left(\varepsilon,\mathbf{r}\right) \to \hat{\phi}\left(\varepsilon,\mathbf{r}\right), & K\left(\varepsilon\right) \to \hat{K}\left(\varepsilon\right), & L_n^{(N)} \to \hat{L}_n^{(N)}, \\ \chi\left(\varepsilon,\mathbf{r}\right) \to \hat{\chi}\left(\varepsilon,\mathbf{r}\right), & G\left(\varepsilon\right) \to \hat{G}\left(\varepsilon\right), & E^{(M)} \to \hat{E}^{(M)}. \end{array} \tag{117}$$

Remember, that the substitutions for $K\left(\varepsilon\right)$ and $G\left(\varepsilon\right)$ do not correspond to a similarity transformation.

As long as we only consider $\hat{T}\left(\varepsilon\right)$-transformations which are independent of N, the step-down relation (80) holds for the transformed NMTOs and for its transfer matrices, because the derivation merely made use of (57), which transforms into (109). This shows that $\hat{E}^{(0)} - \varepsilon_0$ equals $-\hat{K}_0 = -K_0\hat{T}_0$, as expected, but that: $\left\langle \hat{\chi}^{(0)} \mid \hat{\chi}^{(-1)} \right\rangle = 1$ does not hold. The hatted version of (85) therefore only holds for $N \geq 1$. For $N = 0$:

$$\left\langle \hat{\chi}^{(0)} \mid \mathcal{H} - \varepsilon_0 \mid \hat{\chi}^{(0)} \right\rangle = -\hat{T}_0^\dagger K_0 \hat{T}_0 = \hat{T}_0^\dagger \left(\hat{E}^{(0)} - \varepsilon_0 \right) \equiv \hat{H}^{(0)} - \varepsilon_0. \tag{118}$$

The expressions for the transformed NMTO in terms of divided differences of *transformed* kinked partial waves are the hatted versions of (86) and (88). One should remember that the divided difference, $\hat{\phi}\left([0..M],\mathbf{r}\right)$, is a linear combination of the $M + 1$ functions $\phi_0\left(\mathbf{r}\right)\hat{T}_0, .., \phi_M\left(\mathbf{r}\right)\hat{T}_M$, and hence, a linear combination of the $M + 1$ divided differences: $\phi_0\left(\mathbf{r}\right), .., \phi\left([0..M],\mathbf{r}\right)$. This is the generalization of the property: $d\phi\left(\varepsilon,\mathbf{r}\right)\hat{T}\left(\varepsilon\right)/d\varepsilon\big|_{\varepsilon_\nu} = \dot{\phi}\left(\mathbf{r}\right)\hat{T} + \phi\left(\mathbf{r}\right)\dot{\hat{T}}$, used in the 2nd-generation LMTO formalism. Explicitly:

$$\hat{\phi}\left([0...M],\mathbf{r}\right) = \sum_{n=0}^{M} \frac{\phi_n\left(\mathbf{r}\right)\hat{T}_n}{\prod_{m=0,\neq n}^{M}\left(\varepsilon_n - \varepsilon_m\right)} \tag{119}$$

$$= \sum_{m=0}^{M} \phi\left([m..M],\mathbf{r}\right)\hat{T}\left[0..m\right] = \phi\left([0...M],\mathbf{r}\right)\hat{T}_0 + .. + \phi_M\left(\mathbf{r}\right)\hat{T}\left[0...M\right].$$

The transformed versions of the results (102), (103) are complicated, unless $\hat{T}\left(\varepsilon\right)$ is independent of ε. In that case, the right-hand sides just have $K\left(\varepsilon\right)$ substituted by $\hat{T}^\dagger K\left(\varepsilon\right)\hat{T} \equiv \bar{K}\left(\varepsilon\right)$.

Usually $\left\langle \hat{\phi}\left[0..M\right] \mid \hat{\phi}\left[0..N\right] \right\rangle \neq \left\langle \hat{\phi}\left[0..N\right] \mid \hat{\phi}\left[0..M\right] \right\rangle$, unless $\hat{T}\left(\varepsilon\right) = \hat{T}$, or the matrix is diagonal; $\left\langle \hat{\phi}\left[0..M\right] \mid \hat{\phi}\left[0..N\right] \right\rangle = \left\langle \hat{\phi}\left[0..N\right] \mid \hat{\phi}\left[0..M\right] \right\rangle^\dagger$ of course always holds.

5   Hamiltonian Energy Matrices and Orthonormal Sets

Having seen that an energy-dependent, linear transformation (107) of the MTO set does *not* change the Hilbert space spanned by the set of energy-independent NMTOs, but merely the individual basis functions, we now turn to the objective of finding a representation in which the energy matrices $\hat{E}^{(M)}$ – but not necessarily the Green matrix $\hat{G}\left(\varepsilon\right)$ – are *Hermitian*. The energy matrices will then be the two-center Hamiltonians entering expressions like (1) for the orbitals. From (85), we obviously want:

$$\hat{E}^{(M)} - \varepsilon_M = \left\langle \hat{\chi}^{(M)} \mid \mathcal{H} - \varepsilon_M \mid \hat{\chi}^{(M)} \right\rangle \equiv \hat{H}^{(M)} - \varepsilon_M \tag{120}$$

for $1 \leq M \leq N$, and since this condition leads to the *near-orthonormality condition* (108), it guides the way to make *one* of the NMTO sets – let us call it the Lth – *orthonormal*.

In order to solve the N *near-orthonormality* conditions for the Hamiltonian matrices, we first insert the transformed version of expression (83) for the inverse of the Mth divided difference of the Green matrix in terms of the transfer matrices and $\hat{H}^{(0)} - \varepsilon_0$, defined by (118),

$$-\hat{G}\,[0...M]^{-1} = \hat{T}_0^{-1\dagger}\left(\hat{H}^{(0)} - \varepsilon_0\right)\left(\hat{H}^{(1)} - \varepsilon_1\right)..\left(\hat{H}^{(M)} - \varepsilon_M\right), \qquad (121)$$

into the transformed version of expression (95) for the Hamiltonian in terms of the $2M$th Hermite divided difference of the original Green matrix $G(\varepsilon)$. We then use (120) and notice that one factor $\hat{H}^{(M)} - \varepsilon_M$ cancels out so that the equation may be solved for this highest transfer matrix:

$$\hat{H}^{(M)} - \varepsilon_M = \left[\begin{array}{c}\left(\hat{H}^{(M-1)} - \varepsilon_{M-1}\right)..\left(\hat{H}^{(1)} - \varepsilon_1\right)\left(\hat{H}^{(0)} - \varepsilon_0\right) \\ \times\hat{T}_0^{-1}\left(-G\,[[0..M-1]\,M]\right)\hat{T}_0^{-1\dagger} \\ \times\left(\hat{H}^{(0)} - \varepsilon_0\right)\left(\hat{H}^{(1)} - \varepsilon_1\right)..\left(\hat{H}^{(M-1)} - \varepsilon_{M-1}\right)\end{array}\right]^{-1}$$

for $M \geq 1$. Solving recursively for the transfer matrices, and including (118) at the top, we obtain the following results:

$$\hat{H}^{(0)} - \varepsilon_0 \;=\; -\hat{T}_0^{\dagger}\,G\,[[\,]\,0]^{-1}\,\hat{T}_0$$

$$\hat{H}^{(1)} - \varepsilon_1 \;=\; -\hat{T}_0^{-1}\,G\,[[\,]\,0]\,G\,[[0]\,1]^{-1}\,G\,[[\,]\,0]\,\hat{T}_0^{-1\dagger}$$

$$\hat{H}^{(2)} - \varepsilon_2 \;=\; -\hat{T}_0^{\dagger}\,G\,[[\,]\,0]^{-1}\,G\,[[0]\,1]\,G\,[[01]\,2]^{-1}\,G\,[[0]\,1]\,G\,[[\,]\,0]^{-1}\,\hat{T}_0$$

$$\hat{H}^{(M)} - \varepsilon_M \;=\; -\hat{T}_0^{(-1)^M(\dagger)^{M+1}}\,G\,[[\,]\,0]^{(-1)^{M+1}} \,...\, G\,[[0..M-1]\,M]^{-1}$$
$$...G\,[[\,]\,0]^{(-1)^{M+1}}\,\hat{T}_0^{(-1)^M(\dagger)^M}, \qquad (122)$$

where for reasons of systematics we have used the notation (154):

$$G\,[[\,]\,0] = G\,[0] = G_0 = K_0^{-1},$$

explained in the Appendix.

The divided differences (121) of the transformed Green matrix are needed for specification of the transformation via (110), the orbitals via (115), or the transformed kinked partial waves via (111), and are seen to be given by:

$$\hat{G}\,[0]^{-1} \;=\; G\,[[\,]\,0]^{-1}\,\hat{T}_0$$

$$\hat{G}\,[01]^{-1} \;=\; -G\,[[0]\,1]^{-1}\,G\,[[\,]\,0]\,\hat{T}_0^{-1\dagger} \qquad (123)$$

$$\hat{G}\,[012]^{-1} \;=\; G\,[[01]\,2]^{-1}\,G\,[[0]\,1]\,G\,[[\,]\,0]^{-1}\,\hat{T}_0$$

$$\hat{G}\,[0...M]^{-1} \;=\; (-)^M\,G\,[[0..M-1]\,M]^{-1} \,...\, G\,[[\,]\,0]^{(-1)^{M+1}}\,\hat{T}_0^{(-1)^M(\dagger)^M}.$$

Since we originally had the $N + 1$ matrices $\hat{T}_0...\hat{T}_N$ at our disposal and have used N to satisfy the near-orthonormality conditions, we have one, $\hat{T}_0$, left. This –

and thereby implicitly also the other $\hat{T}_n$'s – may now be chosen equal to a matrix, $\check{T}_0$, which makes the Lth set *orthonormal*. Note that whereas the transformation $\hat{T}(\varepsilon)$ did not depend on the order of any basis set, the transformation $\check{T}(\varepsilon)$ does; it depends on L.

Let us first discuss whether the transformation (98) to an orthonormalized NMTO set may at all be arrived at by an energy-dependent linear transformation of the kinked partial waves: According to (115), othonormality of the Lth set happens for any transformation $\check{T}(\varepsilon)$ which satisfies: $(\check{T}^{-1}G)[0...L] = (-G[[0..L]])^{1/2}$, where $G[[0..L]]$ is the $(2L+1)$st Hermite divided difference (152) of the original Green matrix. Hence, this is a linear equation between the $L+1$ values of the matrix $\check{T}(\varepsilon)^{-1}$ at the first $L+1$ mesh points, and it is therefore plausible that it may be used to fix $\check{T}_0$.

The better way of writing this equation is, like for the Hamiltonian matrix, to insert (121) for $\hat{G}[0..L]^{-1}$ into the transformed version of expression (94) for the overlap matrix. As a result:

$$\left\langle \hat{\chi}^{(L)} \mid \hat{\chi}^{(L)} \right\rangle = \left(\hat{H}^{(L)} - \varepsilon_L \right) .. \left(\hat{H}^{(1)} - \varepsilon_1 \right) \left(\hat{H}^{(0)} - \varepsilon_0 \right) \times \qquad (124)$$
$$\hat{T}_0^{-1} \left(-G[[0..L]] \right) \hat{T}_0^{-1\dagger} \left(\hat{H}^{(0)} - \varepsilon_0 \right) \left(\hat{H}^{(1)} - \varepsilon_1 \right) .. \left(\hat{H}^{(L)} - \varepsilon_L \right)$$

$$= -\hat{T}_0^{(-1)^L(\dagger)^{L+1}} G[[\,]0]^{(-1)^{L+1}} ..G[[0..L]]..G[[\,]0]^{(-1)^{L+1}} \hat{T}_0^{(-1)^L(\dagger)^L}.$$

We see that the equation $\left\langle \hat{\chi}^{(L)} \mid \hat{\chi}^{(L)} \right\rangle = 1$, in contrast to the equation: $\left\langle \hat{\chi}^{(M)} \mid \mathcal{H} - \varepsilon_M \mid \hat{\chi}^{(M)} \right\rangle = \hat{H}^{(M)} - \varepsilon_M$, is *quadratic* in *all* Hamiltonians, and therefore can only be solved by taking the square root of a matrix.

Hence, our strategy is to choose a $\hat{T}_0$, which makes the *non-orthonormality*,

$$\left\langle \hat{\chi}^{(L)} \mid \hat{\chi}^{(L)} \right\rangle - 1 \equiv \hat{O}^{(L)}, \qquad (125)$$

so small, that we may use an expansion like (100) to find $\check{T}_0$ and the corresponding Hamiltonians $\check{H}^{(M)}$. Of these, $\check{H}^{(L)}$ equals the variational Hamiltonian (99) with N substituted by L, and its eigenvalues are therefore correct to order $2L+1$. Expression (124) now tells us that:

$$\hat{T}_0^{(-1)^{L+1}(\dagger)^{L+1}} \left\langle \hat{\chi}^{(L)} \mid \hat{\chi}^{(L)} \right\rangle \hat{T}_0^{(-1)^{L+1}(\dagger)^L} = \check{T}_0^{(-1)^{L+1}(\dagger)^{L+1}} \check{T}_0^{(-1)^{L+1}(\dagger)^L},$$

which may be solved to yield:

$$\check{T}_0 = \hat{T}_0 \sqrt{1 + \hat{O}^{(L)}}^{(-1)^{L+1}} = \hat{T}_0 \begin{cases} 1 + \frac{1}{2}\hat{O}^{(L)} - \frac{1}{8}\left(\hat{O}^{(L)}\right)^2 + .. \\ 1 - \frac{1}{2}\hat{O}^{(L)} + \frac{3}{8}\left(\hat{O}^{(L)}\right)^2 - .. \end{cases} \qquad (126)$$

Here, the upper result is for L odd and the lower for L even. Since $\hat{O}^{(L)}$ will be chosen small, and for $L > 1$ is usually of order $(\varepsilon_i - \varepsilon_1)(\varepsilon_i - \varepsilon_0)$ as we shall argue in (137) and (144), this transformation preserves the RL-character of each

NMTO. The Hamiltonian matrix (122) is seen to transform like the overlap matrix (124) with M substituted for L and, as a consequence,

$$\check{H}^{(M)} - \varepsilon_M =$$
$$\sqrt{1 + \hat{O}^{(L)}}^{(-1)^{L-M+1}} \left(\hat{H}^{(M)} - \varepsilon_M \right) \sqrt{1 + \hat{O}^{(L)}}^{(-1)^{L-M+1}} . \qquad (127)$$

Similarly, from (123):

$$\check{G}\,[0...M]^{-1} = \hat{G}\,[0...M]^{-1} \sqrt{1 + \hat{O}^{(L)}}^{(-1)^{L-M+1}} . \qquad (128)$$

A procedure for computing $[1 + O]^{\pm \frac{1}{2}}$, which is more robust than the matrix Taylor series (126), is included in our codes [60].

Choosing $\hat{T}_0$. Since the near-orthonormality conditions (108) merely fix the *geometrical average* $\langle \hat{\chi}^{(M)} \mid \hat{\chi}^{(M-1)} \rangle$ of successive sets, the nearly orthonormal scheme (122)-(124) only makes sense if the transformation $\hat{T}_0$ of the kinked partial waves at ε_0 is chosen in such a way that the non-orthonormality $\hat{O}^{(0)}$ is small compared with the unit matrix. The nearly-orthonormal scheme alone, does not make the orthonormalization integrals $\langle \hat{\chi}^{(M)} \mid \hat{\chi}^{(M)} \rangle$ converge towards the unit matrix, but make them behave like:

$$\left\langle \hat{\chi}^{(M)} \mid \hat{\chi}^{(M)} \right\rangle \sim \left\langle \hat{\chi}^{(0)} \mid \hat{\chi}^{(0)} \right\rangle^{(-1)^M} .$$

This alternates with fluctuations depending on the size of $\langle \hat{\chi}^{(0)} \mid \hat{\chi}^{(0)} \rangle$.

The first thing to do is therefore to *renormalize* the MTOs in such a way that $\hat{T}_0^{a\dagger} \left\langle |\phi_{RL}^a|^2 \right\rangle \hat{T}_0^a = 1$, instead of (47). Hence, the first choice is:

$$\hat{T}_0^a = \left(\dot{k}_0^a \right)^{-\frac{1}{2}} \qquad (129)$$

where $\dot{k}_0^a$ is the energy-*ind*ependent *diagonal* matrix with elements

$$\left\langle |\phi_{RL}^a(\varepsilon_0)|^2 \right\rangle = \dot{K}_{RL,RL}^a(\varepsilon_0) \equiv \dot{k}_{RL,RL}^a(\varepsilon_0) . \qquad (130)$$

Another choice is to start with a *Löwdin orthonormalized* 0th-order set:

$$\hat{T}_0^a = \left(\dot{k}_0^a \right)^{-\frac{1}{2}} \sqrt{1 + O^a}^{-1} \qquad (131)$$

where O^a is the non-orthonormality of the 0th-order, renormalized MTO set:

$$O^a \equiv \left(\dot{k}_0^a \right)^{-\frac{1}{2}} \dot{K}_0^a \left(\dot{k}_0^a \right)^{-\frac{1}{2}} - 1. \qquad (132)$$

This choice therefore corresponds to taking $L = 0$.

Test case: GaAs. We have tested this orthonormalization method for GaAs using the minimal Ga spd As sp basis set and going all the way up to $L = 3$, that is, to a CMTO basis with the properties that $\breve{H}^{(3)} = \langle \breve{\chi}^{(3)} | \mathcal{H} | \breve{\chi}^{(3)} \rangle$ and $\langle \breve{\chi}^{(3)} | \breve{\chi}^{(3)} \rangle = 1$, so that $\breve{H}^{(3)}$ is a 7th-order Hamiltonian. $\breve{H}^{(2)}$ and $\breve{H}^{(1)}$ are of lower order, however, and neither of the three Hamiltonians commute.

We diagonalized $\breve{H}^{(L)}$ for $L = 1, 2, 3$ and compared with the band structures obtained with the corresponding non-orthonormal variational method discussed in Sect. 3.2. Both starting choices (129) and (131) were tried, and both gave fast convergence of the square-root expansions. The first choice which only requires evaluation of a square root at the last stage (127) but whose non-orthonormality $\hat{O}^{(L)}$ is larger, was found to be the fastest [24].

Aleph-representation. The renormalization (129) is of the same nature as – but simpler than (due to lack of energy dependence) – the one performed in Subsection 2.3, where we went from phase-shift normalization to screening-sphere normalization. That diagonal transformation was given by (45) for the screened spherical waves, by (46) and (47) for the 0th-order MTOs, and by (49) for the KKR matrix. Since we distinguished between those two normalizations by using respectively Greek and Latin superscripts for the screening, e.g. α and a, and since it is irrelevant, whether one arrives at a nearly orthonormal representation from quantities normalized one-or-another way, it is logical to label quantities having the integral normalization (129) by *Hebraic* superscripts, e.g. $\aleph$ as corresponding to the same screening as α and a. Although not diagonal, and therefore influencing the shape of the kinked partial waves, also the Löwdin orthonormalization (131) is an energy-*in*dependent *similarity* transformation, and so is any of the following transformations:

$$\phi^\aleph (\varepsilon, \mathbf{r}) \equiv \phi^a (\varepsilon, \mathbf{r}) \, \hat{T}_0^a \qquad \chi^{\aleph(N)} (\varepsilon, \mathbf{r}) \equiv \chi^{a(N)} (\varepsilon, \mathbf{r}) \, \hat{T}_0^a$$

$$K^\aleph (\varepsilon) \equiv \hat{T}_0^{a\,\dagger} K^a (\varepsilon) \, \hat{T}_0^a \qquad G^\aleph (\varepsilon) \equiv \hat{T}_0^{a\,-1} G^a (\varepsilon) \, \hat{T}_0^{a\,-1\dagger} \tag{133}$$

with $\hat{T}_0^a$ arbitrary. From the latter energy-independent similarity transformation of $G(\varepsilon)$, the *non*-Hermitian matrices $L_n^{(N)}$ and $E^{(N)}$, which are given in terms of $G(\varepsilon)$ by respectively (73) and (81), are seen to transform like:

$$L_n^{\aleph(N)} = \hat{T}_0^{a\,-1} L_n^{a(N)} \, \hat{T}_0^a \quad \text{and} \quad E^{\aleph(M)} = \hat{T}_0^{a\,-1} E^{a(M)} \, \hat{T}_0^a . \tag{134}$$

This – (133)-(134) – has all concerned an energy-independent similarity transformation of *un*-hatted quantities.

In order to ensure that the *hatted* quantities are independent of which representation – a or $\aleph$ – we start out from, e.g.

$$\hat{\phi}^\aleph (\varepsilon, \mathbf{r}) = \hat{\phi}^a (\varepsilon, \mathbf{r}) = \phi^a (\varepsilon, \mathbf{r}) \, \hat{T}^a (\varepsilon) = \phi^\aleph (\varepsilon, \mathbf{r}) \, \hat{T}^\aleph (\varepsilon)$$

and

$$\hat{G}^\aleph (\varepsilon) = \hat{G}^a (\varepsilon) = \hat{T}^a (\varepsilon)^{-1} G^a (\varepsilon) \, \hat{T}^a (\varepsilon)^{-1\dagger} = \hat{T}^\aleph (\varepsilon)^{-1} G^\aleph (\varepsilon) \, \hat{T}^\aleph (\varepsilon)^{-1\dagger}$$

where, from the latter, it follows that

$$\hat{L}_n^{\aleph(N)} = \hat{L}_n^{a(N)} \quad \text{and} \quad \hat{E}_n^{\aleph(M)} = \hat{E}_n^{a(M)},$$

it suffices to satisfy the relation:

$$\hat{T}^{\aleph}(\varepsilon) \equiv \hat{T}_0^{a\,-1}\hat{T}^a(\varepsilon), \quad \text{which leads to}: \quad \hat{T}_0^{\aleph} = 1. \tag{135}$$

In conclusion, under the substitution $a \to \aleph$, all previous equations remain valid, and the factors $\hat{T}_0^{\aleph}$ may be deleted.

The virtue of this notation is that, once we have decided upon the normalization and the screening, we can drop the superscripts; and this is what we shall do: From now on, and throughout the remainder of this paper, un-hatted quantities, i.e. the kinked partial waves, the kink and the Green matrices, and the Lagrange and energy matrices, are all supposed to have the integral (ortho)normalization (129) or (131), that is, they are all in the Aleph-representation. All equations derived previously are then unchanged, and $\hat{T}_0$ may be dropped.

Accuracies of Hamiltonians. The accuracies of the Hamiltonians depend on the sizes of the corresponding non-orthonormalities. Specifically, since the residual error of the one-electron energy after use of the variational principle (5) for the set $\hat{\chi}^{(M)}(\mathbf{r})$,

$$\hat{H}^{(M)}v_i = \varepsilon_i v_i + (\varepsilon_i - \varepsilon_M)\,\hat{O}^{(M)}v_i,$$

is proportional to $(\varepsilon_i - \varepsilon_0)^2 .. (\varepsilon_i - \varepsilon_M)^2$, neglect of the non-orthonormality, leads to the error:

$$\delta\hat{\varepsilon}_i^{(M)} = (\varepsilon_i - \varepsilon_M)\,\hat{O}_{ii}^{(M)} + \mathcal{O}\left\{(\varepsilon_i - \varepsilon_0)^2 .. (\varepsilon_i - \varepsilon_M)^2\right\}, \tag{136}$$

where $\hat{O}_{ii}^{(M)} \equiv v_i^\dagger \hat{O}^{(M)}v_i$ and $\mathcal{O}$ means at the order of. The goal should thus be to reduce the non-orthonormality to:

$$\hat{O}_{ii}^{(M)} = \mathcal{O}\left\{(\varepsilon_i - \varepsilon_0)^2 .. (\varepsilon_i - \varepsilon_{M-1})^2 (\varepsilon_i - \varepsilon_M)\right\}$$

because in that case, the error from non-orthonormality will be of the same order as that of the residual error. This can usually only achieved for $M = L$.

The order of the non-orthonormality may be found by use of the difference function:

$$\hat{\chi}^{(M)}(\mathbf{r}) - \hat{\chi}^{(M-1)}(\mathbf{r}) = \hat{\phi}([01],\mathbf{r})\left(\hat{H}^{(M)} - \hat{H}^{(M-1)}\right) + \hat{\phi}([012],\mathbf{r})$$

$$\times \left\{ \begin{array}{c} \left(\hat{H}^{(M-1)} - \varepsilon_1\right)\left(\hat{H}^{(M)} - \varepsilon_0\right) \\ - \left(\hat{H}^{(M-2)} - \varepsilon_1\right)\left(\hat{H}^{(M-1)} - \varepsilon_0\right) \end{array} \right\} + ..,$$

obtained from (88) and where we should take $\hat{H}^{(m)} \equiv 0$ if $m < 1$. As a result:

$$
\begin{aligned}
\hat{O}^{(M)} &= \left\langle \hat{\chi}^{(M)} \mid \hat{\chi}^{(M)} - \hat{\chi}^{(M-1)} \right\rangle \\
&= \left\langle \hat{\phi}_0 \mid \hat{\phi}\,[01] \right\rangle \left(\hat{H}^{(M)} - \hat{H}^{(M-1)} \right) \\
&\quad + \left(\hat{H}^{(M)} - \varepsilon_0 \right) \left\langle \hat{\phi}\,[01] \mid \hat{\phi}\,[01] \right\rangle \left(\hat{H}^{(M-1)} - \varepsilon_1 \right) \left(\hat{H}^{(M)} - \varepsilon_0 \right) \\
&\quad + \left\langle \hat{\phi}_0 \mid \hat{\phi}\,[012] \right\rangle \left\{ \begin{array}{c} \left(\hat{H}^{(M-1)} - \varepsilon_1 \right) \left(\hat{H}^{(M)} - \varepsilon_0 \right) \\ - \left(\hat{H}^{(M-2)} - \varepsilon_1 \right) \left(\hat{H}^{(M-1)} - \varepsilon_0 \right) \end{array} \right\} + ..
\end{aligned}
\tag{137}
$$

which is usually of order $\left(\hat{H}^{(M-1)} - \varepsilon_1 \right) \left(\hat{H}^{(M)} - \varepsilon_0 \right)$ when $M > 1$.

To evaluate integrals like $\left\langle \hat{\phi}_0 \mid \hat{\phi}\,[01] \right\rangle$ we must transform to the original representation using (119) and then use (102). In this way we get:

$$
\left\langle \hat{\phi}_0 \mid \hat{\phi}\,[01] \right\rangle = \langle \phi_0 \mid \phi\,[01] \rangle + \langle \phi_0 \mid \phi_1 \rangle \hat{T}\,[01] = K\,[[0]\,1] + K\,[01]\,\hat{T}\,[01]. \tag{138}
$$

Remember, that we are using the Aleph-normalization (133), because this influences the right-hand sides. For a condensed mesh, (138) reduces to:

$$
\left\langle \hat{\phi} \mid \dot{\hat{\phi}} \right\rangle = \left\langle \phi \mid \dot{\phi} \right\rangle + \dot{K}\dot{\hat{T}} = \frac{\ddot{K}}{2!} + \dot{K}\dot{\hat{T}}.
$$

We shall conclude this study of the accuracy of the Hamiltonians in Eq. (145) below.

6   Connecting Back to the ASA Formalism

What remains to be demonstrated is that the NMTO sets, $\chi^{(N)}\,(\mathbf{r})$, $\hat{\chi}^{(N)}\,(\mathbf{r})$, and $\check{\chi}^{(N)}\,(\mathbf{r})$, of which the two former are based on Löwdin-orthonormalized kinked partial waves at the first mesh point (131), and the last corresponds to the $L=1$-set being orthonormal, are the generalizations to overlapping MT-potentials, arbitrary N, and discrete meshes of the well-known LMTO-ASA sets given in the Overview by respectively (1), (8), and (9).

Since in the present paper we have not made use of the ASA, but merely a MT-potential – plus redefinition of the partial waves followed by a Löwdin-orthonormalization – we merely need to show that the formalism developed above reduces to the one given in the Overview for the case $N=1$ and a condensed mesh. In order to bridge the gap between the new and old formalisms, a bit more will be done though.

$N = 0,\ L = 0.$ For the 0th-order set we have:

$$
\chi^{(0)}\,(\mathbf{r}) = \hat{\chi}^{(0)}\,(\mathbf{r}) = \phi_0\,(\mathbf{r}) = \hat{\phi}_0\,(\mathbf{r}).
$$

All un-hatted quantities in the present section will correspond to using kinked partial waves, transformed to be orthonormal at this first mesh point, ε_0. That is: All un-hatted quantities are in the Aleph-representation (133)-(135) with $\hat{T}_0^a$ given by (131). In this representation all previously derived relations hold, and in addition:

$$\hat{T}_0 = 1 \quad \text{and} \quad \dot{K}_0 = 1. \tag{139}$$

Relating back to the Overview, this means that instead of the ASA-relation (13), we have (133) with $\hat{T}_0^a$ given by (131). The latter is the proper definition of $\dot{K}_0^{a\,-1/2}$, now that $\dot{K}_0^a = \langle \dot{\phi}_0^a \mid \dot{\phi}_0^a \rangle$ is no longer diagonal. We now see that the un-hatted quantities used in the Overview were, in fact, in the Aleph representation.

The overlap and Hamiltonian matrices for the 0th-order set are thus:

$$\left\langle \chi^{(0)} \mid \chi^{(0)} \right\rangle \qquad = \langle \phi_0 \mid \phi_0 \rangle = \left\langle \hat{\chi}^{(0)} \mid \hat{\chi}^{(0)} \right\rangle = \left\langle \hat{\phi}_0 \mid \hat{\phi}_0 \right\rangle = 1$$

$$\left\langle \chi^{(0)} \mid \mathcal{H} - \varepsilon_0 \mid \chi^{(0)} \right\rangle = \left\langle \hat{\chi}^{(0)} \mid \mathcal{H} - \varepsilon_0 \mid \hat{\chi}^{(0)} \right\rangle = H^{(0)} - \varepsilon_0 = -\dot{K}_0, \tag{140}$$

and with the 0th-order set being orthonormal, the Hamiltonian is variational. Hence, $H^{(0)} = \hat{H}^{(0)}$ is the *first*-order, two-center, TB Hamiltonian of the 3rd-generation scheme.

$N = 1, \; L = 0.$ For the LMTO set we have:

$$\chi^{(1)}(\mathbf{r}) = \phi_0(\mathbf{r}) + \phi([01],\mathbf{r}) \left(E^{(1)} - \varepsilon_0 \right) \;\rightarrow\; \phi(\mathbf{r}) + \dot{\phi}(\mathbf{r}) \left(H^{(0)} - \varepsilon_\nu \right),$$

where $E^{(1)}$ – as given by (90) – is seen to become the Hermitian, *first*-order Hamiltonian $H^{(0)}$ given by (140) if the mesh condenses. This proves (1).

The Hamiltonian and overlap matrices were given in respectively (96) and (97), and using now $\dot{K} = 1$ together with (102), we see that for a condensed mesh

$$\left\langle \chi^{(1)} \mid \mathcal{H} - \varepsilon_1 \mid \chi^{(1)} \right\rangle \;\rightarrow\; -\dot{G}^{-1} \frac{\ddot{G}}{2!} \dot{G}^{-1} \;=\; -K + K \frac{\ddot{K}}{2!} K$$

$$= H^{(0)} - \varepsilon_\nu + \left(H^{(0)} - \varepsilon_\nu \right) \left\langle \phi \mid \dot{\phi} \right\rangle \left(H^{(0)} - \varepsilon_\nu \right)$$

and

$$\left\langle \chi^{(1)} \mid \chi^{(1)} \right\rangle \;\rightarrow\; -\dot{G}^{-1} \frac{\dddot{G}}{3!} \dot{G}^{-1} \;=\; 1 - K \frac{\ddot{K}}{2!} - \frac{\ddot{K}}{2!} K + K \frac{\dddot{K}}{3!} K$$

$$= 1 + \left(H^{(0)} - \varepsilon_\nu \right) \left\langle \dot{\phi} \mid \phi \right\rangle + \left\langle \phi \mid \dot{\phi} \right\rangle \left(H^{(0)} - \varepsilon_\nu \right)$$

$$+ \left(H^{(0)} - \varepsilon_\nu \right) \left\langle \dot{\phi} \mid \dot{\phi} \right\rangle \left(H^{(0)} - \varepsilon_\nu \right),$$

which are exactly (7). Merely $\left\langle \phi \mid \dot{\phi} \right\rangle$ is not a *diagonal* matrix of radial integrals like in the ASA.

The nearly orthonormal LMTO set is:

$$\hat{\chi}^{(1)}(\mathbf{r}) = \hat{\phi}_0(\mathbf{r}) + \hat{\phi}([01],\mathbf{r})\left(\hat{H}^{(1)} - \varepsilon_0\right),$$

and the two conditions: $\left\langle \hat{\chi}^{(0)} \mid \hat{\chi}^{(0)} \right\rangle = 1 = \left\langle \hat{\chi}^{(1)} \mid \hat{\chi}^{(0)} \right\rangle$, therefore lead to:

$$\left\langle \hat{\phi}[01] \mid \hat{\phi}_0 \right\rangle = 0 = \left\langle \hat{\phi}_0 \mid \hat{\phi}[01] \right\rangle, \quad \text{and} \quad \left\langle \hat{\phi}_1 \mid \hat{\phi}_0 \right\rangle = 1 = \left\langle \hat{\phi}_0 \mid \hat{\phi}_1 \right\rangle.$$

Of these matrix equations, the first means that *any* $\hat{\phi}_{RL}([01],\mathbf{r})$ is *orthogonal* to *any* $\hat{\phi}_{R'L'}(\varepsilon_0,\mathbf{r})$. As a consequence, the *leading term* of the non-orthonormality (137) *vanishes*. The non-orthonormality of this LMTO set is then:

$$\hat{O}^{(1)} = \left(\hat{H}^{(1)} - \varepsilon_0\right)\left\langle \hat{\phi}[01] \mid \hat{\phi}[01] \right\rangle\left(\hat{H}^{(1)} - \varepsilon_0\right), \tag{141}$$

which by use of (136) shows that the errors of the $\hat{H}^{(1)}$-eigenvalues are:

$$\delta\hat{\varepsilon}_i^{(1)} \approx \left\langle \hat{\phi}[01] \mid \hat{\phi}[01] \right\rangle_{ii} (\varepsilon_i - \varepsilon_1)(\varepsilon_i - \varepsilon_0)^2. \tag{142}$$

This is one order better than the error $\propto (\varepsilon_i - \varepsilon_0)^2$ obtained by diagonalization of $H^{(0)}$, but one order worse than the error $\propto (\varepsilon_i - \varepsilon_1)^2 (\varepsilon_i - \varepsilon_0)^2$ obtained variationally using the LMTO set. Hence, $\hat{H}^{(1)}$ is a *second*-order Hamiltonian. From (122):

$$\hat{H}^{(1)} - \varepsilon_1 = -G_0 G[[0]1]^{-1} G_0 \rightarrow -G\left[\frac{\ddot{G}}{2!}\right]^{-1} G =$$

$$\left(1 - K\frac{\ddot{K}}{2!}\right)^{-1}(-K) = \left[1 + \left(H^{(0)} - \varepsilon_\nu\right)\langle \phi \mid \phi \rangle\right]^{-1}\left(H^{(0)} - \varepsilon_\nu\right),$$

which for a condensed mesh is exactly (8).

For the transformation (115) from the χ to the $\hat{\chi}$-set, we get by use of (123):

$$G[01]\,\hat{G}[01]^{-1} = -G[01]\,G[[0]1]^{-1} G_0$$

$$\rightarrow -\dot{G}\left[\frac{\ddot{G}}{2!}\right]^{-1} G = G^2\left[\frac{\ddot{G}}{2!}\right]^{-1} G = \left[1 + \langle \phi \mid \phi \rangle\left(H^{(0)} - \varepsilon_\nu\right)\right]^{-1}$$

which – since from (102): $\left\langle \dot{\phi} \mid \phi \right\rangle = \left\langle \phi \mid \dot{\phi} \right\rangle$ – is exactly (8).

The transformation (119) of the kinked partial waves is most easily found by using the orthogonality of $\hat{\phi}_0(\mathbf{r})$ and $\hat{\phi}([01],\mathbf{r})$ together with (138). For a condensed mesh, the result is simple:

$$\dot{\hat{\phi}}(\mathbf{r}) = \dot{\phi}(\mathbf{r}) + \phi(\mathbf{r})\dot{T} = \dot{\phi}(\mathbf{r}) - \phi(\mathbf{r})\left\langle \phi \mid \dot{\phi} \right\rangle = \dot{\phi}(\mathbf{r}) - \phi(\mathbf{r})\frac{\ddot{K}}{2!},$$

and well known – see Eqs. (8) and (12). For a *discrete mesh*, things look more complicated in K-language: From (138),

$$\hat{T}[01] = -K[01]^{-1}K[[0]1] = -K[01]^{-1}\frac{1-K[01]}{\varepsilon_0-\varepsilon_1},$$

where the 2nd equation has been obtained by use of (154): $F[[0]1] = \frac{\dot{F}_0-F[01]}{\varepsilon_0-\varepsilon_1}$, together with: $\dot{K}_0 = 1$. For (119) we thus obtain:

$$\begin{aligned}
\hat{\phi}([01],\mathbf{r}) &= \phi([01],\mathbf{r}) + \phi_1(\mathbf{r})\hat{T}[01] \\
&= \phi([01],\mathbf{r})\left(1 + (\varepsilon_1-\varepsilon_0)\hat{T}[01]\right) + \phi_0(\mathbf{r})\hat{T}[01] \\
&= \phi([01],\mathbf{r})K[01]^{-1} + \phi_0(\mathbf{r})\hat{T}[01] \\
&= \left\{\phi([01],\mathbf{r}) + \phi_0(\mathbf{r})\hat{T}[01]K[01]\right\}K[01]^{-1} \\
&= \left\{\phi([01],\mathbf{r}) - \phi_0(\mathbf{r})K[[0]1]\right\}K[01]^{-1} \quad (143)
\end{aligned}$$

where from (102): $K[[0]1] = \langle \phi_0 \mid \phi[01]\rangle$ is the equivalent to the usual radial integral and the new factor $K[01]$ in the transformation is caused by the presence of $\phi_1(\mathbf{r})$ rather than $\phi_0(\mathbf{r})$ on the right-hand side of the top line in (143).

In order to complete the identification of the nearly-orthonormal LMTO representation for a discrete mesh with the ASA version (8) and (12), we need an explicit expression for the third parameter, which is the matrix entering the non-orthonormality (141). With the help of (143), and remembering that $\hat{\phi}_0(\mathbf{r})$ and $\hat{\phi}([01],\mathbf{r})$ are orthogonal, we get:

$$\begin{aligned}
\left\langle \hat{\phi}[01] \mid \hat{\phi}[01]\right\rangle &= K[01]^{-1}\left\langle \phi[01] \mid \hat{\phi}[01]\right\rangle \\
&= K[01]^{-1}\left(\langle\phi[01] \mid \phi[01]\rangle - K[[0]1]^2\right)K[01]^{-1} \\
&= K[01]^{-1}\left(K[[01]] - K[[0]1]^2\right)K[01]^{-1} \\
&\rightarrow \left\langle \dot{\hat{\phi}} \mid \dot{\hat{\phi}}\right\rangle = \frac{\dddot{K}}{3!} - \left[\frac{\ddot{K}}{2!}\right]^2,
\end{aligned}$$

where, in the third equation, we have used (102).

TB parametrization For tight-binding parametrizations of many bands over a relatively wide energy range, it is usually important to have as few parameters as possible. Our experience [61,20] for the occupied and lowest excited bands of semiconductors and transition metals is that the off-diagonal elements of $\langle\phi_0 \mid \phi_1\rangle = K[01]$, $\langle\phi_0 \mid \phi[01]\rangle$, and $\left\langle\hat{\phi}[01] \mid \hat{\phi}[01]\right\rangle$ may be neglected. This is in the spirit of the ASA. We therefore need to tabulate only those few diagonal elements, together with the single TB matrix $H^{(0)}$. These quantities may then be used to construct for instance the Hamiltonian and overlap matrices $\langle\chi^{(1)} \mid \mathcal{H} - \varepsilon_1 \mid \chi^{(1)}\rangle$ and $\langle\chi^{(1)} \mid \chi^{(1)}\rangle$. This is like in the ASA, but now, we neither need this approximation nor a condensed mesh.

$N > 1$, $L = 0$. The nearly-orthonormal QMTO set is:

$$\hat{\chi}^{(2)}\left(\mathbf{r}\right) = \hat{\phi}_0\left(\mathbf{r}\right) + \left\{\hat{\phi}\left(\left[01\right],\mathbf{r}\right) + \hat{\phi}\left(\left[012\right],\mathbf{r}\right)\left(\hat{H}^{(1)} - \varepsilon_1\right)\right\}\left(\hat{H}^{(2)} - \varepsilon_0\right)$$

with the non-orthonormality:

$$\begin{aligned}\hat{O}^{(2)} = \left\langle\hat{\chi}^{(2)} \mid \hat{\chi}^{(2)} - \hat{\chi}^{(1)}\right\rangle &= \left\langle\hat{\phi}_0 \mid \hat{\phi}\left[012\right]\right\rangle\left(\hat{H}^{(1)} - \varepsilon_1\right)\left(\hat{H}^{(2)} - \varepsilon_0\right) + \\ &+ \left(\hat{H}^{(2)} - \varepsilon_0\right)\left\langle\hat{\phi}\left[10\right] \mid \hat{\phi}\left[01\right]\right\rangle\left(\hat{H}^{(2)} - \hat{H}^{(1)}\right) + \dots\end{aligned}$$

This – together with (136) – shows that the eigenvalue errors of $\hat{H}^{(2)}$ are:

$$\delta\hat{\varepsilon}_i^{(2)} \approx \left\langle\hat{\phi}_0 \mid \hat{\phi}\left[012\right]\right\rangle_{ii}\left(\varepsilon_i - \varepsilon_2\right)\left(\varepsilon_i - \varepsilon_1\right)\left(\varepsilon_i - \varepsilon_0\right),$$

which means, that $\hat{H}^{(2)}$ is a *second*-order Hamiltonian like $\hat{H}^{(1)}$, but different from it. In general, for $N > 1$, the leading non-orthonormality is:

$$\hat{O}^{(N)} \approx \left\langle\hat{\phi}_0 \mid \hat{\phi}\left[012\right]\right\rangle\left(\hat{H}^{(N-1)} - \varepsilon_1\right)\left(\hat{H}^{(N)} - \varepsilon_0\right), \qquad (144)$$

as seen from (137). This means that $\hat{H}^{(N)}$ remains a 2nd-order Hamiltonian when $N > 1$, and that its eigenvalue errors are:

$$\delta\hat{\varepsilon}_i^{(N)} \approx \left\langle\hat{\phi}_0 \mid \hat{\phi}\left[012\right]\right\rangle_{ii}\left(\varepsilon_i - \varepsilon_N\right)\left(\varepsilon_i - \varepsilon_1\right)\left(\varepsilon_i - \varepsilon_0\right). \qquad (145)$$

This is much inferior to the variational estimate obtainable with an NMTO basis. Moreover, the same result would have been obtained had we started out from the cheaper, renormalized scheme based on (129). Hence, with the present scheme only the Hamiltonians $H^{(M)}$ with $M \sim L$, have eigenvalues which are accurate approximations to the one-electron energies.

$N = 1$, $L = 1$. We finally use the general procedure (125)-(128) to orthonormalize the nearly-orthonormal LMTO set considered above. The small parameter – the non-orthonormality $\hat{O}^{(L=1)}$ – is thus given by (141).

The transformation from the nearly to the completely orthonormal set is obtained from (128), with $L = M = 1$, as:

$$\check{\chi}^{(1)}\left(\mathbf{r}\right) = \hat{\chi}^{(1)}\left(\mathbf{r}\right)\hat{G}\left[01\right]\check{G}\left[01\right]^{-1} = \hat{\chi}^{(1)}\left(\mathbf{r}\right)\left[1 + \hat{O}^{(1)}\right]^{-\frac{1}{2}},$$

which is the generalization to discrete meshes and (overlapping) MT-potentials of the first equation (9). The resulting, orthonormal LMTO set is:

$$\check{\chi}^{(1)}\left(\mathbf{r}\right) = \check{\phi}_0\left(\mathbf{r}\right) + \check{\phi}\left(\left[01\right],\mathbf{r}\right)\left(\check{H}^{(1)} - \varepsilon_0\right),$$

with the *third*-order Hamiltonian obtained from (127) with $L = M = 1$ as:

$$\check{H}^{(1)} - \varepsilon_1 = \left[1 + \hat{O}^{(1)}\right]^{-\frac{1}{2}}\left(\hat{H}^{(1)} - \varepsilon_1\right)\left[1 + \hat{O}^{(1)}\right]^{-\frac{1}{2}}.$$

This is the second ASA equation (9).

For the transformation of the kinked partial waves, we have from (126):

$$\check{\phi}_0(\mathbf{r}) = \hat{\phi}_0(\mathbf{r}) \left[1 + \hat{O}^{(1)}\right]^{\frac{1}{2}}$$

and putting all of this together, we may obtain:

$$\check{\phi}([01], \mathbf{r}) \approx \hat{\phi}([01], \mathbf{r}) - \hat{\phi}_0(\mathbf{r}) \left(\hat{H}^{(1)} - \varepsilon_0\right) \left\langle \hat{\phi}[01] \mid \hat{\phi}[01] \right\rangle,$$

which is a new result. Finally, we may check that:

$$\left\langle \check{\chi}^{(1)} \mid \check{\chi}^{(0)} \right\rangle = \left\langle \check{\phi}_0 \mid \check{\phi}_0 \right\rangle + \left(\check{H}^{(1)} - \varepsilon_0\right) \left\langle \check{\phi}[01] \mid \check{\phi}_0 \right\rangle =$$
$$1 + \hat{O}^{(1)} - \left(\check{H}^{(1)} - \varepsilon_0\right) \left\langle \hat{\phi}[01] \mid \hat{\phi}[01] \right\rangle \left(\check{H}^{(1)} - \varepsilon_0\right) \left\langle \check{\phi}_0 \mid \check{\phi}_0 \right\rangle \approx 1.$$

7 Outlook

Of the new developments described above, only the use of overlapping MT-potentials and efficient computation of total energies and forces from TB-LMTO-ASA charge densities were planned. Those parts turned out to be the hardest and still await their completion. But on the way, we did pick up a number of beautiful and useful instruments. Now that we have an accordion for playing Schrödinger, maybe Poisson can be learned as well.

8 Acknowledgments

It is a pleasure to thank Mark van Schilfgaarde for drawing our attention to a weakness in the 3rd-generation tangent-LMTO scheme. This triggered the development of the general and robust NMTO method. To make it presentable, took longer than expected, and we are most grateful to Hugues Dreyssé and all other contributors to this book for their patience and encouragement.

9 Appendix: Classical Polynomial Approximations

Lagrange and Newton interpolation. In these interpolation schemes, a function $f(\varepsilon)$ is approximated by that *polynomial* of Nth degree, $f^{(N)}(\varepsilon)$, which coincides with the function at the $N+1$ energies, $\varepsilon_0, \varepsilon_1, .., \varepsilon_N$, forming the *mesh*. The error is proportional to $(\varepsilon - \varepsilon_0)(\varepsilon - \varepsilon_1)..(\varepsilon - \varepsilon_N)$.

The expression for the approximating polynomial in terms of the $N+1$ values of the function, $f(\varepsilon_n) \equiv f_n$, with $n = 0, 1, .., N$, is:

$$f^{(N)}(\varepsilon) = \sum_{n=0}^{N} f_n \, l_n^{(N)}(\varepsilon), \quad \text{where} \quad l_n^{(N)}(\varepsilon) \equiv \prod_{m=0, \neq n}^{N} \frac{\varepsilon - \varepsilon_m}{\varepsilon_n - \varepsilon_m} \tag{146}$$

is the *Lagrange* polynomial of Nth degree. It has nodes at all mesh points, except at the nth, where it takes the value 1. Since Lagrange interpolation is exact for all functions ε^M with $M \leq N$, the Lagrange polynomials satisfy the sum rules: $\varepsilon^M = \sum_{n=0}^{N} (\varepsilon_n)^M l_n^{(N)}(\varepsilon)$, for $M = 0, ..., N$.

The *same* approximating polynomial may be expressed as a *divided difference* – or *Newton* – series:

$$f^{(N)}(\varepsilon) = \sum_{M=0}^{N} f[0,..,M] \prod_{n=0}^{M-1} (\varepsilon - \varepsilon_n) \tag{147}$$

$$= f[0] + f[0,1](\varepsilon - \varepsilon_0) + .. + f[0...N](\varepsilon - \varepsilon_{N-1}) .. (\varepsilon - \varepsilon_1)(\varepsilon - \varepsilon_0),$$

where the square parentheses denote divided differences as defined in the following table:

$$\varepsilon_0 \quad f_0 \equiv f[0]$$
$$\frac{f[0]-f[1]}{\varepsilon_0-\varepsilon_1} \equiv f[0,1]$$
$$\varepsilon_1 \quad f_1 \equiv f[1] \qquad\qquad \frac{f[0,1]-f[1,2]}{\varepsilon_0-\varepsilon_2} \equiv f[0,1,2]$$
$$\frac{f[1]-f[2]}{\varepsilon_1-\varepsilon_2} \equiv f[1,2]$$
$$\varepsilon_2 \quad f_2 \equiv f[2]$$

In general, that is:

$$f[m, m+1, .., n, n+1] \equiv \frac{f[m, m+1, ., n] - f[m+1, ., n, n+1]}{\varepsilon_m - \varepsilon_{n+1}}, \tag{148}$$

where $m \leq n$. Note that the two energies in the denominator are those which refer to the mesh points *not* common to the two divided differences in the nominator. Also, note their order, which defines the sign. A divided difference, $f[0...M]$, is thus a *linear combination* of $f_0, f_1, ..., f_M$. The divided differences entering (147) are those descending along the upper string in the table, but other forms are possible. Besides, the order of the energies need not be monotonic. In fact, *all* divided differences of degree $M + 1$ involving M specific mesh points are identical. This means that the *order* of the *arguments* in $f[0, 1, ., M - 1, M]$ is *irrelevant,* as may be seen explicitly from expression (149) below. When we have a long string of arguments, we usually order them after increasing mesh number, for simplicity of notation.

We may express any divided difference, $f[0..M]$, entering the Newton form (147) as a linear combination of the f_n's with $n \leq M$, and thereby establish the relation to the Lagrange form (146). To do this, we apply both Newton and Lagrange interpolation to a function, which we take to be that Mth degree polynomial, $f^{(M)}(\varepsilon)$, which coincides with $f(\varepsilon)$ at the first $M + 1$ mesh points. This is allowed, because $f[0..M]$ is independent of the f_n's with $n > M$. In this way, we get the identity:

$$f^{(M)}(\varepsilon) = \sum_{m=0}^{M} f[0..m] \prod_{n=0}^{m-1} (\varepsilon - \varepsilon_n) = \sum_{n=0}^{M} f_n l_n^{(M)}(\varepsilon)$$

and taking now the *highest derivative*, we obtain the important relation:

$$f\left[0...M\right] = \sum_{n=0}^{M} \frac{f_n}{\prod_{m=0,\neq n}^{M} \left(\varepsilon_n - \varepsilon_m\right)}. \tag{149}$$

The inverse relation, that is the expression for f_n in terms of divided differences for a (sub)mesh containing ε_n, is of course just the Newton series (147) evaluated at the mesh point ε_n.

In order to factorize $(\phi G)\left[0...N\right]$ in expression (70) for the NMTO, we shall need to express the Nth-order divided difference of a *product function, $f\left(\varepsilon\right) g\left(\varepsilon\right)$*, in terms of divided differences on the same mesh of the individual functions. Since the product is local in energy, we start by expressing its divided difference in the Lagrange form (149):

$$(fg)\left[0...N\right] = \sum_{n=0}^{N} \frac{f_n g_n}{\prod_{m=0,\neq n}^{N} \left(\varepsilon_n - \varepsilon_m\right)}.$$

For $f\left(\varepsilon\right)$ we may choose to use the divided differences in the upper, descending string of the table. We therefore use (147) to express f_n in terms of the divided differences on the $(0..n)$-part of the mesh and thereafter reorder the summations:

$$(fg)\left[0...N\right] = \sum_{n=0}^{N} \sum_{M=0}^{N} f\left[0..M\right] \prod_{m'=0}^{M-1} \left(\varepsilon_n - \varepsilon_{m'}\right) \frac{g_n}{\prod_{m=0,\neq n}^{N} \left(\varepsilon_n - \varepsilon_m\right)}$$

$$= \sum_{M=0}^{N} f\left[0..M\right] \sum_{n=0}^{N} \frac{\prod_{m'=0}^{M-1} \left(\varepsilon_n - \varepsilon_{m'}\right)}{\prod_{m=0,\neq n}^{N} \left(\varepsilon_n - \varepsilon_m\right)} g_n.$$

Since $\prod_{m'=0}^{M-1} \left(\varepsilon_n - \varepsilon_{m'}\right) = 0$ for $n < M$,

$$\sum_{n=0}^{N} \frac{\prod_{m'=0}^{M-1} \left(\varepsilon_n - \varepsilon_{m'}\right)}{\prod_{m=0,\neq n}^{N} \left(\varepsilon_n - \varepsilon_m\right)} g_n = \sum_{n=M}^{N} \frac{\prod_{m'=0}^{M-1} \left(\varepsilon_n - \varepsilon_{m'}\right)}{\prod_{m=0,\neq n}^{N} \left(\varepsilon_n - \varepsilon_m\right)} g_n$$

$$= \sum_{n=M}^{N} \frac{g_n}{\prod_{m=M,\neq n}^{N} \left(\varepsilon_n - \varepsilon_m\right)} = g\left[M..N\right],$$

according to (149). We have thus proved the *binomial formula:*

$$(fg)\left[0...N\right] = \sum_{M=0}^{N} f\left[0..M\right] g\left[M..N\right], \tag{150}$$

which expresses the Nth divided difference of a product on the $(0...N)$-mesh as a sum of products of divided differences on respectively the $(0..M)$- and $(M..N)$-parts of the mesh, with M being the only point in common. Hence, this formula is in terms of the divided differences descending forwards along the upper string for f, and the divided differences descending backwards along the lower string for g, but this is merely one of many possibilities. For the special case: $g\left(\varepsilon\right)=\varepsilon$, we get the useful result:

$$\left(\varepsilon f\right)\left[0...N\right]\;=\;f\left[0..N-1\right]\;+\;\varepsilon_N f\left[0...N\right].\tag{151}$$

Since the numbering of the points is irrelevant, we could of course have singled out *any* of the $N+1$ points, not merely the last.

Newton interpolation has the conceptual advantage over Lagrange interpolation that the 1st divided differences, $f\left[n-1,n\right]$, are the slopes of the chords connecting points $n-1$ and n, and hence approximations to the 1st derivatives, the 2nd divided differences, $f\left[n-1,n,n+1\right]$, are 'local' approximations to $\frac{1}{2!}$ times the 2nd derivatives, and so on, as expressed by (71). For the mesh condensing onto the one energy, ε_ν, Newton *interpolation* becomes *Taylor expansion,* which is of course simpler. An example of this is the binomial expression for the Nth derivative of a product: For a discrete mesh, there are many alternatives to (150), but for a condensed mesh, there is only one expression.

Hermite interpolation. It will turn out that the NMTO Hamiltonian and overlap matrices are best understood and computed using the formalism of *Hermite* interpolation. Here, one seeks the polynomial of degree $M+N+1$ which fits not only the values, f_n, at the $N+1$ points, but also the *slopes*, $\dot{f}_n$, at a subset of $M+1$ points. We shall number the points in such a way, that the $M+1$ points are the *first*. This polynomial is:

$$f^{(M+N+1)}\left(\varepsilon\right)=$$

$$\sum_{n=0}^{M}\left[f_n+\left(\dot{f}_n-f_n\left(\sum_{m=0,\neq n}^{M}\frac{2}{\varepsilon_n-\varepsilon_m}+\sum_{m=M+1}^{N}\frac{1}{\varepsilon_n-\varepsilon_m}\right)\right)\left(\varepsilon-\varepsilon_n\right)\right]$$

$$\times l_n^{(M)}\left(\varepsilon\right)l_n^{(N)}\left(\varepsilon\right)\;+\;\sum_{n=M+1}^{N}f_n\,l_n^{(M+1)}\left(\varepsilon\right)l_n^{(N)}\left(\varepsilon\right).$$

For those interested in *why* this is so, here are the arguments: The product of Lagrange polynomials

$$l_n^{(M)}\left(\varepsilon\right)l_n^{(N)}\left(\varepsilon\right)=\prod_{m=0,\neq n}^{M}\left(\frac{\varepsilon-\varepsilon_m}{\varepsilon_n-\varepsilon_m}\right)^2\prod_{m=M+1}^{N}\frac{\varepsilon-\varepsilon_m}{\varepsilon_n-\varepsilon_m},$$

with $0\leq n\leq M$, is of degree $M+N$. At a mesh point, $\varepsilon=\varepsilon_{n'}$, this product has value 1 when $0\leq n'=n\leq M$, value 0 and slope 0 when $0\leq n'\neq n\leq M$, and value 0 when $M<n'\leq N$. Since the slope is:

$$\left(\sum_{m=0,\neq n}^{M}\frac{2}{\varepsilon-\varepsilon_m}+\sum_{m=M+1}^{N}\frac{1}{\varepsilon-\varepsilon_m}\right)l_n^{(M)}\left(\varepsilon\right)l_n^{(N)}\left(\varepsilon\right),$$

the polynomial of degree $M + N + 1$:

$$\left(1 - (\varepsilon - \varepsilon_n)\left(\sum_{m=0,\neq n}^{M} \frac{2}{\varepsilon_n - \varepsilon_m} + \sum_{m=M+1}^{N} \frac{1}{\varepsilon_n - \varepsilon_m}\right)\right) l_n^{(M)}(\varepsilon)\, l_n^{(N)}(\varepsilon),$$

with $0 \leq n \leq M$, has value 1 and slope 0 if $\varepsilon = \varepsilon_n$. If $\varepsilon = \varepsilon_{n'} \neq \varepsilon_n$, it has value 0 and slope 0 when $0 \leq n' \leq M$, and value 0 and some slope when $M < n' \leq N$. The polynomial of degree $M + N + 1$:

$$(\varepsilon - \varepsilon_n)\, l_n^{(M)}(\varepsilon)\, l_n^{(N)}(\varepsilon),$$

with $0 \leq n \leq M$, vanishes at *all* mesh points, has slope 1 for $\varepsilon = \varepsilon_n$, slope 0 for $\varepsilon = \varepsilon_{n'} \neq \varepsilon_n$ when n' and $0 \leq n' \leq M$, and some slope when $M < n' \leq N$. Finally, the product:

$$l_n^{(M+1)}(\varepsilon)\, l_n^{(N)}(\varepsilon) = \prod_{m=0}^{M} \left(\frac{\varepsilon - \varepsilon_m}{\varepsilon_n - \varepsilon_m}\right)^2 \prod_{m=M+1,\neq n}^{N} \frac{\varepsilon - \varepsilon_m}{\varepsilon_n - \varepsilon_m},$$

with $M < n \leq N$, is a polynomial of degree $M + N + 1$. For $\varepsilon = \varepsilon_{n'}$ it has value 0 and slope 0 if $0 \leq n' \leq M$, value 0 and some slope if $M < n' \neq n \leq N$, and value 1 and some slope if $M < n' = n \leq N$.

What we shall really need is, like in (149), $\frac{1}{(M+N+1)!}$ times the *highest derivative* of the polynomial $f^{(M+N+1)}(\varepsilon)$. Calculated as the coefficient to the highest power of ε, this *Hermite divided difference* is:

$$\frac{f^{(M+N+1)}_{(M+N+1)}}{(M+N+1)!} = \sum_{n=0}^{M} \frac{\dot{f}_n - f_n\left(\sum_{n'=0,\neq n}^{M} \frac{2}{\varepsilon_n - \varepsilon_{n'}} + \sum_{n'=M+1}^{N} \frac{1}{\varepsilon_n - \varepsilon_{n'}}\right)}{\displaystyle\prod_{m=0,\neq n}^{M}(\varepsilon_n - \varepsilon_m)^2 \prod_{m=M+1}^{N}(\varepsilon_n - \varepsilon_m)}$$

$$+ \sum_{n=M+1}^{N} \frac{f_n}{\displaystyle\prod_{m=0}^{M}(\varepsilon_n - \varepsilon_m)^2 \prod_{m=M+1,\neq n}^{N}(\varepsilon_n - \varepsilon_m)}$$

$$= \lim_{\epsilon \to 0} f\,[0.....M+N+1] \equiv f\,[[0...M]\,..N]. \tag{152}$$

In the last line, we have indicated that the Hermite divided difference may be considered as the divided difference for the folded and paired mesh:

$$\varepsilon_0\ \varepsilon_{N+1} \quad \varepsilon_1\ \varepsilon_{N+2} \quad \cdot\cdot\quad\cdot\cdot\quad \varepsilon_M\ \varepsilon_{M+N+1} \quad\cdot\quad\cdot\quad \varepsilon_N$$

in the limit that the energy differences, $\epsilon_n \equiv \varepsilon_{n+N+1} - \varepsilon_n$, between the pairs tend to zero. In analogy with the notation for the divided differences, we have denoted the $(M + N + 1)$st Hermite divided difference: $f\,[[0...M]\,..N]$, which means that the mesh points listed inside *two* square parentheses have both f_n and $\dot{f}_n$ associated with them, whereas those listed inside only *one* square parenthesis have merely f_n. Like for the divided differences, the order of the arguments inside a square parenthesis is irrelevant, but for long strings we usually choose the order of increasing n. For a condensed mesh,

$$f\,[[0...M]..N] \quad \rightarrow \quad \frac{\overset{(M+N+1)}{f}}{(M+N+1)!}. \tag{153}$$

As examples of Hermite divided differences we have:

$$f\,[[0]] = \dot{f}_0 \qquad\qquad f\,[[0]\,1] = \frac{\dot{f}_0 - f[01]}{\varepsilon_0 - \varepsilon_1}$$

$$f\,[[01]] = \frac{\dot{f}_0 - 2f[0,1] + \dot{f}_1}{(\varepsilon_0 - \varepsilon_1)^2} \qquad\qquad f\,[[\,]\,0..N] = f\,[0..N] \tag{154}$$

In the NMTO formalism the Hermite divided difference (152) comes in the disguise of the following double sum (92):

$$\sum_{n=0}^{N} \sum_{n'=0}^{M} \frac{f\,[n,n']}{\prod_{m=0,\neq n}^{N} (\varepsilon_n - \varepsilon_m) \prod_{m'=0,\neq n'}^{M} (\varepsilon_{n'} - \varepsilon_{m'})}, \tag{155}$$

which may, in fact, be viewed as a divided difference (149) – albeit in two dimensions – but that brings little simplification. So let us prove that (152) and (155) are identical: First of all, the $\dot{f}_n$-terms of the double sum (155) are those for which $n = n'$, and they obviously equal those of the single sum (152). Secondly, the f_n-terms in (155) are:

$$\sum_{n=0}^{N} \sum_{n'=0,\neq n}^{M} \frac{f_n \,(\varepsilon_n - \varepsilon_{n'})^{-1} + f_{n'}\,(\varepsilon_{n'} - \varepsilon_n)^{-1}}{\prod_{m=0,\neq n}^{N} (\varepsilon_n - \varepsilon_m) \prod_{m=0,\neq n'}^{M} (\varepsilon_{n'} - \varepsilon_m)} =$$

$$\sum_{n=M+1}^{N} \frac{f_n}{\prod_{m=0,\neq n}^{N} (\varepsilon_n - \varepsilon_m)} \sum_{n'=0}^{M} \frac{(\varepsilon_n - \varepsilon_{n'})^{-1}}{\prod_{m=0,\neq n'}^{M} (\varepsilon_{n'} - \varepsilon_m)} + \tag{156}$$

$$\sum_{n=0}^{M} \frac{f_n}{\prod_{m=0,\neq n}^{N} (\varepsilon_n - \varepsilon_m)} \sum_{n'=0,\neq n}^{M} \frac{(\varepsilon_n - \varepsilon_{n'})^{-1}}{\prod_{m=0,\neq n'}^{M} (\varepsilon_{n'} - \varepsilon_m)} +$$

$$\sum_{n=0}^{M} \frac{f_n}{\prod_{m=0,\neq n}^{M} (\varepsilon_n - \varepsilon_m)} \sum_{n'=0,\neq n}^{N} \frac{(\varepsilon_n - \varepsilon_{n'})^{-1}}{\prod_{m=0,\neq n'}^{N} (\varepsilon_{n'} - \varepsilon_m)}.$$

Now, according to (149),

$$\sum_{n'=0}^{M} \frac{\frac{1}{\varepsilon_n - \varepsilon_{n'}}}{\prod_{m=0,\neq n'}^{M} (\varepsilon_{n'} - \varepsilon_m)} = \frac{1}{\varepsilon_n - \varepsilon}\,[0...M] \tag{157}$$

is the Mth divided difference of the single-pole function $1/(\varepsilon_n - \varepsilon)$, provided that n is *not* on the mesh $0...M$. For the sum where n *is* on the mesh – but the $n'=n$-term is excluded – we have:

$$\sum_{n'=0,\neq n}^{M} \frac{\frac{1}{\varepsilon_n - \varepsilon_{n'}}}{\prod_{m=0,\neq n'}^{M} (\varepsilon_{n'} - \varepsilon_m)} = \sum_{n'=0,\neq n}^{M} \frac{\frac{-1}{(\varepsilon_n - \varepsilon_{n'})^2}}{\prod_{m=0,\neq n,\neq n'}^{M} (\varepsilon_{n'} - \varepsilon_m)}$$

$$= \frac{-1}{(\varepsilon_n - \varepsilon)^2} [0..n-1, n+1..M]. \qquad (158)$$

This result also holds if M is named N, and is therefore relevant for both of the last terms in (156). We then need simpler expressions for the divided differences of the single- and double-pole functions. Guided by the results:

$$\frac{1}{M!} \frac{d^M}{d\varepsilon^M} \frac{1}{\varepsilon_i - \varepsilon} = \frac{1}{(\varepsilon_i - \varepsilon)^{M+1}}, \qquad \frac{1}{M!} \frac{d^M}{d\varepsilon^M} \frac{1}{(\varepsilon_i - \varepsilon)^2} = \frac{M+1}{(\varepsilon_i - \varepsilon)^{M+2}},$$

for the derivatives, we postulate that for a discrete mesh,

$$\frac{1}{\varepsilon_i - \varepsilon} [0...M] = \frac{1}{\prod_{m=0}^{M} (\varepsilon_i - \varepsilon_m)}, \qquad \frac{1}{(\varepsilon_i - \varepsilon)^2} [0...M] = \frac{\sum_{n=0}^{M} \frac{1}{\varepsilon_i - \varepsilon_n}}{\prod_{m=0}^{M} (\varepsilon_i - \varepsilon_m)}. \qquad (159)$$

For $M=0$, these expressions obviously reduce to the correct results, $(\varepsilon_i - \varepsilon_0)^{-1}$ and $(\varepsilon_i - \varepsilon_0)^{-2}$. For $M > 0$, our conjectures inserted on the right-hand side of (148) and subsequent use of (149) yield:

$$\frac{\frac{1}{\varepsilon_i - \varepsilon} [0..M-1] - \frac{1}{\varepsilon_i - \varepsilon} [1..M]}{\varepsilon_0 - \varepsilon_M} = \frac{1}{\prod_{m=0}^{M} (\varepsilon_i - \varepsilon_m)} = \frac{1}{\varepsilon_i - \varepsilon} [0...M],$$

$$\frac{\frac{1}{(\varepsilon_i - \varepsilon)^2} [0..M-1] - \frac{1}{(\varepsilon_i - \varepsilon)^2} [1..M]}{\varepsilon_0 - \varepsilon_M} = \frac{\sum_{n=0}^{M} \frac{1}{\varepsilon_i - \varepsilon_n}}{\prod_{m=0}^{M} (\varepsilon_i - \varepsilon_m)} = \frac{1}{(\varepsilon_i - \varepsilon)^2} [0...M],$$

which are obviously correct too. Hence, equations (159) have been proved.

Using finally (159) in (157) and (158), and right back in (156), leads to the f_n-terms in (152). We have therefore demonstrated that:

$$\sum_{n=0}^{N} \sum_{n'=0}^{M} \frac{f[n, n']}{\prod_{m=0,\neq n}^{N} (\varepsilon_n - \varepsilon_m) \prod_{m'=0,\neq n'}^{M} (\varepsilon_{n'} - \varepsilon_{m'})} = f[[0...M]..N]. \qquad (160)$$

The final expression needed for the NMTO formalism, is one for the Hermite divided difference of the product-function $\varepsilon f(\varepsilon)$. For this we can use (151) applied to the folded and paired mesh. As a result:

$$(\varepsilon f) [[0...M]..N] = f[[0..M-1]..N] + \varepsilon_M f[[0...M]..N]. \qquad (161)$$

Since the numbering of the points is irrelevant, we could of course have singled out *any* of the $M+1$ points, not merely the last.

References

1. O.K. Andersen and O. Jepsen, Phys. Rev. Lett. **53**, 2571 (1984). O.K. Andersen, O. Jepsen, and D. Glötzel in *Highlights of Condensed Matter Theory*, eds. F. Bassani, F. Fumi, and M.P. Tosi (North-Holland, New York 1985).
2. O.K. Andersen, O. Jepsen and M. Sob, in *Lecture Notes in Physics: Electronic Band Structure and Its Applications*, eds. M. Yussouff (Springer-Verlag, Berlin, 1987).
3. S. Frota-Pessoa, Phys. Rev. B **36**, 904 (1987); P. R. Peduto, S. Frota-Pessoa, and M. S. Methfessel, Phys, Rev. B **44**, 13283 (1991).
4. H.J. Nowak, O.K. Andersen, T. Fujiwara, O. Jepsen and P. Vargas, Phys. Rev. B **44**, 3577 (1991).
5. S.K. Bose, O. Jepsen, and O.K. Andersen, Phys. Rev. B **48**, 4265 (1993).
6. P. Vargas in *Lectures on Methods of Electronic Structure Calculations*, edited by V. Kumar, O.K. Andersen, and A. Mookerjee (World Scientific Publishing Co., Singapore, 1994), pp. 147-191.
7. T. Saha, I. Dasgupta, and A. Mookerjee, J. Phys.: Condens. Matter **6**, L245 (1994).
8. I. A. Abrikosov and H. L. Skriver, Phys. Rev. B **47**, 16532 (1993).
9. A. V. Ruban, I. A. Abrikosov, D. Ya. Kats, D. Gorelikov, K. W. Jacobsen, and H. L. Skriver, Phys. Rev. B **49**, 11383 (1994).
10. I. Turek, V. Drchal, J. Kudrnovsky, M. Sob, and P. Weinberger, *Electronic Structure of Disordered Alloys, Surfaces, and Interfaces* (Kluwer Academic Publishers, Boston/London/Dordrecht, 1997).
11. P. Weinberger, I. Turek, and L. Szunyogh, Int. J. Quant. Chem. **63**, 165 (1997).
12. W.R.L. Lambrecht and O.K. Andersen, Surface Science **178**, 256-263 (1986).
13. H.L. Skriver and N.M. Rosengaard, Phys. Rev. B **43**, 9538 (1991); .N.M. Rosengaard and H.L. Skriver, Phys. Rev. **50**, 4848 (1994).
14. M. van Schilfgaarde and F. Herman, Phys. Rev. Lett. **71**, 1923 (1993).
15. O.K. Andersen, Solid State Commun. **13**, 133 (1973); O.K. Andersen, Phys. Rev. B **12**, 3060 (1975); O. Jepsen, O.K. Andersen, and A.R. Mackintosh, Phys. Rev. **12**, 3084 (1975).
16. P. Hohenberg and W. Kohn, Phys. Rev. **136,** B 864 (1964); W. Kohn and L.J. Sham, Phys. Rev. **140,** A1133 (1965).
17. D. Glötzel, B. Segall, and O.K. Andersen, Solid State Commun. **36**, 403 (1980).
18. W.R.L. Lambrecht and O.K. Andersen, Phys. Rev. B **34**, 2439 (1986).
19. O.K. Andersen, O. Jepsen, and G. Krier in *Lectures on Methods of Electronic Structure Calculations*, edited by V. Kumar, O.K. Andersen, and A. Mookerjee (World Scientific Publishing Co., Singapore, 1994), pp. 63-124.
20. O.K. Andersen, C. Arcangeli, R.W. Tank, T. Saha-Dasgupta, G. Krier, O. Jepsen, and I. Dasgupta in *Tight-Binding Approach to Computational Materials Science*, Eds. L. Colombo, A. Gonis, P. Turchi, Mat. Res. Soc. Symp. Proc. Vol. 491 (Materials Research Society, Pittsburgh, 1998) pp 3-34. An earlier, and for certain aspects more complete account of the formalism may be found in [19].
21. O.K. Andersen, A.V. Postnikov, and S. Yu. Savrasov, in *Applications of Multiple Scattering Theory to Materials Science*, eds. W.H. Butler, P.H. Dederichs, A. Gonis, and R.L. Weaver, MRS Symposia Proceedings No. 253 (Materials Research Society, Pittsburgh, 1992) pp 37-70.
22. T. Saha-Dasgupta, O. K. Andersen, G. Krier, C. Arcangeli, R.W. Tank, O. Jepsen, and I. Dasgupta (unpublished).

23. T. F. A. Müller, V. Anisimov, T. M. Rice, I. Dasgupta, and T. Saha-Dasgupta, Phys. Rev, B **57**, R12655 (1998).
24. T. Saha-Dasgupta and O. K. Andersen (unpublished).
25. R. Hughbanks and R. Hoffmann, J. Am. Chem. Soc. **105**, 3528 (1983).
26. R. Dronskowski and P.E. Blöchl, J. Phys. Chem. **97**, 8617 (1993).
27. D. Johrendt, C. Felser, O. Jepsen, O.K. Andersen, A. Mewis, and J. Rouxel, J. Solid State Chem. **130**, 254 (1997).
28. F. Boucher and R. Rousseau, Inorg. Chem. **37**, 2351 (1998).
29. R. Car and M. Parrinello, Phys. Rev. Lett. **55**, 2471 (1985).
30. The Stuttgart TB-LMTO program. http://www.mpi-stuttgart.mpg.de
31. O. Jepsen and O.K. Andersen, Z. Phys. B **97**, 35 (1995).
32. J.M. Wills (unpublished); M. Alouani, J.M. Wills, and J.W. Wilkins, Phys. Rev. B **57**, 9502 (1998); J.M. Wills and B.R. Cooper, Phys. Rev. B **36**, 3809 (1987), D.L. Price and B.R. Cooper, Phys. Rev. B **39**, 4945 (1989).
33. S.Y. Savrasov, Phys. Rev. B **54**, 16470 (1996).
34. M. Methfessel, C.O. Rodriguez, and O.K. Andersen, Phys. Rev. B **40**, 2009 (1989).
35. M. Methfessel, Phys. Rev. **38**, 1537 (1988).
36. M. Springborg and O.K. Andersen, J. Chem. Phys. **87**, 7125 (1986).
37. K.H. Weyrich, Solid State Commun. **54**, 975 (1985).
38. F. Casula and F. Herman, J. Chem. Phys. **78**, 858, (1983).
39. O. Gunnarsson, J. Harris, and R.O. Jones, Phys. Rev. B **15**, 3027 (1977).
40. O.K. Andersen and R.G. Woolley, Mol. Phys. **26**, 905 (1973). R.V. Kasowski and O.K. Andersen, Solid State Commun. **11**, 799 (1972).
41. J. Korringa, Physica **13**, 392 (1947); W. Kohn and J. Rostoker, Phys. Rev. **94**, 1111 (1954); F.S. Ham and B. Segal, Phys. Rev. **124**, 1786 (1961).
42. C. Arcangeli and O.K. Andersen (unpublished).
43. C. Arcangeli, R.W. Tank, and O.K. Andersen (unpublished).
44. O.K. Andersen, Z. Pawlowska and O. Jepsen, Phys. Rev. B **34**, 5253 (1986).
45. L. Vitos, J. Kollar, and H.L. Skriver, Phys. Rev. B **49**, 16694 (1994).
46. A. Savin, O. Jepsen, J. Flad, O.K. Andersen, H. Preuss, and H.G. von Schnering, Angew. Chem. **104**, 186 (1992); Angew. Chem. Int. Ed. Engl. **31**, 187 (1992).
47. R.W. Tank, O. K. Andersen, G. Krier, C. Arcangeli, and O. Jepsen (unpublished).
48. R.W. Tank, C. Arcangeli, G. Krier, O. K. Andersen, and O. Jepsen, in *Properties of Complex Inorganic solids*, eds. Gonis *et al.* (Plenum, New York, *1997)* pp *233-237.*
49. R.W. Tank, C. Arcangeli, and O.K. Andersen (unpublished).
50. I.V. Solovyev, A.I. Liechtenstein, V.A. Gubanov, V.P. Antropov, O.K. Andersen, Phys. Rev. B **43**, 14414-422 (1991).
51. O. K. Andersen and T. Saha-Dasgupta (unpublished).
52. O.K. Andersen, in *Computational Methods in Band Theory,* eds. P.M. Marcus, J.F. Janak, and A.R. Williams (Plenum, 1971) p.178. O. K. Andersen and R.V. Kasowski, Phys. Rev. B **4**, 1064 (1971).
53. R. W. Hamming, *Numerical Methods for Scientists and Engineers (*McGraw-Hill, New York *1962).*
54. V. Anisimov, J. Zaanen, and O.K. Andersen, Phys.Rev. B**44**, 943-954 (1991), A.I. Liechtenstein, J. Zaanen, and V.I. Anisimov, Phys. Rev. B **52**, R5467 (1995).
55. O.K. Andersen, Europhysics News **12**, 5, 1 (1981); in *The Electronic Structure of Complex Systems,* eds. W. Temmerman and P.Phariseau (Plenum 1984) p. 11-66.
56. O. Gunnarsson, O. Jepsen, and O.K. Andersen, Phys. Rev. B **27**, 7144 (1983).
57. O.K. Andersen and O. Jepsen (unpublished).
58. R. Zeller, P.H. Dederichs, B. Ujfalussy, L. Szunyogh, and P. Weinberger, Phys. Rev. B **52**, 8807 (1995).

59. J.C. Slater, Phys. Rev. **51,** 846 (1937)
60. C. Arcangeli and O.K. Andersen (unpublished).
61. T. Saha-Dasgupta, O. K. Andersen, C. Arcangeli, R.W. Tank, and O. Jepsen (unpublished).

From ASA Towards the Full Potential

J. Kollár[1], L. Vitos[1,2], and H. L. Skriver[3]

[1] Research Institute for Solid State Physics,
   H-1525 Budapest, P.O.Box 49, Hungary
[2] Condensed Matter Theory Group, Physics Department,
   Uppsala University, S-75121 Uppsala, Sweden
[3] Center for Atomic-scale Materials Physics and Department of Physics,
   Technical University of Denmark, DK-2800 Lyngby, Denmark

Abstract. To combine the simplicity and efficiency of atomic-sphere approximation (ASA) based electronic structure calculations and the accuracy of full potential techniques, we have developed a *full charge-density* (FCD) method. In this method the charge density is obtained from the output of self-consistent linear muffin-tin orbitals (LMTO) ASA calculations, the Coulomb energy is calculated exactly from the complete, nonspherically symmetric charge density defined within nonoverlapping, space-filling Wigner-Seitz cells, and the exchange-correlation energy is evaluated by means of the local density approximation or the generalized gradient approximation applied to the complete charge-density. The kinetic energy is obtained as the ASA kinetic energy corrected for the nonspherically symmetric charge-density by a gradient expansion. The integration over the Wigner-Seitz cell is carried out by means of the *shape truncation function* technique, which is also discussed in detail. The FCD technique retains most of the simplicity and computational efficiency of the LMTO-ASA method, while several tests for bulk metals and surfaces show that the accuracy of the method is similar to that of full potential methods.

1 Introduction

As a consequence of the rapidly increasing computational facilities followed by the development of computer codes, the *ab initio* electronic structure methods are able to treat more and more complicated problems, closely related to applications, with sufficiently high accuracy. During the last two decades the linear muffin-tin orbitals (LMTO) method [1,2,4,9,31–33,35,51] has been one of the most commonly used method in electronic structure calculations. In particular, due to its simplicity and extreme computational efficiency it has been extensively used in total-energy calculations for close-packed high-symmetry systems where the atomic sphere approximation (ASA) may be applied with sufficient accuracy. However, if the local arrangements of atoms deviate strongly from spherical symmetry or the atoms change their positions away from high symmetry positions the ASA cannot be applied. Thus, although the LMTO-ASA may be used to calculate, for instance, the electronic pressure, it cannot in its conventional implementations yield forces and, if uncorrected, the ASA breaks down when used to calculate elastic shear moduli. To increase the number of systems to which the LMTO method may be applied, including systems with low symmetry, one has developed a number of full-potential (FP) LMTO techniques [10–15]. These techniques are of course highly accurate but lack the efficiency of the LMTO-ASA method. Hence, they may be used in static but not in molecular dynamics calculations and they cannot be used in many applications.

According to the theorem of Hohenberg and Kohn [6], there exists a unique energy functional which is variational in the density. Hence, if the functional is evaluated with a trial density close to the exact ground state density, the error in the total energy is only of second order in the difference between the trial density and the ground state density. This variational property means that in many cases one can achieve the required accuracy simply by evaluating the total energy functional using an appropriate trial density and thus avoiding the most time consuming part of the calculation, the self-consistent iterations. In order to do this, one has to answer the following questions: How does one construct densities which applied in the true functional yield total energies of sufficient accuracy? In the context of the LMTO method one has the related question: How does one evaluate the true functional rather than the approximate ASA functional? It is the purpose of the present paper to provide one answer to these questions.

In the following, we describe and test an efficient technique for total energy calculations based on the LMTO-ASA method in the tight-binding (TB) representation [9,32,33,51]. According to this, we use the complete, non-spherically symmetric charge density generated in self-consistent ASA calculations to evaluate the true energy functional. We have developed the new technique during the last years and used it successfully in many applications [17–19], [20–24]. In the first version of the method we used the uncorrected ASA kinetic energy for the total energy calculation and only the electrostatic and exchange-correlation terms of the energy functional were evaluated from a complete non-spherical charge density. In many applications this approximation proved to be sufficiently accurate, like e.g. in studying the ground state atomic volumes of open crystal structures such as the α-phases of the light actinides [19–21]. It has turned out, however, that although the ASA kinetic energy is often a suitable approximation, it does not, for instance, yield sufficiently accurate total energies for the small orthorhombic and tetragonal deformations needed in calculations of elastic constants [22]. Therefore we improved the kinetic energy calculation beyond the ASA and thereby take the remaining step towards constructing the true energy functional.

Finally, we point out that although in its present form our *full charge density* (FCD) total energy calculation is based on the conventional LMTO-TB method, it can be an even more promising technique in many applications if it is combined with the recently developed *exact muffin tin orbitals* method by Andersen *et al.* [25]. The new method gives more accurate interstitial charge density and kinetic energy, and thus the accuracy of the FCD total energy can be substantially improved. The development of such a new FCD technique is in progress.

The structure of the paper is the following: In Section II we set up the FCD total energy and in two subsections we discuss the details of the kinetic and exchange-correlation energy, as well as the Coulomb energy calculations. We put emphasis on the calculation of the kinetic energy correction to the ASA and the technique used in the determination of the intercell Coulomb (Madelung) energy. In Section III we describe the construction of the charge density in a one-center form using a general muffin-tin orbitals formalism. The expressions valid on the LMTO basis are given in the Appendix C. In the calculation of the integrals over the Wigner-Seitz cell we used the *shape truncation function*, or simply *shape function technique*, which is discussed in Section IV. Some important calculational details faced to actual applications are presented and discussed through several examples in Section V. The accuracy of the FCD method is demonstrated in comparison with full potential and experimental results. Finally the paper is ended with the Conclusions.

2   Energy Functional

Within density functional theory the total energy of the system may be decomposed in the form [6]

$$E[n] \equiv G[n] + F[n], \tag{1}$$

where $G[n]$ is a universal functional consisting of the kinetic energy $T[n]$ of the non-interacting system and the exchange-correlation energy $E_{xc}[n]$, i.e.,

$$G[n] \equiv T[n] + E_{xc}[n], \tag{2}$$

and $F[n]$ is the Coulomb contribution to the total energy

$$F[n] \equiv \int v(\mathbf{r})n(\mathbf{r})d\mathbf{r} + \frac{1}{2}\int\int\frac{n(\mathbf{r})n(\mathbf{r}')}{|\mathbf{r} - \mathbf{r}'|}d\mathbf{r}d\mathbf{r}'. \tag{3}$$

Here, $v(\mathbf{r})$ is an external potential. The total charge density $n(\mathbf{r})$ may be given by the sum

$$n(\mathbf{r}) = \sum_R n_R(\mathbf{r}_R) \tag{4}$$

over lattice positions R of atomic-centred charge densities $n_R(\mathbf{r}_R)$ defined within space filling, non-overlapping cells Ω_R, which in turn may be written in the one-centre form [20]

$$n_R(\mathbf{r}_R) = \sum_L n_{RL}(r_R)Y_L(\hat{\mathbf{r}}_R), \tag{5}$$

where L is short-hand notation for (l, m), $\mathbf{r}_R = \mathbf{r} - \mathbf{R}$, and Y_L is a real harmonic. These atomic centred charge densities are normalized within the cells and the total charge density is continuous and continuously differentiable in all space.

In the following we assume that the total energy functional, in accordance with (4) may be partitioned into cell-contributions, i.e. $E[n] = \sum_R E_R[n]$, where $E_R[n] \equiv G_R[n] + F_R[n]$. Due to the non-local character of the interactions these functionals depend on the total density (4). However, in the local density as well as the generalized gradient approximation to the exchange-correlation and kinetic energies G_R depends only on n_R.

2.1   Kinetic and Exchange-Correlation Energy

In the self-consistent ASA based methods both the Schrödinger equation and the Poisson equation are solved within the spherical symmetric approximation for the charge density and the Wigner-Seitz cell. When self-consistency has been reached the Coulomb energy may be evaluated exactly by solving Poisson's equation for the proper charge density and atomic polyhedron. However, the one-electron energies and, hence, the kinetic energy, will reflect the approximation used in the solution of Schrödinger's equation. Therefore, in order to achieve the required accuracy of the total energy we correct the ASA kinetic energy. There are several reasons why a correction to the ASA kinetic energy of the kind presented here has not been previously attempted. First of all, in most LMTO calculations the electrostatic and exchange-correlation terms have been evaluated from a spherically symmetric charge density and, hence, there is no need for a more accurate kinetic energy. Secondly, the kinetic energy, which is obtained from the Kohn–Sham equations [7] as

$$T^{\mathrm{ASA}} = \sum_j^{\mathrm{occ}} \epsilon_j - \int n^{\mathrm{ASA}}(r) v_{\mathrm{eff}}^{\mathrm{ASA}}(r) d\mathbf{r}, \tag{6}$$

where ϵ_j are the one-electron energies, $n(r)$ the electron density, and the $v_{eff}(r)$ the effective potential, is variational in the potential, and it has often been assumed that the ASA kinetic energy is in fact sufficiently accurate. Finally, to improve on the ASA kinetic energy one would need to know an explicit kinetic energy functional, e.g. in the form of a gradient expansion. However, in view of the relatively slow convergence of the known kinetic energy gradient expansions it is not obvious that this would in fact lead to the required accuracy.

The solution to this impasse is to evaluate the main contribution to the kinetic energy in the ASA and then apply an approximate functional form to evaluate the difference between the ASA and the true kinetic energy. This remainder is presumably small and may be obtained with sufficient accuracy by a gradient expansion. A similar approach based on Hartree-Fock densities has been used in atomic calculations by DePristo and Kress [27]. The procedure is closely related to the modern gradient correction to local density functional theory and as we shall demonstrate the corrected FCD method has the accuracy of the full potential methods while retaining most of the simplicity and efficiency of the LMTO-ASA.

In principle, the exchange-correlation energy can be evaluated directly from the total charge density. However, in many cases it is useful to use an expansion arround a uniform or spherical symmetric charge density. Therefore, in the following we consider the universal functional $G[n]$ and we will separate the kinetic and exchange-correlation terms later. The energy density g, corresponding to the functional $G_R[n_R]$, is defined as

$$G_R[n_R] \equiv \int_{\Omega_R} g([n_R], \mathbf{r}_R) d\mathbf{r}_R \tag{7}$$

which may, within the density-gradient approximation, be expressed as [28]

$$\begin{aligned}
g([n_R], \mathbf{r}_R) &\equiv t([n_R], \mathbf{r}_R) + \epsilon_{xc}([n_R], \mathbf{r}_R) n_R(\mathbf{r}_R) \\
&= t(n_R, |\nabla n_R|^2, ...) + \epsilon_{xc}(n_R, |\nabla n_R|^2, ...) n_R(\mathbf{r}_R) \tag{8} \\
&\equiv g([n_R]),
\end{aligned}$$

where t and $\epsilon_{xc} n$ are the kinetic and exchange-correlation energy densities, respectively. For charge densities which deviate weakly from spherically symmetry $g([n_R])$ may be represented by a Taylor series around the sperically symmetric charge density $n_R^0(r_R) \equiv \frac{1}{\sqrt{4\pi}} n_{R0}(r_R)$, i.e.,

$$\begin{aligned}
g([n_R]) = g([n_R^0]) &+ \tilde{n}_R(\mathbf{r}_R) \left. \frac{\partial g([n_R])}{\partial n_R} \right|_{n_R^0} + \nabla \tilde{n}_R(\mathbf{r}_R) \left. \frac{\partial g([n_R])}{\partial \nabla n_R} \right|_{n_R^0} \\
&+ \frac{1}{2} \tilde{n}_R(\mathbf{r}_R)^2 \left. \frac{\partial^2 g([n_R])}{\partial n_R^2} \right|_{n_R^0} + \frac{1}{2} (\nabla \tilde{n}_R(\mathbf{r}_R))^2 \left. \frac{\partial^2 g([n_R])}{\partial (\nabla n_R)^2} \right|_{n_R^0} \\
&+ \tilde{n}_R(\mathbf{r}_R) \nabla \tilde{n}_R(\mathbf{r}_R) \left. \frac{\partial^2 g([n_R])}{\partial n_R \partial \nabla n_R} \right|_{n_R^0} + ... \tag{9}
\end{aligned}$$

where $\tilde{n}_R(\mathbf{r}_R) \equiv n_R(\mathbf{r}_R) - n_R^0(r_R)$. As a result, the universal functional may be expanded in the following form

$$G_R[n_R] = G_R^0[n_R^0] + G_R^1[\tilde{n}_R, n_R^0] + G_R^2[\tilde{n}_R^2, n_R^0] + ..., \tag{10}$$

which may be used to calculate the total energy provided one knows the energy density functions and the corresponding gradients. Unfortunately, this is not the case and one must resort to approximations.

Within modern density functional theory the problem is solved, as far as the exchange-correlation energy $E_{xc;R}[n_R]$ is concerned, by means of the local density approximation (LDA) or generalized gradient approximation (GGA) [29] which yield analytic expressions that may easily be applied in conjunction with the full LMTO charge density. Thus, only the kinetic energy $T_R[n_R]$ remains to be accurately evaluated. Here, the problem is that neither the Kohn–Sham equation (6) in the ASA nor a straight density gradient expansion of the kinetic energy based on the explicit analytic expressions given, for instance, in Ref. [28] have sufficient accuracy when used seperately. However, as we shall show in the following one may by a combination of the two techniques in the form of a density-gradient correction to the ASA obtain kinetic energies with the desired accuracy.

We proceed by isolating the lowest order terms in (7 - 10) which may be evaluated in the ASA and the "small terms" which may be evaluated using a suitable functional form. In the ASA the kinetic energy is obtained from the Kohn–Sham one-electron equations in the form (6) which depends only on the spherical average of the charge density, because the effective one-electron ASA potential is spherically symmetric. Hence, viewed as a functional of an arbitrary density, equation (6) would give the same value for any non-spherically symmetric charge density having the spherical average n_R^0. It may therefore be identified as the kinetic energy belonging to the charge density n_R^0. Thus, we write

$$T_R^0[n_R^0] \approx T_R^{ASA}[n_R^{ASA}] + \Delta[n_R^0, n_R^{ASA}] \qquad (11)$$

where T_R^{ASA} is the kinetic energy obtained in the ASA from a spherical symmetric self-consistent calculation and the second term is a "small" shape-correction connected with the fact that the kinetic energy $T_R^0[n_R^0]$ corresponding to the spherically symmetric charge density n_R^0 is defined within the Wigner-Seitz cell at $\mathbf{R}$ while the ASA kinetic energy is defined inside the corresponding atomic sphere. Within the LMTO-ASA method the kinetic energy may be expressed by means of the ASA Hamiltonian H^{ASA} and the one-electron wave functions $\psi_j(\mathbf{r}_R)$ as [30]

$$T_R^{ASA} = \sum_j^{occ} \int_{S_R} \psi_j^*(\mathbf{r}_R) \, H^{ASA} \, \psi_j(\mathbf{r}_R) \, d\mathbf{r}_R$$
$$- \int_{S_R} n_R^{ASA}(r_R) \, v_{eff}([n_R^{ASA}], r_R) \, d\mathbf{r}_R, \qquad (12)$$

where $v_{eff}([n_R^{ASA}], r_R)$ is the effectiv one-electron potential, S_R the atomic Wigner-Seitz radius, and $n_R^{ASA}(r_R)$ the ASA charge-density normalized within the atomic sphere which is equivalent to $n_R^0(r_R)$ inside of the cell and sphere. This form may include the so-called combined correction [33,30]. The shape-correction term in (11) may be obtained as

$$\Delta[n_R^0, n_R^{ASA}] = \int_{\Omega_R} t([n_R^0]) \, d\mathbf{r}_R - \int_{S_R} t([n_R^{ASA}]) \, d\mathbf{r}_R, \qquad (13)$$

where the first integral is performed within the Wigner-Seitz cell, Ω_R, and the second one within the atomic sphere.

The shape correction and the kinetic energy part of the higher order terms in (10), i.e., those of first and second order in $\tilde{n}_R$ and $\nabla \tilde{n}_R$, are evaluated by means of a semi-local kinetic energy density functional. In Appendix A we present some functionals used in the actual applications.

For strongly anisotropic electron densities, like in the case of surfaces, the expansion (10) is not convergent. In this case the exchange-correlation energy has to be evaluated by a direct three dimensional integration of the LDA or GGA functional, while the non-spherical kinetic energy correction, i. e. the higher order terms in (10), is calculated as

$$\int_{\Omega_R} t([n_R]) \, d\mathbf{r}_R - \int_{\Omega_R} t([n_R^0]) \, d\mathbf{r}_R. \tag{14}$$

This procedure is much more time consuming than the Taylor expansion (10).

In order to show the effect of the correction terms to the kinetic energy we consider an orthorhombic shear deformation [22,31] of bcc Mo, which can be used to determine the c_{44} elastic constant. In Fig. 1 we show the total energy, with and without the kinetic energy corrections, versus the relative deformation parameter d.

The energy determined from the experimental shear elastic constant [31] is also shown. We observe that using the ASA kinetic energy the calculated c_{44} is negative and, hence, the bcc structure of Mo will be unstable against such an orthorhombic distortion. Obviously, the ASA kinetic energy is not sufficiently accurate to render the bcc structure of Mo stable. It is only when the kinetic energy correction is applied that a positive c_{44} is obtained which is in fact very close to the measured c_{44} value.

2.2   Coulomb Energy

In this section our aim is to calculate the Coulomb energy of a given charge distribution in a solid with arbitrary symmetry. We devide the total electrostatic contribution belonging to the cell at $\mathbf{R}$ into the intracell, that depends only on the electron density in the cell, and intercell terms, i.e.

$$F_R[n] = F_R^{\text{intra}}[n_R] + F_R^{\text{inter}}[Q]. \tag{15}$$

The intracell energy

$$F_R^{\text{intra}}[n_R] \equiv \int_{\Omega_R} \left[-\frac{Z_R}{r_R} + \frac{1}{2} \int_{\Omega_R} \frac{n_R(\mathbf{r}'_R)}{|\mathbf{r}_R - \mathbf{r}'_R|} d\mathbf{r}'_R \right] n_R(\mathbf{r}_R) d\mathbf{r}_R, \tag{16}$$

where Z_R is the atomic number, may be determined by solving the $l-$dependent Poisson equation or by numerical integration using, for instance, the shape function technique [2,17,32]. Denoting by $\tilde{n}_{RL}(r_R)$ the $Y_L(\hat{\mathbf{r}}_R)$ projection of $n_R(\mathbf{r}_R)$

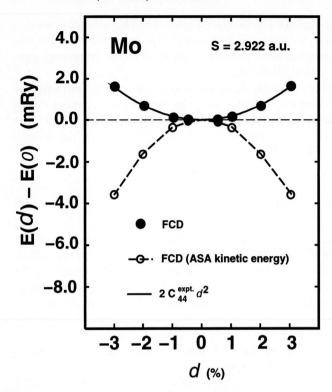

Fig. 1. Change of the total energy of Mo for orthorhombic shear deformation as a function of the relative deformation parameter d.

on a spherical surface of radius r_R that lies inside the cell and performing the angular integrations the expression (16) for the intracell energy can be brought to the form

$$F_R^{\text{intra}}[n_R] = \frac{\sqrt{4\pi}}{S} \sum_L \int_0^{S_R^c} \tilde{n}_{RL}(r_R) \left[\left(\frac{r_R}{S} \right)^l P_{RL}(r_R) \right.$$
$$\left. + \left(\frac{r_R}{S} \right)^{-l-1} Q_{RL}(r_R) \right] r_R^2 dr_R \tag{17}$$

where

$$P_{RL}(r_R) = \frac{\sqrt{4\pi}}{2l+1} \int_{r_R}^{S_R^c} \tilde{n}_{RL}(r_R') \left(\frac{r_R'}{S} \right)^{-l-1} (r_R')^2 dr_R' \tag{18}$$

and

$$Q_{RL}(r_R) = \frac{\sqrt{4\pi}}{2l+1} \int_0^{r_R} \tilde{n}_{RL}(r_R') \left(\frac{r_R'}{S} \right)^l (r_R')^2 dr_R' - \delta_{L,(0,0)} Z_R. \tag{19}$$

Here S_R^c stands for the radius of the sphere circumscribed to the Wigner-Seitz cell Ω_R at $\mathbf{R}$, and S is the average Wigner-Seitz radius. The explicit form for the $\tilde{n}_{RL}(r_R)$ function will be established later using the shape functions.

In Fig. 2 the intracell Hartree energy of fcc Cu is plotted relative to its converged value as a function of $l_{\max}$ used in (17). As may be seen from the figure the energy difference of 0.3 mRy obtained for $l_{\max} = 8 - 11$ is reduced below 0.1 mRy already for $l_{\max} = 12$. In the actual calculations we found that for a wide range of structures a reasonable accuracy of the intracell energy given by (17) can be achieved by performing the summation over l up to $l_{\max} = 8 - 14$.

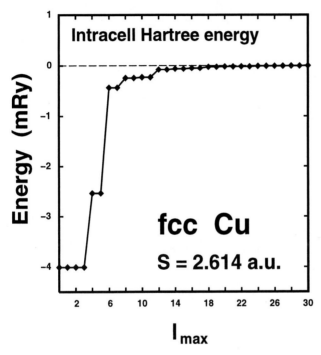

Fig. 2. The convergence test for the intracell Hartree energy of fcc Cu. The results are plotted relative to their converged value as a function of the maximal l value used in (17) or in (43).

The intercell interaction energy belonging to the cell at $\mathbf{R}$ may be written in the following form [17,33]

$$
F_R^{\text{inter}}[Q] = -\frac{1}{2S} \sum_{R'L} \frac{1}{2l+1} \left(\frac{b_{RR'}}{S}\right)^l Y_L(\hat{b}_{RR'}) \sum_{L',L''} Q_{RL'} C_{L',L''}^L
$$
$$
\times \frac{4\pi(2l''-1)!!}{(2l-1)!!(2l'-1)!!} \delta_{l'',l+l'} \sum_{L'''} S_{L'',L'''}(\mathbf{R'} - \mathbf{R} + b_{RR'}) Q_{R'L'''}, \qquad (20)
$$

and it can be completely described in terms of the multipole moments defined as

$$QRL = \frac{\sqrt{4\pi}}{2l+1} \int_{\Omega_R} n_R(\mathbf{r}_R) \, Y_L(\hat{\mathbf{r}}_R) \left(\frac{r_R}{S}\right)^l d\mathbf{r}_R - \delta_{L,(0,0)} Z_R$$
$$= Q_{RL}(S_R^C). \tag{21}$$

In (20) $S_{L,L'}(\mathbf{R})$ are the conventional LMTO structure constants and $\mathbf{R'}$ runs over the lattice vectors. The *displacement vector* $\mathbf{b}_{RR'}$ has to be chosen in such a way that the circumscribed spheres of the cells in question do not intersect each other (here we assume that the directions of $\mathbf{b}_{RR'}$ and $\mathbf{R'} - \mathbf{R}$ coincide)

$$|\mathbf{R'} - \mathbf{R}| + b_{RR'} > S_R^c + S_{R'}^c. \tag{22}$$

For the cells with nonoverlapping bounding spheres we can choose $b_{RR'} = 0$ so the equation (20) reduces to the well-known form

$$F_R^{\text{inter,no}}[Q] = -\frac{1}{2S} \sum_L Q_{RL} \sum_{\mathbf{R}_{no},L'} S_{L,L'}(\mathbf{R}_{no}) Q_{R_{no}L'} \tag{23}$$

where $\mathbf{R}_{no}$ runs over the cells with nonoverlapping bounding spheres.

For neighbouring cells with overlapping bounding spheres $b_{RR'}$ according to the inequality (22) has to be nonzero. Furthermore, as was pointed out by Gonis *et al.* [33], the outer sum (over l) is conditionally convergent and for a fixed value of the upper limit of the summation over l', l''', it may start to diverge above a certain value of l thus defining a range of convergence of the summation. It was also shown that the range of convergence for the summation in (20) depends sensitively on the choice of $b_{RR'}$ [33]. For larger values of $b_{RR'}$ the sum converges more slowly, but the range of convergence is wider. Therefore we expect an optimal value for $b_{RR'}$ to exist , which is large enough to satisfy (22) and to ensure a wide range of convergence but at the same time the summations over l converge rapidly allowing us to use a reasonably low $l_{\max}$ value in the actual calculations. The choice $b_{RR'} = |\mathbf{R'} - \mathbf{R}|$ proposed by Gonis *et al.* [33] is suitable for cubic systems [17,18]. However, for a crystal with lower symmetry it may happen that the convergency ranges for different neighbours do not even overlap with this choice of $b_{RR'}$ (e.g. for a tetragonal lattice with large c/a ratio). We have solved this problem by choosing the displacement vector equal to the radius of the circumscribed sphere S_R^c.

On the basis of a simple model [34], described in Appendix B, we have shown that the displacement vector is related to the circumscribed spheres radii as

$$b_{RR'} + |\mathbf{R'} - \mathbf{R}| = (1 + \alpha)(S_R^c + S_{R'}^c), \tag{24}$$

where α is a small positive number which determines the ratio between the maximal orbital quantum numbers included in the outer and inner summations, i. e.

$$l_{\max}/l'_{\max} \simeq \alpha. \tag{25}$$

In order to ensure similar accuracy for different neighbours we have to choose $\alpha = const$. We can see that the limit where α tends to zero corresponds to the lower limit of $b_{RR'} + |\mathbf{R}' - \mathbf{R}|$ in the inequality (22), but in this case the inner summation should go to infinity. For finite $l'_{\max}$, $b_{RR'} + |\mathbf{R}' - \mathbf{R}|$ has to be chosen according to (24) above its lower limit to assure the convergency of the inner summation for any l. With decreasing α, the number of terms in the outer sum decreases. On the other hand, with decreasing α, we need more and more multipole moments in the calculation of the inner sum. By choosing the coordinate system with z axis pointing in the direction of the neighboring atom $\mathbf{R}'$ the summations in (20) can be evaluated efficiently even for very high $l_{\max}$ values. Therefore, it is preferable to choose relatively high α values in order to minimize the computer time needed for the calculation of the multipole moments.

In order to test the accuracy of the intercell energy term evaluated by Eq. (20) and using for the displacement vectors the recipe given in (24) and (25) we calculated the Coulomb energy of a homogeneous charge distribution of several lattices with different symmetry. In these calculations we took $\alpha = 0.5$ and $l_{\max} = 20 - 22$. The scaled Coulomb energies $-E^{\mathrm{Coulomb}}/(Z^2/S)$, i.e., the average Madelung constants are plotted in Fig. 3 as a function of $l_{\max}$ for cubic, tetragonal, $\alpha - Np$ and $\alpha - Pu$ structures. The exact results obtained by the Ewald technique are also indicated in the figure. For the intracell Coulomb energy given in (17) and for the intercell Coulomb energy for cells with nonoverlapping bounding spheres (23) the converged values were used for any $l_{\max}$.

We can see from the figure that the results converge to the exact values smoothly in each case, indicating that using (24) the sums over l converge simultaneously for different neighbouring cells. The relative deviations from the exact values are less than 0.03% for $l_{\max} = 20$ in each case.

3   Construction of the Charge Density

It was shown by Andersen *et al.* [9,35] that even for open structures such as the diamond structure one may obtain good charge densities by means of an LMTO-ASA potential. In their approach, however, the output charge density is given in a multi-centre form which requires double lattice summations and is therefore less suitable in total energy calculations. Our aim is to rewrite the output ASA charge density in a one-centre form (5). This expression is simple to evaluate and well suited for the integration in the Wigner-Seitz cell at $\mathbf{R}$.

The muffin-tin (MT) orbitals for the low l orbital quantum numbers are defined as [35]

$$\chi_{RL}(\epsilon, \mathbf{r}_R) = Y_L(\hat{\mathbf{r}}_R) \begin{cases} \varphi_{Rl}(\epsilon, r_R) + P_{Rl}(\epsilon) j_l(\kappa, r_R) & \text{for } r_R \leq S_R \\ n_l(\kappa, r_R) & \text{for } r_R > S_R, \end{cases} \tag{26}$$

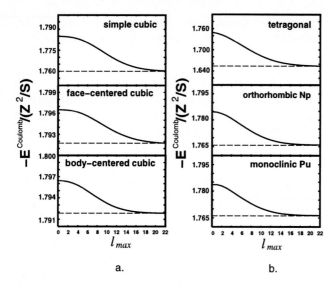

Fig. 3. The scaled average Coulomb energy per cell of a homogeneous charge distribution (Madelung constant) of (a) simple, face-centered and body-centered cubic and (b) tetragonal, $\alpha - Np$ ($oP8$) and $\alpha - Pu$ ($mP16$) structures as a function of l_{max}. For the tetragonal lattice $c/a = 1.5$ was used. The exact values for the Madelung constants obtained by the Ewald procedure are indicated by dashed lines.

where $\varphi_{Rl}(\epsilon, r_R)$ is solution of the radial Schrödinger equation, $\kappa^2 \equiv \epsilon - v_0$ is the "kinetic energy in the interstitial region", j_l and n_l are, respectively, regular and irregular solutions at the origin of the radial wave equation for the constant potential v_0 (the muffin-tin zero), and $P_{Rl}(\epsilon)$ is the potential function. In a standard self-consistent calculation only the s, p, d(and f) partial waves are included in the basis set, therefore in (26) $l \leq l_{max} = 2$(or 3).

The trial wave function for the energy ϵ is set up as the linear combination of the MT orbitals

$$\psi(\epsilon, \mathbf{r}) = \sum_{RL} \chi_{RL}(\epsilon, \mathbf{r}_R) u_{RL}. \tag{27}$$

The expansion coefficients u_{RL}^j as well as the energies ϵ_j are determined from the condition that the wave function $\psi(\epsilon, \mathbf{r})$ should be a solution of the Schrödinger equation for the muffin-tin potential

$$v_R(\mathbf{r}_R) = \begin{cases} v_R(r_R) & \text{if } r_R \leq S_R \\ v_0 & \text{if } r_R > S_R \end{cases}. \tag{28}$$

In order to set up algebraic formulation of this condition we expand the tails of the $n_L(\kappa, \mathbf{r}_R) \equiv n_l(\kappa, r_R)Y_L(\hat{\mathbf{r}}_R)$ envelope functions arround the site $\mathbf{R}'$ in terms of the $j_{L'}(\kappa, \mathbf{r}_{R'}) \equiv j_{l'}(\kappa, r_{R'})Y_{L'}(\hat{\mathbf{r}}_{R'})$ functions, i.e.

$$n_L(\kappa, \mathbf{r}_R) = -\sum_{L'} j_{L'}(\kappa, \mathbf{r}_{R'}) \, S_{R'L', RL}(\kappa). \qquad (29)$$

Here the expansion coefficients are the well known structure contants. It is very important to note that for each l the l' summation in (29) goes to infinity, and, because at high orbital quantum numbers the centrifugal potential becomes dominant, all the high l' terms from the right hand side of (29) are solutions of the Schrödinger equation.

For the energies ϵ_j where nontrivial solution of the secular or tail cancellation equation exists the multi-centre form of the wave function (27) inside the MT sphere at $\mathbf{R'}$ reduces to the one-centre form

$$\psi(\epsilon_j, \mathbf{r}_{R'}) \sim \sum_{L'} \varphi_{R'L'}(\epsilon_j, \mathbf{r}_{R'}) u^j_{R'L'} \quad \text{if } r_{R'} \leq S_{R'}. \qquad (30)$$

This is the expression that is used during the self-consistent iterations. In (30) two significant approximations have been made: i) the use of the incorrect basis functions in the overlap region, and ii) the neglect of the high l' terms in the expansion (29).

In the case of overlapping spheres, there is, in the region of overlap, an uncancelled remainder which is the superposition of the functions

$$f_{RL}(\epsilon, \mathbf{r}_R) \equiv \chi_{RL}(\epsilon, \mathbf{r}_R) - n_L(\kappa, \mathbf{r}_R) = f_{Rl}(\epsilon, r_R) Y_L(\hat{\mathbf{r}}_R) \Theta_R(\mathbf{r}_R) \qquad (31)$$

coming from the neighbouring sites [35]. Here,

$$f_{Rl}(\epsilon, r_R) = \varphi_{Rl}(\epsilon, r_R) + P_{Rl}(\epsilon) j_l(\epsilon, r_R) - n_l(\epsilon, r_R), \qquad (32)$$

and $\Theta_R(\mathbf{r}_R)$ is a step function, which is 1 inside and zero outside the MT sphere at $\mathbf{R}$. Thus, in the case of overlapping spheres centered at $\mathbf{R}$ the wave function at $\mathbf{R'}$, i.e. $r_{R'} \leq S_{R'}$, should be corrected as [35]

$$\psi(\epsilon_j, \mathbf{r}_{R'}) \sim \sum_{L'} \varphi_{R'L'}(\epsilon_j, \mathbf{r}_{R'}) u^j_{R'L'} + \sum_{R} \sum_{L'} f_{RL'}(\epsilon_j, \mathbf{r}_R) u^j_{RL'}. \qquad (33)$$

In Ref. [20] we describe a method whereby the f function can be included in the one-centre form of the charge density. In many systems we found that this correction is negligible compared to the effect of the high l' terms from the expansion (29). Therefore, here we focus on these terms and present a technique how the high tail components can be taken into account when the one-centre charge density is constructed.

Including the high l' terms in (29) the approximate wave function can be written in the following simple one-centre form

$$\psi(\epsilon_j, \mathbf{r}_{R'}) = \sum_{L'} \varphi_{R'L'}(\epsilon_j, \mathbf{r}_{R'}) u^j_{R'L'} - {\sum_{L'}}' j_{L'}(\kappa_j, \mathbf{r}_{R'}) u^j_{R'L'}, \qquad (34)$$

where the second summation includes only terms with $l' > l_{\max}$. This expression is valid inside and outside of the MT spheres as well. When $r_{R'} > S_{R'}$ the partial waves $\varphi_{R'l'}$ has to be substituted by the proper free electron solution $n_{l'} - P_{R'l'} j_{l'}$. In (34) we have extended the definition of the expansion coefficients for the high l' quantum numbers, i.e.,

$$u^j_{R'L'} \equiv \sum_{RL} S_{R'L',RL}(\kappa_j) u^j_{RL} \quad \text{for} \quad l' > l_{\max} \text{ and } l < l_{\max}. \qquad (35)$$

In this expression the off-diagonal ($l' > l$) structure constants are involved. In the conventional, unscreened, MTO method the calculation of these matrix elements is obvious. In the case of tight-binding representation the high-low subblock of the structure constants is constructed using the so called "blowing-up" technique [33]. The screening parameters for the high l indices are set to zero, and therefore the internal l'' summation in the Dyson equation [33], written for the high-low subblock,

$$S^\alpha_{R'L',RL}(\kappa) = S^0_{R'L',RL}(\kappa) + \sum_{R''L''} S^0_{R'L',R''L''}(\kappa) \alpha_{R''L''} S^\alpha_{R''L'',RL}(\kappa) \qquad (36)$$

is truncated at $l''_{\max} = l_{\max}$. Therefore, having the low-low subblock of the screened structure constant and using the high-low unscreened matrix elements, $S^0_{R'L',RL}$ we can determine by simple matrix multiplication the high-low block of the screened structure constants.

The one-centre form of the charge density can be obtained from Eq. (34) and (35). Using a compact notation for the radial functions, i.e.

$$\phi_{Rl}(\epsilon, r_R) = \begin{cases} \varphi_{Rl}(\epsilon, r_R) & \text{if } l \le l_{\max} \\ -j_l(\kappa, r_R) & \text{if } l > l_{\max}, \end{cases} \qquad (37)$$

we obtain the following expression for the partial components of the full charge density

$$n_{RL}(r_R) = \sum_{L''L'} C^L_{L''L'} \sum_j^{occ.} \phi_{Rl''}(\epsilon_j, r_R) u^{j*}_{RL''} u^j_{RL'} \phi_{Rl'}(\epsilon_j, r_R), \qquad (38)$$

where $C^L_{L''L'}$ are the real Gaunt numbers. In (38) both l'' and l' summations includes all the terms up to $l'_{max} > l_{max}$. Similar expressions valid for the LMTO basis set are given in Appendix C. In practical applications we found that for an accurate representation of the non-spherical charge density l'_{max} should be about $8 - 12$ depending on the structure.

When both the overlap correction (33) and the high l' tails (34) are included in the charge density besides the terms from (38) or (64) and those from Ref. [20] one has to include the cross terms, $f_{RL''}j_{L'}$ and $j_{L''}f_{RL'}$ as well.

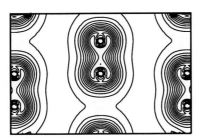

Fig. 4. Charge density contours (in electron/a.u.^{3}) of the *hexagonal* graphite within an atomic layer (left) and in a plane perpendicular to the layers (right). The density was calculated by the TB-LMTO using the one enter form (64).

In Fig. 4 we present charge density contour plots for hexagonal graphite calculated at the theoretical equilibrium volume using Eq. (64) given in Appendix C. From a comparison of the present plots withthose calculated using a full-potential method [36], we can conclude that the effect of the non-spherical potential terms, neglected in the present self-consistent LMTO calculation, have minor effects on the valence charge distribution. The ASA potential describes very well the strongly covalent double humped character of the C-C bonds in the hexagonal graphite. Futher examples of the application of the above technique for the calculation of charge densities are given in Refs. [20,37].

Finally we note that, because the high l' tails and the shape of the cell are neglected during the self-consistent iterations, the total charge density given by (5) and (38) will not be exactly normalized within the Wigner-Seitz cell. This difficulty can be overcome by a simple spherical symmetric renormalization of the charge density within the cells.

4   Shape Function Technique

In order to calculate the different energy terms discussed earlier, we need a technique to integrate over the W-S cell. To carry out this integration we use the so-called *shape function technique* [2,17,32] which was recently implemented by Drittler *et al.* [48] in their development of a full-potential Korringa-Kohn-Rostoker multiple scattering method. An alternative scheme to treat interstitial integrals has been proposed by Savrasov and Savrasov [15]. Here, we apply the original version of the method and define the following shape function

$$\sigma_R(\mathbf{r}_R) = \begin{cases} 1 & \text{for } \mathbf{r}_R \in \Omega_R \\ 0 & \text{otherwise} \end{cases} \tag{39}$$

which can be expanded in terms of real harmonics

$$\sigma_R(\mathbf{r}_R) = \sum_L \sigma_{RL}(r_R) Y_L(\hat{\mathbf{r}}_R). \tag{40}$$

Here $\sigma_{RL}(r_R)$ are the partial components of the shape function. By means of the shape function any integral over the W-S cell may be transformed into an integral over the sphere which circumscribes the cell, i.e.

$$\int_{\Omega_R} n_R(\mathbf{r}_R) \, K[n] \, d\mathbf{r}_R = \int_{S_R^c} \sigma_R(\mathbf{r}_R) \, n_R(\mathbf{r}_R) \, K[n] \, d\mathbf{r}_R. \tag{41}$$

Here $K[n]$ can be an arbitrary functional of electron density, for example, the kinetic and exchange-correlation energy density from Eqns. (10,13,14), or the Coulomb potential from Eq. (16), etc. The quantity $\sigma_R(\mathbf{r}_R) \, n_R(\mathbf{r}_R)$ can be expanded in terms of real harmonics

$$\sigma_R(\mathbf{r}_R) \, n_R(\mathbf{r}_R) \equiv \sum_L \tilde{n}_{RL}(r_R) Y_L(\hat{\mathbf{r}}_R), \tag{42}$$

where the partial radial functions represent the $Y_L(\hat{\mathbf{r}}_R)$ projections of the charge density on a spherical surface that lies inside the Wigner-Seitz cell. These functions are used in Section II in the evaluation of the Coulomb energy. In terms of the partial components of the shape function and of the charge density they can be expressed as

$$\tilde{n}_{RL}(r_R) = \sum_{L',L''} C^L_{L',L''} n_{RL'}(r_R) \sigma_{RL''}(r_R), \tag{43}$$

where $C^L_{L',L''}$ are the real Gaunt coefficients.

4.1   Evaluation of the Shape Functions

In order to determine the partial components of the shape function (39)

$$\sigma_{RL}(r_R) = \int \sigma_R(\mathbf{r}_R) Y_L(\hat{\mathbf{r}}_R) d\hat{\mathbf{r}}_R, \tag{44}$$

we divide the polyhedron, generated by the well-known Wigner-Seitz approach, into N_t tetrahedra, and we choose a local coordinate system for each tetrahedron. Thus the two dimensional surface integral in (44) may be performed only for the non-equivalent tetrahedra of number N_n, and the total shape function is obtained as

$$\sigma_{RL}(r_R) = \sum_t^{N_n} \sum_i^{N_e(t)} \sum_{m'} D^l_{m\,m'}(\alpha_i, \beta_i, \gamma_i) \sigma^t_{Rlm'}(r_R), \tag{45}$$

where $D^l_{m\,m'}$ are the matrix elements of finite rotations defined, for example, in Ref. [39], and $\alpha_i, \beta_i, \gamma_i$ are the Euler angles of the local coordinate system associated with the tetrahedron i. In (45) $N_e(t)$ denotes the total number of tetrahedra of type t, i.e. $\sum_t^{N_n} N_e(t) = N_t$.

The angular integration for each non-equivalent tetrahedron was performed by integrating analytically over θ and numerically over ϕ, as described by Stefanou et al. [40]. In this way we managed to achieve both high accuracy and efficiency, and thus develop a general algorithm for determining the shape function from the neighboring lattice vectors for an arbitrary structure.

In Fig. 5 we show the $L = (0,0), (4,0), (4,4), (10,0)$ and $(14,0)$ partial components of the shape function for the bcc structure as a function of r in units of the lattice constant a. Apart from the spherical component $\sigma_{(0,0)}(r)$, all the other terms are zero inside the inscribed sphere and outside the circumscribed sphere. The partial components of the shape function and their derivatives have kinks at the points where the sphere of radius r passes through a face, an edge or a corner of the polyhedron. In order to ensure the required accuracy in the radial numerical integration we take these points as mesh points.

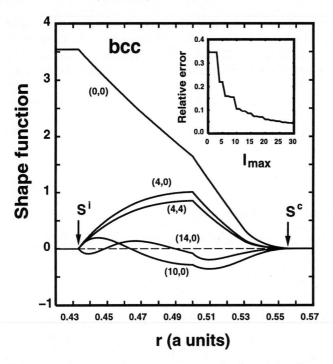

Fig. 5. The partial components of the *bcc* shape function for $L = (0,0), (4,0), (4,4), (10,0)$ and $(14,0)$ as functions of the radius. In the insert the relative error $(V_{\Omega-S^i} - \sum_l^{l_{max}} d_l)/V_{\Omega-S^i}$ (see Eq. (46)) is shown for different l_{max} values.

The partial wave expansion of the shape function (40) oscillates strongly and its convergence towards the step function is rather slow. The insert in Fig. 5 shows the relative error of the volume between the Wigner-Seitz cell and the inscribed sphere, calculated as

$$V_{\Omega-S^i} = \int_{S^c} \sigma(\mathbf{r})\,\sigma(\mathbf{r})\,d\mathbf{r} - \int_{S^i} \sigma(\mathbf{r})\,\sigma(\mathbf{r})\,d\mathbf{r}$$

$$= \sum_L \int_{S^i}^{S^c} \sigma_L^2(r)\,r^2\,dr \equiv \sum_l d_l. \qquad (46)$$

As it can be seen from the figure the relative error for $l_{max} = 30$ is still very high (4.2%). However, the partial components exhibit several oscillations within the interval $r \in (S^i, S^c)$ and the number of oscilations increases with the orbital quantum numbers. Therefore, the quantities derived from the shape function by integration like (41) are well behaved, as it can be seen from Fig. 2, where the Hartree energy of *fcc* Cu is shown as a function of l_{max} used in (17) or equivalalently in (43). The figure illustrates that a resonable accuracy is achieved already for $l_{max} = 8 - 12$.

5   Discussion

In this section we review and discuss through several examples some important calculational details related to the FCD total energy calculations. In the self-consistent procedure we solve the scalar-relativistic Dirac equation using either the hamiltonian or the Green's function formalism. In the former description the so called combined correction term [31,35,9] is included. The core electrons are treated within the frozen core approximation, and the semi-core states are included in a second energy panel. In this panel all the l chanels, except that corresponding to the semi-core states, are downfolded [33,51]. In the upper energy panel the inactive chanels are dow-folded. This procedure accounts correctly for the important weak hybrization in the occupied parts of the bandstructure and reduces the rank of the eigenvalue problem to that of the number of active orbitals. Moreover, in this way we manage to avoid the numerical difficulties that arises by fixing the expansion parameter $\epsilon_{\nu\ Rl}$ in a different position than the center of the occupied part of the corresponding chanel.

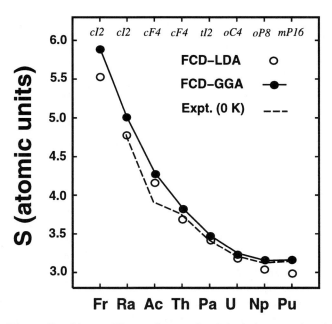

Fig. 6. Equilibrium Wigner-Seitz radii of the light actinides obtained by the full charge density method. The calculations are performed in the LDA or the GGA for the crystallographic α-phases indicated in Pearson notation at the top of the figure. The measured room temperature values are corrected to $T = 0$K using the measured mean thermal expansion coefficients.

As an illustration of this problem we consider the light actinides in the low temperature α phases. The uncorrected ASA fails [37] in the case of these open systems and, therefore, a more accurate and, at the same time, efficient approach is needed. In Fig. 6 we show the equilibrium Wigner-Seitz radii for the light actinides calculated by the present method using either LDA [41,42] or GGA [29] exchange-correlation energy functionals. The crystallographic α phases are indicated in the top of the figure. To obtain fully converged binding energy curves with a proper minimum for the light actinide metals in the FCD method the $6p$ states must be included in the first energy panel and the $7p$ states in a second energy panel. In the first panel we down-folded the s, d and f states and in the second only the p states. For all the active and inactive orbitals the ϵ_ν was fixed at the center of the occupied part of the corresponding band. We have found that the GGA results are on average only 1.3% larger than the zero temperature experimental values. This slight overestimate of the equilibrium volume is a common feature of the GGA functional observed in the case of the transition metals as well [22]. We note that the agrement between the LDA results and the experimental values is very good at the beginning of the series but the experimental trend, reproduced by the GGA, is not properly described by the LDA functional. In order to see the effect of the inactive $7p$ band we fixed ϵ_ν $7p$ at the center of the $7p$ band, thereby reducing the theoretical LDA atomic radius for Th by 2%. Throughout the actinide series the $7p$ band ascends and the errors related to fixing ϵ_ν become smaller. Hence, an incorrect parametrization of the LMTO basis can alter the theoretical trend provided by the FCD method.

The number of orbitals or basis functions that should be considered in the solution of Schrödinger's equation for a crystal potential is one of the basic problems in any band structure calculation method. A low cutoff in the orbital quantum number, i.e., a low l_{max} in (26), prefered in the secular equation, does not always quarantee basis set convergency. Regarding the present method, in several applications we found that the total energy evaluated form the total charge density represents a better basis-set convergence compared to the ASA total energy. In other words, in the case of the transition metals, for example, the inclusion of the f orbitals has a relatively small effect on the FCD total energy, while the ASA total energy can be affected by about 40% [17,18]. This is illustrated in Fig. 7, where we plotted the surface energies of the $4d$ transition metals obtained for the close packed fcc (111) surface by means of the LMTO-ASA and FCD methods. The two sets of results correspond to s, p, d and s, p, d, f basis sets. The isotropic experimental values are taken from Ref. [43]. The ASA surface energies and the surface full charge densities, needed for the FCD surface energies, were calculated by means of the self-consistent surface Green's function technique implemented by Skriver and Rosengaard [4]. Details of this calculation can be found in Refs. [17,23]. As it can be seen from the figure, the two sets of FCD results are very close to each other, and in fact they follow the experimental trend. In the ASA calculation the inclusion of the f orbitals has a strong negative effect in the total energy of surface layers. This is due to the fact that in a conventional ASA method the tail expansion (29) is truncated at the same l_{max} as is used for the muffin-tin orbitals. In the FCD technique the truncation in Eq. (29) is always higher than the one in Eq. (26).

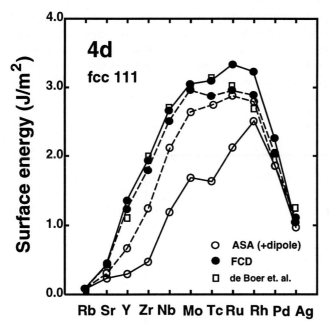

Fig. 7. Comparison of the surface energies obtained using the full charge density LMTO and the LMTO-ASA methods for the *fcc* (111) surface of the 4*d* element. The dashed lines connect the result obtained on the *s, p, d* basis set, and the solid lines those obtained by inclusion of the *f* orbitals. The experimental results by de Boer *et al.* are also shown.

The kinetic and exchange-correlation energy functionals used within the FCD technique involve the functional derivatives of the energy densities, see Eq. (9). The first order functional derivative of the exchange-correlation energy, for example, is the well known exchange-correlation potential, μ_{xc}, etc. Some of the parametrized density functionals, however, have different analytical forms for different densities or density gradients, which matches continuously. As a result, they may display an artificial discontinuity in the second or higher derivatives with respect to density or gradient and cannot be applied in conjunction with (9) and (10) which requires simple analytic representations of the exchange-correlation and kinetic energies. Therefore, in the actual applications the exchange-correlation energy is evaluated using either the LDA functional by Perdew and Wang [45], or the GGA functional by Perdew *et. al* [29,46], while the kinetic energy correction, Eqns. (10-13), is evaluated using either the gradient expansion (47) or the Pade's form (52).

In the FCD technique Poisson's equation is solved exactly and, hence, the Coulomb energy is calculated exactly from the total, non-spherical charge density. Moreover, within the LDA or GGA the exchange-correlation energy is also evaluated exactly, and the kinetic energy is corrected for the nonspherical effects. However, the ASA part of the spherical kinetic energy contribution (11) reflects the accuracy of the ASA potential, e.g., how well the spherical part of the full-potential is described by the ASA potential inside the atomic sphere. In open structures with large overlap between the atomic spheres the ASA does not work well and one has to introduce empty spheres into the structure to reduce the overlap and maintain the accuracy of the ASA. As an example of such a system we consider *hexagonal* graphite, where we include 8 empty spheres in the unit cell in addition to the 4 C spheres.

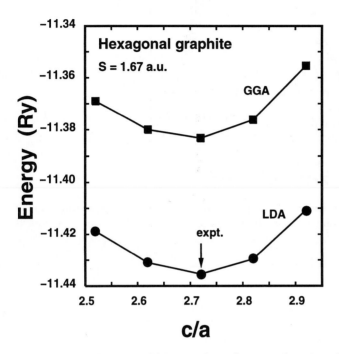

Fig. 8. The total energy of *hexagonal* graphite as a function of the c/a ratio obtained by means of the full charge density LMTO method.

The contour plot of the graphite charge density is shown in Fig. 4. Including 8 empty spheres the true charge density, obtained by the full potential method [36], is equally well reproduced in the C-C direction, in the centre of the hexagonal arrangement of the C atoms and between the atomic layers as well. The equilibrium Wigner-Seitz radius and bulk modulus obtained by the present calculation are 1.601 a.u. and 3.32 Mbar within the LDA, and 1.618 a.u. and 3.07 Mbar within the GGA, respectively. These numbers are only in moderate agreement with the experimental values, 1.67 a.u. and 2.86 Mbar [36], and with the full potential results from Ref. [47], 1.68 a.u. and 2.36 Mbar. However, when the c/a ratio is considered, our LDA value shown in Fig. 8 is in very good agreement with both the experimental, 2.72 and full potential value, 2.77 from Ref. [36]. The c/a ratio in the present work was calculated by minimizing the total energy with respect to the c/a at constant volume, like in Ref. [36], and, therefore, it reflects only partially the C-C interlayer bonding.

6   Conclusions

We have presented a full charge density technique based on the complete charge density from a self-consistent LMTO calculation employing a spherically symmetric ASA potential. In the calculations, besides the exact Coulomb and exchange-correlation terms within LDA or GGA, we include a correction to the ASA kinetic energy, which means that we evaluate the true functional rather than an ASA functional. The technique has been tested through several calculations for systems where the conventional ASA method fails [18,20,22,23,37]. The comparison with the experimental values and with full potential results shows that the FCD technique has an accuracy similar to that of a full potential description, while the required computational effort is not significantly larger than in conventional spherically symmetric LMTO-ASA calculations.

7   Acknowledgements

This work was supported by the research project OTKA 016740 and 023390 of the Hungarian Scientific Research Fund. The Swedish Natural Science Research Council is acknowledged for financial support. Center for Atomic-scale Materials Physics is sponsored by the Danish National Research Foundation.

8   Appendix A

The starting point for the kinetic energy correction is the density-gradient expansion of the noninteracting kinetic energy functional [28]

$$T[n] = T^{(0)}[n] + T^{(2)}[n] + T^{(4)}[n] + ... \tag{47}$$

with

$$T^{(2k)} = \int t^{(2k)}(\mathbf{r})d\mathbf{r}. \tag{48}$$

Here, $t^{(2k)}$ is a kinetic energy density which (in atomic Ry units) has the following explicit forms

$$t^{(0)} = \frac{3}{5}(3\pi^2)^{2/3} n^{5/3}, \tag{49}$$

$$t^{(2)} = \frac{1}{36} \frac{(\nabla n)^2}{n}, \tag{50}$$

$$t^{(4)} = \frac{1}{370(3\pi^2)^{2/3}} \left[\left(\frac{\nabla^2 n}{n} \right)^2 - \frac{9}{8} \left(\frac{\nabla^2 n}{n} \right) \left(\frac{\nabla n}{n} \right)^2 + \frac{1}{3} \left(\frac{\nabla n}{n} \right)^4 \right]. \tag{51}$$

for $k = 0, 1, 2$. This expansion is valid for slowly varying or high densities, i.e. when $s \equiv |\nabla n|/n^{4/3} \ll 1$. For high values of s the expansion diverges and, therefore, other approximate functional forms should be considered. In the locally truncated expansion suggested in Ref. [48] the number of terms included in the series is determined in each point in space by a local criterion based on the properties of an asymptotic series. In practice, this means that in regions of high gradients the functional reduces to the Thomas-Fermi form, which, obviously, represents a poor approximation to the real kinetic energy density. In Ref. [49] we showed that the local inclusion of the von Weizsäcker term, $t_W = 9t^{(2)}$, rather than the Thomas-Fermi term, gives resonable accurate results in the large gradient limit as well. The parametrized form of this kinetic energy functional is

$$t_{3,2} = t^{(0)} \frac{1 + 0.95x + 9ax^3}{1 - 0.05x + ax^2}, \tag{52}$$

where $x = (5/27)s^2$. The parameter a was determined using the exact Kohn–Sham kinetic energies for the jellium surface [50], and we found $a = 0.396$ [49].

9 Appendix B

To find a reasonable choice for the displacement vector introduced in Eq. (20) we have calculated the electrostatic interaction energy of two truncated spheres of radius S^c with a uniform charge distribution n_0, separated by a distance d [34]. In the case of overlapping bounding spheres, $d < 2S^c$, the intercell energy (20) is given by

$$F^{\text{inter}}[\{Q_l\}] = \sum_l \left(\frac{b_d}{b_d + d} \right)^l \sum_{l',l''} Q_{l'}$$
$$\times \frac{(l + l' + l'')!}{l! l'! l''!} \frac{1}{(b_d + d)^{l' + l'' + 1}} Q_{l''}. \tag{53}$$

For nonoverlapping spheres, $d \geq 2S^c$, this expression reduces to the form

$$F^{\text{inter},no}[\{Q_l\}] = \sum_{l,l'} Q_l \frac{(l+l')!}{l!l'!} \frac{1}{d^{l+l'+1}} Q_{l'} \qquad (54)$$

where Q_l stands for the l-th multipole moment

$$Q_l = \frac{2\pi n_0}{l+3}(S^c)^{l+3} \int_{-1}^{\frac{d}{2S^c}} P_l(\mathrm{x})dx + \frac{2\pi n_0}{l+3} \int_{\frac{d}{2S^c}}^{1} \left(\frac{d}{2x}\right)^{l+3} P_l(\mathrm{x})dx. \qquad (55)$$

Here $P_l(x)$'s are the Legendre polynomials. We mention that owing to the axial symmetry of the system the multiple moments vanish for $m \neq 0$.

The necessary condition for the convergency of the outer sum in (53) is the convergency of the inner sums over l' and l'' for each value of l. The diagonal terms of the inner sum in (53):

$$A_{l'l'} = Q_{l'}^2 \frac{(l+2l')!}{l!(l'!)^2} \frac{1}{(b_d+d)^{2l'+1}} \qquad (56)$$

have a maximum around $l' = \tilde{l}$ for $l, l' \gg 1$

$$\tilde{l}(l) = l\frac{\frac{Q_{\tilde{l}+1}}{Q_{\tilde{l}}}}{b_d + d - 2\frac{Q_{\tilde{l}+1}}{Q_{\tilde{l}}}}. \qquad (57)$$

From (54) we can estimate the upper limit for the ratio of the multipole moments $\frac{Q_{l+1}}{Q_l}$ since this sum is always convergent for $d \geq 2S^c$ (similar result can be obtained from (55)). Therefore for higher l values we have

$$\frac{Q_{l+1}}{Q_l} < S^c \qquad (58)$$

and for $\tilde{l}$ we obtain

$$\tilde{l}(l) < l\frac{S^c}{b_d + d - 2S^c} \equiv \frac{1}{2\alpha}l. \qquad (59)$$

Thus we see that the individual terms in the inner sum show a maximum, which (strictly speaking its upper bound) is proportional to l and the coefficient α depends on b_d. In order to ensure the required accuracy of the inner sum it is reasonable to assume that the summation should be carried out at least up to $l'_{max} \simeq \frac{1}{\alpha}l > 2\tilde{l}(l)$ for any l; it is easy to show that in this case for the neglected terms in the inner summmations $\frac{A_{l'l'}}{A_{\tilde{l}\tilde{l}}} < 1\%$. This should hold for the largest value of l as well, i.e. $l_{max} = \alpha l'_{max}$, and we obtain the relation

$$b_d + d = (1 + \alpha)\, 2S^c. \tag{60}$$

The validity of this choise of the displacement vector should hold for realistic systems as well because in (58) for the ratio of the multipole moments we use an estimate which is independent on the shape of the cell and on the charge distribution. We note that Eq. (24) is a generalization of (60).

On the basis of (60) we can explain the existence of the range of convergence in the outer summation. For a fixed value of l'_{max} it is obvious that with increasing l, we start to neglect significant terms in the inner summation if $\frac{1}{\alpha}l > l'_{max}$ which may lead to the divergence of the outer sum. Therefore an upper limit of the range of convergence in l may be defined as $l'_{max}\left(\frac{b_d + d}{2S^c} - 1\right)$. Thus with increasing b_d the range of convergence becomes wider in accordance with the observation by Gonis *et al.* ([33]).

10   Appendix C

In the linear muffin-tin orbitals method [1,2,31,4,35,32,33,51,9] the lower part of the tail functions are substituted (taking into account the normalization function) by the first energy derivative of the partial wave, $\dot{\varphi}^\gamma$ calculated at the center of the energy range of interest, ϵ_ν. In this way the LMTO's become energy independent up to the second order and the summation over the occupied states in (38) can be evaluated separately. The energy dependence of the envelope functions within the LMTO formalism is fixed at $\kappa^2 = 0$. In the following we use the nearly orthogonal representation (γ) but the relations can easily be generalized to other representations.

The second order expression for the wave function (34) that includes the high tail contributions becomes

$$\psi(\epsilon_j, \mathbf{r}_{R'}) = \sum_{L'} [\varphi_{R'L'}(\mathbf{r}_{R'}) + \frac{1}{2}(\epsilon_j - \epsilon_{\nu\, R'l'})^2 \ddot{\varphi}_{R'L'}(\mathbf{r}_{R'})] u^j_{R'L'}$$
$$+ \sum_{L'} \tilde{j}^\gamma_{R'L'}(\mathbf{r}_{R'}) \tilde{u}^j_{R'L'}, \tag{61}$$

where $\ddot{\varphi}$ denotes the second order energy derivative. The first summation includes terms up to the $l_{max} \leq 2$(or 3), while the second summation includes all the high l' tail components as well. In (61) we have introduced a compact notation for the tail functions

$$\tilde{j}^\gamma_{R'l'}(r_{R'}) = \begin{cases} \dot{\varphi}^\gamma_{R'l'}(r_{R'}) & \text{if } l' \leq l_{max}, \\ j_{l'}(r_{R'}) & \text{if } l' > l_{max} \end{cases}, \tag{62}$$

and for the eigenvectors

$$\tilde{u}^j_{R'L'} = \begin{cases} (\epsilon_j - \epsilon_{\nu\,R'l'})u^j_{R'L'} & \text{if } l' \leq l_{max}, \\ -\sum_{RL} S^\gamma_{R'L',RL} \frac{1}{\sqrt{\frac{S}{2}\dot{P}^\gamma_{Rl}}} u^j_{RL} & \text{if } l' > l_{max} \end{cases} \tag{63}$$

In (61) the partial waves, their energy derivatives and the energy derivative of the potential function are calculated at $\epsilon = \epsilon_\nu$. Now we can write the expression for the partial components of the full charge density,

$$\begin{aligned} n_{RL} = \sum_{L''L'} C^L_{L''L'} [\varphi_{Rl''}\varphi_{Rl'}m^0_{RL''L'} + \varphi_{Rl''}\tilde{j}^\gamma_{Rl'}m^1_{RL''L'} \\ + \frac{1}{2}\varphi_{Rl''}\ddot{\varphi}_{Rl'}m^2_{RL''L'} + + \tilde{j}^\gamma_{Rl''}\varphi_{Rl'}(m^1_{RL'L''})^* \\ + \frac{1}{2}\ddot{\varphi}_{Rl''}\varphi_{Rl'}(m^2_{RL'L''})^* + \tilde{j}^\gamma_{Rl''}\tilde{j}^\gamma_{Rl'}m^3_{RL''L'}], \end{aligned} \tag{64}$$

where for simplicity we have neglected the radial variable r_R, and

$$m^0_{RL''L'} = \sum_j^{occ.} u^{j*}_{RL''}u^j_{RL'} \qquad \text{if } l'', l' \leq l_{max},$$

$$m^1_{RL''L'} = \sum_j^{occ.} u^{j*}_{RL''}\tilde{u}^j_{RL'} \qquad \text{if } l'' \leq l_{max},$$

$$m^2_{RL''L'} = \sum_j^{occ.} u^{j*}_{RL''}(\epsilon_j - \epsilon_{\nu\,Rl'})^2 u^j_{RL'} \quad \text{if } l'', l' \leq l_{max},$$

$$m^3_{RL''L'} = \sum_j^{occ.} \tilde{u}^{j*}_{RL''}\tilde{u}^j_{RL'} \qquad .$$

In (64) the l'' and l' summations include the high tail components as well, however, for these high indices the φ and $\ddot{\varphi}$ should be set to zero.

The high-low block of the structure constant S^γ can be calculated using the Dyson equation

$$S^\gamma_{R'L',RL} = S^\alpha_{R'L',RL} + \sum_{R''L''} S^\alpha_{R'L',R''L''}(\gamma_{R''l''} - \alpha_{R''l''})S^\gamma_{R''L'',RL}, \tag{65}$$

where $\gamma_{R''l''} = \alpha_{R''l''} = 0$ for $l'' > l_{max}$, and the high-low subblock of the tight-binding LMTO structure constant S^α is obtained from (36) written for $\kappa^2 = 0$.

References

1. O.K. Andersen, Solid State Commun. **13**, 133 (1973).

2. O.K. Andersen and R.G. Wolley, Mol. Phys. **26**, 905 (1973).
3. O. K. Andersen, Phys. Rev. B **12**, 3060 (1975).
4. A.R. William, J. Kübler, and C.D. Gelatt, Phys. Rev. B **19**, 6094 (1979).
5. H.L. Skriver, *The LMTO Method* (Springer-Verlag, Berlin, 1984).
6. O.K. Andersen and O. Jepsen, Phys. Rev. Lett. **53**, 2571 (1984).
7. O.K. Andersen, O. Jepsen, and D. Glötzel, in *Highlights of Condensed-Matter Theory* , edited by F. Bassani, F. Fumi, and M. P. Tosi (North Holland, New York, 1985).
8. W.R.L Lambrecht and O.K. Andersen, Phys. Rev. B **34**, 2439 (1986).
9. O.K. Andersen, Z. Pawlowska, and O. Jepsen, Phys. Rev. B **34**, 5253 (1986).
10. G.W. Fernando, B.R. Cooper, M.V. Ramana, H. Krakauer, and C.Q. Ma, Phys. Rev. Lett. **56**, 2299 (1986).
11. J.M. Wills and B.R. Cooper, Phys. Rev. B **36**, 3809 (1987).
12. M. Springborg and O.K. Andersen, J. Chem. Phys. **87**, 7125 (1987).
13. M. Methfessel, Phys. Rev. B **38**, 1537 (1988).
14. M. Methfessel, C.O. Rodriguez, and O.K. Andersen, Phys. Rev. B **40**, 2009 (1989).
15. S. Savrasov and D. Savrasov, Phys. Rev. B **46**, 12181 (1992).
16. P. Hohenberg and W. Kohn, Phys. Rev. **136B** 864 (1964).
17. L. Vitos, J. Kollár and H. L. Skriver, Phys. Rev. B **49**, 16694 (1994).
18. J. Kollár,L. Vitos, and H. L. Skriver, Phys. Rev. B **49**, 11288 (1994).
19. J. Kollár, L. Vitos and H. L. Skriver, NATO ASI Series, Vol.41, ed. P. A. Sterne, A. Gonis and A. A. Borovoi, p.97 (1998)
20. L. Vitos, J. Kollár, and H. L. Skriver, Phys. Rev. B **55**, 4947 (1997).
21. J. Kollár, L. Vitos and H. L. Skriver, Phys. Rev. B **55**, 15353 (1997).
22. L. Vitos, J. Kollár, and H. L. Skriver, Phys. Rev. B **55**, 13521 (1997).
23. L. Vitos, A.V. Ruban, H. L. Skriver and J. Kollár, Surface Sci. **411**, 186 (1998)
24. L. Vitos, A.V. Ruban, H. L. Skriver and J. Kollár, Phil. Mag. **78**, 487 (1998)
25. O.K. Andersen, O. Jepsen, and G. Krier, in *Methods of Electronic Structure Calculations* , edited by V. Kumar, O.K. Andersen, and A. Mookerjee (World Scientific, Singapore, 1994), p. 63.
26. W. Kohn and L.J. Sham, Phys. Rev. **140** A1133 (1965).
27. A.E. DePristo and J.D. Kress, Phys. Rev. A **35**, 438 (1987).
28. R.M. Dreizler and E.K.U. Gross, Density Functional Theory, (Springer-Verlag, 1990).
29. J. D. Perdew, in *Electronic Structure of Solids*, edited by P. Ziesche and H. Eschrig, Academic Verlag, Berlin, p. 11 (1991).
30. O.K. Andersen, O. Jepsen and M. Sob, in Electronic Band Structure and its Applications, ed. M. Yussouff (Springer Lecture Notes, 1987).
31. P. Söderlind, O. Eriksson, J.M. Wills and A.M. Boring, Phys. Rev. B **48**, 5844, (1993)
32. J. van W. Morgan, J. Phys. C, **10**, 1181 (1977)
33. A. Gonis, E. C. Sowa and P. A. Sterne, Phys. Rev. Lett. **66**, 2207 (1991)
34. L. Vitos and J. Kollár, Phys. Rev. B **51**, 4074, (1995).
35. O.K. Andersen, A.V. Postnikov and S.Yu. Savrasov, in Applications of Multiple Scattering Theory to Materials Science, Eds. W.H. Butler, P.H. Dederichs, A. Gonis and R.L. Weaver, MRS Symp. Proc. p.37 (1992)
36. R. Ahuja, S. Auluck, T. Trygg, J. M. Wills, O. Eriksson, and B. Johansson, Phys. Rev. B **51**, 4813 (1995).
37. L. Vitos, J. Kollár, and H. L. Skriver in NATO ASI Series B:Physics, Stability of Materials, ed. A. Gonis, P.E.A. Turchi and J. Kudrnovsky, p. 393, Plenum Press, New York (1996).

38. B. Drittler, M. Weinert, R. Zeller and P. H. Dederichs, Solid State Commun. **79**, 31, (1991)
39. A.R. Edmonds, Angular Momentum in Quantum Mechanics, Ed. by E. Wigner and R. Hofstadter, Princeton University Press, Princeton, (1957)
40. N. Stefanou, H. Akai and R. Zeller, Comput. Phys. Comun. **60**,231 (1990)
41. J. Perdew and A. Zunger, Phys. Rev. B **23**, 5048 (1981).
42. D. M. Ceperley and B. J. Alder, Phys. Rev. Lett. **45**, 566 (1980).
43. F. R. de Boer, R. Boom, W. C. M. Mattens, A. R. Miedema, and A. K. Niessen, *Cohesion in Metals* (North-Holland, Amsterdam, 1988).
44. H.L. Skriver and N. M. Rosengaard, Phys. Rev. B **46**, 7157 (1992).
45. J. Perdew and Y. Wang, Phys. Rev. B **45**, 13244 (1992).
46. J.P. Perdew, K. Burke, and M. Ernzerhof, Phys. Rev. Lett. **77**, 3865 (1996).
47. M. T. Yin, Marvin. L. Cohen, Phys. Rev. B **29**, 6996 (1984).
48. E.W. Pearson, and R.G. Gordon, J. Chem. Phys. **82**, 881 (1985)
49. L. Vitos, H. L. Skriver, and J. Kollár, Phys. Rev. B **57**, 12611 (1998).
50. N. D. Lang and W. Kohn, Phys. Rev. B **1**, 4555 (1970).

A Full-Potential LMTO Method
Based on Smooth Hankel Functions

M. Methfessel[1], M. van Schilfgaarde[2], and R. A. Casali[3]

[1] Institute for Semiconductor Physics, Walter-Korsing-Str. 2, D-15230 Frankfurt (Oder), Germany

[2] SRI International, 333 Ravenswood Avenue, Menlo Park, California 94025 (Present Address: Sandia National Laboratories, Livermore, California 94551)

[3] Department of Physics, Universidad Nacional del Nordeste, 3400 Corrientes, Argentina

Abstract. The paper presents a recently developed full-potential linear muffin-tin orbital (FP-LMTO) method which does not require empty spheres and can calculate the forces accurately. Similar to previous approaches, this method uses numerical integration to calculate the matrix elements for the interstitial potential, which is the limiting step for any FP-LMTO approach. However, in order to reduce the numerical effort as far as possible, we use a newly introduced basis consisting of "augmented smooth Hankel functions" which play the role of the LMTO envelope functions. After presenting the basics of the approach, we report the results of numerical test for typical condensed-matter systems. The calculations show that good accuracy can be reached with an almost minimal basis set. These features of the method open the way to efficient molecular dynamics studies and simulated-annealing calculations to optimize structures while retaining the advantages of the LMTO method.

1 Introduction

The linear muffin-tin orbital (LMTO) method [1] has played a very successful role among the various techniques for solving the density-functional equations [2] for a condensed-matter system. Two characteristic features of this approach are (i) the use of atom-centered basis functions of well-defined angular momentum, constructed out of Hankel functions, and (ii) the use of augmentation to introduce atomic detail into the basis functions in the vicinity of each nucleus. Overall, the rationale behind this approach is to construct basis functions which closely resemble the actual wavefunctions from the very beginning. The consequence is that a comparatively small basis set already leads to a converged total energy, hopefully giving rise to substantial reductions in the computation time and the storage requirements.

In general term, when using a sophisticated basis in this manner, it is not *a priori* clear that these savings will actually be the realized. The price for a more sophisticated basis set is an increased computational effort in some steps of the calculation. This extra effort may or may not cancel out the gain due to the reduced basis size. For the LMTO method, the balance is no doubt positive if the atomic-sphere approximation (ASA) [3] is used. Hereby the single-electron potential is modeled by a superposition of spherical potentials inside overlapping space-filling spheres. Where this approximation is applicable, the LMTO-ASA method is presumably the most efficient procedure available for solving the density-functional equations to a reasonably high degree of accuracy. However, a "full-potential" treatment which goes beyond the ASA is needed for many systems of interest. Typical examples are the total energy changes associated with phonons distortions and atomic relaxations, say at a surface or around an impurity. Furthermore, for low-symmetry situations the LMTO-ASA method becomes unwieldy as the question of "empty spheres" to improve the packing fraction becomes more important. Finally, since the energies associated with such distortions are not reliable, the question of the calculation of the forces on the atoms does not even arise. However, the forces are precondition for simulated annealing and molecular dynamics studies in the spirit of the Car-Parrinello [4] method.

A number of different approaches have been developed in the past to go beyond the ASA in the LMTO method. Their common element is that the potential is treated correctly within the existing LMTO basis set. Unfortunately, this has lead to a substantial increase in the computational effort. In the interest of efficiency, we have explored and implemented a novel approach in which the LMTO basis functions themselves are modified in a controlled manner. This has two benefits: the basis set can be even smaller, and the effort for a numerical integration of the potential matrix elements is reduced. At the same time, a reformulation of the augmentation procedure is required since the standard structure-constant expansion cannot be used for the modified functions. Turning this into an advantage, we have used a "projection" description which is partly based on previous developments due to Blöchl [5] and Vanderbilt [6] but which includes some new features. Among other benefits, the separation into a "smooth" and atomic "local" terms is cleaner, leading to a straightforward expression for the forces.

Overall, the following criteria were considered imperative when developing this variant of the LMTO method:

- The forces must be calculated.
- All reasonable geometries should be handled without the need for empty spheres.
- It must be possible to improve the accuracy systematically by turning up the various convergence parameters.

Within these requirements, it was tried as far as possible to maintain the characteristic advantages of computational efficiency and small memory demands.

In the rest of this paper, the modified LMTO envelope functions are discussed in Sect. 2.2. The reformulated augmentation procedure and the force theorem are described in Section 2.3. Finally, we present some practical experience with the resulting method in Section 3. Specifically, we investigate the dependence of the results on the various convergence parameters and compare to calculations with other methods for a number of realistic systems.

2 Description of the Method

This section presents the basic ideas behind the method. After a discussion the central role of the interstitial potential matrix elements, the modified LMTO-like envelope functions (smooth Hankel functions) are presented. Finally, the reformulated augmentation procedure and the forces are discussed.

2.1 The Central Role of the Interstitial Potential Integrals

In a density-functional calculation, the three main tasks are (i) to solve the Poisson equation and to add on the exchange-correlation potential, making the effective potential felt by an electron, (ii) to solve the single-particle Schrödinger equation, and (iii) to add together the squared moduli of the wavefunctions to accumulate the new output density. In most cases, the solution of Schrödinger's equation is the most difficult and the most expensive step of these. In fact, if the potential is to be treated without any further approximation (*i.e.*, in a "full-potential" method), the main bottlenecks usually turn out to be in two closely related substeps:

- To set up the Hamiltonian matrix, we require the matrix elements of the effective potential $V_{\text{eff}}(r)$ for the basis functions $\chi_i(r)$:

$$V_{ij} = \int \chi_i^*(r) V_{\text{eff}}(r) \chi_j(r) \, dr \; . \tag{1}$$

- To obtain the output density, we must sum over the squared moduli of the wave functions $\psi_n(r)$:

$$n^{\text{out}}(r) = \sum_{n\text{occ}} w_n \left| \psi_n(r) \right|^2 \; . \tag{2}$$

Since the wavefunction is a linear combination of the basis functions, the real task when making the output density is to express the product of any two basis functions $\chi_i^* \chi_j$ in a form suitable for further handling. Specifically, we must be able to evaluate the integral of this product times the potential, since this is one important contribution in the total energy. Furthermore, the representation must "close the loop" so that the output density can be mixed with the input density and the result fed into the next iteration. In practice, both of the steps above are essentially the same problem, namely the evaluation of the potential integrals of (1).

In augmentation approaches such as the LMTO and linear augmented plane-wave (LAPW) methods [1], space is partitioned into atom-centered muffin-tin spheres and an interstitial region. The potential integrals split up accordingly into the integrals over the two types of region. The contributions from the muffin-tin spheres can be calculates in a straightforward and reasonably efficient manner since the potential as well as the basis functions can be expressed using one-center expansions. That is, each function is written in polar coordinates around the relevant sphere center as a radial part times a spherical harmonic.

Thus, the term which requires most attention is in fact only the interstitial contribution to the potential matrix element:

$$V_{ij}^{(\mathrm{IR})} = \int_{\mathrm{IR}} H_i^*(\boldsymbol{r}) V(\boldsymbol{r}) H_j(\boldsymbol{r}) \, d\boldsymbol{r} \ . \tag{3}$$

Here the functions $H_i(\boldsymbol{r})$ denote the envelope functions, *i.e.*, the analytic functions which will be augmented inside the atomic spheres to obtain the final basis functions $\chi_i(\boldsymbol{r})$, and IR denotes the interstitial region. This matrix element is always problematic, independent of the representation chosen for the interstitial potential. In fact, a major part of devising a viable method is to select a representation for the density and the potential which makes it possible to compute these matrix elements reasonably efficiently. The efficiency of the LMTO-ASA method is clear: by making the atomic spheres space filling and neglecting the interstitial potential, the most demanding computational step is eliminated, leaving only terms which can be evaluated in a compact and effective manner.

Basically there are two ways to proceed in a full-potential approach. First, the potential in the interstitial region can be expanded in some suitable set of auxiliary functions. For example, a set of atom-centered functions can be used, which can lead to a very compact representation. Substituting the expansion under the integral in (3) leads to a sum of integrals, each over a product of three terms. Unfortunately, such integrals can almost never be evaluated in closed form (the notable exception is when all terms are gaussians). In our context, two of the terms are LMTO envelopes, and no reasonable choice of a potential expansion considered to date leads to a closed form.

Alternatively, the interstitial potential can be specified by tabulating it on a regular mesh which extends through the unit cell. Equivalently, the coefficients of the Fourier expansion of the potential can be given. Although such a numerical representation requires more data to specify the potential, there are some considerable advantages. First, since the exchange-correlation potential must be evaluated point by point, a mesh representation is needed at some stage in any case. Second, by using a regular mesh the Poisson equation can be solved easily using a fast Fourier transform. Third, the additional step of fitting the output density or the effective potential to the set of auxiliary functions is avoided. Finally, enhancements such as gradient corrections can be implemented more easily on a regular mesh than in a more complicated representation. For these reasons, a real-space mesh representation of the interstitial potential (as well as the interstitial density) will be used in the following. In effect, we have opted for a hybrid treatment in which the wavefunctions are represented using a carefully constructed atom-centered basis set but the interstitial potential and density are given as numerical tabulations.

At this point, a comparison to different existing methods is in order. The generally used procedure is to extend the basis functions and the interstitial potential smoothly through the atomic spheres in some manner, to integrate (3) over the complete unit cell using these smooth functions, and finally to subtract off the unwanted contributions inside the spheres in conjunction with the augmentation step. In the full-potential LAPW method, all the involved functions are simply plane waves and the integral over the cell can be written down immediately. In the various FP-LMTO approaches, the smooth extension must be explicitly constructed for the sphere on which the function is the centered. This can be done by matching an analytical expression (such as a polynomial) at the sphere radius [7,8] or by using analytical functions similar to ours with a suitable choice of the parameters [9]. In either case, care must be taken that the resulting function strictly equals the standard LMTO envelope in the interstitial region. The alternative approach followed in this paper is to "bend over" the envelope functions already in the interstitial. As discussed next, this reduces the basis size and at the same time permits a coarser mesh for the numerical integration. We note that a completely different approach to the interstitial potential matrix elements is to re-expand the product of any two envelopes as a sum of auxiliary atom-centered basis functions. In this way the integrals over three factors in (3) are changed to a sum of two-center integrals. The expansion can be obtained approximately by fitting on the surfaces of the muffin-tin spheres [10] or more accurately for molecules by tabulating the results of a careful numerical fit [11]. Furthermore, an approach suitable for molecules and polymers has been developed which handles integrals over the product of three Hankel functions by expanding two of the terms around the site of the third [12].

2.2   Smooth Hankel Functions

In the following, we present the advantages of an LMTO-like basis consisting of augmented smoothed Hankel functions. The standard LMTO envelope function is a Hankel function of a (usually) zero or negative energy parameter times a spherical harmonic. This object will be denoted as a "solid Hankel function" in the following. It solves the Schrödinger equation for a flat potential, decays exponentially at large distances if the energy parameter is negative, and has a singularity at the site where it is centered. The essence of our modification is to remove the singularity. The resulting "smooth Hankel function" is smooth and analytic in all parts of space. When such a function is used to construct the basis, the parameters can (and should) be chosen so that the functions deviate from the unsmoothed variants already outside the central atomic sphere. As will be explained below, this speeds up the calculation for two separate reasons: the basis can be smaller, and numerical integration can be done using a coarser mesh.

Basic Properties

The smooth Hankel functions (discussed in detail in Refs. [13,14]) are shown in Fig. 1 for angular momentum 0, 1, and 2. For large radii, the smooth function to each angular momentum equals the corresponding standard Hankel function, showing the same exponential decay proportional to $\exp(-\kappa r)$, as specified by the negative energy parameter $\epsilon = -\kappa^2$. At smaller radii, the function bends over gradually until it finally approaches r^l close to $r = 0$. When multiplied by the spherical harmonic $Y_L(\hat{r})$, the result is analytic in all parts of space.

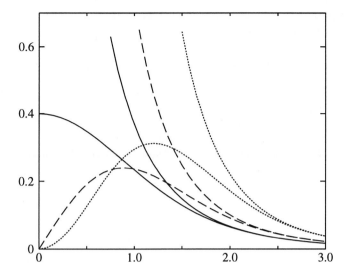

Fig. 1. Comparison of smooth and standard Hankel functions for $l=0$ (continuous lines), $l=1$ (dashed), and $l=2$ (dotted lines). The energy ϵ equals -1 and the smoothing radius R_{sm} equals 1.0. For large radii, the smooth and standard functions coincide. Near the origin, the smooth function bends over gradually until it enters as r^l whereas the standard function has a singularity proportional to $1/r^{l+1}$.

Of some importance is the parameter R_{sm}, denoted as the "smoothing radius" associated with the function. For practical purposes, the standard Hankel function and its smooth variant are equal where the gaussian $\exp(-r^2/R_{sm}^2)$ is negligible, say for $r > 3R_{sm}$ When R_{sm} is increased, the deviation from the standard function starts at a larger value of r and the resulting function is more strongly smoothed. Specifically, the value at $r = 0$ for $\ell = 0$ becomes smaller as the former singularity is washed out more and more.

A central distinction to the standard LMTO envelopes is that two separate parameters determine the shape of each function. More exactly, the energy parameter determines the exponential decay at large radii, and the smoothing radius determines how strongly the function has been smoothed. Consequently, in order to tune the function to mimic the true wavefunction as far as possible, both of these parameters should be adjusted.

As a basis set, these functions combine many of the advantages of Hankel functions and gaussians. In fact, a smooth Hankel function is a convolution of these two types of functions. Due to the exponential decay at large radii, they constitute a numerically more stable and more compact basis than pure gaussians. In contrast to the standard Hankel functions, they have a smooth nonsingular shape near the origin. Furthermore, many important quantities (such as two-center integrals) can be evaluated analytically.

Formally, the smooth Hankel functions are defined in the following way. The usual Hankel function for angular momentum zero is $h_0(r) = e^{-\kappa r}/r$ where κ

defines the decay at large radii. As a function of $r = |r|$ in three-dimensional space, h_0 satisfies the differential equation

$$(\Delta + \epsilon)h_0(r) = -4\pi\delta(r) \tag{4}$$

where $\epsilon = -\kappa^2$ is the energy associated with the function, here always taken to be negative. Thus, $\Delta + \epsilon$ applied to h_0 is zero everywhere except at $r = 0$, where a delta function arises from the $1/r$ singularity of h_0. Expressed differently, $h_0(r)$ is the response of the operator $\Delta + \epsilon$ to a delta-function source term.

To change this standard Hankel function into a "smooth Hankel function" the infinitely sharp delta function is smeared out into a gaussian:

$$(\Delta + \epsilon)h_0(r) = -4\pi g_0(r) . \tag{5}$$

By defining a suitable normalization of the modified source term $g_0(r) = C \exp(-r^2 / R_{sm}^2)$, the smooth Hankel approaches the standard function for large r. As r becomes smaller and reaches the range where $g_0(r)$ is non-negligible, the function $h_0(r)$ now bends over smoothly and behaves as a constant times r^l for $r \to 0$.

We will also need smooth Hankel functions for higher angular momenta in order to construct basis functions for the s, p, d... states. These can be constructed by applying the differential operator $\mathcal{Y}_L(-\nabla)$ defined as follows. The spheric harmonic polynomial $\mathcal{Y}(r) = r^l Y_L$ is a polynomial in x, y, and z, for example $C(x^2 - y^2)$. By substituting the partial derivatives $-\partial_x$, $-\partial_y$, and $-\partial_z$ for x, y, and z, respectively, the required operator is obtained in a straightforward manner. Applying this operator to the delta function yields point dipoles, quadrupoles and so on, and applying it to $g_0(r)$ yields smeared-out gaussian versions of these. Thus the L-th smooth Hankel functions is $H_L(r) = \mathcal{Y}_L(-\nabla)h_0(r)$ and satisfies the differential equation

$$(\Delta + \epsilon)H_L = -4\pi G_L(r) = -4\pi\mathcal{Y}_L(-\nabla)g_0(r) . \tag{6}$$

A number of important quantities can be calculated analytically for these functions, including the overlap integral and the kinetic energy expectation value between any two functions. They can also be expanded around some point in the unit cell. For further details, see Ref. [14].

Advantages of Smooth-Hankel Envelopes

The first reason for using the smooth-Hankel basis functions is that this reduces the size of the basis set, leading to a substantial gain in efficiency. To make this plausible, note that the standard LMTO basis functions are in fact not optimal as a basis for representing the crystal or molecular wave functions. The main problem is that they are "too steep" in the interstitial region close to the muffin-tin sphere on which they are centered. This is illustrated in Fig. 2. The standard Hankel functions solves Schrödinger's equation for a flat potential. However, when approaching a nucleus the true crystal potential is not flat but decreases

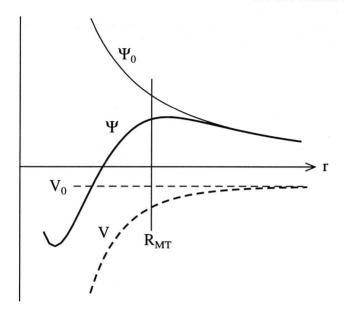

Fig. 2. Sketch to explain why smooth Hankel functions lead to an improved basis. For the flat potential V_0, the solution of the radial Schrödinger equation Ψ_0 is a standard Hankel function with a singularity at the origin. As the true potential V starts to feel the attractive nuclear potential, the correct wavefunction Ψ bends over. This behavior already starts outside the muffin-tin radius and is built into the smooth Hankel functions.

as it feels the attractive nuclear potential. The curvature of the wavefunction equals the potential minus the energy which therefore becomes negative. In response, the wavefunction bends over and changes from exponential to oscillatory behavior. By using smooth Hankel functions, this typical form is inherent in each basis function.

This effect can be appreciated by inspecting the way in which the standard LMTO basis functions combine to describe a crystal wavefunction. Generally the basis set must include some slowly decaying functions together with others which are considerably more localized. In the course of the calculation these combine with opposite signs, in this way modeling the required change of curvature. Using smooth Hankel functions as envelopes, these already have the correct behavior and some of the additional localized functions can be left away.

In practice, the amount of gain depends on the type of atom. For the important angular momenta, a tripled basis can often be replaced by a doubled set. Less important channels such as the d states in an sp atom can be described by one radial function instead of two. An overall reduction by a factor of almost two is sometimes possible. In the order(N^3) steps, the computation time in such a favorable case divides by eight.

The second major advantage to using smooth Hankel functions instead of the standard LMTO envelopes is that the time-consuming matrix elements (3) for the interstitial potential can be calculated more efficiently. As described above, the integrals can be obtained by integrating over the complete unit cell using a regular mesh and subsequently subtracting the contributions inside the spheres. The danger when calculating three-dimensional integrals using a mesh is that the computational effort can easily dominate all other steps. To keep the effort manageable, it is of high priority to make the integrands as smooth as possible. This can be done by using smooth Hankels as envelopes. As an example, consider silicon with a muffin-tin radius of 2.2 bohr. For the standard LMTO basis, the smoothing must be noticeable inside the MT sphere only, demanding a smoothing radius no larger than 0.6 to 0.7 bohr. Outside the central sphere, the smooth and conventional Hankel functions are then identical to acceptable precision. The required integration mesh spacing is approximately 0.35 bohr. If we permit the functions to bend over outside the MT sphere, we find that the optimal basis functions have a smoothing radius of about 1.4 bohr. For these functions, the integration mesh can be twice as coarse. Consequently the number of mesh points and the computational effort are divided by eight.

At this point it should already be mentioned that in the final implementation, the matrix elements of the smooth potential are actually calculated in reciprocal space. While this is at first sight equivalent to a real-space integration, the important difference is that a different reciprocal-space cutoff can be used for each function. For integrals involving envelopes with a large smoothing radius, a small cutoff is adequate. This property is very important for systems such as an oxygen impurity in silicon. The O basis functions demand a fine underlying real-space mesh through the unit cell, but this is effectively only used for the O functions. The effort for integrals between Si functions can be calculated with the same effort as without the presence of the oxygen atom.

Altogether, a modified basis using "smooth Hankel functions" combines two major advantages. Since these are more similar to the final wavefunctions, adequate convergence can be attained using a smaller basis set. Secondly, since each function is smoother, the mesh used to evaluate the potential integral can be coarser. These effects combine to a substantial saving in computer time. As an estimate (admittedly for an extremely favorable case), the integration mesh can be twice as coarse, leading to a saving of about $(1/2)^3 = 1/8$ in the order-3 steps. The basis set will contain something like one-half of the functions needed without extra smoothing, giving another factor of approximately 1/8. Together, the computer time is divided by 64. While the overall gain will not be as large for many systems, a speedup by a factor between 10 to 20 is realistic.

Analytical Two-Center Integrals

In the preceding, it was explained how the smooth Hankel functions help to evaluate the matrix elements of the interstitial potential more efficiently. In the course of a calculation, we also require the matrix elements of the kinetic energy operator and the overlap integrals. A major advantage of the smooth Hankel functions is that these integrals can be evaluated analytically. In fact, most two-center integrals involving these functions as well as gaussians can be obtained in basically the same way. The idea is to use Parseval's equality:

$$\int f_1(\boldsymbol{r})^* f_2(\boldsymbol{r}) \, d\boldsymbol{r} = \frac{1}{(2\pi)^3} \int \hat{f}_1^*(\boldsymbol{q}) \hat{f}_2(\boldsymbol{q}) \, d\boldsymbol{q} \tag{7}$$

and the explicit expression for the Fourier transform of a smooth Hankel function $H_L(\boldsymbol{r})$ located at some site $\boldsymbol{R}$:

$$\hat{H}_L(\boldsymbol{q}) = \frac{-4\pi}{\epsilon - q^2} e^{\gamma(\epsilon - q^2)} \mathcal{Y}_L(-i\boldsymbol{q}) e^{-i\boldsymbol{q} \cdot \boldsymbol{R}} \tag{8}$$

where $\gamma = R_{\text{sm}}^2/4$ is one-fourth of the squared smoothing radius. When two such expressions are multiplied together, the result can be readily written as a sum of terms of the same basic form, combined with additional powers of q^2 arising from the Clebsch-Gordon factorization of $\mathcal{Y}_L(-i\boldsymbol{q})^* \mathcal{Y}_{L'}(-i\boldsymbol{q})$. Since the phase factor in the product is $\exp[i\boldsymbol{q} \cdot (\boldsymbol{R}_1 - \boldsymbol{R}_2)]$, the final result is that the desired integral can be expressed analytically as a sum of smooth Hankel functions, evaluated for the connecting vector between the two sites. Hereby the extra powers of q^2 mean that the (only slightly more complicated) functions $\Delta H_L, \Delta^2 H_L \ldots$ are also needed. Furthermore, the resulting expression for the two-center integral is equally valid for molecular and Bloch-summed functions. In the later case, the only difference is that Bloch functions are also substituted in the final analytical expression. In addition to the overlap integrals considered here, integrals involving any power of the kinetic energy operator $-\Delta$ as well as Coulomb integrals can be calculated analytically in a similar manner. Another related application is to obtain the coefficients of a local expansion around another site, as described next.

Expansion Around a Site

In order to do perform the augmentation in practice, one of the steps which is needed is to expand a smooth Hankel function around some point in the unit cell. Far away from the center, the smooth Hankel function equals the unsmooth variant and the well-known structure constant expansion for the standard Hankel functions could be used. On the central sphere, the function is given explicitly by its definition. It is for sites close to the central sphere, such as nearest-neighbor atoms, where something new is needed. Here the function generally starts to bend over and the standard expansion does not apply.

This problem is solved as follows. We define a family of higher-order gaussians $G_{kL}(\boldsymbol{r})$ by applying differential operators to the seed function $g_0(r) = C \exp(-r^2/R_{\text{sm}}^2)$:

$$G_{kL}(\boldsymbol{r}) = \Delta^k \mathcal{Y}_L(-\nabla) g_0(r) . \tag{9}$$

We can construct biorthogonal polynomials to these functions, i.e., a set of polynomials $P_{kL}(\boldsymbol{r})$ with the property

$$\int G_{kL}(\boldsymbol{r}) P_{k'L'}(\boldsymbol{r}) \, d\boldsymbol{r} = \delta_{kk'} \delta_{LL'} . \tag{10}$$

In fact, it turns out that P_{kL} is just G_{kL} divided by $g_0(r)$ times a normalization constant. To expand an arbitrary function $f(r)$ as a sum of the P_{kL}, each coefficient can be calculated by the integral over $f(r)$ times the corresponding gaussian:

$$f(r) = \sum_{kL} A_{kL} P_{kL}(r) \tag{11}$$

where

$$A_{kL} = \int f(r) G_{kL}(r) \, dr \; . \tag{12}$$

This expansion, when truncated to some low value of k, is considerably more accurate than, for example, a Taylor series. This is because the expansion converges smoothly towards $f(r)$ in the range where $g_0(r)$ is large as more terms are included. When $f(r)$ is a smooth Hankel function centered anywhere in space, the integrals defining expansion coefficients can be done analytically. This supplies the desired local expansion.

The expansion is used in several different steps, most prominently to augment the envelope functions. Note that in this procedure, there are two distinct parameters which influence the accuracy of the expansion. By choosing a cutoff $p_{\max}$ for the terms in the expansion, the radial function is represented as a polynomial of order $p_{\max}$. The range over which the expansion is accurate is determined by the smoothing radius R_{sm} of the projection gaussians G_{kL}. When R_{sm} is chosen larger, the expansion can be used over a larger part of space but will not be as accurate overall for the same value of $p_{\max}$. Choosing R_{sm} in the vicinity of one-third of the muffin-tin radius will usually give a reasonable expansion within the muffin-tin sphere.

2.3   Augmentation

In the following, we describe the modified augmentation procedure used in the method. In general terms, the pseudopotential formulation and augmentation are two competing approaches to introduce atomic detail into the wavefunction near the nuclei. When a pseudopotential formulation is used, this is implicit: although only smooth functions are manipulated during a calculation, the true wavefunctions could be derived from these in a well-defined manner. When augmentation is used, the basis functions are explicitly constructed to show the required strongly-varying and oscillatory character close to an atom. The first step is to cut space into atomic spheres and an interstitial region. Throughout the interstitial region, the basis functions are equal to suitable smooth "envelope functions" which in the present case are the smooth Hankel functions introduced above. Inside each atomic sphere, every envelope function is replaced by a numerical solution of the Schrödinger equation. Specifically, in the well-known linear methods [1], numerical solutions of the Schrödinger equation in the spheridized potential and their energy derivatives are combined to match smoothly to the envelope function at the sphere boundary. Overall, this procedure amounts to a piecewise definition of the basis function in different parts of space.

Comparing the two approaches, the norm-conserving pseudopotential formulation [15] has a number of advantages, once the initial effort of constructing the pseudopotential has been completed. Angular momentum cutoffs are generally low and an expression for the force is easy to obtain. In contrast, since augmentation works by cutting out the part of space near the nucleus and treating it separately, quantities must sometimes be expanded to relatively high angular momenta. Due to the complexity of the augmentation procedure, it is often difficult to derive a valid force theorem. In practice, the augmentation and pseudopotential approaches have some similarity. Both methods expand a set of smooth basis functions by angular momentum around the different sites, then operate on the different angular momentum components independently. This suggests that a more unified description should be possible.

With the aim of a practical formulation for all-electron methods which has the simplicity and transferability of the pseudopotential approach, the formulation of augmentation described in the following shares some aspects with the projector augmented-wave (PAW) method [5] and has some similarity with Vanderbilt's ultrasmooth pseudopotentials [6]. As in the PAW method, "additive augmentation" is used to reduce angular-momentum cutoffs for the representation of the wavefunctions, charge density, and potential. However, the present approach eliminates the need to construct pseudopartial waves and projector functions. Also, completeness of the partial-wave expansion is not an issue, the force theorem here seems to be simpler, and orthogonality to the core states is automatic.

Augmented Basis Functions

As will be described in more detail below, the crystal potential will be written in the following form:

$$V(r) = \tilde{V}_0(r) + \sum_\nu \left\{ V_{1\nu}(r) - \tilde{V}_{2\nu}(r) \right\} \tag{13}$$

Here $\tilde{V}_0$ is a smooth potential extending through the unit cell, tabulated on a real-space mesh; $V_{1\nu}$ is the true potential inside the atomic sphere ν, given as an expansion in spherical harmonics times numerical radial functions, and $\tilde{V}_{2\nu}$ is the smooth mesh potential expanded in the same way. The tilde over a potential term indicates that compensating charges modeled by gaussians entered into the electrostatic contribution. One sees that the potential is expressed as a smooth function $\tilde{V}_0$ which is augmented by adding a local term $V_{1\nu} - \tilde{V}_{2\nu}$ inside each sphere. The relevant point is that the local expansions of $V_{1\nu}$ and $\tilde{V}_{2\nu}$ can be truncated at the (same) low angular-momentum cutoff l_L with small loss of accuracy. To obtain a smooth overall potential, $V_{1\nu}$ and $\tilde{V}_{2\nu}$ should have the same values and slopes at the sphere surface. The higher l-components of the two functions are closely similar and can be left away in the difference $V_{1\nu} - \tilde{V}_{2\nu}$.

This does not mean that the higher angular momentum components are set to zero in the total potential; instead, they are carried by the smooth mesh potential $\tilde{V}_0$.

After identifying the form in which the potential is specified, the next step is to calculate the Hamiltonian and overlap matrix elements for the basis functions $\chi_i(r)$, obtained by augmenting the (as yet unspecified) smooth envelope functions $F_i(r)$. The required integrals over the unit cell are

$$H_{ij} = \int \chi_i^*(r)[-\Delta + V(r)]\chi_j(r)\, dr \tag{14}$$

$$S_{ij} = \int \chi_i^*(r)\chi_j(r)\, dr\ . \tag{15}$$

At this stage, the envelopes could be the smooth Hankel functions, plane waves, or some other set of smooth functions extending through the unit cell.

In order to perform the augmentation, the first step is to project out local information about the envelopes near a chosen site. This can be done using some local set of radial functions, denoted by P_{kL}, used to expand the i-th envelope as

$$F_i(r) = \sum_{kL} C_{kL}^{(i)} P_{kL}(r)\ . \tag{16}$$

The notation anticipates that we will later use the polynomials of Section 2.2, but in this context the P_{kL} are general functions with well-defined angular momentum and sufficient radial degrees of freedom.

To augment, we first construct functions $\tilde{P}_{kL}$ which are augmented versions of the separate P_{kL}. Assuming P_{kL} is given in the form

$$P_{kL}(r) = p_{kl}(r)Y_L(\hat{r}) \tag{17}$$

then the augmented version is defined as

$$\tilde{P}_{kL}(r) = \tilde{p}_{kl}(r)Y_L(\hat{r}) = \left[A_{kl}\phi_l(r) + B_{kl}\dot{\phi}_l(r)\right]Y_L(\hat{r}) \tag{18}$$

where $\tilde{p}_{kl}(r)$ equals the contents of the square brackets. In the standard way, $\phi_l(r)$ and $\dot{\phi}_l(r)$ are a specific solution of the radial Schrödinger equation and its energy derivative, respectively. The coefficients A_{kl} and B_{kl} are chosen so that $\tilde{p}_{kl}$ and p_{kl} have the same value and derivative at the muffin-tin radius R_{mt}. The augmented envelope function then is

$$\tilde{F}_i(r) = F_i(r) + \sum_{kL} C_{kL}^{(i)}\left\{\tilde{P}_{kL}(r) - P_{kL}(r)\right\}\ . \tag{19}$$

At first sight, this expression seems slightly nonsensical. Using (16), the first term should cancel against the last, leaving only a sum over the $\tilde{P}_{kL}$. The relevant feature is that, when the sums over k and L are truncated (as will be assumed from here on), the result is still close to the complete sum. Exactly as in the case of the potential, we start with a smooth function containing all components up to infinity and replace only a few of the lower terms by numerical functions.

Overlap and Kinetic Energy Matrix Elements

The next task is to calculate overlap integrals for the augmented envelope functions. These integrals are expressed as follows:

$$\int \tilde{F}_i^* \tilde{F}_j \, d\mathbf{r} = \int F_i^* F_j \, d\mathbf{r} + \sum_{kk'L} C_{kL}^{(i)*} \sigma_{kk'l} C_{k'L}^{(j)} \tag{20}$$

where

$$\sigma_{kk'l} = \int_S \left\{ \tilde{P}_{kL} \tilde{P}_{k'L} - P_{kL} P_{k'L} \right\} d\mathbf{r} \tag{21}$$

is an integral over the atomic sphere. For simplicity in the notation, it was assumed that there is one atom in the unit cell and that P_{kL} is real. Note that $\sigma_{kk'l}$ depends only on l, not on $L=(l,m)$. In (20), the integral is evaluated by first calculating the overlap between the smooth envelopes as an integral over the whole unit cell. Then local information about the envelopes is projected out in the form of the coefficient vectors $C^{(i)}$ and $C^{(j)}$. Finally, the product $C^{(i)\dagger} \sigma C^{(j)}$ is added, where σ is a small symmetric matrix characterizing the atom at this site.

It should be pointed out that equation (20) is not formally equal to the integral over $\tilde{F}_i^* \tilde{F}_j$ when these functions are given by (19). Instead, the straightforward integral over the product would lead to a large number of unwieldy cross terms. The point is that (20) gives the correct result when all sums are taken to infinity. For finite cutoffs, the result is a good approximation to that which would be obtained if no truncation were done. As before, this follows because the discarded terms in the integrals over $\tilde{P}_{kL} \tilde{P}_{k'L}$ and $P_{kL} P_{k'L}$ are similar. Therefore the truncation errors cancel to a large extent when the the difference between the two contributions to σ is taken. Thus, the question of leaving away the cross terms in the expression for this and similar matrix elements is purely a convergence issue: we have an expression which is correct when all terms to infinity are included, but which must be truncated to a suitable finite sum for a practical calculation. Consequently it is of high priority to arrange matters for rapid convergence.

As a second comment, let us assume for a moment that the radial augmentation functions ϕ_l and $\dot{\phi}_l$ are kept frozen throughout the calculation. This is usually a reasonable procedure in practice, although it was not done in the calculations presented further on. Then the augmented expansion functions $\tilde{P}_{kL}$ are also invariant and consequently the local matrix $\sigma_{kk'l}$ is completely independent of the environment. The formulation thus begins to approach that of a unique and transferable pseudopotential.

Analogous to the overlap integral, the kinetic energy integrals are given as follows:

$$\int \tilde{F}_i^* [-\Delta] \tilde{F}_j \, d\mathbf{r} = \int F_i^* [-\Delta] F_j \, d\mathbf{r} + \sum_{kk'L} C_{kL}^{(i)*} \tau_{kk'l} C_{k'L}^{(j)} \tag{22}$$

where

$$\tau_{kk'l} = \int_S \left\{ \tilde{P}_{kL}[-\Delta]\tilde{P}_{k'L} - P_{kL}[-\Delta]P_{k'L} \right\} d\mathbf{r} \ . \tag{23}$$

Although it is not immediately apparent, the local kinetic energy matrix $\tau_{kk'l}$ is also symmetric. When the operator $-\Delta$ is moved from the second to the first function under each integral, two surface terms over the sphere boundary arise. These cancel because $\tilde{P}_{kL}$ and P_{kL} match in value and slope. As in the case of the overlap, $\tau_{kk'l}$ is independent of the environment if the radial augmentation functions ϕ_l and $\dot{\phi}_l$ are kept frozen during the calculation.

Potential Matrix Elements

The potential matrix element is somewhat more complicated. The electrostatic potential inside a given sphere depends not only on the density inside this sphere, but also on the density in all other parts of the unit cell. This is one reason why it is not easy to separate out a transferable local potential from the overall eigenvalue problem, even though this is a fundamental feature of the pseudopotential approach.

In precise terms, the potential is an auxiliary function needed to minimize the total energy respective to the trial density. Taken times some density variation and integrated, the potential should give the first-order response of the electrostatic and exchange-correlation energy for the given variation of the trial density. As described below, when evaluating the electrostatic energy, compensating gaussians are added to the smooth mesh density to make a "pseudodensity" with the correct multipole moments in the spheres. This construction enters into the total energy and, as a consequence, gives rise to certain terms in the potential matrix elements. To formulate this properly, write the total crystal density in a way similar to the potential (13) as

$$n(\mathbf{r}) = n_0(\mathbf{r}) + \left\{ \rho_1(\mathbf{r}) - \rho_2(\mathbf{r}) \right\} \tag{24}$$

where n_0 is a smooth function on the real-space mesh and ρ_1, ρ_2 are true and smooth local terms defined only inside an atomic sphere. (For the case of several atoms per unit cell, the term in braces is replaced by a sum over the spheres.) If the local sphere density $\rho_1 - \rho_2$ has multipole moments q_M, the compensated mesh density is

$$\tilde{n}_0(\mathbf{r}) = n_0(\mathbf{r}) + \sum_M q_M G_M(\mathbf{r}) \tag{25}$$

where G_M is a gaussian of moment unity with angular momentum M, localized inside the muffin-tin sphere except for a negligible tail.

Similar to the other matrix elements, the potential energy integral can be written in the form

$$\int \tilde{F}_i^* V \tilde{F}_j \, d\mathbf{r} = \int F_i^* \tilde{V}_0 F_j \, d\mathbf{r} + \sum_{kk'LL'} C_{kL}^{(i)*} \pi_{kk'LL'} C_{k'L'}^{(j)} \ . \tag{26}$$

Again, the integral is first evaluated for the smooth mesh functions, after which the projection coefficients are combined with a small local matrix to add on the contribution of an atomic site. Here, the local potential matrix $\pi_{kk'LL'}$ is diagonal in L only if the sphere potential terms V_1 and $\tilde{V}_2$ are taken as spherical. To derive an expression for $\pi_{kk'LL'}$ we inspect the changes in the electrostatic and exchange-correlation energies due to a variation of the charge density. Presenting only the results, the final expression involves the multipole moments of $\tilde{P}_{kL}\tilde{P}_{k'L'} - P_{kL}P_{k'L'}$:

$$Q_{kk'LL'M} = \int_S \left\{ \tilde{P}_{kL}\tilde{P}_{k'L'} - P_{kL}P_{k'L'} \right\} r^m Y_M(\hat{r}) \, d\boldsymbol{r} \ . \tag{27}$$

The matrix $\pi_{kk'LL'}$ turns out to be the sum of two contributions. The first term involves the smooth potential $\tilde{V}_0$ for the compensated mesh density:

$$\pi_{kk'LL'}^{\mathrm{mesh}} = \sum_M Q_{kk'LL'M} \int \tilde{V}_0 G_M \, d\boldsymbol{r} \ . \tag{28}$$

The second term involves the true and smooth local potentials, V_1 and $\tilde{V}_2$:

$$\pi_{kk'LL'}^{\mathrm{local}} = \int_S \left\{ \tilde{P}_{kL}V_1\tilde{P}_{k'L'} - P_{kL}\tilde{V}_2 P_{k'L'} \right\} d\boldsymbol{r}$$
$$- \sum_M Q_{kk'LL'M} \int_S \tilde{V}_2 G_M \, d\boldsymbol{r} \ . \tag{29}$$

Of course, 28 will cancel to a large extent against the last term in 29 since $\tilde{V}_2$ should be a local representation of the mesh potential $\tilde{V}_0$. It is necessary to include all the terms as described in order to minimize the total energy exactly.

These expressions have moved the situation quite a bit towards the desired environment-independence of all local atomic terms. In (29), the true potential V_1 is felt by the true partial density $\tilde{P}_{kL}\tilde{P}_{k'L'}$ while the smooth potential $\tilde{V}_2$ is felt by smooth partial density $P_{kL}P_{k'L'} + \sum Q_{kk'LL'M}G_M$. These two partial densities have the same multipole moments; this was essentially the definition of the quantity $Q_{kk'LL'M}$. If we modify the boundary conditions for the electrostatic potential inside the sphere in arbitrary way, this just adds the same linear combination of the harmonic functions $r^m Y_M(\boldsymbol{r})$ to both V_1 and $\tilde{V}_2$ (see below). It follows that the result of (29) does not depend on the electrostatic boundary conditions on the sphere, which can therefore be set to zero. The result is that (29) defines a quantity which can be calculated completely and unambiguously with only the density inside the sphere as input. However, this does not yet make $\pi_{kk'LL'}^{\mathrm{local}}$ a fixed quantity when the $\tilde{P}_{kL}$ are kept frozen (as was the case for the overlap and kinetic energy matrix elements). The reason is that the local sphere density $\rho_1 - \rho_2$ can change in the course of a calculation, which will lead to changes in V_1 and $\tilde{V}_2$. Keeping the local sphere density frozen is thus an additional approximation, albeit reasonably plausible, which must be assumed in order to get full independence of π^{local} from the environment. While this was not done in the current method, further investigations in this direction would be interesting.

Finally, we discuss the significance of the term $\pi_{kk'LL'}^{mesh}$ given by (28). Augmentation modifies the charge density inside a muffin-tin sphere, including the total sphere charge as well as the higher multipole moments. In this sense our procedure is much more similar to Vanderbilt's ultrasmooth pseudopotential approach [6] than to the usual norm-conserving formulation. In the smooth representation of the problem, the sphere charge and the higher moments are mapped onto gaussians of the correct normalization. The term in (28) describes the interaction of these gaussians with the smooth mesh potential.

Construction of the Output Density

To complete the self-consistency loop, the output density must be constructed:

$$n^{out}(\boldsymbol{r}) = \sum_n w_n |\psi_n(\boldsymbol{r})|^2 \tag{30}$$

where the sum runs over the occupied eigenstates and the w_n are occupation numbers including spin degeneracy. Each wavefunction is a linear combination of the basis functions in the form

$$\psi_n(\boldsymbol{r}) = \sum_i T_{in} \tilde{F}_{in}(\boldsymbol{r}) \tag{31}$$

and the output density is a sum over the products $\tilde{F}_i^* \tilde{F}_j$:

$$n^{out}(\boldsymbol{r}) = \sum_{ij} \left\{ \sum_n w_n T_{in}^* T_{jn} \right\} \tilde{F}_i^*(\boldsymbol{r}) \tilde{F}_j(\boldsymbol{r}) . \tag{32}$$

The output density should be expressed in the same form as the input density in (24). That is, it should be given by a smooth mesh density n_0^{out} together with separate true and smooth local contributions $\rho_{1\nu}^{out}$ and $\rho_{2\nu}^{out}$ for each sphere $\nu..$ In view of the preceding discussion of the augmentation process, the product of two augmented basis functions should be calculated as

$$\tilde{F}_i^* \tilde{F}_j = F_i^* F_j + \sum_{kk'LL'} C_{kL}^{(i)*} \left\{ \tilde{P}_{kL} \tilde{P}_{k'L'} - P_{kL} P_{k'L'} \right\} C_{k'L'}^{(j)} \tag{33}$$

where we have again assumed one atom per unit cell for notational simplicity. The integral over this quantity gave the overlap matrix, and the integral over this quantity times the potential gave the potential matrix element. In the latter case, the first term interacted with the smooth mesh potential $\tilde{V}_0$, the second with the local true potential V_1, and the third with the local smooth potential V_2 in the sphere. Correspondingly, the accumulated sums over the first, second, and third terms produce n_0^{out}, $\rho_{1\nu}^{out}$, and $\rho_{2\nu}^{out}$, respectively.

2.4   Representation of the Density and Potential

In the method presented here, some effort was taken to formulate matters such that low angular-momentum cutoffs can be used. As already mentioned above, quantities defined throughout the crystal are represented as a smooth function which extend through the whole unit cell, plus contributions which are non-zero inside the muffin-tin spheres. This "additive augmentation" formulation has advantages over a piecewise definition, which would *replace* the smooth function by another quantity inside a sphere. It is the key to good accuracy at low angular-momentum cutoffs because it permits the smooth interstitial function to supply the higher angular momentum components.

To represent the valence density, we start with the smooth function $n_0(r)$ tabulated on the real-space mesh. For each atom ν we carry about a "true" density, which is to be added on, and a "smooth" contribution, which is to be subtracted and should equal the local expansion of $n_0(r)$ except for the angular-momentum cutoff. The potential is written in a similar way. Collecting together the corresponding expressions from above, we have

$$n(r) = n_0(r) + \sum_{\nu} \left\{ \rho_{1\nu}(r) - \rho_{2\nu}(r) \right\} \tag{34}$$

and

$$V(r) = \tilde{V}_0(r) + \sum_{\nu} \left\{ V_{1\nu}(r) - \tilde{V}_{2\nu}(r) \right\} . \tag{35}$$

Here $\rho_{1\nu}$ includes the core and the nucleus, and $\rho_{2\nu}$ includes their smooth "pseudo" versions, modeled by localized gaussians. Each local contribution $\rho_\nu = \rho_{1\nu} - \rho_{2\nu}$ is a sum over various angular momentum components, is non-zero only inside the corresponding MT sphere, and goes to zero smoothly as it approaches the MT radius R_{mt}^ν. As already emphasised, even if ρ_ν is truncated to a low angular momentum (possibly only to its spherical part) the full density $n(r)$ includes contributions for all L up to infinity. At high angular momentum, the radial part for any smooth function approaches a constant times r^l. It is pointless to include these components explicitly in the local representation, since they are already contained in $n_0(r)$.

A special situation arises if the core states are extended enough to spill out of the muffin-tin sphere. In such a case, the smooth core density is written as the sum of a gaussian and a single smooth Hankel function of angular momentum zero, whereby the latter term describes the spilled-out tail core density. In the "frozen overlapped core approximation" (FOCA) [16] the smooth crystal core density is obtained by overlapping these atomic contributions. This means that a further parameter enters, namely the smoothing radius which defines this core Hankel function. Whereas a large smoothing radius makes it possible to use a coarser real-space mesh to represent the overlapped core density, generally the correct core density outside the sphere can be modeled more accurately using a smaller value of the smoothing radius.

The potential is made by solving the Poisson equation for the input density and adding the exchange-correlation potential. This is described next.

Electrostatic Energy and Potential

At the start of each iteration, the crystal density is available in the form given in (34) above. To make the electrostatic potential, first a smooth "pseudodensity" is constructed which equals the given density in the interstitial region and has the correct multipole moments in all spheres. This is done by adding localized gaussians g_ν to the smooth mesh density with the same multipole moments as the local contributions $\rho_{1\nu} - \rho_{2\nu}$ within the spheres:

$$\tilde{n}_0 = n_0 + \sum_\nu g_\nu = n_0 + \sum_{\nu M} q_{\nu M} G_{\nu M} \tag{36}$$

The electrostatic potential of the smooth mesh density $n_0(r)$ is made by transforming to reciprocal space by fast Fourier transform, dividing the coefficients by the squares of the reciprocal vectors, and transforming back. The electrostatic potential due to the gaussian terms includes the contribution of the nuclei and is handled analytically in order to avoid the need for a higher plane wave cutoff. Together, this produces an electrostatic potential $\tilde{V}_0^{es}(r)$ which is valid throughout the interstitial region and extends smoothly through the spheres.

Next, the electrostatic potential inside the spheres is determined. To recover the true density from the smooth pseudodensity, the following quantity must be added at each site ν:

$$\rho_{1\nu} - (\rho_{2\nu} + g_\nu) = \rho_{1\nu} - \tilde{\rho}_{2\nu} . \tag{37}$$

By construction, this local contribution has multipole moments which are zero. We solve the Poisson equation twice to obtain the "true" local potential $V_{1\nu}^{es}$ and the "smooth" local potential $\tilde{V}_{2\nu}^{es}$:

$$\Delta V_{1\nu}^{es} = -8\pi \rho_{1\nu} \tag{38}$$
$$\Delta \tilde{V}_{2\nu}^{es} = -8\pi (\rho_{2\nu} + g_\nu) . \tag{39}$$

The source terms $\rho_{1\nu}$ and $\tilde{\rho}_{2\nu} = \rho_{2\nu} + g_\nu$ have the same multipole moments, and we are really only interested in the difference $V_{1\nu}^{es} - \tilde{V}_{2\nu}^{es}$. Thus, the boundary conditions when solving the Poisson equation are here irrelevant, as long as the same set is used in both equations, and can be set to zero.

Finally, the electrostatic energy is obtained by adding together a smooth mesh term and local terms for all sites:

$$\int \tilde{n}_0(r) \tilde{V}_0^{es}(r)\, dr + \sum_\nu \left\{ \int_{S_\nu} \rho_{1\nu} V_{1\nu}^{es}\, dr - \int_{S_\nu} \tilde{\rho}_{2\nu} \tilde{V}_{2\nu}^{es}\, dr \right\} . \tag{40}$$

As a general note on the energy integrals, expression (40) for the electrostatic energy involves an important convergence issue similar to the one discussed for the augmentation procedure. If Equation (34) for the density is taken seriously, it should be possible to add together $\rho_{1\nu}$ and $-\rho_{2\nu}$ for each site from the very beginning. However, in our approach the program explicitly carries about both functions in separate arrays, making it possible to evaluate the electrostatic energy as described. In fact, the sum in (40) is not strictly the exact electrostatic energy for the density in (34) for any finite angular-momentum cutoff; again, a straightforward evaluation would lead to a large number of problematic cross terms.

In the description used here, there is a "true" and a "smooth" part to the problem which are kept separate. When evaluating energy integrals like the one above, the true density is not allowed to interact with the smooth potential and vice versa. The choice between the two possible expressions for an energy integral is only relevant for the rate of convergence with the angular-momentum cutoff, and not for the final outcome at convergence. When all angular-momentum components up to infinity are included, the distinction between the different ways to calculate the integral disappears. In the end, the main consequences of the formulation as in (40) are:

- In the resulting energy integrals, the higher angular momenta are supplied by the smooth density $n_0(r)$. That is, we start with an expression which contains all angular momentum components up to infinity in each sphere, then replace a few of the lower momentum terms.
- The aforementioned unwieldy cross terms do not have to be evaluated.
- The augmentation procedure described above cleanly separates into a smooth part plus local contributions, whereby each local term is independent of the environment of the atom.

Exchange-Correlation Energy and Potential

In the same spirit, the exchange-correlation energy is calculated by integrating a smooth function over the whole unit cell, then replacing the smooth contribution by the true one inside each MT sphere. Things are simpler here than for the electrostatic energy because no additional terms are needed to correct the multipole moments. Thus, the smooth exchange-correlation potential is made by evaluating it point-by-point for the smooth mesh density $n_0(r)$. Adding this to the electrostatic potential for the compensated density gives the total mesh potential

$$\tilde{V}_0(r) = \tilde{V}_0^{es}(r) + \mu_{xc}(n_0(r)) \tag{41}$$

Inside each sphere, two different exchange-correlation potentials are made and added respectively to the true and smooth local electrostatic potentials:

$$V_{1\nu}(r) = V_{1\nu}^{es}(r) + \mu_{xc}(\rho_{1\nu}(r)) \tag{42}$$

$$\tilde{V}_{2\nu}(r) = \tilde{V}_{2\nu}^{es}(r) + \mu_{xc}(\rho_{2\nu}(r)) \tag{43}$$

to produce the final potentials seen by the true and smooth densities at this site. The exchange-correlation energy is made by constructing the energy density ϵ_{xc} in exactly the same way and then adding together the smooth mesh term

$$\int n_0 \epsilon_{xc}(n_0) \, dr \tag{44}$$

and the sum over the local contributions

$$\int_{S_\nu} \rho_{1\nu} \epsilon_{\mathrm{xc}}(\rho_{1\nu})\, d\boldsymbol{r} - \int_{S_\nu} \rho_{2\nu} \epsilon_{\mathrm{xc}}(\rho_{2\nu})\, d\boldsymbol{r} \ . \tag{45}$$

Since $\epsilon_{\mathrm{xc}}(\rho)$ is a nonlinear function of the density, it mixes together all angular momentum components up to infinity. That is, $\epsilon_{\mathrm{xc}}(\rho_L + \rho_K)$ is not the same as $\epsilon_{\mathrm{xc}}(\rho_L) + \epsilon_{\mathrm{xc}}(\rho_K)$ and we should really include all L terms, even when making something as fundamental as the spherical part of $\epsilon_{\mathrm{xc}}(\rho)$. By evaluating the exchange-correlation energy as described, the interaction with the higher angular momenta is taken over by the smooth mesh contribution. When $\rho_{1\nu}$ and $\rho_{2\nu}$ are truncated to the same low angular momentum, the errors in the two integrands $\rho_{1\nu}\epsilon_{\mathrm{xc}}(\rho_{1\nu})$ and $\rho_{2\nu}\epsilon_{\mathrm{xc}}(\rho_{2\nu})$ are similar. Thus, while the two integrals in (45) converge rather sedately with the L cutoff separately, their difference converges rapidly.

Force Theorem

The force on an atom is defined as the negative gradient of the total energy respective to the corresponding atomic coordinates:

$$\boldsymbol{F}_\mu = -\nabla_\mu E_{\mathrm{SC}}(\boldsymbol{R}_1, \dots, \boldsymbol{R}_\mu, \dots, \boldsymbol{R}_N) \ . \tag{46}$$

Here E_{SC} is the full self-consistent total energy, which depends only on the atomic positions. In terms of small differences, the system is made self-consistent at two slightly different geometries and the total energies are compared. A force theorem is a closed expression, making it possible to calculate the forces $\boldsymbol{F}_\mu$ without explicitly shifting the atoms. Since the self-consistency process mixes together the various energy terms, a straightforward differentiation of the total energy expression is a strenuous (and not always feasible) way to obtain such a theorem.

A more convenient way to derive a force theorem is as follows [11]. Assume we want to calculate the forces at the geometry $\boldsymbol{P}^{(0)} = (\boldsymbol{R}_1^{(0)}, \dots, \boldsymbol{R}_N^{(0)})$, for which the self-consistent density is known. For this purpose we make an arbitrary guess for the way in which the density responds as the atoms are moved. That is, we define a density $\tilde{n}_P(\boldsymbol{r})$ for each geometry $\boldsymbol{P} = (\boldsymbol{R}_1, \dots, \boldsymbol{R}_N)$ which conserves the total charge and which approaches the correct self-consistent density smoothly as $\boldsymbol{P} \to \boldsymbol{P}^{(0)}$. From the variational properties of the energy functional it follows that

$$\nabla_\mu E_{\mathrm{SC}}(\boldsymbol{R}_1, \dots, \boldsymbol{R}_N) = \nabla_\mu \tilde{E}(\boldsymbol{R}_1, \dots, \boldsymbol{R}_N) \tag{47}$$

where $\tilde{E}$ is defined as the Harris energy [17] evaluated for the guessed density:

$$\tilde{E}(\boldsymbol{R}_1, \dots, \boldsymbol{R}_N) = E_{\mathrm{H}}[\tilde{n}_P(\boldsymbol{r})] \ . \tag{48}$$

It follows that the forces can also be obtained by differentiating the auxiliary function $\tilde{E}$. This is a considerably easier task than differentiation of the self-consistent energy because the Harris energy can be written down explicitly as function of the density. Furthermore, since the guessed function $\tilde{n}_P(r)$ can be chosen freely, different force theorems can be obtained depending on this choice. Sensibly, $\tilde{n}_P(r)$ should be defined in a way which makes it easy to perform the differentiation of $\tilde{E}$.

In the present formalism, the charge density is defined by a smooth mesh density $n_0(r)$ together with true and smooth local terms $\rho_{1\nu}(r)$ and $\rho_{2\nu}(r)$ associated with each site ν. If these functions represent the self-consistent density at geometry $P^{(0)}$, a natural choice for the guessed density $\tilde{n}_P(r)$ at a different geometry P is that $\rho_{1\nu}$ and $\rho_{2\nu}$ are carried along rigidly with the moving atoms while n_0 is unchanged.

This choice leads to a simple force theorem, as will be shown next. When the core states are treated separately, the Harris energy takes this form:

$$E_H = \sum \epsilon_n^{\text{val}} - \int n^{\text{val}} V_{\text{eff}} + U + E_{\text{xc}} + \tilde{T}_{\text{core}} \tag{49}$$

where $\tilde{T}_{\text{core}}$ equals $\sum \epsilon_i^{\text{core}} - \int n^{\text{core}} V_{\text{eff}}$ and all eigenvalues are calculated in the effective potential V_{eff}, made from the input density $n^{\text{val}} + n^{\text{core}}$. Integrals (including the implicit ones in U and E_{xc}) are all assembled in a similar way out of contributions from the smooth mesh density and from the spheres.

The aim is to derive the first-order change δE_H for the density change defined above. It reasonably straightforward that the last two terms give no contribution. Furthermore, all integrals over the spheres do not contribute. This is because the total-energy terms and the augmentation matrices for each atomic sphere are independent of the environment, depending only on the (here invariant) densities $\rho_{1\nu}$ and $\rho_{2\nu}$. On the other hand, a number of terms arise because the electrostatic potential on the mesh changes by some amount $\delta \tilde{V}_0$ as the compensating gaussians g_ν move along with the atoms. Each compensating gaussian g_ν can be split into a valence part g_ν^{val} and a smooth representation of the core and nucleus g_ν^{cn}. The compensated mesh density then equals

$$\tilde{n}_0 = n_0 + \sum_\nu g_\nu = \left[n_0 + \sum_\nu g_\nu^{\text{val}} \right] + \sum_\nu g_\nu^{\text{cn}} \tag{50}$$

where the term in brackets is the smooth representation of the valence density.

The contributions to δE_H from the first three terms of (49) are obtained as follows:

- By first-order perturbation theory, each eigenvalue ϵ_n^{val} changes by

$$\delta \epsilon_n^{\text{val}} = < C_n | \delta H - \epsilon_n^{\text{val}} \delta S | C_n > \tag{51}$$

where H and S are the Hamiltonian and overlap matrices and C_n is the column eigenvector. By summing over the occupied states and inspecting how $\tilde{V}_0$ enters into the Hamiltonian, the outcome is

$$\delta \sum \epsilon_n^{\text{val}} = \int \delta \tilde{V}_0 \left[n_0 + \sum_\nu g_\nu^{\text{val}} \right] + \delta^R \sum \epsilon_n^{\text{val}} . \tag{52}$$

Here δ^R refers to the eigenvalue change when the potential $\tilde{V}_0$ and the augmentation matrices entering into H and S are kept frozen.

- The change in the second term is

$$-\delta \int n^{\text{val}} V_{\text{eff}} = -\int \left[n_0 + \sum_\nu g_\nu^{\text{val}} \right] \delta \tilde{V}_0 - \int \tilde{V}_0 \, \delta g_\nu^{\text{val}} . \tag{53}$$

- The change in the third term is

$$\delta U = \int \tilde{V}_0 \left[\delta g_\nu^{\text{val}} + \delta g_\nu^{\text{cn}} \right] . \tag{54}$$

By summing these three contributions, the final force theorem is obtained:

$$\delta E_{\text{H}} = \int \tilde{V}_0 \, \delta g_\nu^{\text{cn}} + \delta^R \sum_n \epsilon_n^{\text{val}} . \tag{55}$$

Here the first term describes the force of the smooth density on the gaussian lumps which represent the core and the nucleus at each site. The second term is a generalized Pulay [18] term. It describes the eigenvalue shifts for a changing geometry but invariant smooth mesh potential and augmentation matrices. However, these augmentation matrices are used at the shifted atomic positions. This term can be evaluated in a straightforward manner since it mainly involves the gradients of the various quantities (such as the expansion coefficients of the envelope functions) which are needed to assemble the Hamiltonian and overlap matrices.

3   Tests of the Method

We have performed a number of tests to check the validity of the method. This we do in two ways. First, we investigate the sensitivity of the total energy to the various parameters at our disposal, i.e. the charge density and augmentation l-cutoffs l_ρ and l_a, the augmentation polynomial cutoff k_a, the smoothed density spacing, the treatment of the core, and the smoothing radii for the charge density, augmentation, and core, which we label respectively R_{sm}^ρ, R_{sm}^a, and R_{sm}^c. Second, we compare our results to other local-density calculations for several materials systems. Lastly, we present rules of thumb for the choice of basis.

3.1   Dependence on $l-$ and k-Cutoffs

Because the augmentation here shares a lot in common with a pseudopotential formulation, we expect that the l-cutoffs in the augmentation l_a and explicit representation of the charge density l_ρ to converge as efficiently. Indeed we find (see Fig. 3) that for $l_a=2$ the total energy is converged to $\sim 10^{-3}$ Ry/atom, and to $\sim 10^{-4}$ Ry/atom for $l_a=3$, even for the transition metal Ti. For a fixed l_a, a somewhat faster convergence in l_ρ was found: even $l_\rho=0$ was quite adequate

Table 1. Parameters for baseline reference in the tests for GaN, Ti, and Se. R_{mt}, R_{sm}^c and R_{sm}^ρ are in units of the average Wigner-Seitz radius. As fractions of touching sphere radii, the R_{mt} are 0.99, 0.97, and 0.96, respectively

	R_{mt}	R_{sm}^c	R_{sm}^ρ	k_a	l_a	l_ρ	mesh	basis
GaN	0.69	0.22	0.17	4	4	4	20	$spd \times 3$
(small basis)								$spd \times 2$
Ti	0.88	0.28	0.19	4	4	4	24	$spd \times 2 + p$
(small basis)								$spd \times 2$
(minimal basis)								spd
Se	0.60	0.20	0.15	4	4	4	24	$spd \times 2 + sp$

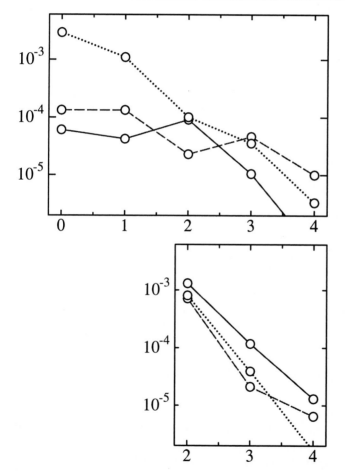

Fig. 3. Errors in the total energy, in Ry/atom, as a function of the l-cutoffs l_ρ for fixed $l_a = 5$ (top panel) and l_a, with $l_\rho = l_a$ (bottom panel). Solid line: GaN, dashed line: Ti; dotted line: Se.

for GaN and Ti, but this may be an artifact of the relatively high symmetry of those lattices.

The polynomial cutoff k_a in the P_{kl} expansion is another parameter whose convergence was checked. Because the tails have a rather weak energy dependence, we would expect that already after two terms ($k_a=1$) the total energy should be reasonable. Fig. 4 shows that the energy is converged to $\sim 10^{-3}$Ry/atom for $k_a=1$ (though the Ti calculation produced nonsensical results in that case), and $\sim 10^{-4}$Ry/atom for $k_a=2$. It is evident that the convergence of k_a will be sensitive to the augmentation smoothing R_{sm}^a, i.e. the smoothing radius for which one projects tails of the envelope functions into polynomials inside the augmentation site. The larger one makes R_{sm}^a, the more broadly dispersed the errors in the polynomial expansion; the optimal choice of R_{sm}^a distributes the errors most evenly throughout the augmentation sphere. From experience we have found that $R_{sm}^a \sim R_{mt}/3$ is approximately the smoothing for which the most rapid convergence with k_a is attained.

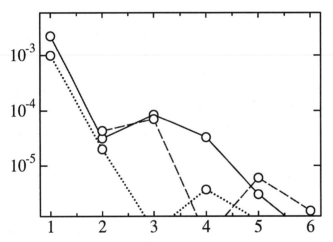

Fig. 4. Errors in the total energy, in Ry/atom, as a function of the k-cutoff. Solid line: GaN, dashed line: Ti; dotted line: Se.

3.2   Dependence on MT and Smoothing Radii

One critical test of the theory's validity is its dependence of the total energy on muffin-tin radius R_{mt}, because by changing R_{mt} one changes the representation of the basis. If the basis is complete, and all the cutoffs (l_a, l_ρ, k_a and mesh spacing) and smoothing radii (R_{sm}^c, R_{sm}^ρ) are set fine enough, there should be no dependence on R_{mt}. Indeed, we find this to be essentially true for the cases we studied (ZB GaN, Se, and Ti). Parameters for which the sensitivity was checked in the greatest detail were the core smoothing radius R_{sm}^c, and truncation of the basis. Using the reference parameters in Table 1, the total energy was found to be constant for all three tests to within ~1 mRy/atom when the MT radius was varied from ~0.85×R_T, to ~1.1×R_T, where R_T is the touching-sphere radius. Fig. 5 illustrates the dependence for GaN and Ti, for both the baseline case and when one parameter was changed relative to it.

That the energy is independent of R_{mt} for $R_{mt} < R_T$ is a demonstration that the errors can be well controlled, and the energy converged to a high precision as with the LAPW technique. That the energy is is independent of R_{mt} for $R_{mt} > R_T$ is less obvious, since spheres overlap and wave functions have different, and inconsistent representations. Previous implementations of full-potential, augmented-wave programs do not show this independence, and we attribute this to the fact that both the augmentation and the local contributions to the potential go smoothly and differentiably to zero at the MT boundary; thus the overlapping regions are "doubly counted" with a very small weight.

Fig. 5 shows that the energy is rather sensitive to the core smoothing radius. This is not surprising, since the core energy is very large and small changes in the core density can produce significant effects. But, energy differences between different structures should be significantly smaller than this. Also, as expected, there is a much stronger dependence on the total energy when the basis is made small. (This error can be reduced by re-optimizing the wave function parameters for each new choice of MT radius. No attempt was made to do this.) As one shrinks R_{mt}, the interstitial volume increases and throws greater weight into the interstitial representation. The figure also shows that R_{sm}^ρ must be small enough to really properly confine the projection of the MT potential into the augmentation sphere, or about 0.25×R_{mt}. R_{sm}^c can be set set to about 0.4×R_{mt}, negligible loss in accuracy, and up to about 0.6×R_{mt}, with a loss in absolute accuracy of ~1 mRy/atom, provided R_{mt} is kept close to touching. This is quite adequate for most applications, since such error will cancel when comparing relative energy differences. Some information about the MT dependence in Se is discussed in the comparison to other LDA results.

3.3   Dependence of the Total Energy on Basis

Having the extra degrees of freedom in the basis is one key advantage the present scheme has over conventional LMTO approaches. The two degrees of freedom, the energy ϵ and smoothing radius R_{sm} can be adjusted for each orbital. As we show here, a minimal basis of *spd* orbitals produces ground state energies within a few mRy of the completely converged LDA energy in close-packed systems; similar convergence can be achieved with an *spdsp* basis in open structures.

As of yet, there is no automatic prescription for finding the optimum choice of these parameters suitable to the solid, and we offer some general rules of thumb based on practical experience. One obvious choice is to determine R_{sm} and ϵ from the free-atomic wave functions. This turns to be a reasonable choice, particularly for the compact orbitals such as transition metal *d* orbitals or deep states such as the Ga *d* semicore levels.

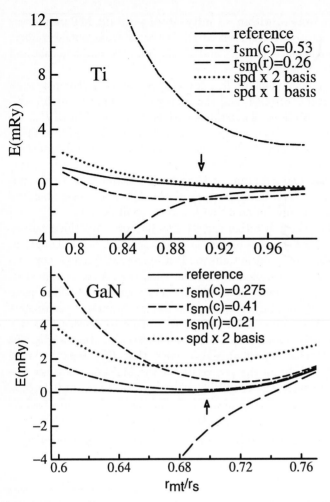

Fig. 5. Dependence of the total energy, in Ry/atom on muffin-tin radius. Reference parameters are provided in Table 1. Top: Ti; bottom: ZB GaN. Arrows mark MT radius for touching spheres. The large R_{sm}^c (short dashed line) correspond to $R_{sm}^c \sim 0.6 \times R_T$; the reference R_{sm}^c (solid lines) correspond to $\sim 0.32 \times R_T$. The large R_{sm}^ρ (long dashed line) correspond to $R_{sm}^\rho \sim 0.3 \times R_T$; the reference R_{sm}^ρ correspond to $R_{sm}^\rho \sim 0.25 \times R_T$. Calculations repeated using the perturbation approach to the core-valence interaction are indistinguishable from the ones shown.

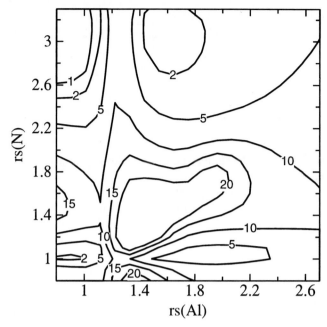

Fig. 6. Contours of the deviation of the total energy from the optimum value, in mRy/cell, as functions of R_{sm} for Al and N. All the orbitals on each atom were given the same R_{sm}. There are four minima, at $(R_{sm}(Al), R_{sm}(N)) \approx (0.95,1), (0.85,3), (1.9,1)$, and $(1.6,3)$. All the minima are within 4 mRy of each other.

Interestingly, the extended sp orbitals usually have two very different optimum values of R_{sm}. This is illustrated for AlN in Fig. 6. Some rules of thumb, gathered from experience for a number of cases, are:

- For localized, narrow-band orbitals, the values of R_{sm} and ϵ fit to the free atom are close to optimum.
- Reasonable choices of ϵ are $-1 < \epsilon < 0$. For wide-band orbitals, the best choice of ϵ is near zero.
- For sp orbitals, the "small" optimum R_{sm} is usually $\approx 2/3\ R_{mt}$. The "large" optimum R_{sm} is $\approx 1.5\ R_{mt}$ and mostly resembles a gaussian orbital in the near R_{mt}.
- For close-packed structures, a minimal basis consisting of the nine spd orbitals (16 for rare earths) is sufficient to produce a total energy with ~ 5mRy/atom or so of the totally converged LDA result. The error is slightly larger for heavier elements, unless f orbitals are included.
- For open structures, a basis consisting of thirteen $spdsp$ orbitals usually produces a result with a similar accuracy. In that case, choosing one set of sp orbitals with a "small" R_{sm}, and the the other with a large R_{sm} seems to work well. There is an occasional exception to this rule; for example, the deep Ga d states in GaN require two sets of d orbitals to reach this accuracy.

- The fastest convergence in k_a was achieved for R_{sm}^a, is $\sim 1/3 \times R_{mt}$; using that value, $k_a=2$ or 3 was adequate for all systems we studied.

3.4   Comparison with Other Density-Functional Calculations

First, we investigate the cases of elemental Se and Te. These form an open, low-symmetry crystal with approximately 90-degree bond angles. The electronic structure is approximately described by pure atomic p orbitals linked together in one-dimensional chains, with a strong σ bond connecting neighbors in the chain and with a weak interaction between the chains. The weak inter-chain interaction combined σ-like character of the bond make the stiffness of the crystal weak with respect to angular distortions, and thus pose a rather stringent test for the local-density approximation, and the low symmetry and open structure make a a good test for an atom-centered method's method's ability to reproduce the converged local-density result. The crystal structure of Se and Te is hexagonal with three atoms per unit cell, and may be specified by the a and c parameters of the hexagonal cell, and one internal displacement parameter u. Using a plane-wave pseudopotential approach, Dal Corso and Resta[19] have calculated the bonding in some detail, and they have also calculated the equilibrium structure of Se for both the LDA and adding gradient corrections.

Table 2 shows our LDA structures agrees well with that of Corso and Resta. For both calculations, the LDA predicts reasonably well the strong intra-chain bond length, but rather poorly the inter-chain bond length. The table also shows the dependence of the structural parameters and total energy on the size of basis and the MT radius. The reference basis consisted of one spd group with $R_{sm}=1.3$, $\epsilon=-0.1$ Ry and an additional sp group with $R_{sm}=1.4$, $\epsilon=-1$ Ry, for 13 orbitals. Adding a second d and a third s orbital lowers the energy by $2/3$ mRy per atom; adding an f orbital lowers the energy by an additional 2.5 mRy per atom. Thus, the 13-orbitals basis comes to within ~ 3 mRy/atom of a totally converged energy. Also, the table illustrates the dependence on R_{MT}: the lattice parameters vary weakly and the total energy varies by ~ 1 mRy/atom when R_{MT} is changed from -10 to 10 percent of touching for the minimal basis; the dependence is slightly weaker with a large basis.

Next, we examine in some detail the behavior of the (Al,Ga,In)N compounds, because it involves several useful tests of the method's validity. The deep Ga and In d levels must be included in the valence for a well-converged calculation. (Pseudopotential results that pseudize the d electrons differ considerably from those that do not.) Even though the states are deep, it turns out that they must treated carefully. For example, to produce accurate heats of mixing of the GaInN alloy it was found that two sets of Ga and In d orbitals were required.

Our next case concerns the [111] γ-surface of Ni. The [111] glide planes are connected with dislocation motion, and therefore are of significant practical interest. To model the γ surface with the periodic boundary conditions necessitated by the present computational approach, we employ a supercell of 8 [111] planes of Ni atoms, with the fault between the fourth and fifth planes. The first and 8th layers were separated by a vacuum of two layers. Lattice vectors in the plane of

Table 2. Crystal structure properties of Se. Lattice parameters a and c are in atomic units, as are intrachain bond length d_1 and interchain bond length d_2; u is an internal displacement parameter as described in Ref. [19] and E is the binding energy relative to the free atom, in Ry

	a	c	u	d_1	d_2	E
[a], R_{mt}=2.11	7.45	9.64	0.258	4.63	5.83	-0.7121
[a], R_{mt}=2.32	7.40	9.57	0.266	4.67	5.75	-0.7147
[a], R_{mt}=2.45	7.40	9.57	0.267	4.68	5.73	-0.7152
[a], R_{mt}=2.58	7.40	9.59	0.268	4.69	5.72	-0.7149
[b], R_{mt}=2.11	7.40	9.58	0.259	4.61	5.78	-0.7152
[b], R_{mt}=2.32	7.41	9.61	0.264	4.66	5.76	-0.7167
[c], R_{mt}=2.32	7.42	9.57	0.264	4.67	5.77	-0.7261
[d], R_{mt}=2.32	7.38	9.60	0.270	4.71	5.70	-0.7131
[e], R_{mt}=2.32	7.40	9.59	0.267	4.72	5.72	-0.7131
Ref. [19]	7.45	9.68	0.256	4.61	5.84	
Experiment	8.23	9.37	0.228	4.51	6.45	

[a] 13 orbital basis
[b] 21 orbital basis
[c] $21+f$ orbital basis
[d] $l_\rho=0$
[e] $l_a=2\ l_\rho=2$

the fault were $[1\bar{1}0]$ and $[11\bar{2}]$, so that each layer had two atoms. The total energy was calculated as a function of the translation τ along $[11\bar{2}]$. It was assumed that the four layers separating the fault and the vacuum was large enough that there was no coupling between the fault and the surface. The thickness of the vacuum itself was checked and the energy shown in Fig. 7 change was negligible when the vacuum thickness was doubled. l_a and l_ρ were taken to be 3, R^c_{sm} to be $0.4 \times R_{mt}$.

Fig. 7 shows the energy of the Ni γ surface, without allowing any atomic relaxation. It was calculated using a 9-orbital minimal basis of *spd* orbitals of energy -0.1 Ry and smoothing radius R_{sm} 2.0 a.u. and 1.0 a.u. for *sp* and *d* orbitals, respectively (dark circles), and then repeated for a basis enlarged by an *spd* set of energy -1 Ry, R_{sm}=1.3 a.u. (light circles). As the Figure shows, the addition of the larger basis made a 5% correction to the barrier. The BZ was integrated by sampling, using 27 irreducible points. A conservative mesh of $15 \times 24 \times 120$ divisions was used, corresponding to a mesh spacing of $\sim$0.31 a.u.; the inset in Fig. 7 shows the error in the barrier as a function of the mesh spacing. It is seen that an acceptable error was obtained for a mesh spacing of $\sim$0.5 a.u..

Table 3. Lattice properties of III-nitrides. Lattice constant a is in Å. All data are for the wurtzite structure, except for rows marked a (ZB). Data marked OLD-FP are calculations from the present work, using the LDA and the exchange-correlation functional of Barth and Hedin [34]. Data marked GGA-LMTO employ gradient-corrected functional of Perdew [35]. Data marked PWPP are LDA results using a plane-wave basis and a pseudopotential with the Ga $3d$ and In $4d$ in the valence [33]. $\Delta E_{\mathrm{ZB-WZ}}$ is the ZB-WZ energy difference per atom pair, in meV

		AlN	GaN	InN
Small basis	a	3.096	3.163	3.525
	c/a	1.600	1.629	1.606
	u	0.3818	0.3775	0.3805
	a (ZB)	4.353	4.481	4.961
	$\Delta E_{\mathrm{ZB-WZ}}$	54	14	23
Large basis	a	3.095	3.160	3.509
	c/a	1.601	1.629	1.613
	u	0.3818	0.3766	0.3794
	a (ZB)		4.478	4.947
	$\Delta E_{\mathrm{ZB-WZ}}$	51	12	18
Old-FPLMTO	a	3.091	3.160	3.528
	c/a	1.602	1.626	1.611
	u	0.381	0.377	0.380
	a (ZB)	4.345	4.464	4.957
LAPW[h]	a	3.098	3.170	3.546
	c/a	1.601	1.625	1.612
	u	0.382	0.377	0.379
	a (ZB)	4.355	4.476	4.964
	$\Delta E_{\mathrm{ZB-WZ}}$	45	11	21
PWPP[f]	a	3.084	3.162	3.501
	c/a	1.604	1.626	1.619
	u	0.381	0.377	0.378
	a (ZB)	4.342	4.460	4.932
	$\Delta E_{\mathrm{ZB-WZ}}$	36	22	20
Experiment	a	3.11[d]	3.190[d]	3.544[e]
	c/a	1.601[d]	1.626[d]	1.613[e]
	u	0.382[d]	0.377[d]	
	a (ZB)	4.38 [a]	4.52-4.55[b]	4.98[c]

[a] Ref. [24]
[b] Refs. [25–29], taken from Ref. [23]
[c] Ref. [22]
[d] Ref. [30]
[e] Ref. [21]
[f] Structural properties taken from Ref. [33]; $\Delta E_{\mathrm{ZB-WZ}}$ taken from Ref. [20].
[g] Taken from Ref. [31]
[h] Taken from Ref. [32]

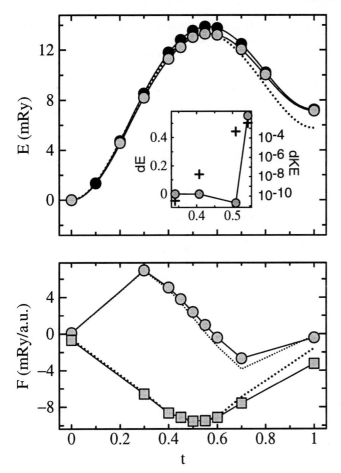

Fig. 7. The total energy for a γ surface of [111] Ni, as as function of translation τ/4[11$\bar{2}$], as described in the text. Upper panel: total energy (mRy/unit cell) for a nine-orbital basis (dark circles), an 18-orbital basis (light circles) and for a pseudopotential calculation (dotted line), using VASP. Inset: circles show error in the peak energy (mRy) as a function of mesh spacing in a.u.. Crosses show corresponding error in numerical integration of the kinetic energy of the d orbital on the mesh. Lower panel: x (circles) and y (squares) force on the atoms at the interface. Dotted lines show VASP result.

This corresponds to an error in the numerical integration of kinetic energy of $\sim 10^{-6}$ Ry.

These results were compared to a pseudopotential calculation using Vanderbilt's [6] ultrasoft potentials; this latter calculation was performed with the VASP code [36]. The VASP code used the same BZ integration, and the default mesh (16×24×108 divisions). Those results are illustrated by the dotted lines in Fig. 7. There is good agreement with VASP throughout except for a modest discrepancy for τ ~ 1. The source of the discrepancy is not clear; it would seem that in the all-electron case, the remaining sources of error (truncation in basis, truncation in l, smoothing of the core) all appear to be controlled.

4   Summary

In this paper, we have presented a newly developed full-potential LMTO method with the following properties:

- The potential is treated without any shape approximation, based on a muffin-tin type of geometry consisting of non-overlapping atomic spheres and an interstitial region.
- Empty spheres are not needed for any system.
- The results can be systematically improved by increasing the convergence parameters.
- The forces are as accurate derivatives of the total energy.

Similar to some previous FP-LMTO approaches, the interstitial density and potential are represented on a real-space mesh while the wavefunctions are assembled from atom-centered basis functions with well-defined angular momenta. In order to reduce the basis size further and to make the evaluation of integrals more efficient, we have introduced a new type of function which plays the role of the smooth envelope. These "smooth Hankel functions" are analytic in all parts of space and equal the standard Hankel functions at large radii. Closer to the origin, they bend over smoothly and are nonsingular. To introduce atomic detail into these envelope functions, a modified augmentation scheme was employed which combines some features of the projector-augmented wave method and the pseudopotential approach. By systematically using additive augmentation for the density, the potential, and the wavefunctions, the approach allows low angular momentum cutoffs throughout. Tests were performed to assess the quality of the method. Results have shown that state-of-the-art convergence can indeed be attained with low angular momentum cutoffs and small basis sets. Since the forces are calculated and the potential is treated without any shape approximation, the method presented here can be used to relax atomic structures and to study the energy changes associated with perturbations of the atoms around their equilibrium sites.

References

1. O.K. Andersen, Phys. Rev. B **12**, 3060 (1975).
2. P. Hohenberg and W. Kohn, Phys. Rev. **136**, B6864 (1964); W. Kohn and L.J. Sham, Phys. Rev. **140**, A1133 (1965); R.O. Jones and O. Gunnarsson, Rev. Mod. Phys. **61**, 689 (1989).
3. D. Glötzel, B. Segall, and O.K. Andersen, Solid State Commun. **36**, 403 (1980); A.K. McMahan, Phys. Rev. B **30**, 5835 (1984).
4. R. Car and M. Parrinello, Phys. Rev. Lett. **55**, 2471 (1985).
5. P.E. Blöchl, Phys. Rev. B **50**, 17 953 (1994).

6. D. Vanderbilt, Phys. Rev. B **41**, 7892 (1990).
7. K.H. Weyrich, Phys. Rev. B **37**, 10269 (1988).
8. J. Wills, unpublished.
9. S.Y. Savrasov, Phys. Rev. B **54**, 16470 (1996).
10. M. Methfessel, Phys. Rev. B **38**, 1537 (1988); M. Methfessel, C.O. Rodriguez, and O.K. Andersen, Phys. Rev. B **40**, 2009 (1989).
11. M. Methfessel and M. van Schilfgaarde, Phys. Rev. B **48**, 4937 (1993).
12. M. Springborg and O.K. Andersen, J. Chem. Phys. **87**, 7125 (1987).
13. M. Methfessel, PhD thesis, Katholieke Universiteit Nijmegen (1986).
14. E. Bott, Diplomarbeit, Technical University Darmstadt (1997); E. Bott, M. Methfessel, W. Krabs, and P.C. Schmidt, J. Math. Phys. **39**, 3393 (1998).
15. G.B. Bachelet, D.R. Haman, and M. Schlüter, Phys. Rev. B **26**, 4199 (1982).
16. M. Methfessel, *NFP Manual* (Institute for Semiconductor Physics, Frankfurt (Oder), 1997).
17. J. Harris, Phys. Rev. B **31**, 1770 (1985); W.M.C. Foulkes and R. Haydock, Phys. Rev. B **39**, 12 520 (1989).
18. P. Pulay, Mol. Phys. **17**, 197 (1969).
19. A. Dal Corso and R. Resta, Phys. Rev. B**50**, 4327 (1994).
20. C.-Y. Yeh, Z. W. Lu, S. Froyen and A. Zunger, Phys. Rev. B**46**, 10086 (1992).
21. K. Osamura, S. Naka and Y. Murakami, J. Appl. Phys. **46**, 3432 (1975).
22. S. Strite, D. Chandrasekhar, D. J. Smith, J. Sariel, N. Teraguchi and H. Morkoç, J. Crys. Growth **127**, 204 (1993).
23. H. Morkoç, S. Strite, G. B. Gao, M. E. Lin, B. Sverdlov and M. Burns, J. Appl. Phys. **76**, 1363 (1994).
24. P. Petrov, E. Mojab, R. C. Powell and J. E. Greene Appl. Phys. Lett. **60** 2491 (1992).
25. M. J. Paisley, Z. Sitar, J. B. Posthill and R. F. Davis, J. Vac. Sci. Technol. A**7** 1701 (1989).
26. S. Strite, J. Ruan, Z. Li, A. Salvador, H. Chen D. J. Smith, W. Y. Choyke and H. Morkoç, J. Vac. Sci. Technol. B **9**, 1924 (1991).
27. R. C. Powell, G. A. Tomasch, Y.-W. Kim, J. A. Thornton and J. E. Green, Mater. Res. Soc. Symp. Proc. **162**, 525 (1990).
28. T. Lei, M. Fanciulli, R. J. Molnar, T. D. Moustakas, R. J. Graham and J. Scanlon, Appl. Phys. Lett. **59**, 944 (1991).
29. M. Mizuta, S. Fujieda, Y. Matsumoto and T. Kawamura, Jpn. J. Appl. Phys. **25**, L945 (1986).
30. H. Schulz and K. H. Thiemann, Sol. State Commun. **23**, 815 (1977).
31. M. van Schilfgaarde, A. Sher and A.-B. Chen "Theory of AlN, GaN, InN and Their Alloys," J. Crystal Growth **178**, 8 (1997).
32. S.- H.- Wei, private communication.
33. A. F. Wright and J. S. Nelson, Phys. Rev. B**51**, 7866 (1995); A. F. Wright and J. S. Nelson, Phys. Rev. B**50**, 2159 (1994).
34. U. von Barth and L. Hedin, J. Phys. C**5**, 1629 (1972).
35. J. P. Perdew, K. Burke, and M. Ernzerhof, Phys. Rev. Lett. **77**, 3865 (1996).
36. G. Kresse and J. Hafner, Phys. Rev. B**47**, 558 (1993); G. Kresse, Thesis, Technische Universität Wien 1993; G. Kresse and J. Furthmüller, Comput. Mat. Sci. 6, 15-50 (1996); G. Kresse and J. Furthmüller, Phys. Rev. B**54**, 11169 (1996).

Full-Potential LMTO
Total Energy and Force Calculations

J. M. Wills[1], O. Eriksson[2], M. Alouani[3], and D. L. Price[4]

[1] Los Alamos National Laboratory, Los Alamos, NM 87545, USA
[2] Uppsala University, Uppsala, Sweden
[3] IPCMS, 23 rue du Loess, 67037 Strasbourg, France
[4] University of Memphis, Memphis, TN 38152, USA

Abstract. The essential features of a full potential electronic structure method using Linear Muffin-Tin Orbitals (LMTOs) are presented. The electron density and potential in the this method are represented with no inherent geometrical approximation. This method allows the calculation of total energies and forces with arbitrary accuracy while sacrificing much of the efficiency and physical content of approximate methods such as the LMTO-ASA method.

1   Introduction

This paper describes a particular implementation of a full-potential electronic structure method using Linear Muffin-Tin Orbitals (LMTO's) [2,10] as basis functions. There have been several "FP-LMTO" implementations [3–7]. The one described here has not been published in detail, although calculations performed with this method have been reported for quite some time [3]. There are many aspects to an electronic structure method. This paper is focussed on those aspects which enable a full potential treatment. Relatively small details pertaining to full-potential methods will be discussed while larger details having to do with, for example, relativity will not be.

The emphasis of a variational full-potential method is somewhat different from that of a method such as the LMTO-ASA method. The emphasis of the former is on the completeness of the basis while in the latter it is in the physical content (and interpretability) of the basis. These concepts are, of course, intimately related, but the emphasis is different.

The exposition here is for an infinite system periodic in three dimensions. This method has been implemented for two-dimensional systems,[8] but that will not be discussed here.

Notation

Papers on electronic structure methods unavoidably carry a high overhead in functional symbols and indices. It is simplest to define here, without motivation, the special symbols and functions that will be used in this paper, for future reference. These special functions (although not necessarily the symbols used here) have been used extensively in LMTO documentation and are largely due to Andersen.[10]

Spherical harmonics:

$$\mathcal{Y}_{\ell m}(\hat{\boldsymbol{r}}) \equiv i^{\ell} Y_{\ell m}(\hat{\boldsymbol{r}}) \tag{1}$$

$$C_{\ell m}(\hat{\boldsymbol{r}}) \equiv \sqrt{\frac{4\pi}{2\ell+1}}\, Y_{\ell m}(\hat{\boldsymbol{r}}) \tag{2}$$

$$\mathcal{C}_{\ell m}(\hat{\boldsymbol{r}}) \equiv i^{\ell} C_{\ell m}(\hat{\boldsymbol{r}}) \tag{3}$$

where Y is a spherical harmonic.[9]

Bessel functions:

$$\mathcal{K}_{\ell}(\kappa, r) \equiv -\kappa^{\ell+1} \begin{cases} n_{\ell}(\kappa r) - i j_{\ell}(\kappa r) & \kappa^2 < 0 \\ n_{\ell}(\kappa r) & \kappa^2 > 0 \end{cases} \tag{4}$$

$$\mathcal{K}_L(\kappa, \boldsymbol{r}) \equiv \mathcal{K}_{\ell}(\kappa, r)\mathcal{Y}_L(\hat{\boldsymbol{r}}) \tag{5}$$

$$\mathcal{J}_{\ell}(\kappa, r) \equiv j_{\ell}(\kappa r)/\kappa^{\ell} \tag{6}$$

$$\mathcal{J}_L(\kappa, \boldsymbol{r}) \equiv \mathcal{J}_{\ell}(\kappa, r)\mathcal{Y}_L(\hat{\boldsymbol{r}}) \tag{7}$$

where L denotes ℓm and n_{ℓ} and j_{ℓ} are spherical Neumann and Bessel functions, respectively.

Geometry: For computational purposes, the crystal is divided into non-overlapping spheres surrounding atomic sites (*muffin-tin spheres*) where the charge density and potential vary rapidly and the *interstitial* region between the spheres, where the charge density and potential vary slowly. This is the *muffin-tin* geometry used as an idealized potential and charge density in early electronic structure methods (KKR and APW). Here, the division is a computational one, and does not restrict the final shape of the charge density or potential. In the muffin-tin spheres, the basis functions, electron density, and potential are expanded in spherical waves; in the interstitial region, the basis functions, electron density, and potential are expanded in Fourier series.

There are many relevant considerations in choosing muffin-tin radii. Assuming all expansions are taken to convergence, the density and potential depend on the muffin-tin radii only through the dependence of basis functions on the radii. As discussed below, basis functions have a different functional form inside the muffin-tin spheres, and the choice of muffin-tin radius affects this crossover. Hence, assuming the Hamiltonian is the same inside and outside the spheres (the treatment of relativity may affect this as discussed below), the muffin-tin radii are variational parameters and the optimum choice minimizes the total energy. If the basis is large enough however (suitably complete within and without the spheres), the energy is insensitive to the choice of radii. A reasonable choice results from choosing radii that are within both the minimum in charge density and the maximum in potential along a line between nearest neighbors. Relativistic effects are usually taken into account only in the muffin-tin spheres, in which case the Hamiltonian depends on the radii; hence when relativistic effects are important, the radii are not variational parameters.

In what follows, lattice positions are vectors $R = Rn$, integer multiples of a basis R. Atomic positions in the unit cell are denoted by τ. A set of atomic positions invariant under the point group of the lattice are said to be of the same symmetry type, t. Similarly, in the reciprocal lattice, vectors are $g = Gn$ for the reciprocal basis $G = 2\pi R^{-T}$. Brillouin zone (or reciprocal unit cell) vectors are denoted by k.

Symmetric functions: Within the muffin-tin region, functions invariant are expressed in harmonic series. If $f(r)$ is such a function, at site τ

$$f(r)\Big|_{r_\tau < s_\tau} = \sum_h f_{ht}(r_\tau) D_{ht}(\mathcal{D}_\tau \hat{r}_\tau) \qquad (8a)$$

$$D_{ht}(\hat{r}) = \sum_m \alpha_{ht}(m) C_{\ell_h m}(\hat{r}) \qquad (8b)$$

In Equation (8a), $\mathcal{D}_\tau$ is a transformation to a coordinate system local to site τ; the local coordinates of sites of the same type are related by an element of the crystal point group that takes one site into another. Expressed in this way, the functional form of D_{ht} (Equation (8b)) depends only on symmetry type.

In the interstitial region, symmetric functions are expressed in Fourier series:

$$f(r)\Big|_{r \in \mathcal{I}} = \sum_S f(S) D_S(r) \qquad (9a)$$

$$D_S(r) = \sum_{g \in S} e^{ig \cdot r} \qquad (9b)$$

The sum in Equation (9a) is over symmetry stars S of the reciprocal lattice.

2 Basis Set

2.1 Interstitial

In the interstitial region (symbolically $\mathcal{I}$) between the muffin-tin spheres, bases are Bloch sums of spherical Hankel or Neumann functions:

$$\psi_i(k, r)\Big|_{r \in \mathcal{I}} = \sum_R e^{ik \cdot R} K_{\ell_i}(\kappa_i, |r - \tau_i - R|) \mathcal{Y}_{\ell_i m_i}(\mathcal{D}_{\tau_i}(r - \tau_i - R)) \qquad (10)$$

The rotation $\mathcal{D}_\tau$ in (10) takes the argument into a coordinate system local to each site τ. The purpose of this will be made evident later. The function on the right hand side of Equation (10) is sometimes called the envelope function.

Notice the parameters, specifying a basis function, inherent in this definition. They are the site τ in the unit cell on which the spherical wave is based, the angular momentum parameters ℓ and m of the spherical wave with respect to its parent cell, and the kinetic energy κ^2 of the basis in the interstitial region. The angular momentum parameters specifying the basis set are chosen to represent the atomic states from which crystal eigenstates are derived. In the LMTO-ASA, it is usual to include ℓ bases one higher than the highest relevant band. In the method described here, this is rarely necessary, possibly because of the multiplicity of bases with the same angular momentum parameters. It is usual to use "multiple κ" basis sets, having all parameters except the tail parameter the same.

There appears to be no simple algorithm for choosing a good set of interstitial kinetic energy parameters. Schemes such as bracketing the relevant energy spectrum have been proposed.[7] The optimum set would minimize the total energy. This can be done but is time consuming even for relatively simple systems. It seems, however that parameter sets obtained in this way for simple systems in representative configurations can give good results when used for related systems over a broad pressure range. Thus good sets are arrived at through some experimentation. The choice can be important as it's possible to pick a set of parameters that will give very bad results, and the parameter set used in any new calculation should be always checked for stability.

2.2   Muffin Tins

In the muffin-tin spheres, bases are linear combinations of spherical waves matching continuously and differentiably to the envelope function at the muffin-tin sphere. The envelope function $\mathcal{K}$ may be expanded in a series of spherical Bessel functions about any site except it's center. A basis function on a muffin-tin sphere in the unit cell at $\boldsymbol{R} = 0$ is therefore

$$
\begin{aligned}
\psi_i(\boldsymbol{k}, \boldsymbol{r})\Big|_{r_\tau = s_\tau} = &\sum_R e^{i\boldsymbol{k}\cdot\boldsymbol{R}} \sum_L \mathcal{Y}_L(\mathcal{D}_\tau \hat{\boldsymbol{r}}_\tau)\Big(\ \mathcal{K}_\ell(\kappa_i, s_\tau)\delta(R,0)\delta(\tau,\tau_i)\delta(L,L_i) \\
+ & \qquad\qquad\qquad\qquad\qquad \mathcal{J}_L(\kappa,s_\tau)B_{L,L_i}(\kappa_i, \boldsymbol{\tau}-\boldsymbol{\tau}'-\boldsymbol{R})\Big) \\
= &\sum_L \mathcal{Y}_L(\mathcal{D}_\tau \hat{\boldsymbol{r}}_\tau)\Big(\qquad \mathcal{K}_\ell(\kappa_i, s_\tau)\delta(\tau,\tau_i)\delta(L,L_i) \\
+ & \qquad\qquad\qquad\qquad\qquad \mathcal{J}_L(\kappa,s_\tau)B_{L,L_i}(\kappa_i, \boldsymbol{\tau}-\boldsymbol{\tau}',\boldsymbol{k})\Big)
\end{aligned}
$$
(11)

where $r_\tau \equiv \boldsymbol{r} - \boldsymbol{\tau}$ and B is equivalent to the KKR structure constant. [10] The unitary transformation applied to B rotates components into site-local coordinates from the left and right.

Equation (50) is compactly expressed by defining a two-component row vector K so that

$$
K_\ell(\kappa, r) = (\mathcal{K}_\ell(\kappa, r), \mathcal{J}_\ell(\kappa, r))
$$
(12)

and a two component column vector S so that

$$
S_{L,L'}(\kappa, \boldsymbol{\tau}-\boldsymbol{\tau}', \boldsymbol{k}) = \begin{pmatrix} \delta(\tau,\tau')\delta(L,L') \\ B_{L,L'}(\kappa, \boldsymbol{\tau}-\boldsymbol{\tau}', \boldsymbol{k}) \end{pmatrix} \quad .
$$
(13)

Then the value of a basis function on a muffin-tin boundary is expressed simply as

$$\psi_i(\boldsymbol{k}, \boldsymbol{r})\Big|_{r_\tau = s_\tau} = \sum_L \mathcal{Y}_L(\mathcal{D}_\tau \hat{\boldsymbol{r}}_\tau) K_\ell(\kappa_i, s_\tau) S_{L, L_i}(\kappa_i, \boldsymbol{\tau} - \boldsymbol{\tau}', \boldsymbol{k}) \qquad (14)$$

The radial part a basis function inside a muffin-tin sphere is a linear combination of atomic like functions ϕ and their energy derivatives $\dot{\phi}$ [10,2] matching continuously and differentiably to the radial function K in Equation (14). Collecting ϕ and $\dot{\phi}$ in a row vector

$$U(e, r) \equiv \left(\phi(e, r), \dot{\phi}(e, r) \right) \quad,$$

a simple case of this matching condition may be expressed as $U(e, s)\Omega(e, \kappa) = K(\kappa, s)$ and $U'(e, s)\Omega(e, \kappa) = K'(\kappa, s)$, where Ω is a matrix of order 2.

The use of these radial functions in the method described here is different than that used by most other methods, however. For the broadest utility, a basis set must be flexible enough to describe energy levels derived from atomic states having different principle quantum numbers but the same angular momentum quantum number. For example, describing the properties of elemental actinides at any pressure requires a basis with both $6p$ and $7p$ character. Similarly, an adequate calculation of the structural properties of transition metal oxides requires both semi-core and valence s and p states on the transition metal ions. The description of the evolution of core states from localized to itinerant under pressure also requires multiple principle quantum numbers per ℓ value. It is usual in LMTO-based methods to perform calculations for the eigenstates and eigenvalues of "semi-core" and valence states separately, using a different basis set, with a single set of energy parameters $\{e_\ell\}$, for each "energy panel". This approach fails when energy panels overlap, and has the disadvantage that the set of eigenvectors is not an orthogonal set. The problem of "ghost bands" also arises.[2]

In the method described here, bases corresponding to multiple principle quantum numbers are contained within a single, *fully hybridizing* basis set. This is accomplished simply by using functions ϕ and $\dot{\phi}$ calculated with energies $\{e_{n\ell}\}$ corresponding to different principal quantum numbers n to describe the radial dependence of a basis in the muffin-tin spheres. The Hamiltonian matrix for an actinide, for example, will have elements $\langle \psi_{6p} | H | \psi_{7p} \rangle$ and the overlap matrix elements $\langle \psi_{6p} | \psi_{7p} \rangle$, We may formally express the radial part of basis i in a muffin-tin sphere by the function $f(r) = \sum_n a_i(n\ell)U(e_{n\ell}, s)\Omega(e_{n\ell}, \kappa_i)$ but in practice it is sufficient to restrict the coefficients by $a_i(n\ell) = \delta(n, n_i)$ so that the basis set (although not eigenvectors) will have pure principal quantum number "parentage". This method of expanding the energy range of a basis set has been used (and reported) extensively. Representative calculations in which this method was essential are described in Reference [11].

Thus another parameter specifying a basis function is the set of energy parameters $\{e_{t\ell}\}$ that will be used to calculate the radial basis functions $\phi_{t\ell}$ and

$\dot{\phi}_{t\ell}$ used to express the basis function in muffin-tin spheres of each symmetry type. A basis function in a muffin-tin sphere is therefore

$$\psi_i(\boldsymbol{k},\boldsymbol{r})\Big|_{r_\tau < s_t} = \sum_{L}^{\ell \leq \ell_m} \mathrm{U}_{tL}(e_i, \mathcal{D}_\tau \boldsymbol{r}_\tau)\varOmega_{t\ell}(e_i,\kappa_i)\mathrm{S}_{L,L_i}(\kappa_i, \boldsymbol{\tau} - \boldsymbol{\tau}', \boldsymbol{k}) \quad (15)$$

where e_i means "use the energy parameter $e_{n\ell}$ corresponding to the principal quantum number specified for basis i" and

$$\mathrm{U}_{tL}(e,\boldsymbol{r}) \equiv \mathcal{Y}_L(\hat{r})\mathrm{U}_{t\ell}(e,r) \quad . \quad (16)$$

The necessary cutoff in angular momentum has now been made explicit. The 2×2 matrix $\varOmega$ matches U to K continuously and differentiably at the muffin-tin radius. Specifically, $\varOmega$ is specified by

$$\begin{pmatrix} \phi_{t\ell}(e,s_t) & \dot{\phi}_{t\ell}(e,s_t) \\ \phi'_{t\ell}(e,s_t) & \dot{\phi}'_{t\ell}(e,s_t) \end{pmatrix} \varOmega_{t\ell}(e,\kappa) = \begin{pmatrix} \mathcal{K}_\ell(\kappa,s_t) & \mathcal{J}_\ell(\kappa,s_t) \\ \mathcal{K}'_\ell(\kappa,s_t) & \mathcal{J}'_\ell(\kappa,s_t) \end{pmatrix} \quad (17)$$

In principle, and as programmed, each $(\tau\ell\kappa)$ basis can use its own unique energy set. It is more usual to use a common energy set for a set of basis states giving rise to bands of similar energy within the scope of a particular calculation. The configuration of the basis shown in Table 1 for example uses a set of energies for "semi-core" $6s$ and $6p$ bases, and another set of energies to represent "valence" bases. The calculation of energies in an energy parameter set is discussed below.

A parameter introduced in (15) is the angular momentum cut-off ℓ_m. In most cases, a converged total energy is achieved with values $\ell_m \sim 6 - 8$. Note that since a basis set generally contains functions based on spherical waves with $\ell \leq 3$, the KKR structure constant in (13) is rectangular.

Table 1. Parameters for typical basis set for an elemental actinide: parent angular momentum parameter (ℓ), energy set for radial expansions (e-set), and the index of the kinetic energy in the interstitial region(κ-index). A typical set of κ^2 values, corresponding to the kinetic energy indices, is given at the bottom of the table.

n	ℓ	e-set	κ-index	n	ℓ	e-set	κ-index	n	ℓ	e-set	κ-index
6	s	1	1	7	s	2	3	6	d	2	3
6	s	1	2	7	s	2	4	6	d	2	5
6	p	1	1	7	s	2	5	5	s	2	3
6	p	1	2	7	p	2	3	5	s	2	5
				7	p	2	4				
				7	p	2	5				

κ^2:	1:	-1.96582916	3:	-3.44402161
	2:	$-.193652690$	4:	-1.56582916
			5:	$.331719550$

3   Matrix Elements

3.1   Muffin-Tin Matrix Elements

The potential in a muffin-tin at τ has an expansion in linear combinations of spherical harmonics invariant under that part of the point group leaving τ invariant:

$$V(r)\Big|_{r_\tau < s_t} = \sum_h v_{ht}(r_\tau) D_{ht}\big(\mathcal{D}_\tau \hat{r}_\tau\big) \tag{18a}$$

$$D_{ht}(\hat{r}) = \sum_m \alpha_{ht}(m) \mathcal{C}_{\ell_h m}(\hat{r}) \quad . \tag{18b}$$

The utility of referring bases and potentials in muffin-tin spheres to site-local coordinates is apparent in (18a). If the site local coordinates of sites are constructed so that $\mathcal{D}_{\tau'} = \mathcal{D}_\tau \mathcal{Q}^{-1}$ for some $\mathcal{Q}$ such that $\mathcal{Q}\tau = \tau'$, then the harmonic functions D_{ht} depend only on the symmetry type, rather than on each site. The normalization for the spherical harmonic in (18a) ($\mathcal{C} = \sqrt{4\pi/(2\ell+1)}\mathcal{Y}$) is chosen so that $v_{ht}(r)$ is the potential when $\ell_h = 0$.

Combining (15) and (18a), the potential matrix is

$$\langle \psi_i | V | \psi_j \rangle \Big|_{mt} = \sum_\tau \sum_L S^\dagger_{L,L_i}(\kappa_i, \tau - \tau_i, k) \tag{19}$$

$$\times \left(\sum_h \sum_{L'} \Omega^T_{t\ell}(e_i, \kappa_i) \langle U^T_{t\ell}(e_i) | v_{ht} | U_{t\ell}(e_j) \rangle \Omega_{t\ell}(e_j, \kappa_j) \right.$$

$$\left. \langle L | D_{ht} | L' \rangle S_{L,L_j}(\kappa_j, \tau - \tau_j, k) \right) \quad .$$

The matrix element of the D_{ht} is a sum over Gaunt coefficients:

$$\langle L | D_{ht} | L' \rangle = \sum_{m_h} \alpha_{ht}(m_h) \mathcal{G}\big(\ell', m'; \ell, m; \ell_h, m_h\big)$$

$$\mathcal{G}\big(\ell', m'; \ell, m; \ell_h, m_h\big) = \int \mathcal{Y}_{\ell'm'} \mathcal{Y}^*_{\ell m} \mathcal{C}_{\ell_h m_h}$$

In electronic structure methods using muffin-tin orbitals, the muffin-tin energy parameters $\{e_\ell$ are usually taken from "ℓ-projected average energies". With multiple energy sets, this is a reasonable choice provided that the basis set, which uses separate sets, gives rise to bands well separated in energy. The ℓ-projected charge, integrated over a muffin-tin sphere, is a sum over cross terms between energy sets

$$Q_\ell = \sum_{ij} Q_\ell(e_i, e_j)$$

and must be made diagonal in some approximation for the resulting energy- and ℓ-projected energies and charges to be representative.

Another criterion, particularly useful for states using different sets not well separated in energy or for states not having significant occupation is to maximize the completeness of the basis. To accomplish this, the energy parameter for the low energy state $e_\ell(1)$ can be set to a set of projected energy averages, and the energy parameters for the same ℓ in higher energy sets may be chosen so that the radial function has one more node and the same logarithmic derivative at the muffin-tin radius, hence

$$\int_0^s r^2 dr \; \phi_\ell(e_1, r)\phi_\ell(e_i, r) = 0 \quad , \quad i > 1 \quad . \tag{20}$$

Although this usually generates energy parameters out of the range of occupied states (since the logarithmic derivative of semi-core states is usually large in magnitude and negative), this choice seems to give a total energy close to the minimum with respect to this parameter. This is an example of the difference mentioned in the introduction in emphasis between an accurate "basis-set" method and a method motivated by a physical model.

The convergence of the harmonic expansion of the potential in a muffin-tin sphere (18a) depends, of course, on the basis, atomic constituents, and geometry. Using harmonics through $\ell_{h_{max}} = 6$ is usually sufficient, and it has never been necessary to go beyond $\ell_{h_{max}} = 8$.

3.2 Interstitial Matrix Elements

Overlap and Kinetic Energy: The interstitial overlap matrix can be easily obtained from an integral over the interstitial surface (the only non-zero contributions, in a crystal periodic in three dimensions, come from the surfaces of the muffin-tin spheres) and the kinetic energy is proportional to the overlap:

$$\int_\mathcal{I} \psi_i^\dagger(r)\psi_j(r) = -(\kappa_j^2 - \kappa_i^2)^{-1} \int_\mathcal{I} (\psi_i^\dagger \nabla^2 \psi_j - (\nabla^2 \psi_i^\dagger)\psi_j)$$

$$= (\kappa_j^2 - \kappa_i^2)^{-1} \sum_\tau s_t^2 \int d\Omega_\tau W(\psi_i^\dagger, \psi_j) \tag{21}$$

where $W(f, g) = fg' - f'g$. Basis functions on muffin-tin spheres are given in (14), hence

$$\langle \psi_i | \psi_j \rangle \Big|_\mathcal{I} = \sum_\tau s_t^2 \sum_L S_{L,L_i}^\dagger(\kappa_i, \tau - \tau_i, k)$$

$$\times \quad \frac{W\big(K_\ell^T(\kappa_i, s_t), K_\ell(\kappa_j, s_t)\big)}{\kappa_j^2 - \kappa_i^2} S_{L,L_j}(\kappa_j, \tau - \tau_j, k) \tag{22}$$

In the limit $\kappa_j^2 \to \kappa_i^2$, the evaluation of (21) requires the derivative with respect to κ^2 of the structure constant.

Potential Matrix Elements: The greatest difference between LMTO-based full-potential methods is in the way the matrix elements of the potential are calculated over the interstitial region. The method being described here uses a Fourier representation of basis functions and the interstitial potential to calculate these matrix elements. Other approaches for computing these elements are described in the literature. [4,5]

A Fourier transform of the basis functions described in Section 2 would be too poorly convergent for practical use. However, the evaluation of the interstitial potential matrix requires only a correct treatment of basis functions and potential in the interstitial region. This degree of freedom can be used to design "pseudo basis-set", equal to the true basis in the interstitial region although not in the muffin-tin spheres, and have a Fourier transform which converges rapidly enough for practical use. We define this pseudo basis set by

$$\tilde{\psi}_i(\mathbf{k},\mathbf{r})\Big|_{r\in\mathcal{I}} = \sum_R e^{i\mathbf{k}\cdot\mathbf{R}}\tilde{\mathcal{K}}_{\ell_i}(\kappa_i,|\mathbf{r}-\boldsymbol{\tau}_i-\mathbf{R}|)i^\ell Y_{\ell_i m_i}(\mathbf{r}-\boldsymbol{\tau}_i-\mathbf{R}) \qquad (23a)$$

$$\tilde{\mathcal{K}}_\ell(\kappa,r) \equiv \mathcal{K}_\ell(\kappa,r), \quad r > s, \quad s \leq s_\tau \qquad (23b)$$

Since rapid Fourier convergence is the criterion for constructing the pseudo-basis, it is useful to consider the Fourier integral of a Bloch function with wave-number $\mathbf{k}$:

$$\tilde{\psi}(\mathbf{g}) = -\frac{1}{V_c(|\mathbf{k}+\mathbf{g}|^2 - \kappa^2)} \int_{V_c} d^3r\, e^{-i(\mathbf{k}+\mathbf{g})\cdot\mathbf{r}}\left(\nabla^2+\kappa^2\right)\tilde{\psi}(\mathbf{r}) \qquad (24)$$

where V_c is the unit cell volume. Equation (24) is obtained by casting $\nabla^2+\kappa^2$ on the plane wave then doing two partial integrations; surface terms vanish due to periodicity. From (24) it is evident that the Fourier integral of a pseudo-basis satisfying the first criterion (equal to the true basis in the interstitial region) may be obtained from integral over muffin-tin spheres. If in addition, the pseudo-basis is different from a Hankel function only in it's parent sphere, the Fourier integral is a finite integral over a single muffin-tin sphere. The problem then is to find a function $\tilde{\psi}$ such that $(\nabla^2+\kappa^2)\tilde{\psi}$ has a rapidly convergent Fourier integral, vanishes outside a radius less than or equal to the parent muffin-tin radius for the basis, and has a value and slope equal to $\mathcal{K}$ at this radius.

A good choice for such a function is obtained by solving

$$\left(\nabla^2 + \kappa^2\right)\tilde{\mathcal{K}}_\ell(\kappa,r)\mathcal{Y}_L(\hat{r}) = -c_\ell\left(\frac{r}{s}\right)^\ell\left[1 - \left(\frac{r}{s}\right)^2\right]^n \mathcal{Y}_L(\hat{r})\Theta(s-r) \qquad (25)$$

for a radius $s < s_{t_i}$, and with with c_ℓ chosen to match on to $\mathcal{K}$ at s. This is easily done analytically. The resulting Fourier transform is

$$\tilde{\psi}_i(\mathbf{k}+\mathbf{g}) = \frac{4\pi}{V_c}\frac{Y_{L_i}(\mathbf{k}+\mathbf{g})e^{-i(\mathbf{k}+\mathbf{g})\cdot\boldsymbol{\tau}_i}}{(|\mathbf{k}+\mathbf{g}|^2 - \kappa_i^2)}|\mathbf{k}+\mathbf{g}|^{\ell_i}\frac{\mathcal{J}_N(|\mathbf{k}+\mathbf{g}|,s)}{\mathcal{J}_N(\kappa_i,s)} \qquad (26)$$

where $N = \ell_i+n_i+1$. The subscript i has been purposely left off N and s (see below).

These coefficients converge like $1/g^{n+4}$, provided $\mathcal{J}_N(|\mathbf{k+g}|, s)$ achieves it's large argument behavior, and n can be chosen to optimize convergence. Weinert [12] used an analogous construction as tool to solve Poisson's equation. He proposed a criterion for the convergence of the Fourier series (26) which amounts to choosing the exponent n in Equation (26) so that $|\mathbf{k+g}_{max}|s$ would be greater than the position of the first node of $\mathcal{J}_{\ell+n+1}$. We find this criterion to be useful provided anisotropy in reciprocal space is accounted for. This is accomplished by using the minimum reciprocal lattice vector on the surface of maximal reciprocal lattice vectors, rather than simply using g_{max}.

Notice that this criterion is a criterion for $N = \ell+n+1$. The basis Fourier components are simplified, and the amount of information stored reduced, by simply using a single argument for all bases; *i.e.* all bases use the same value of N. It is also possible to use a single radius s, less than or equal to the smallest muffin-tin radius, since the only requirement is on the pseudo bases in the interstitial region. In practice, a few radii are desirable if large and small atoms are present in the same calculation, since small radii give less convergent Fourier coefficients. In any event, no more than a few radii are necessary to handle systems with many atoms. Notice also that local coordinates have been left out of (26). The resulting potential matrix may be easily rotated to local coordinates at the end of the calculation.

As expressed in (26), the Fourier components are products of phases $e^{-i(\mathbf{k+g})\cdot\boldsymbol{\tau}}$, which scale like the number of atoms squared (the size of the reciprocal lattice grid grows linearly with the number of atoms), and a function of lattice vectors and a few parameters, which scales linearly with the number of atoms. The phase factors are simple to calculate by accumulation and need not be stored.

The potential in the interstitial region is similarly obtained from a "pseudo-potential" $\tilde{V}$ that equals the true potential in the interstitial region and has rapidly converging Fourier coefficients:

$$V(\mathbf{r})\Big|_{\mathcal{I}} = \tilde{V}(\mathbf{r})\Big|_{\mathcal{I}} \tag{27a}$$

$$\tilde{V}(\mathbf{r}) = \sum_{\mathcal{S}} \tilde{V}(\mathcal{S}) D_{\mathcal{S}}(\mathbf{r}) \tag{27b}$$

$$D_{\mathcal{S}} = \sum_{g \in \mathcal{S}} e^{i\mathbf{g}\cdot\mathbf{r}} \tag{27c}$$

The sum in Equation (27b) is over stars $\mathcal{S}$ of the reciprocal lattice.

Integrals over the interstitial region are performed by convoluting the potential with an interstitial region step function and integrating over the unit cell:

$$\langle \psi_i|V|\psi_j\rangle_{\mathcal{I}} = \langle \tilde{\psi}_i|\tilde{V}|\tilde{\psi}_j\rangle_{\mathcal{I}} = \langle \tilde{\psi}_i|\theta_{\mathcal{I}}\tilde{V}|\tilde{\psi}_j\rangle_c \quad .$$

The potential matrix element is calculated by convoluting the convoluted potential with a basis, and performing a direct product between convoluted and unconvoluted bases. If basis functions are calculated n^3 reciprocal lattice vectors, the interstitial potential will be calculated on $(2n)^3$ vectors. The convolution is exact if it is carried out on a lattice containing $(4n)^3$ vectors. The size of the set of reciprocal lattice vectors necessary to converge the total energy using this treatment of the interstitial region varies from between $\sim 150 - 300$ basis plane waves per atom, depending on the smoothness of the potential and the convergence required.

Another way of integrating over the interstitial region, more usual in site-centered methods, is to integrate Fourier series over the unit cell and subtract the muffin-tin contributions with pseudo-bases and pseudo-potential expressed as an expansion in spherical waves. The convolution has an advantage in acting with a single representation, and, given a finite representation for bases and potential, the convolution may be done exactly.

Empty spheres are never used with this scheme. Bases, and the charge density and potential are calculated as accurately as necessary using the scheme described above and a basis set expanded with tail parameters and energy sets has proven to be flexible enough to accurately describe the contribution of the electronic states in the interstitial region.

4   Charge Density

When a solution to the wave equation at every physical energy is available, the charge density may be obtained from a set of energy-dependent coefficients. The spherically symmetric charge density in a muffin-tin sphere, coupled with an $\ell - projected$ density of states, is an example. In a variational calculation, as is being described here, all that is available is a (variational) solution to the wave equation at a set of discreet energies, and the charge density must be obtained simply from the square of the eigenvectors, or equivalently from expectation values of occupation numbers.

Having calculated a set of eigenvalues and eigenvectors $\mathcal{A}$ of the generalized eigenvalue problem, the charge density in the interstitial region is

$$\tilde{n}(r)\Big|_{\mathcal{I}} = \sum_{\mathcal{S}} \tilde{n}(\mathcal{S})D_{\mathcal{S}}(r) \tag{28a}$$

$$\tilde{n}(\mathcal{S}) = \frac{1}{N_{\mathcal{S}}} \sum_{g \in \mathcal{S}} \sum_{nk} w_{nk} \frac{1}{V_c} \int_{V_c} d^3r\, e^{-ig\cdot r} \Big| \sum_i \tilde{\psi}_i(k, r)\mathcal{A}_i(nk)\Big|^2 \tag{28b}$$

where $N_{\mathcal{S}}$ is the number of vectors in the reciprocal lattice star $\mathcal{S}$. The square of the wave function is obtained by convoluting the Fourier components of ψ with $\mathcal{A}$, Fourier transforming, and taking the modulus.

In the muffin-tin spheres the charge density is

$$n(r)\Big|_{r_\tau < s_t} = \sum_h n_{ht}(r_\tau) D_{ht}(\mathcal{D}_\tau r_\tau) \tag{29a}$$

$$n_{ht}(r) = \sum_{e\ell} \sum_{e'\ell'} U_{t\ell'}(e_{i'}, r) M_{ht}(e\ell, e'\ell') U^T_{t\ell}(e_i, r) \tag{29b}$$

$$M_{ht}(e\ell, e'\ell') = \frac{2\ell_h + 1}{4\pi} \sum_{m_h, mm'} \alpha^*_{ht}(m_h) \mathcal{G}\big(\ell, m; \ell', m'; \ell_h, m_h\big) \tag{29c}$$

$$\times \sum_{nk} w_{nk} \mathcal{V}_{\tau\ell m}(e) \mathcal{V}^\dagger_{\tau\ell'm'}(e')$$

$$\mathcal{V}_{\tau\ell m}(e) = \sum_i \delta(e, e_i) \Omega_{t\ell}(e, \kappa_i) S_{\ell m, \ell_i m_i}(\kappa_i, \tau - \tau_i, k) \mathcal{A}_i(nk) \tag{29d}$$

The process of calculation is evident in the sequence of equations.

5   Core States

Core states, even spherically symmetric complete shells, contribute non-muffin-tin components to the interstitial region and to muffin-tin spheres surrounding other sites. Whether it is essential to include this contribution depends on the size of the contribution, and any sizable contribution implies that there are states being treated as localized which aren't localized within the scope of the calculation. Nevertheless, confining states to the core is often useful, and including the core contribution to the full potential is not difficult. One possibility, the one used in this method, is to fit the part of the core electron density to a linear combination of Hankel functions, and expand this density in the interstitial region as a Fourier serie and in the muffin-tin spheres in a harmonic series, in the same way the basis functions are treated.

6   Potential

6.1   Coulomb Potential

The Coulomb potential is obtained by first calculating the Coulomb potential in the interstitial region, then, using the value of the interstitial potential on the muffin-tin sphere, calculating the potential in the spheres by a numerical Coulomb integral of the muffin-tin electron density for each harmonic.

   The interstitial Coulomb potential is calculated in a way similar to that suggested by Weinert [12]. Express the electron density as

$$n(r) = \tilde{n}(r) + \sum_{R\tau} (n(r) - \tilde{n}(r))\Theta(s_t - r_\tau) \tag{30}$$

where $\tilde{n}$ is the squared modulus of the pseudo-eigenvectors, which is equal to the true electron density in the interstitial region. The first term on the right-hand side of (30) has, by construction, a convergent Fourier series. The second term is confined to muffin-tin spheres. To calculate the Coulomb potential in the interstitial region, this term may be replaced by any density also confined to the muffin-tin spheres and having the same multipole moments. If a charge density satisfies these requirements and also has a convergent Fourier series, the Coulomb potential in the interstitial region may be easily calculated from the combined Fourier series. Such a charge density can be constructed in a similar way to that detailed for the pseudo-bases. Construct a pseudo charge-density satisfying

$$\tilde{n}^{(p)}(\boldsymbol{r}) = \sum_{R\tau}\sum_{h} \tilde{n}^{(p)}(ht, r_{R\tau}) D_{ht}(\mathcal{D}_{\tau}\hat{\boldsymbol{r}}_{R\tau}) \tag{31a}$$

$$\tilde{n}_{ht}^{(p)}(\boldsymbol{r}) = c_{ht}\left(\frac{r}{s_t}\right)^{\ell_h}\left(1-\left(\frac{r}{s_t}\right)^2\right)^n \Theta(s_t-r) \tag{31b}$$

$$0 = \int_{\tau} d^3r\, r_{\tau}^{\ell} D_{ht}^*(\mathcal{D}_{\tau}\hat{\boldsymbol{r}}_{\tau})\big(\tilde{n}^{(p)}(\boldsymbol{r})-n(\boldsymbol{r})+\tilde{n}(\boldsymbol{r})\big) \quad . \tag{31c}$$

This charge density has Fourier components

$$\tilde{n}^{(p)}(\boldsymbol{r}) = \sum_{\tau}\sum_{h} e^{-i\boldsymbol{g}\cdot\boldsymbol{\tau}}(-i)^{\ell_h} D_{ht}(\mathcal{D}_{\tau}\boldsymbol{g})\frac{4\pi}{V_c}\frac{(Q_{ht}\{n\}-Q_{ht}\{\tilde{n}\})}{s^{\ell_h+n+1}}$$
$$\times \frac{(2(\ell_h+n+1)+1)!!}{(2\ell_h+1)!!}g^{\ell_h}\mathcal{J}_{\ell_h+n+1}(g,s_t) \tag{32}$$

where the multipole moments Q are defined by

$$Q_{ht}\{n\} = \frac{2\ell_h+1}{4\pi}\int_{s_t>r_{\tau}} r_{\tau}^{\ell_h} D_{ht}(\hat{\boldsymbol{r}}_{\tau})n(\boldsymbol{r})\, d^3r_{\tau} \tag{33}$$

The Fourier components $\tilde{n}^{(p)}(\boldsymbol{r})$ converge like $1/g^{n+2}$ provided $j_{\ell+n+1}$ attains it's asymptotic form. The exponent n is chosen using the same considerations as for the pseudo-basis set.

The Coulomb potential in the interstitial region is then given by

$$V_c(\boldsymbol{r})\Big|_{\mathcal{I}} = \tilde{V}_c(\boldsymbol{r})\Big|_{\mathcal{I}}$$
$$= \sum_{g\neq 0}\frac{4\pi e^2\big(\tilde{n}(g)+n^{(p)}(g)\big)}{g^2}e^{i\boldsymbol{g}\cdot\boldsymbol{r}} \tag{34}$$

From the Coulomb potential in the interstitial region follows the Coulomb potential on the surface of the muffin-tin spheres. The coulomb Potential inside the muffin-tin spheres is

$$V^{(c)}(\boldsymbol{r})\Big|_{r_{\tau}<s_t} = \sum_{h} D_{ht}(\mathcal{D}_{\tau}\hat{\boldsymbol{r}}_{\tau})\left[e^2\int_0^{s_t}\frac{r_<^{\ell_h}}{r_>^{\ell_h+1}}\frac{4\pi r'^2 n_h(r)}{2\ell_h+1}dr'\right.$$
$$\left.+\left(V_h^{(c)}(s)-\frac{e^2}{s^{\ell_h+1}}\int_0^s\frac{4\pi r'^{\ell_h+2}n_h(r')}{2\ell_h+1}dr'\right)\left(\frac{r}{s}\right)^{\ell_h}\right] \tag{35}$$

where

$$V_{ht}^{(c)}(s_t) \equiv \frac{2\ell_h+1}{4\pi} \int_{r_\tau=s_t} d\hat{r} D_{ht}^*(\mathcal{D}_\tau \hat{r}) V^{(c)}(r) \tag{36}$$

is the harmonic component of the potential on a sphere boundary.

6.2   Density Gradients

Gradients of the electron density are needed for the evaluation of gradient corrected density functionals. These functionals depend on invariants (with respect
to the point group) constructed from density gradients (*e.g.* $|\nabla n|^2$). This re-
duces computation significantly in the muffin-tin spheres, for if f and g are
invariant functions (*i.e.* $f(r) = \sum_h f_h(r)D_h(\hat{r})$), and $d = \nabla f \cdot \nabla g$, then $d(r) = \sum_h d_h(r)D_h(\hat{r})$ with

$$\frac{4\pi r^2}{2\ell_h+1} d_h(r) = \sum_{h,h'} \sum_{k,k'=\pm 1} f_h^{(k)}(r) g_{h'}^{(k')}(r) I(kk'; hh') \tag{37}$$

where the set of parameters I is easily calculable from $3j$ and $6j$ coefficients and
integrals over the harmonic functions D_h, and

$$f_h^{(k)} = \frac{4\pi}{2\ell+1} \begin{cases} rf' - \ell_h & k = 1 \\ rf' + \ell_h + 1 & k = -1 \end{cases} \tag{38}$$

and similarly for g.

Gradients of the interstitial charge density, represented as a Fourier series,
are poorly represented by differentiating the series term by term. A stable representation of the density gradient that converges well is obtained by defining the
derivative as the difference between adjacent grid points, divided by twice the
grid spacing as suggested by Lanczos.[13] This is equivalent to differentiating,
term by term, the Lanczos-damped series for the charge density.

7   All-Electron Force Calculations

7.1   Symmetry

The set of internal forces acting on the atomic sites of a crystal is a symmetric,
discrete function of atomic coordinates and has a spherical expansion on the
crystal sites with the same coefficients as continuous symmetric functions (8a)
and (8b). Since forces are vectors, their representation has $\ell = 1$, and if a site
has no invariant harmonics with $\ell = 1$, there is no force on that site. So the force
on an atomic site may be expressed as

$$f(\tau) = \sum_{h:\ell_h=1} f_{ht} \sum_m \alpha_m \hat{e}_m \, \mathcal{U}_\tau \tag{39}$$

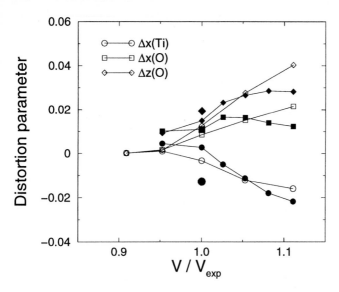

Fig. 1. The deviation of the internal coordinates of rhombohedral $BaTiO_3$ from ideal, calculated using all-electron force calculations as a function of volume with both LDA (open symbols) and GGA (filled symbols) exchange-correlation functionals. The grey filled symbols are experimental points[16]. The LDA equilibrium volume is $.958\ V_{exp}$; the GGA volume is $1.037\ V_{exp}$. The energy was also minimized with respect to the rhombohedral angle at each volume.

where the coefficients α are as in (8b), the $\hat{\mathbf{e}}_m$ are spherical unit vectors, [14] and $\mathcal{U}_\tau$ is the transformation to local coordinates for spherical vectors. A force calculation is, as much as possible, a calculation of the set $\{f_{ht}\}$; The size of this set is often much smaller than three times the number of atoms. The displacements of atoms allowed by symmetry also have the form of (39):

$$\delta\tau = \sum_{h:\ell_h=1} \delta\tau_{ht} \sum_m \alpha_m \hat{\mathbf{e}}_m\, \mathcal{U}_\tau \qquad (40)$$

Minimizing the energy with respect to the atomic positions is a process of finding the set $\{\delta\tau_{ht}\}$ that gives $f_{ht} = 0$.

7.2   Force Calculations

The calculation of forces in an all-electron method has been nicely described by Yu *et al.* [15] for the LAPW method. In addition to the terms discussed in that paper, a force calculation using a site-centered basis has the additional, and significant, complication that the bases depend on atomic position not only through augmentation but also through parentage.

The contributions to the total force on a site in an all-electron calculation follow directly from a derivative of the LDA total energy with respect to atomic positions. The terms listed by Yu *et al.* are 1) a "Helmann-Feynman" term,

$\partial E/\partial\boldsymbol{\tau}$, which accounts for the explicit dependence of the energy functional on atomic positions, 2) an "Incomplete Basis Set" (IBS) term, which arises when derivatives of basis functions aren't contained in the space covered by the basis set, 3) a core-correction term, arising because core states are calculated using only the spherical average of the potential, and 4) a muffin-tin term, a surface term arising from the change in integration boundaries when atoms are moved and the discontinuity of the second derivative of basis functions across muffin-tin boundaries. There are two other terms to consider. The first arises when a calculation isn't fully self-consistent, and has the form $-\int_{V_c}(V_{\text{out}}-V_{\text{in}})dn(\boldsymbol{r})/d\boldsymbol{\tau}$, where V_{out} and V_{in} are output and input potentials. The second term arises from the way in which Brillouin zone integrals are done. Whether by quadrature or linear interpolation, the result is a set of weights (occupations) multiplying quantities evaluated at discrete Brillouin zone points. The terms listed above do not take into account the change of weights with atomic positions.

The evaluation of the IBS term in a method using site-centered bases is significantly more involved than in the LAPW method. This term has the form

$$\boldsymbol{F}_{\text{IBS}} = -\sum_{nk} w_{nk} \sum_{ij} \mathcal{A}^*_{i,nk} \left(\left\langle \psi_i \middle| H - e_{nk} \middle| d\psi_j/d\boldsymbol{\tau} \right\rangle \right.$$
$$\left. + \left\langle d\psi_i/d\boldsymbol{\tau} \middle| H - e_{nk} \middle| \psi_j \right\rangle \right) \mathcal{A}_{j,nk} \qquad (41)$$

where the $\mathcal{A}$ are eigenvectors. Both LAPW and LMTO methods have a dependence on atomic positions through augmentation (the expansion of the basis set in atomic-like spherical waves) in the muffin-tin spheres, and both methods have an implicit dependence of basis functions on atomic positions through self-consistency, a term largely ignored and usually negligible. A site-centered basis, however, depends on atomic positions also through it's parent site (the site it's centered on). The contribution from augmentation is fairly easily accounted for at the density stage of a calculation, after integrals over the Brillouin zone have been done. The parent contribution, however, requires evaluation at the part of the calculation where eigenvalues and vectors are obtained, which makes its calculation time consuming.

There are four types of contributions to $d\psi/d\boldsymbol{\tau}$:

$$-\frac{d}{d\boldsymbol{\tau}}\psi_i(\boldsymbol{k},\boldsymbol{r}) = i\left(\delta^{(1)}_{\tau} + \delta^{(2)}_{\tau} + \delta^{(3)}_{\tau} + \delta^{(4)}_{\tau}\right)\psi_i(\boldsymbol{k},\boldsymbol{r}) \qquad (42)$$

$$\delta^{(1)}_{\tau}\psi_i(\boldsymbol{k},\boldsymbol{r}) \equiv \Theta(\boldsymbol{r}\in\mathcal{I})\delta(\tau_i,\tau)\,\hat{\boldsymbol{p}}\psi_i(\boldsymbol{k},\boldsymbol{r}) \qquad (43)$$

$$\delta^{(2)}_{\tau}\psi_i(\boldsymbol{k},\boldsymbol{r}) \equiv \delta(\tau_i,\tau)\sum_{\tau'L}\Theta(s_{t'}-r_{\tau'})U_{t'L}(e_i,\boldsymbol{r}_{\tau'})\Omega_{t'\ell}(e_i,\kappa_i)$$
$$\times \begin{pmatrix} 0 \\ -i\boldsymbol{\nabla}_{\tau}B_{L,L_i}(\kappa_i,\boldsymbol{\tau}'-\boldsymbol{\tau}_i,\boldsymbol{k}) \end{pmatrix} \qquad (44)$$

$$\delta^{(3)}_{\tau}\psi_i(\boldsymbol{k},\boldsymbol{r}) \equiv \Theta(s_t-r_\tau)\sum_{L}\hat{\boldsymbol{p}}U_{tL}(e_i,\boldsymbol{r}_\tau)\Omega_{t\ell}(e_i,\kappa_i)S_{L,L_i}(\kappa_i,\boldsymbol{\tau}-\boldsymbol{\tau}_i,\boldsymbol{k}) \qquad (45)$$

$$\delta_\tau^{(4)}\psi_i(\boldsymbol{k},\boldsymbol{r}) \equiv -\Theta(s_t-r_\tau)\sum_L U_{tL}(e_i,\boldsymbol{r}_\tau)\Omega_{t\ell}(e_i,\kappa_i)$$

$$\times \begin{pmatrix} 0 \\ -i\boldsymbol{\nabla}_\tau B_{L,L_i}(\kappa_i,\boldsymbol{\tau}-\boldsymbol{\tau}_i,\boldsymbol{k}) \end{pmatrix} \tag{46}$$

where $\hat{\boldsymbol{p}}$ is the momentum operator $-i\boldsymbol{\nabla}$. The first two terms, Equations (43) and (44), are parent terms, changes in a basis due to a change in the site the basis is centered on. The first term, Equation (43), is the derivative of the wave function in the interstitial region (Equation (10) with respect to its parent site. Since the gradient of a solution to the Helmholtz equation is a solution to the Helmholtz equation, matrix elements $\langle\psi_i|\hat{\boldsymbol{p}}\psi_j\rangle_{\mathcal{I}}$ and $\langle\psi_i|-\nabla^2|\hat{\boldsymbol{p}}\psi_j\rangle_{\mathcal{I}}$ are calculated as integrals over the surface of the muffin-tin spheres. As in Equation (22), when interstitial region tail parameters are the same, the evaluation requires κ^2 derivatives of structure functions. Working out this contribution proceeds as in Equation (22), although arriving at a finite form requires identities such as

$$\sum_\mu \hat{e}_\mu \mathcal{U}_{\tau_b} \left(B_{\ell_a m_a,\ell_b-1\,m_b-\mu}(\kappa_b,\boldsymbol{\tau}_a-\boldsymbol{\tau}_b,\boldsymbol{k})\mathcal{G}(\ell_b-1,\,m_b-\mu;\,\ell_b,\,m_b;\,1,\,\mu)\,\kappa_b^2 \right.$$

$$\left. - B_{\ell_a m_a,\ell_b+1\,m_b-\mu}(\kappa_b,\boldsymbol{\tau}_a-\boldsymbol{\tau}_b,\boldsymbol{k})\mathcal{G}(\ell_b+1,\,m_b-\mu;\,\ell_b,\,m_b;\,1,\,\mu) \right)$$

$$= \sum_\mu \hat{e}_\mu \mathcal{U}_{\tau_a} \left(B_{\ell_a+1\,m_a+\mu,\ell_b m_b}(\kappa_b,\boldsymbol{\tau}_a-\boldsymbol{\tau}_b,\boldsymbol{k})\mathcal{G}(\ell_a,\,m_a;\,\ell_a+1,\,m_a+\mu;\,1,\,\mu) \right.$$

$$\left. - B_{\ell_a-1\,m_a+\mu,\ell_b m_b}(\kappa_b,\boldsymbol{\tau}_a-\boldsymbol{\tau}_b,\boldsymbol{k})\mathcal{G}(\ell_a,\,m_a;\,\ell_a-1,\,m_a+\mu;\,1,\,\mu)\,\kappa_b^2 \right) \tag{47}$$

Potential matrix elements $\langle\psi_i|V|\psi_j\rangle$ are calculated using Fourier series as in Sect. 3.2 with gradients taken as discussed after equation (38).

The second term, equation (42), is the analog of the first term in the muffin-tin spheres; i.e., this term is the derivative of a basis with respect to its parent site evaluated in the muffin-tin spheres. This term requires the gradient with respect to atomic positions of the structure function B. This gradient is easily obtained from the structure function itself:

$$B'_{\ell m,\ell' m'}(\kappa,\boldsymbol{\tau}-\boldsymbol{\tau}',\boldsymbol{k}) \equiv \left.\frac{\partial}{\partial u}B_{\ell m,\ell' m'}(\kappa,\boldsymbol{u},\boldsymbol{k})\right|_{\boldsymbol{u}=\boldsymbol{\tau}-\boldsymbol{\tau}'}$$

$$\equiv \sum_\mu i\hat{e}_\mu \mathcal{U}_\tau B'^{(\mu)}_{\ell m,\ell' m'}(\kappa,\boldsymbol{\tau}-\boldsymbol{\tau}',\boldsymbol{k})$$

$$B'^{(\mu)}_{\ell m,\ell' m'}(\kappa,\boldsymbol{\tau}-\boldsymbol{\tau}',\boldsymbol{k}) = \left(\mathcal{G}(\ell,\,m;\,\ell+1,\,m+\mu;\,1,\,\mu)B_{\ell+1\,m+\mu,\ell' m'}(\kappa,\boldsymbol{\tau}-\boldsymbol{\tau}',\boldsymbol{k}) \right.$$

$$\left. -\kappa^2\mathcal{G}(\ell,\,m;\,\ell-1,\,m+\mu;\,1,\,\mu)B_{\ell-1\,m+\mu,\ell' m'}(\kappa,\boldsymbol{\tau}-\boldsymbol{\tau}',\boldsymbol{k}) \right)$$

$$\boldsymbol{\tau}-\boldsymbol{\tau}'\neq 0 \tag{48}$$

If convergence with respect to ℓ on the left hand side of the structure function is sufficient for the energy, terms in $\ell_{\max}+1$ in Equation (48) may be neglected

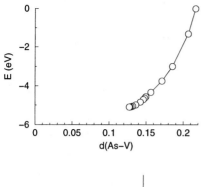

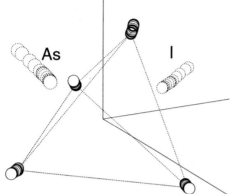

Fig. 2. Relaxation of a silicon 65 atom supercell containing a vacancy, a Si interstitial, and an As interstitial. Of the 106 internal coordinates in this cell, 104 were allowed to relax (2 coordinates were fixed to fix the center of mass of the crystal). The calculation used a simple Broyden's method to zero atomic forces.

in evaluating forces. As stated above, the evaluation of these terms is somewhat time consuming.

Examples of the use of forces for structural relaxation are given in Figures 1 and 2. Figure 1 shows deviations from ideal lattice positions calculated for rhombohedral $BaTiO_3$ as a function of volume compared to experiment [16]. The rhombohedral angle was also relaxed at each volume in this calculation. The Ti coordinate is a displacement along [111]. The oxygen displacements Δx are along face diagonals while Δz is toward the cell center. These calculations included Ti $3s$ and $3p$ and Ba $5s$ and $5p$ along with the usual valence bases in a single, fully hybridizing basis. At convergence, forces on internal coordinates were less than 1 mRy/Bohr. Figure 2 is a calculation of structural relaxation of As-vacancy-interstitial complex in Si. To a sixty-four atom Si supercell was added an As impurity at a tetrahedral interstitial position and a Si interstitial at an exchange position both surrounding a vacancy. The crystal, far from equilibrium, was then allowed to relax. Two internal coordinates (of a total of 106) were fixed to fix the center of mass of the crystal. The energy was minimized with respect to the other 104 internal coordinates by zeroing the forces (to with 1 mRy/Bohr). The forces were zeroed using a simple Broyden's method.

8   Conclusion

In this article we have described our highly accurate full-potential LMTO method for solving the Kohn-Sham equations. In particular, we have shown that by dividing the crystal space into non-overlapping "muffin-tin" spheres and an interstitial region, we can compute the charge density or the potential without any shape approximation, thus eliminating any need for empty spheres which are necessary in other LMTO implementations when the crystal is not closely packed. Another feature of our implementation is that we can describe multiple principle quantum numbers within a single, *fully hybridized* basis set. This is accomplished simply by using functions ϕ and $\dot{\phi}$ calculated with energies $\{e_{n\ell}\}$ corresponding to different principal quantum numbers n to describe the radial dependence of a basis in the muffin-tin region. In the interstitial region our method uses "multiple κ" basis sets, for a better description of the interstitial charge density. Highly accurate charge density can be obtained by systematically increasing the number of variational parameters κ for each angular momentum of the basis set.

The potential in a muffin-tin sphere at τ has an expansion in linear combinations of spherical harmonics invariant under that part of the point group that leaves atomic positions invariant. The evaluation of the interstitial potential matrix only requires a correct treatment of basis functions (and potential) in the interstitial region. We have used this degree of freedom to design "pseudo-basis functions", equal to the true basis functions in the interstitial region and are smooth functions in the muffin-tin region, with the requirement that their Fourier transforms converge rapidly enough for practical use.

The set of internal forces acting on the atomic sites of a crystal is a symmetric, discrete function of atom coordinates and has a spherical expansion on the crystal sites with the same coefficients as continuous symmetric functions. The total force on a site is given by the derivative of the LDA total energy with respect to the atomic position. Our implementation of the forces is in many ways similar to that of Yu *et al.* for the LAPW method [15]. Because our basis set is a site-centered one, we are required to compute additional terms, which can be time consuming. These contributions to the forces are non existant in plane-wave based methods, such as the pseudo-potential method. In addition to the "Helmann-Feynman" term, which accounts for the explicit dependence of the energy functional A on atomic positions, the other contributions are: (1) an "Incomplete Basis Set" term, (2) a core-correction term, (3) a surface term arising from the change in integration boundaries when atoms are moved, (4) a term which arises when the calculation isn't fully self-consistent, and (5) a term arising from the way in which the Brillouin zone integrals are performed. We have showed that the forces are accurate enough to relax atomic structures. As examples, forces have been used to optimize the internal coordinates of rhombohedral $BaTiO_3$ as a function of volume and the geometry of a 65 atom As, vacancy, and interstitial defected Si supercell. Where experimental results are available, good agreement is obtained.

9   Acknowledgments

One of us J.M.W would like to thank the Université Louis Pasteur for his IPCMS stay. M.A and O.E collaboration is partially supported by the TMR network 'Interface Magnetism' of the European Commission (Contract No. EMRX-CT96-0089) .

References

1. O. K. Andersen, Phys. Rev. B **12**, 3060 (1988), and references therein.
2. H. L. Skriver, *The LMTO method* (Springer-Verlag, Berlin, 1984).
3. J. M. Wills (unpublished); J. M. Wills and B. Cooper, Phys. Rev. B **36**, 389 (1987).
4. M. Springborg and O. K. Andersen, J. Chem. Phys. **87**, 7125, (1987).
5. M. Methfessel, Phys. Rev. B **38**, 1537 (1988).
6. K. H. Weyrich, Phys. Rev. B **37**, 10269 (1987).
7. S. Savrasov and D. Savrasov, Phys. Rev. B **46**, 12181 (1992).
8. M. Tischer, O. Hjortstam, D. Arvanitis, J. Hunter Dunn, F. May, K. Baberschke, J. Trygg, J. M. Wills, B. Johansson, and O. Eriksson, Phys. Rev. Lett. **75**, 1602, (1995); O. Hjortstam, J. Trygg, J. M. Wills, B. Johansson, and O. Eriksson, Phys. Rev. B **53**, 9204, (1996).
9. J. D. Jackson, *Classical Electrodynamics*, John Wiley and Sons, (New York, 1978).
10. J. Korringa, Physica **13**, 392 (1947); W. Kohn and N. Rostoker, Phys. Rev. **94**, 1111 (1954).
11. J. M. Wills, O. Eriksson, and A. M. Boring, Phys. Rev. Lett. **67**, 2215 (1991); M. Alouani, R.C. Albers, J. M. Wills, and M. Springborg, Phys. Rev. Lett. **69**, 3104, (1992), M. Alouani, J. W. Wilkins, R.C. Albers, and J. M. Wills, *ibid.*, **71**, 1415 (1993); M. Alouani and J. M. Wills, Phys. Rev. B **54**, 2480 (1996); R. Ahuja et al., *ibid.* **55**, 4999 (1996).
12. M. Weinert, J. Math. Phys. **22**, 2433 (1980).
13. C. Lanczos, *Applied Analysis*, Dover Publications Inc., New York, 1988.
14. A. R. Edmonds, *Angular Momentum in Quantum Mechanics*, Princeton University Press, Princeton, 1974.
15. R. Yu, D. Singh, and H. Krakauer, Phys. Rev. B **43**, 6411 (1991).
16. G. H. Kwei, A. C. Lawson, S. J. L. Billinge, and S.-W. Cheong, J. Phys. Chem. **97**, 2368 (1993).

Excited States Calculated by Means of the Linear Muffin-Tin Orbital Method

M. Alouani[1] and J. M. Wills[2]

[1] IPCMS, Université Louis Pasteur, 23 Rue du Loess, 67037 Strasbourg, France
[2] Los Alamos National Laboratory, Los Alamos, NM 87545, USA

Abstract. The most popular electronic structure method, the linear muffin-tin orbital method (LMTO), in its full-potential (FP) and relativistic forms has been extended to calculate the spectroscopic properties of materials form first principles, i.e, optical spectra, x-ray magnetic circular dichroism (XMCD) and magneto-optical kerr effect (MOKE). The paper describes an overview of the FP-LMTO basis set and the calculation of the momentum matrix elements. Some applications concerning the computation of optical properties of semiconductors and XMCD spectra of transition metal alloys are reviewed.

1 Introduction

The density functional theory (DFT) of Hohenberg, Kohn, and Sham is the method of choice for describing the ground-state properties of materials [1]. However, in the initial derivation of the DFT, the eigenvalues are Lagrange multipliers introduced to orthogonalize the eigenvectors, which in their turn are used to compute the total energy and the charge density. In this formulation the eigenvalues have therefore no physical meaning and should not be considered as excited states. Nevertheless, the DFT in the local density approximation (LDA) or in its spin resolved local density formulation (LSDA), has been used successfully to compute the excited states, namely, optical and magneto-optical properties, x-ray absorption and magnetic dichroism spectra.

The LDA or LSDA were indeed intended to compute the ground-state properties of materials, and their use during the last two decades has produced an excellent track record in the computation of these properties for a wide variety of materials, ranging from simple metals to complex semiconductor superlattices. However, it is now believed that the DFT can do more than computing the ground state properties. This is because the Kohn–Sham equations could be viewed as deriving form a simplified quasi-particle (QP) theory where the self-energy is local and time averaged, i.e., $\Sigma(\mathbf{r}, \mathbf{r}', t) \approx V_{xc}(\mathbf{r})\delta(\mathbf{r}-\mathbf{r}')\delta(t)$, here $V_{xc}(\mathbf{r})$ is the local exchange and correlation potential as, for example, parameterized by Von Barth and Hedin [2]. Viewed in this way, the KS eigenvalues are then approximate QP energies and could be compared to experimental data. This argument is supported by quasiparticle calculations within the so called GW approximation of Hedin [3] showing that the valence QP energies of semiconductors are in good agreement with these obtained using LDA, and the conduction QP energies differ by approximately a rigid energy shift [4,5]. In the literature this shift is often called "scissors-operator" shift [6].

In the last few years spectroscopy is becoming the standard tool for measuring excited states of materials. Its owes its impressive advances to the availability of synchrotron tunable highly polarized radiation. In particular, the measurement of optical, magneto-optical properties as well as magnetic x-ray dichroism are now becoming routine tasks for probing the structural and magnetic properties of materials. Considerable attention has been focused on transition-metal surfaces and and thin films due to their novel physical properties different from that of bulk materials and due to potential industrial applications such as magneto-optical recording, sensors, or technology based on giant magneto-resistance. In this respect, theory is falling far behind experiment and it is becoming hard to give a basic interpretation of experimental data.

This paper, which is far from being a review paper about calculated excited states, tries to bridge the gap between experiment and theory by describing a rather quantitative method for computing excited states of materials. This method uses the local density approximation and the linear muffin-tin orbital (LMTO) method. In the first part of this paper we introduce the density functional theory and the local density approximation and justify the use of LDA eigenvalues as approximate excited states and relate them to quasiparticle energies. In the second part we describe the construction of the LMTO basis set within an all-electron full-potential approach [7,17] which will be used to determine the momentum matrix elements. We devote the third part to the determination of the momentum matrix elements. In the first part of the application section we present some examples of computation of semiconductors optical spectra [9,10], and leave out the optical properties of metals and magneto-optical properties of materials and refer the reader to Ref. [11,12,9,14–19]. In the second part of the applications we show some examples of x-ray magnetic dichroism calculations [20–22].

2   Density Functional Theory

The density functional method of Hohenberg and Kohn [1] which states that the ground state total energy of a system of N interacting electrons in an external potential V_{ext} is a functional of the electron density $\rho(\mathbf{r})$ does not provide an analytical form of the functional [1]. This method remains numerically intractable without the Kohn and Sham introduction of the so called local density approximation [1] in which the exchange and correlation functional $E_{xc}\{n\}$ appearing in the total energy:

$$E\{n\} = T\{n\} + \frac{e^2}{2} \int d^3r \int d^3r' \frac{n(\mathbf{r})n(\mathbf{r}')}{|\mathbf{r} - \mathbf{r}'|} + E_{xc}\{n\}$$
$$+ \int d^3r \, V_{ext}(\mathbf{r})n(\mathbf{r}) + E(V_{ext}) \tag{1}$$

is given by $E_{xc}\{n\} = \int d^3r \ \epsilon_{xc}(n(\mathbf{r}))n(\mathbf{r})$ where ϵ_{xc} is the exchange-correlation energy of a uniform electron gas of density n. Thus, Kohn and Sham constructed a set of self-consistent single-particle equations:

$$\left(-\nabla^2 + \frac{\delta}{\delta n}(E - T)\right)\psi(\mathbf{r}) = e_i\psi_i(\mathbf{r}) \tag{2}$$

where the density $n(\mathbf{r})$ is given by:

$$n(\mathbf{r}) = \sum_i \theta(e_i < E_F)\psi_i(\mathbf{r})\psi_i^\dagger(\mathbf{r}) \tag{3}$$

and

$$V_{\text{ext}}(\mathbf{r}) = -e^2 \sum_{R\tau} \frac{Z_\tau}{|\mathbf{r} - \tau - \mathbf{R}|} \tag{4}$$

$$E(V_{\text{ext}}) = e^2 \sum_{\tau R} \sum_{\tau' R'} (1 - \delta(R, R')\delta(\tau, \tau')) \frac{Z_\tau Z_{\tau'}}{|\tau + \mathbf{R} - \tau' - \mathbf{R}'|} \tag{5}$$

Instead of the true kinetic energy of the electron gas, Kohn and Sham used the homogeneous electron kinetic energy:

$$\bar{T} \equiv \sum_i \theta(e_i < E_F) \int d^3r \ \psi_i^\dagger(-\nabla^2)\psi_i \tag{6}$$

This use of homogeneous-electron kinetic energy in the Kohn–Sham equations redefined the exchange-correlation function to be:

$$\bar{E}_{xc}\{n\} \equiv \left(E_{xc}\{n\} + T\{n\} - \bar{T}\right) = \int d^3r \ \epsilon_{xc}(n(\mathbf{r}))n(\mathbf{r}) \tag{7}$$

It is then crucial to use a good basis-set for the description of the electronic structure of realistic systems. The augmented plane wave [23] (APW), and the Korringa-Kohn-Rostoker [24] (KKR) methods can be used, in principle, to solve exactly the Kohn–Sham equations, however these methods are numerically involved and their linearization, introduced by Andersen is much preferable. Andersen linearization, has not only made the techniques for solving the band-structure problem transparent by reducing it essentially to the diagonalization of one-electron Hamiltonian, and cuts the cost of computation by at least one order of magnitude. The linearized versions of these two powerful methods are the linear augmented plane wave (LAPW) and linear muffin-tin orbital (LMTO) methods, respectively [17].

In this paper, we will only use the LMTO method to study excited states of solids. The reason for this choice is that the LMTO method is the mostly used method in computational electronic structure. This is due primarily to the use of atomic-sphere approximation (ASA) which made the LMTO method run fast even on today's cheap personal computers. Due to this reduced computational cost, the LMTO ASA method became the method of choice of researchers without access to supercomputers.

3   Quasiparticle Theory and Local-Density Approximation Link

The quasiparticle (QP) electronic structure of an interacting many-body system is described by the single-particle eigenstates resulting from the interaction of this single particle with the many-body electron gas of the system. The single-particle eigenstate energies are the results of solving a Schroedinger like equation containing the non-local and energy-dependent self-energy instead of the exchange-correlation potential appearing in Kohn–Sham like equations:

$$(T + V_H + V_{\text{ext}})\Psi(\mathbf{r}) + \int d^3 r' \Sigma(\mathbf{r}, \mathbf{r}', E)\Psi(\mathbf{r}') = E\Psi(\mathbf{r}).$$

Thus the self-energy Σ contains all many-body effects. Almost all ab-initio QP studies were performed within the so-called GW approximation, where the self-energy Σ is calculated within Hedin's GW approximation. This method consists of approximating the self-energy as the convolution of the LDA self-consistent Green function G and the screened coulomb interaction W within the random-phase approximation. The QP eigenvalues are often obtained using first-order perturbation theory starting from LDA eigenvalues and eigenvectors [4,25]. Although there are early calculations starting from Hartree-Fock [26] or tight-binding [27] methods. Nevertheless, the best results are based on a LDA starting point [4,25,28–30].

Thus the GW predicted optical excitations energies of semiconductors are within 0.1 eV form the experimental results and the surprizing fact is that the QP wave functions are almost identical to these produced within the LDA [4] (the wave function overlap is more than 99%). For a general review of GW calculations see the review by Araysetianwan and Gunnarsson [29] or by Aulbur, Jönsson and Wilkins [30].

It is clear that the quasiparticle Schroedinger equation resembles to the Kohn–Sham equation. Both equations describe a fictitious electron moving in a effective potential. The difference is that the self-energy is nonlocal and energy dependent whereas the LDA potential is local and averaged over time. This resemblance can be further pushed by noticing that the DFT can be used to obtain excitation energies. For example, the ionization energy, I, and the electron affinity, A, are difference between ground state energies:

$$I = E(N - 1) - E(N), \qquad \text{and} \qquad A = E(N) - E(N + 1)$$

where N is the number of electrons of the system. And since the DFT gives the correct ground state energies it should produce, in principle, the correct ionization and electron affinity energies. For metals, the addition or removal of an electron from the system costs the same energy, and hence the ionization energy is equal to the electron affinity. For insulators, the energy gap makes all the difference and hence breaks this symmetry. Thus the energy band gap is given by:

$$E_g = I - A = E(N+1) + E(N-1) - 2E(N)$$

In practice, however, the calculation is often obtained within the LDA and the energy band gap is calculated as the difference between the lowest conduction band and the highest valence band. It was shown by Sham and Schlüter [31] and Perdew and Levy [32] that the calculated energy gap differ from the true band gap by an amount Δ even when the DFT is used without the LDA. The Δ value could range from 50% in the case of silicon to 100% in the case of germanium. For most of the semiconductors, the GW calculations show that the LDA eigenvalues differ form the GW quasiparticle energy by a constant Δ which is almost independent of the k-point. This finding is important and shows that the LDA eigenvalues have some meaning and could be used to calculate excited states. So as stated in the introduction, the Kohn–Sham equations could be viewed as deriving form a simplified quasi-particle (QP) theory where the self-energy is made local and time averaged, i.e., $\Sigma(\mathbf{r}, \mathbf{r}', t) \approx V_{xc}(\mathbf{r})\delta(\mathbf{r}-\mathbf{r}')\delta(t)$. This approximation is certainly good for metals where we have a good data base for excited state calculated within the LDA [11,12,9,14–19] and where the agreement with experiment is good. For semiconductors, this approximation is not bad either, provided we know the value of the discontinuity of the exchange and correlation. Usually, this value is provided by GW calculations or by experiment.

4   The Full-Potential LMTO Basis Set

In this section we describe the LMTO basis-set used to calculate the excited states of solids. We discuss the basis used for an all electron calculation where the potential is not supposed to be spherically symmetric nor of muffin-tin type. The use of a general potential makes the study of open structures possible without having to resort to the so-called "empty-sphere" approximation. To define the basis-set, we divide the space into non overlapping spheres called "muffin-tin" spheres and a region between these spheres which we call interstitial region. Inside the muffin-tin spheres the Schroedinger equation is solved at a fixed energy for each angular momentum ℓ and variational parameter κ (which is defined later). The linearization amounts to the use of a linear combination of the solution $\phi_\ell(e, r)$ of the Schroedinger equation for a fixed energy and its energy derivative $\dot{\phi}_\ell(e, r)$ inside the muffin-tin spheres. These linear combinations matche continuously and differentiably to an envelop function (spherical function) in the interstitial region. The Bloch wave function in the interstitial region is given by a linear combination of these Hankel functions centered at each site:

$$\psi_i(\mathbf{k},\mathbf{r}) = \sum_R e^{i\mathbf{k}\cdot\mathbf{R}} \mathcal{K}_{L_i}(\kappa_i, \mathbf{r}-\tau_i-\mathbf{R}) \tag{8}$$

where i stands for the number of the basis function quantum numbers (these numbers are $\{\tau, L, \kappa, \{e_{\ell t}\}\}$), where τ is the site number, $L = (\ell, m)$ groups the two angular quantum numbers, and $e_{\ell t}$ is the linearization energy for a particular atom type t and angular momentum number ℓ. The envelop functions are defined as $\mathcal{K}_L(\kappa, \mathbf{r}) \equiv \mathcal{K}_\ell(\kappa, r)\mathcal{Y}_L(\hat{\mathbf{r}})$.

$$\mathcal{Y}_{\ell m}(\hat{\mathbf{r}}) \equiv i^\ell Y_{\ell m}(\hat{\mathbf{r}}) \tag{9}$$

$$\mathcal{K}_\ell(\kappa, r) \equiv -\kappa^{\ell+1} \begin{cases} n_\ell(\kappa r), & \text{if } \kappa^2 > 0 \\ n_\ell(\kappa r) - i j_\ell(\kappa r), & \text{if } \kappa^2 < 0, \ (\kappa = i|\kappa|) \end{cases} \tag{10}$$

$$\mathcal{J}_\ell(\kappa, r) \equiv \kappa^{-\ell} j_\ell(\kappa r) \tag{11}$$

Here n_ℓ is the Neumann function and j_ℓ Bessel function for the angular momentum ℓ, and $Y_{\ell m}$ are the spherical harmonics.

To get the differentiability of the wave-function at the boundary of the muffin-tin spheres, we write the envelope function inside the muffin-tin spheres. The envelope function for a muffin-tin sphere τ' is given by:

$$\sum_R e^{i\mathbf{k}\cdot\mathbf{R}} \mathcal{K}_L(\kappa, \mathbf{r}-\tau-\mathbf{R})\Big|_{r_{\tau'}<S_{\tau'}} \tag{12}$$

$$= \sum_{L'} \mathcal{Y}_{L'}(\hat{\mathbf{r}}_{\tau'}) \Big(\mathcal{K}_{\ell'}(\kappa, r_\tau)\delta(\tau, \tau')\delta(L, L') \tag{13}$$

$$+ \mathcal{J}_{\ell'}(\kappa, r_\tau) B_{L',L}(\tau'-\tau, \kappa, \mathbf{k}) \Big)$$

To produce smooth basis functions we require that the basis function is differentiable at the boundary of each muffin-tin sphere, i.e., that a linear combination of ϕ and $\dot{\phi}$ matches continuously and differentiably $\mathcal{K}$ and $\mathcal{J}$ at the boundary of the parent sphere and other spheres, respectively. Using these matching conditions at the muffin-tin spheres, the Bloch wave function inside a muffin-tin sphere τ of the unit cell at the origin is given by [7]:

$$\psi_i(\mathbf{k},\mathbf{r})\Big|_{r_\tau<S_\tau} = \sum_L \mathcal{Y}_L(D_\tau \hat{\mathbf{r}}_\tau) U_\ell(e_{\ell t i}, r_\tau) \Omega(\ell t, e_{\ell t i}\kappa_i) B_{L,L_i}(\tau-\tau_i, \kappa_i, \mathbf{k}) \tag{14}$$

where

$$U_\ell(e, r) \equiv \begin{pmatrix} \phi_\ell(e, r) \\ \dot{\phi}_\ell(e, r) \end{pmatrix} \tag{15}$$

$$\Omega(\ell t, e\kappa) \equiv S_\tau^2 \begin{pmatrix} -W(\mathcal{K}, \dot{\phi}) & -W(\mathcal{J}, \dot{\phi}) \\ W(\mathcal{K}, \phi) & W(\mathcal{J}, \phi) \end{pmatrix} \quad (W(f,g) \equiv fg' - f'g) \qquad (16)$$

$$\mathcal{B}_{L,L_i}(\tau - \tau_i, \kappa_i, \mathbf{k}) \equiv \begin{pmatrix} \delta(\tau, \tau_i)\delta(L, L_i) \\ B_{L,L_i}(\tau - \tau_i, \kappa_i, \mathbf{k}) \end{pmatrix} \qquad (17)$$

To add the spin dependence to the basis-set, the Bloch wave function is multiplied by the eigenvector of the Pauli spin operator $\eta_{\pm 1}$:

$$\psi_\sigma(\mathbf{k}, \mathbf{r}) = \psi(\mathbf{k}, \mathbf{r})\eta_\sigma$$

such that:

$$\hat{\mathbf{n}} \cdot \sigma\eta_{\pm 1} = (\pm 1)\eta_{\pm 1}$$

where η is the quantization axis chosen in advance.

5   Dielectric function

5.1   Dynamical Dielectric Function

Here we give a review of the determination of the dielectric response of a semiconductor due to the application of an electric field. We expend the description of our published work [10] by giving more details concerning the calculation of the momentum matrix elements.

An electromagnetic field of frequency ω, and a wave vector $\mathbf{q} + \mathbf{G}$ interacting with atoms in a crystal produces a response of frequency ω and a wave vector $\mathbf{q} + \mathbf{G}'$ ($\mathbf{G}$ and $\mathbf{G}'$ being reciprocal lattice vectors). The microscopic field of wave vector $\mathbf{q} + \mathbf{G}'$ is produced by the umklapp processes as a result of the applied field $E_0(\mathbf{q} + \mathbf{G}, \omega)$

$$E_0(\mathbf{q} + \mathbf{G}, \omega) = \sum_{\mathbf{G}'} \epsilon_{\mathbf{G}, \mathbf{G}'}(\mathbf{q}, \omega)E(\mathbf{q} + \mathbf{G}', \omega) \qquad (18)$$

where $E(\mathbf{q} + \mathbf{G}, \omega)$ is the total field producing the non-diagonal elements in the microscopic dielectric function $\epsilon_{\mathbf{G}, \mathbf{G}'}(\mathbf{q}, \omega)$. The microscopic dielectric function in the random phase approximation is given by [33]:

$$\epsilon_{\mathbf{G}, \mathbf{G}'}(\mathbf{q}, \omega) = \delta_{\mathbf{G}, \mathbf{G}'} - \frac{8\pi e^2}{\Omega|\mathbf{q} + \mathbf{G}||\mathbf{q} + \mathbf{G}'|} \qquad (19)$$

$$\times \sum_{\mathbf{k}, n, n'} \frac{f_{n', \mathbf{k} + \mathbf{q}} - f_{n, \mathbf{k}}}{E_{n', \mathbf{k} + \mathbf{q}} - E_{n, \mathbf{k}} - \hbar\omega + i\delta}$$

$$\langle n', \mathbf{k} + \mathbf{q}|e^{i(\mathbf{q} + \mathbf{G})\mathbf{r}}|n, \mathbf{k}\rangle\langle n, \mathbf{k}|e^{-i(\mathbf{q} + \mathbf{G}')\mathbf{r}}|n', \mathbf{k} + \mathbf{q}\rangle$$

Here n and n' are the band indexes, $f_{n,\mathbf{k}}$ is the zero temperature Fermi distribution, and Ω is the cell volume. The energies $E_{n,\mathbf{k}}$ and the the crystal wave function $|n,\mathbf{k}\rangle$ are produced for each band index n and for each wave vector $\mathbf{k}$ in the Brillouin zone.

The macroscopic dielectric function in the infinite wave length limit is given by the inversion of the microscopic dielectric function:

$$\epsilon(\omega) = \lim_{\mathbf{q}\to 0} \frac{1}{[\epsilon^{-1}_{\mathbf{G},\mathbf{G'}}(\mathbf{q},\omega)]_{0,0}} \tag{20}$$

$$= \epsilon_{0,0}(\omega) - \lim_{\mathbf{q}\to 0} \sum_{\mathbf{G},\mathbf{G'}\neq 0} \epsilon_{0,\mathbf{G}}(\mathbf{q},\omega) T^{-1}_{\mathbf{G},\mathbf{G'}}(\mathbf{q},\omega) \epsilon_{\mathbf{G'},0}(\mathbf{q},\omega)$$

where $T^{-1}_{\mathbf{G},\mathbf{G'}}$ is the inverse matrix of $T_{\mathbf{G},\mathbf{G'}}$ containing the elements $\epsilon_{\mathbf{G},\mathbf{G'}}$ with $\mathbf{G}$ and $\mathbf{G'} \neq 0$. The first term of this equation is the interband contribution to the macroscopic dielectric function and the second term represent the local-field correction to ϵ. The most recent ab-initio pseudopotential calculation found that the local-field effect reduces the static dielectric function by at most 5% [6]. Previous calculations with the same method have also found a decrease of ϵ_∞ by about the same percentage [4,34]. For insulators the dipole approximation of the imaginary part of the first term of equation (21) is given by [35]:

$$\epsilon_2(\omega) = \frac{e^2}{3\omega^2\pi} \sum_{n,n'} \int d\mathbf{k} |\langle n,\mathbf{k}|\mathbf{v}|n',\mathbf{k}\rangle|^2 f_{n,\mathbf{k}}(1 - f_{n',\mathbf{k}}) \delta(e_{\mathbf{k},n',n} - \hbar\omega), \tag{21}$$

Here $\mathbf{v}$ is the velocity operator, and in the LDA $\mathbf{v} = \mathbf{p}/m$ ($\mathbf{p}$ being the momentum operator), and where $e_{\mathbf{k},n,n'} = E_{n',\mathbf{k}} - E_{n,\mathbf{k}}$. The matrix elements $\langle n\mathbf{k}|\mathbf{p}|n'\mathbf{k}\rangle$ are calculated for each projection $p_j = \frac{\hbar}{i}\partial_j$, $j = x$ or y and z, with the wave function $|n\mathbf{k}\rangle$ expressed in terms of the full-potential LMTO crystal wave function described by equations (14) and (8). The $\mathbf{k}$-space integration is performed using the tetrahedron method [36] with a large number of irreducible $\mathbf{k}$ points the Brillouin zone. The irreducible $\mathbf{k}$-points are obtained from a shifted $\mathbf{k}$-space grid from the high symmetry planes and Γ point by a half step in each of the k_x, k_y, and k_z directions. This scheme produces highly accurate integration in the Brillouin zone by avoiding high symmetry points.

5.2   Momentum Matrix Elements

To calculate these matrix elements we first defined a tensor operator of order one out of the momentum operator $\nabla_0 = \nabla_z = \frac{\partial}{\partial z}$ and $\nabla_{\pm 1} = \mp\frac{1}{\sqrt{2}}(\frac{\partial}{\partial x} \pm i\frac{\partial}{\partial y})$. The muffin-tin part of the momentum matrix elements is calculated using the commutator $[\nabla^2, x_\mu] = 2\nabla_\mu$ so that:

$$\int_{S_\tau} d\mathbf{r}\phi_{\tau\ell'}(r)\, Y_{\ell'm'}(\widehat{\mathbf{r}-\tau})\nabla_\mu\phi_{\tau\ell}(r)Y_{\ell m}(\widehat{\mathbf{r}-\tau}) = -\tfrac{i}{2}G^{1\mu}_{\ell m,\ell',m'}$$

$$\int_0^{S_\tau} r^2 dr \phi_{\tau\ell'}\left(\tfrac{2}{r}\tfrac{d}{dr}r + \tfrac{\ell(\ell+1)-\ell'(\ell'+1)}{r}\right)\phi_{\tau\ell}(r) \tag{22}$$

where $G^{1\mu}_{\ell m,\ell',m'}$ are the usual Gaunt coefficients, and S_τ is the radius of the muffin-tin sphere of atom τ. In the interstitial region the plane-wave representation of the wave function (see equation 8) makes the calculation straightforward, but a special care has to be taken for the removal of the extra contribution in the muffin-tin spheres. However, we find it much easier and faster to transform the interstitial matrix elements as an integral over the surface of the muffin-tin spheres using the commutation relation of the momentum operator and the Hamiltonian in the interstitial region. The calculation of the interstitial momentum matrix elements is then similar to the calculation of the interstitial overlap matrix elements. The $\kappa = 0$ case has been already derived by Chen using the Korringa, Kohn and Rostoker Greens-function method [37]. We have tested that both the plane-wave summation and the surface integration provide the same results.

$$-\nabla^2 \, \mathbf{p}\psi = \kappa^2 \mathbf{p}\psi$$

A Hankel function can be integrated over a volume by knowing its integral over the bounding surface:

$$\int_\mathcal{I} d^3 r \nabla \left(\psi_1^\wedge \nabla p_i \psi_2 - \left(\nabla \psi_1^\wedge \right) p_i \psi_2 \right)$$
$$= \left(\kappa_1^2 - \kappa_2^2 \right) \int_\mathcal{I} d^3 r \psi_1^\wedge p_i \psi_2 \tag{23}$$

The surface of the interstitial consists of the exterior of the muffin-tin spheres and the unit cell boundary.

Over the surface of the muffin tins: the surface area is $S^2 d\Omega$ and the normal to the sphere points inward

$$\left(\kappa_1^2 - \kappa_2^2 \right) \int_\mathcal{I} d^3 r \; \psi_1^\wedge p_i \psi_2 = \tag{24}$$
$$-\sum_\tau S_\tau^2 \int dS \; \left(\psi_1^\wedge \frac{\partial}{\partial r} p_i \psi_2 - \left(\frac{\partial}{\partial r} \psi_1^\wedge \right) p_i \psi_2 \right)$$

At a muffin-tin sphere boundary S_τ the Bloch wave function is given by:

$$\left. \psi_i \left(\mathbf{k}, \mathbf{r} \right) \right|_{S_\tau} = \sum_R e^{i\mathbf{k} \cdot \mathbf{R}} K_{L_i} \left. \left(\kappa_i, \mathbf{r} - \tau_i - \mathbf{R} \right) \right|_{S_\tau} \tag{25}$$
$$= \sum_{\ell m} \mathcal{Y}_{\ell m}(\hat{\mathbf{r}}) K_\ell(\kappa_i, S) B_{\ell m, \ell_i m_i} \left(\tau - \tau_i, \kappa_i, \mathbf{k} \right)$$

where $B_{\ell m, \ell_i m_i} \left(\tau - \tau_i, \kappa_i, \mathbf{k} \right) = \begin{pmatrix} \delta(\tau, \tau_i)\delta(\ell, \ell_i)\delta(m, m_i) \\ B_{\ell m, \ell_i m_i} \left(\tau - \tau_i, \kappa_i, \mathbf{k} \right) \end{pmatrix}$ and $K = \begin{pmatrix} \mathcal{K} \\ \mathcal{J} \end{pmatrix}$

Let W denote the Wronskian $W(f, g) = fg' - f'g$

We define then

$$S^2 W_0 = S^2 W(K_\ell^T(\kappa), K_\ell(\kappa)) = \begin{pmatrix} 0 & 1 \\ -1 & 0 \end{pmatrix}$$

and

$$w_{\tau\ell}\kappa_1, \kappa_2 = S_\tau \frac{W(K_\ell^T(\kappa)), K_\ell(\kappa) - W_0}{\kappa_1^2 - \kappa_2^2}$$

$$\mathbf{p}\psi_i|_\tau = \sum_\mu \hat{\mathbf{e}}_\mu \sum_{\ell m} \Bigg[K_{\ell-1m-\mu} \begin{pmatrix} \kappa_i^2 & 0 \\ 0 & 1 \end{pmatrix} \mathcal{G}(\ell-1, m-\mu; \ell, m; 1, \mu)$$

$$-K_{\ell+1m-\mu} \begin{pmatrix} 1 & 0 \\ 0 & \kappa_i^2 \end{pmatrix} \mathcal{G}(\ell+1, m-\mu; \ell, m; 1, \mu) \Bigg]$$

$$B_{\ell m,\ell_i m_i}(\tau-\tau_i, \kappa_i, \mathbf{k}) \Big) \tag{26}$$

then

$$\langle \psi_f \mathbf{p}\psi_i \rangle_\tau = \tag{27}$$

$$\sum_\tau \sum_\mu \hat{\mathbf{e}}_\mu \sum_{\ell m} \Bigg[B_{\ell-1m-\mu,\ell_f m_f}(\tau-\tau_f, \kappa_f, \mathbf{k}) w_{\tau\ell-1}(\kappa_f, \kappa_i)$$

$$\times \begin{pmatrix} \kappa_i^2 & 0 \\ 0 & 1 \end{pmatrix} \mathcal{G}(\ell-1, m-\mu; \ell, m; 1, \mu)$$

$$-B_{\ell+1m-\mu,\ell_f m_f}(\tau-\tau_f, \kappa_f, \mathbf{k}) w_{\tau\ell+1}(\kappa_f, \kappa_i)$$

$$\times \begin{pmatrix} 1 & 0 \\ 0 & \kappa_i^2 \end{pmatrix} \mathcal{G}(\ell+1, m-\mu; \ell, m; 1, \mu) \Bigg]$$

$$B_{\ell m,\ell_i m_i}(\tau-\tau_i, \kappa_i, \mathbf{k}) \Big) + \Delta(f, i, \kappa_i)$$

where

$$(\kappa_f^2 - \kappa_i^2)\Delta(f, i, \kappa_i) = \sum_\mu \hat{\mathbf{e}}_\mu(\tau_i) \Bigg[\tag{28}$$

$$+B_{\ell_i+1m_i-\mu,\ell_f m_f}^\star(\tau_i-\tau_f, \kappa_f, \mathbf{k}) \mathcal{G}(\ell_i+1, m_i-\mu; \ell_i, m_i; 1, \mu)\kappa^2$$

$$-B_{\ell_i-1m_i-\mu,\ell_f m_f}^\star(\tau_i-\tau_f, \kappa_f, \mathbf{k}) \mathcal{G}(\ell_i-1, m_i-\mu; \ell_i, m_i; 1, \mu)\kappa^2 \Bigg]$$

$$+\sum_\mu \hat{\mathbf{e}}_\mu(\tau_f) \Bigg[\tag{29}$$

$$+B_{\ell_f+1m_i+\mu,\ell_f m_f}(\tau_f-\tau_i, \kappa_i, \mathbf{k}) \mathcal{G}(\ell_f, m_i; \ell_f+1, m_f+\mu; 1, \mu)$$

$$-B_{\ell_f-1m_f-\mu,\ell_i m_i}(\tau_f-\tau_i, \kappa_i, \mathbf{k}) \mathcal{G}(\ell_f, m_f; \ell_f-1, m_f+\mu; 1, \mu)\kappa^2 \Bigg]$$

5.3   Velocity Operator and Sum Rules

Equation (21) can not be used directly to determine the optical properties of semiconductors, when the GW approximation or the scissors operator is used to determine the electronic structure. The velocity operator should be obtained from the effective momentum operator $\mathbf{p}^{\text{eff}}$ which is calculated using the self-energy operator, $\Sigma(\mathbf{r}, \mathbf{p})$, of the system [38]:

$$\mathbf{v} = \mathbf{p}^{\text{eff}}/m = \mathbf{p}/m + \partial\Sigma(\mathbf{r}, \mathbf{p})/\partial\mathbf{p} \tag{30}$$

GW calculations show that the quasiparticle wave function is almost equals to the LDA wave function [4,5]. Based on this assumption, it can be easily shown [38] that in the case of the scissors operator, where all the empty states are shifted rigidly by a constant energy Δ, the imaginary part of the dielectric function is a simple energy shift of the LDA dielectric function towards the high energies by an amount Δ, i.e., $\epsilon_2^{QP}(\omega) = \epsilon_2^{\text{LDA}}(\omega - \Delta/\hbar)$. The real part of the dielectric function is then obtained from the shifted ϵ_2 using Kramers-Kronig relations. The expression of ϵ_∞^{QP} is given by:

$$\epsilon_\infty^{QP} = 1 + \frac{2e^2}{3\omega^2\pi^2}\sum_{n,n'}\int d\mathbf{k} f_{n,\mathbf{k}}(1 - f_{n',\mathbf{k}})\frac{|\langle n, \mathbf{k}|\mathbf{p}|n', \mathbf{k}\rangle|^2}{(e_{\mathbf{k},n',n} + \Delta)e_{\mathbf{k},n',n}^2}, \tag{31}$$

ϵ_∞^{QP} is very similar to $\epsilon_\infty^{\text{LDA}}$ except that one of the interband gap $e_{\mathbf{k},n',n}$ is substituted by the QP interband gap $e_{\mathbf{k},n',n} + \Delta$.

To test for the accuracy of the calculation within the LDA the f-sum rule:

$$\frac{2}{3mn_v}\sum_{\mathbf{k}}\sum_{n,n'} f_{n,\mathbf{k}}(1 - f_{n',\mathbf{k}})\frac{|\langle n, \mathbf{k}|\mathbf{p}|n', \mathbf{k}\rangle|^2}{e_{\mathbf{k},n',n}} = 1, \tag{32}$$

where n_v is the number of valence bands, should be always checked to ensure the accuracy of the calculations.

It is easily seen that the dielectric function ϵ_2^{QP} calculated using the scissors-operator shift does not satisfy the sum rule (ω_P is the free-electron plasmon frequency):

$$\int_0^\infty \omega\epsilon_2(\omega)d\omega = \frac{\pi}{2}\omega_P^2 \tag{33}$$

because (i) ϵ_2^{LDA} satisfies this rule, and (ii) ϵ_2^{QP} is obtained by a simple shift of ϵ_2^{LDA} by the scissors-operator Δ towards higher energies. The non simultaneous satisfaction of both the f-sum rule and the integral sum rule within the scissors approximation shows the limitation of this approximation. While the scissors operator approximation describes nicely the low lying excited states, which is seen in the good determination of the static dielectric function and the low energy structures, i.e. E_1 and E_2, in the imaginary part of the dielectric function, it seems to fail for the description of the higher excited states. This is not surprising because the higher excited states which are free electrons like are most probably

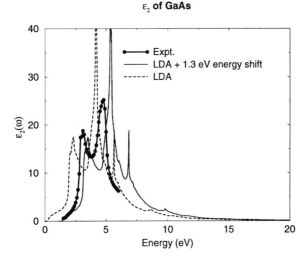

Fig. 1. Calculated Imaginary part of the dielectric function of GaAs at the experimental equilibrium volume both within LDA and shifted by 1.3 eV, compared with the experimental results of Ref. [39]. The experimental E_1 is only slightly underestimated while E_2 is overestimated. Notice that the shifted dielectric function by 1.3 eV, which produces the correct band gap, overestimates the peak positions by about 0.3 eV. Excitonic effect should shift these peaks to lower values in agreement with experiment.

well described within LDA and need no scissors-operator shift. This is supported by the fact that the the energy-loss function, -Imϵ^{-1}, within the LDA has it maximum roughly at the free electron plasmon frequency whereas within the scissors approximation its maximum is shifted to higher energies. For our purpose the scissors-operator shift remains a good approximation for the description of the low-lying excited states of semiconductors and their optical properties.

6   Applications

6.1   Optical Properties

We have used our FP-LMTO method and the formalism outlined above to calculate the optical properties of materials [9–12]. In general our results are often in good agreement with the experimental results. For semiconductors, however, good agreement with experiment is only achieved when the so called scissors-operator shift is used. Figure 1 presents our relativistic calculation of the imaginary part of the dielectric function of GaAs compared to the experimental results of Ref. [39]. The LDA relativistic results underestimates the band gap by about 1.3 eV. When the imaginary part of the dielectric function is shifted to higher energies by 1.3 eV the results the E_1 and E_2 peaks are overestimated in our calculation. One needs to shift the spectrum by less than the band gap as done in Ref. [10] to produce good agreement with experiment. It seems then that the optical band gap is less than the band energy gap (1.5 eV). The optical band gap is produced by interband transitions to the low lying conduction states. Excitonic effects are therefore important and are responsible for the reduction of the energy gap of semiconductors. It is interesting to notice though that the static dielectric function are in good agreement with experiment for GaAs, Si, and Ge when the shift correspond to the energy band gap obtained from photoluminescence [6,10].

More interesting are the wide band-gap materials where the LDA calculated static dielectric function is in good agreement with experiment despite that the band gap is still underestimated by LDA. Correcting the band gap using the scissors operator makes the static dielectric much small than the measured value. As an example of wide gap material, we present in Figure 2 and 3 the imaginary part of the dielectric function of GaN for the cubic (3) and wurtzite structure (B4).

Table 1. Calculated static dielectric function ϵ_∞ for GaN compared to pseudopotential (PP) results and experiment. For the wurtzite structure we have calculated $\epsilon_\infty^\parallel$ for a polarization parallel to the xy plan and $\epsilon_\infty^\perp$ which is perpendicular.

	zinc-blende	wurtzite	
	ϵ_∞	$\epsilon_\infty^\parallel$	$\epsilon_\infty^\perp$
PP	5.74	5.48	5.60
Present work	5.96	5.54	5.65
Expt.		5.35	5.35±0.2

Table 1 shows that our LDA dielectric constant calculations are in agreement with available experimental results and the pseudo-potential (PP) results [40] including local-field effects (an error about our calculation is reported in Ref. [40]; our value for $\epsilon_\infty^\parallel$ is not 4.48 but 5.54 and the PP value should then be 4.48). It is interesting to notice that static dielectric is in good agreement for the for all the nitrides [40] while the band gap is underestimated. The scissors-operator shift fails to explain the static dielectric function of large gap semiconductor. Recently, both local-field effects and electron-hole interaction were included on an ab-initio computation of the dielectric function of few semiconductors [41,42] by extending the semi-empirical Hanke and coworkers approach [26,43] which is based on the solution of the Bethe-Salpeter equation [43]. The excitonic effects seem to improve significantly the agreement between theory and experiment. However for large band-gap semiconductors, such as diamond, the inclusion of the excitonic effects seem to underestimate the optical band gap by as much as 1 eV [42]. It is not clear from these calculations whether the static dielectric function for wide-band gap semiconductors is improved when excitonic effects are included. More theoretical work along these lines is needed to fully understand the dielectric function of wide-gap semiconductors.

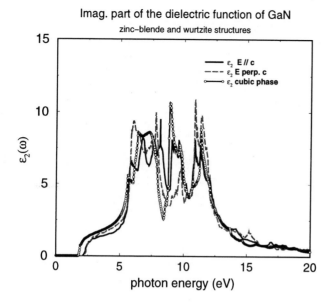

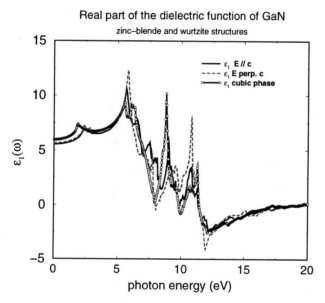

Fig. 2. Calculated imaginary and real parts of the dielectric function of GaN in its cubic and wurtzite forms. The LDA band gap of the cubic phase is 1.8 eV and the wurtzite phase is 2.2 eV.

6.2   Magnetic Circular Magnetic Dichroism

X-ray absorption spectroscopy (XAS) probes selectively each core orbital of each atomic species in a material. Two decades ago the theoretical work of Erskine and Stern show that the x-ray absorption could be used to determine the x-ray magnetic circular dichroism (XMCD) in transition metals when left and right circularly polarized x-ray beams are used [44]. More recently these ideas were implemented experimentally and XAS was used to determine the local magnetic properties of each magnetic atomic orbital in a magnetic compound [45,46]. Thus the circular magnetic x-ray dichroism is an important tool for the investigation of magnetic materials [45–56], especially through the use of sum rules for the direct determination of the local orbital and spin contributions to the total magnetic moment [50].

Thole and co-workers show that the circular-magnetic-x-ray dichroism is related to the magnetic moment of the photo-excited atom when the core electron is excited to the conduction states that are responsible of the magnetic properties of the material. On the theoretical side, Ebert and his co-workers [51,52] have developed a fully-relativistic local-spin-density-approximation approach that was used with success to calculate the XMCD at the K-edge of Fe, the L_3-edge of Gadolinium, and Fe and Co multilayers. Wu et al used slab linear augmented plane wave method to study the $L_{2,3}$ XMCD of Fe [56]. Brouder and co-workers uses Multiple-scattering theory to solve the Schrödinger using spherical potentials and spin-orbit coupling as a perturbation in the final state [53]. Recently Ankudinov and Rehr used a method based on a non-relativistic treatment of propagation based on high order multiple scattering theory and spinor-relativistic Dirac-Fock treatment of the dipole matrix elements to calculate the Fe K edge and Gd L_3 edge XMCD [54].

The calculation of the x-ray absorption for left and right circularly polarized x-ray beams is implemented within the local-density approximation (LDA) by means of all-electron full-relativistic and spin-polarized full-potential linear muffin-tin orbital method (LMTO). The core electrons are spin-polarized and their electronic states are obtained by solving the full-Dirac equation, whereas for the valence electrons the spin-orbit coupling is added perturbatively to the the semi-relativistic Hamiltonian. The total Hamiltonian is then solved self-consistently. To calculate the polarization dependent cross-section we consider the case where the internal field polarizes the spins along the magnetization easy axis. With respect to this axis we defined the left- and right-circular polarization, which correspond to the photon helicity $(+\hbar)$ $(-\hbar)$ respectively and the following dipole interaction: $\hat{e}_{\pm}\mathbf{p} = \frac{1}{\sqrt{2}}(\nabla_x \pm i\nabla_y)$. The absorption cross-section $\mu_{\pm}$ for left $(+)$ and right $(-)$ circular polarized x-ray calculated at the relativistic $j_{\pm}$ $(\ell \pm \frac{1}{2})$ core level and in the dipole approximation is given by:

$$\mu_{\pm}(\omega) = \frac{2\pi}{\hbar} \sum_{m_{j_\pm}} \sum_{n,\mathbf{k}} \langle j_{\pm}m_{j_\pm}|\hat{e}_{\pm}\mathbf{p}|n\mathbf{k}\rangle \langle n\mathbf{k}|\mathbf{p}\hat{e}_{\pm}|j_{\pm}m_{j_\pm}\rangle \delta(\omega - E_{n\mathbf{k}} + E_{j_\pm}) \quad (34)$$

using LDA in conjunction with the relativistic full-potential LMTO technique.

Figure 3. represent the K-edge x-ray absorption of Fe, for left and right circularly polarized light, compared to the experimental results. The agreement at low energy with experiment is good and start degrading at higher energies above the mean absorption peak. It is of interest to point out that the magnetic

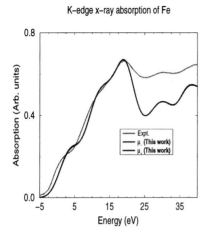

K-edge x-ray absorption of Fe

Legend:
- Expt.
- μ_- (This work)
- μ_+ (This work)

Fig. 3. Calculated x-ray absorption at the K-edge of Fe for left and right circularly polarized light compared to the experimental spectrum. The difference between the two spectra (barely visible on the graph) represents the x-ray magnetic circular dichroism.

x-ray dichroism at the K-edge which is due to the spin polarization and the spin-orbit in the final state is very small in the case of Fe. The difference between the right and left circularly polarization of the light is not even visible on the graph. However, the x-ray magnetic circular dichroism can be measured and Figure 4 shows a good agreement of the calculated dichroic signal with the experimental results of Shütz [46].

At the $L_{2,3}$ edge of $3d$ transition metals the x-ray magnetic dichroism is much important because it is meanly due to the presence of the strong spin-orbit coupling in the initial $2p$ states (in the case of Fe the spin-orbit splitting between the $2p_{3/2}$ and $2p_{1/2}$ is about 13 eV). In Figure 5 we show the calculated x-ray absorption and XMCD at the Co in PtCo ordered alloy [21].

To compare the results with experiment we have to take into account the effect of the core hole and the experimental resolution. This is done by convoluting the calculated spectra by a Lorentzian of widths of 0.9 eV and 1.4 eV for the L_2 and L_3 edges, respectively, in addition a Gaussian broadening of 0.4 eV is added to take into account the experimental resolution. The calculation of the x-ray magnetic circular dichroic signal ignoring the electron-hole recombination effect provides a semi-quantitative agreement with the experimental spectra. Hence, we believe that the core hole effect represented here by a Lorentzian broadening plays a significant role in determining the correct L_3/L_2 branching ratio for $3d$ transition metals. The underestimation of the $L_{2,3}$ branching ratio remains a challenge for theorists and further theoretical development along the line proposed by Schwitalla and Ebert [57] is needed to bring the theory at the level of the experiment.

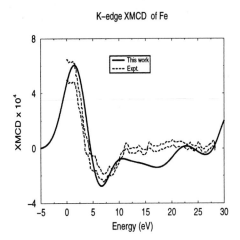

Fig. 4. Calculated x-ray magnetic x-ray dichroism at the K-edge f Fe compared to the experimental spectrum of Shütz[46].

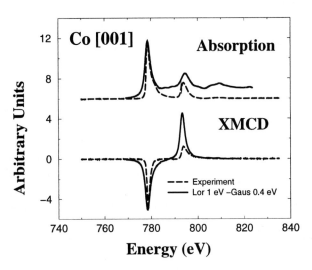

Fig. 5. Calculated x-ray absorption and magnetic x-ray dichroism at the $L_{2,3}$-edge of Fe compared to the experimental spectrum of Grange *et al.*[21].

For the $4d$-transition metals, the core hole is deeper, and the agreement with experiment of the XMCD is satisfactory. Figure 6 shows the calculated XMCD at the site of Pt of the CoPt ordered alloy.

In contrast to what is obtained for Co, the results for the Pt site show a much better agreement with experiment, due to the fact that the core hole effect is

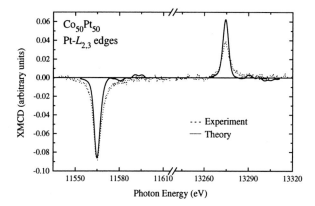

Fig. 6. Calculated x-ray magnetic x-ray dichroism at the $L_{2,3}$-edge of Pt compared to the experimental spectrum of Grange *et al.*[21].

less intense (core hole much deeper than that of Co). For the Pt atom we used both a Lorentzian (1 eV) and a Gaussian (1 eV) to represent the core hole effect and a Gaussian of 1 eV width for the experimental resolution. The experimental and theoretical L_2 and L_3 edges are separated by a spin-orbit splitting of the $2p$ core states of 1709 and 1727 eV respectively. The width of both L_2 and L_3 edges is comparable to experiment, but the calculated L_2 edge is much larger. This produces a calculated integrated branching ratio of 1.49 which is much smaller than the experimental ratio of 2.66. Here again the theory is underestimating the branching ratio.

7   Conclusion

We have reviewed the FP-LMTO method and the implementation of the optical properties and x-ray magnetic dichroism within the local density approximation. We have showed that the momentum matrix elements can be evaluated as a muffin-tin contribution and a surface term. The method has been successfully used to compute the optical properties of metals [11,12], semiconductors [9,10] and magneto-optical properties [11] of transition metals alloys, as well as x-ray magnetic circular dichroism [20–22] with high precision.

For small-gap semiconductors a scissors-operator shift should be used to reproduce the static and dynamic dielectric function [10]. Excitonic effects seem to be important in reproducing the correct optical energy gap [41,42]. For wide-gap semiconductors the local-density approximation (LDA) static dielectric function is in good agreement with experiment and no scissors-operator shift is required despite the underestimation of the band gap by LDA [40].

For the computation of the x-ray magnetic circular dichroism the agreement with experiment is rather good [20–22,51–55]. However, the so called branching ratio is underestimated by the theory. More theoretical work where the electron core-hole interaction is taken into account is needed to bring the theory at the quality level of experiment [57].

Part of this work was done while one of us (M.A) was at Ohio State University and were supported by NSF, grant number DMR-9520319. Supercomputer time was granted by CNUSC (project gem1917) on the IBM SP2 and by the Université Louis Pasteur de Strasbourg on the SGI O2000 supercomputer.

References

1. P. Hehenberg et W. Kohn, Phys. Rev. **136**, B864 (1964); W. Kohn et L. J. Sham, Phys. Rev. **140**, 1133 (1965).
2. U. Von Barth and L Hedin, J. Phys. C **5**, 1629 (1972).
3. L. Hedin, Phys. Rev. **139**, A796 (1965).
4. M. S. Hybertsen and S. G. Louie, Phys. Rev. B **32**, 7005, (1985); *ibid*, **34**, 5390, (1986).
5. R. W. Godby, M. Schlüter, and L. J. Sham, Phys. Rev. B **37**, 10159 (1988).
6. Z. H . Levine and D. Allan, Phys. Rev. B **43**, 4187 (1991); Phys. Rev. Lett. **66**, 41 (1991).
7. See the article by J. M. Wills *et al.* in this book.
8. O. K. Andersen, Phys. Rev. B **12**, 3060 (1975).
9. J. Petalas, S. Logothetidis, S. Boultadakis, M. Alouani, and J. M. Wills, Phys. Rev. B **52**, 8082 (1995).
10. M. Alouani and J. M. Wills, Phys. Rev. B **54**, 2480 (1996).
11. For the calculation of magneto-optical properties of materials see A. Delin, PhD thesis, Uppsala 1998; P. Ravindran *et al.* **59** 15680 (1999).
12. R. Ahuja, S. Auluck, J. M. Wills, M. Alouani, B. Johansson, and O. Eriksson, Phys. Rev. B **55**, 4999 (1997).
13. C. S. Wang and J. Callaway, Phys. Rev. B **9**, 4897 (1979); D. G. Laurent, J. Callaway, and C. S. Wang, Phys. Rev. B **19**, 5019 (1979).
14. Yu. A. Uspenski, E. G. Maksimov, S. U. Rashkeev, and I. I. Mazin, Z. Phys. B **53**, 263 (1983).
15. M. Alouani, J. M. Koch, and M. A. Khan, J. Phys. F **16**, 473 (1986).
16. M. Amiotti *et al.*, Phys. Rev. B **45**, 13285 (1992).
17. S. V. Halilov and Yu A. Uspenskii, J. Phys: Condens. Matter **2**, 6137 (1990); S. V. Halilov, *ibid* **4**, 1299 (1992).
18. P. M. Oppeneer, T. Maurer, J. Sticht, and J. Kübler, Phys. Rev. B **45**, 10924 (1992); T. Kraft, P. M. Oppeneer, V. N. Antonov, and H. Eschrig, *ibid* **52**, 3561 (1995).
19. K. W. Wierman, J. N. Hifiker, R. F. Sabiryanov, S. S. Jaswal, R. D. Kirby, J. A. Woollam, Phys. Rev. B **55**, 3093 (1997).
20. M. Alouani, J.M. Wills, and J.W. Wilkins, Phys. Rev. B **57**, 9502 (1998).
21. W. Grange, I. Galanakis, M. Alouani, M. Maret, J. P. Kappler, A. Rogalev, Phys. Rev. B (accepted).
22. I. Galanakis, S. Ostanin, M. Alouani, H. Dreysse, J. M. Wills, Phys. Rev. B (submitted).
23. J. C. Slater, **51**, 151 (1937); T. L. Loucks, *Augmented plane wave method*, (W. A. Benjamin, Inc., New York 1967)

24. J. Korringa, Physica **13**, 392 (1947); W. Kohn and N. Rostoker, Phys. Rev. **94**, 1111 (1954).

25. R. W. Godby, M. Schlüter, and L. J. Sham, Phys. Rev. Lett. **56**, 2415, (1986); Phys. Rev. B **35**, 4170, (1987); *ibid*, **36**, 6497 (1987), and **37**, 10159 (1987).

26. G. Strinati, H. J. Mattausch, and W. Hanke, Phys. Rev. B **25**, 2867 (1982).

27. P.A. Sterne and J. C. Inkson, J. Phys. C **17**, 1497 (1984).

28. M. Rohlfing, P. Krüger, and J. Pollmann, Phys. Rev. B **48**, 17791 (1993).

29. F. Araysetianwan and O. Gunnarsson, Rep. Prog. Phys. **61**, 237-312 (1998).

30. W. G. Aulbur, L. Jönsson, and J. W. Wilkins, *'Quasiparticle calculations in solids'*, to be published in Solid state Physics; edited by H. Ehrenreich.

31. L. J. Sham and M. Schlüter, Phys. Rev. Lett. **51**, 1888 (1983); Phys. Rev. B **32**, 3883 (1985).

32. J. P. Perdew and M. Levy, Phys. Rev. Lett. **51**, 1884 (1983)

33. S. L. Adler, Phys. Rev. **126**, 413 (1962); N. Wiser, Phys. Rev. **129**, 62 (1963).

34. S. Baroni and R. Resta, Phys. Rev. B **33**, 7017 (1986).

35. H. Ehrenreich and M. L. Cohen, Phys. Rev. **115**, 786 (1959).

36. O. Jepsen et O. K. Andersen, Solid State Commun. **9**, 1763 (1971); G. Lehmann et M. Taut, Phys. Stat. Sol. **54**, 469 (1972).

37. A. B. Chen, Phys. Rev. B **14**, 2384 (1976).

38. R. Del Sole and R. Girlanda, Phys. Rev. B **48**, 11789 (1993).

39. D. E. Aspnes and A. A. Studna, Phys. Rev. B **27**, 985 (1983).

40. J. Chen, Z. H. Levine, and J. W. Wilkins, Appl. Phys. Lett. **66**, 1129 (1995).

41. S. Albrecht, L. Reining, R. Del Sole, and G. Onida, Phys. Rev. Lett. **80**, 4510 (1998).

42. L. X. Benedict, E. L. Shirley, and R. B. Bohn, Phys. Rev. Lett. **80**, 4514 (1998); Phys. Rev. B **57**, R9385 (1998).

43. G. Strinati, H. J. Mattausch, and W. Hanke, Phys. Rev. Lett. **45**, 290 (1980); H. J. Mattausch, W. R. Hanke, and G. Strinati, Phys. Rev. B **27**, 3735 (1983); N. Meskini, H. J. Mattausch, and W. Hanke, Solid Stat. Comm. **48**, 807 (1983); M. del Castello-Mussot and L. J. Sham, Phys. Rev. B **31**, 2092 (1985).

44. J. L Erskine and E. A. Stern, Phys. Rev. B **12**, 5016 (1975).

45. G. van der Laan *et al.*, Phys. Rev. B **34**, 6529 (1986).

46. G. Shütz *et al.*, Phys. Rev. Lett. **58**, 737 (1987).

47. P. Carra and M. Alterelli Phys. Rev. lett. **64**, 1286 (1990).

48. C. T. Chen *et al.*, Phys. Rev. Lett. **75**, 152 (1995) and references therein.

49. P. Carra *et al.*, Phys. Rev. Lett. **66**, 2595 (1991).

50. For atoms: B. T. Thole *et al.*, Phys. Rev. Lett. **68**, 1943 (1992), P. Carra et al., Phys. Rev. Lett **70**, 694 (1993), for solids: A. Ankudinov and J. J. Rehr, Phys. Rev. B **51**, 1282 (1995).

51. H. Ebert, P. Strange, and B. L. Gyorffy, J. Appl. Phys. **63**, 3055 (1988); P. Strange and B. L. Gyorffy, Phys. Rev. B **52**, R13091 (1995).

52. G. Y. Guo *et al.*, Phys. Rev. B **50**, 3861 (1994); E. Ebert, Rep. Prog. Phys. **59** 1665 (1996).

53. Ch. Brouder, M. Alouani, and K. H. Bennamann, Phys. Rev. B **54**, 7334 (1996) and references therein.

54. A. Ankudinov and J. J. Rehr, Phys. Rev. B **56**, 1712 (1997).

55. N. V. Smith, C. T. Chen, F. Sette, and L. F. Mattheiss, Phys. Rev. B **46**, 1023 (1992).

56. R. Wu, D. Wang, and A. J. Freeman, Phys. Rev. lett. **71**, 3581 (1993); R. Wu and A. J. Freeman, Phys. Rev. Lett. **73** 1994 (1994).

57. J. Schwitalla and H. Ebert, Phys. Rev. Lett. **80**, 4586 (1998).

Part II

Magnetic Properties

Fully Relativistic Band Structure Calculations for Magnetic Solids – Formalism and Application

H. Ebert

Institut für Phys. Chemie, Univ. München, Butenandtstr. 5-13, D-81377 München

Abstract. Relativistic effects, in particular the spin-orbit coupling, give rise for magnetic systems to a great number of interesting and technologically important phenomena. The formal and technical aspects of corresponding fully relativistic theoretical investigations are reviewed. The properties of the underlying Dirac equation, set up within the framework of density functional theory (DFT) are discussed together with the Breit-interaction and Brooks' orbital polarization mechanism. As an example for a corresponding band structure method, the Korringa-Kohn-Rostoker (KKR) Green's function method is adopted. In particular, some technical aspects specific to this technique are discussed. The numerous applications that will be presented are primarily meant to demonstrate the many different facets of relativistic – this means in general – of spin-orbit induced effects in magnetic solids. In addition, these also demonstrate the tremendous flexibility of band structure schemes based on the Green's function formalism.

1 Introduction

Relativistic influences on the electronic properties of solids are known for quite a long time. One of the most prominent examples for these is the position of the optical absorption edge of Au. Compared to that of Ag this is higher in energy giving rise to the characteristic yellow colour of Au [1]. Another example is the relativistic contraction of s-type electronic shells, that has important consequences even in chemistry [2]. In the case of Au this leads to the existence of Au^--ions in the compounds CsAu and RbAu, while corresponding Ag-compounds are not found. One of the early hints for the importance of spin-orbit coupling for the electronic band structure of solids stems from X-ray absorption experiments. Namely, it was observed by Cauchois and Manescu [3] that for the L_3-absorption spectrum of Pt there occurs a *white line* at the absorption edge while none was found for the L_2-edge. Mott [4] ascribed this finding to the spin-orbit coupling, that should cause the d-states of Pt above the Fermi-energy to have predominantly $d_{5/2}$-character. As a consequence of this and because of the dipole selection rules $\Delta j = 0, \pm 1$ on expects strong absorption for the L_3-edge but not for the L_2-edge. Another example for the influence of the spin-orbit coupling is supplied by de Haas-van Alphen-experiments on W. Here it was found that the electron- and hole-surfaces do not touch in the Γ-H-direction, as it was predicted from non-relativistic band structure calculations [5]. The mentioned experiments and many others clearly demonstrated the influence of relativistic effects and that way the need to account for them within a corresponding band structure calculation. Nevertheless, it will in general depend on the specific experiment one wants to describe to what extent this has to be done. For example, in the case of the quadrupolar and magnetic hyperfine interaction, that takes place in the vicinity of the nucleus where relativistic influences on the electrons are most pronounced, these will show up even for relatively light elements [6].

For many cases, it is well justified to deal with relativistic effects by introducing corresponding corrections to the Schrödinger equation. One of the standard techniques to derive these is to apply a Foldy-Wouthuysen-transformation to remove the coupling between the large and small component of the Dirac equation [7]. Because of technical problems connected with this approach – in particular its convergence behavior – several alternative schemes have been suggested that in general also lead to a two-component formalism and that in some cases are also somewhat problematic. Among these for example the elimination methods [8–14], the Douglas-Kroll-Heß-transformation [15], or the ZORA-scheme [16] aim to derive an effective Hamiltonian that contains – compared to a non-relativistic Schrödinger-Hamiltonian – relativistic correction terms. In contrast to this, Gesztesy et al. worked out an expansion scheme for the corresponding Green's function [17,18]. In spite of the obvious differences between the various schemes mentioned, they nevertheless all lead to the mass-velocity, the Darwin and the spin-orbit coupling terms [7] as the most important corrections, that are all proportional to $(1/c^2)$. Of course, the explicit expressions supplied by the various approaches differ to some extent. Nevertheless, one may unambiguously call the first two correction terms scalar-relativistic because of their transformation properties [19]. Most important, these imply that the scalar-relativistic corrections leave spin as a good quantum number. Accordingly, it is very simple to include them in a band structure programme that is set up in a non-relativistic way - even for spin-polarized systems. However, one has to note that modifying the basic electronic Hamiltonian, one may have to adopt the expressions for operators representing physical observables [20,21]. The most prominent example is that for the Fermi-contact part of the magnetic hyperfine interaction. Inconsistent scalar-relativistic calculations indicated for example for 3d-transition metals relativistic corrections in the order of 40 % [22,23] while these amount only to about 10 % [24,25]. Another important example in this context is the electron-photon interaction operator. While this implies the selection rule $\Delta m_s = 0$ for the non-relativistic case, the corresponding relativistic corrections involve the spin-orbit coupling operator allowing for that reason for spin-flip optical transitions [21,26–28].

Compared to the scalar-relativistic corrections, accounting for the spin-orbit coupling within a band structure calculation is much more demanding because the corresponding correction operator contains explicitly the spin operator. An efficient way to include its effects within a band structure calculation, that is based on the variational principle, is to add the matrix of the corresponding operator to the Hamiltonian matrix in the secular equation [29,30]. For a paramagnetic system this will increase the time to solve the secular equation by a factor of 8 because the dimension of the matrices is doubled. This computational effort can be reduced to some extent by applying the second variational technique [31]. Because in both cases the basis functions are set up only in a non- or scalar-relativistic level, it was often argued that this procedure will lead to problems if the spin-orbit coupling is very strong. However, recently it could be demonstrated that one can apply it without problems even to compounds containing Pb, for which the spin-orbit coupling for the p-states is quite high [32]. Of course, the most reliable way to deal with all relativistic effects is to start from the Dirac equation. In fact, for more or less any of the standard band structure methods corresponding versions have been developed during the last 30 years (for example APW [33], OPW [34], KKR [35,36], ASW [37]). Dealing with paramagnetic solids these methods do not require more computer time than accounting for spin-orbit coupling in the variational step, because the dimension of the Hamiltonian matrix is just the same. In spite of this, the later approach seems to be much more popular because it allows one to discuss the final results in familiar terms.

In contrast to the scalar-relativistic corrections, the spin-orbit coupling has many far-reaching consequences for the qualitative aspects of the electronic structure. The reason for this is the lowering in symmetry that is caused by the coupling of the spin and orbital degrees of freedom, that leads – among others – to a removal of energetic degeneracies. Another consequence of spin-orbit coupling is the occurrence of physical phenomena, that cannot be described within a non- or scalar-relativistic framework. A very well-known example for this is the so-called Fano-effect [38,39], that denotes the finding that one obtains a spin-polarized photo-electron current even for a paramagnetic solid [40] if one uses circularly polarized light. Of course, the spin polarization gets just reversed if the helicity of the radiation is reversed. For a magnetic solid, however, this symmetry is broken. A direct consequence of this broken symmetry are the magneto-optical Kerr-effect [41,42] in the visible regime of light and the circular magnetic X-ray dichroism [42,43] at higher photon energies. Of course, there are many other phenomena in magnetic solids, that are caused by the common occurrence of spin-orbit coupling and spin-polarization, as for example the galvano-magnetic phenomena [44], the magneto-crystalline anisotropy [45], orbital contributions to the hyperfine fields and magnetic moments [24] or electric field gradients in cubic solids [46].

The first band structure calculations aiming to calculate spin-orbit induced properties in magnetic solids have been done by Callaway and coworkers [29,26,47]. These authors and later on many other authors [48–56] accounted for spin-orbit coupling in the variational step with the unperturbed Hamilton matrix describing a spin-polarized system. The basis functions used in this approach depend only on the orbital angular momentum quantum number l but carry no information on the spin-orbit coupling. This does not apply to the scheme suggested by the author [57] for which the role of the exchange splitting and the spin-orbit coupling are interchanged compared to the approach of Callaway and others. In particular this means that four-component basis functions are used that are obtained as solutions to the Dirac equation for a spin-averaged potential, while the spin-dependent part is accounted for in the variational step. An alternative way to include the influence of spin-orbit coupling already in the wave functions has been used by McLaren and Victora [58]. These authors adopted the formalism suggested by Koellling and Harmon [10] that works with two component wave functions in a (l, m_l, m_s)-representation and that leads to coupled sets of radial differential equations if the spin-orbit coupling term is included. A very similar scheme has been used by Akai [59], who restricted the effect of the spin-orbit coupling to within a spin subsystem keeping spin as a good quantum number that way. To avoid the numerical effort in solving the coupled set of radial differential equations, Ankudinov [60] suggested an approximate scheme that is exact for vanishing spin-orbit coupling or vanishing exchange splitting and interpolates between these two extreme cases. Because the later three schemes account for spin-orbit coupling and exchange splitting already in calculating the wave functions, they can be used straightforwardly as a starting point for multiple scattering theory.

To deal with all relativistic effects and magnetism – at least for transition metals this means first of all spin polarization – on the same level, it was suggested already in the 1970s to work on the basis of the appropriate Dirac equation. Dealing with exchange and correlation within the framework of density functional theory this leads in a rather natural way to current density functional theory (CDFT) [61] instead of the conventional spin density functional theory (SDFT) [62]. However, because of the many unsolved problems connected with this general scheme a relativistic version of spin density functional theory has been suggested [63,64]. Instead of dealing with the resulting Dirac-Hamiltonian for spin-polarized systems, Richter and Eschrig [65] suggested to use the corresponding squared Dirac Hamiltonian and developed a corresponding spin-polarized relativistic LCAO-band structure method. The first step to start from the spin-polarized relativistic Dirac equation itself has been done already 20 years ago by Doniach and Sommers, who derived the corresponding coupled radial Dirac equation [66]. The problem has later been investigated in more detail by Feder et al. [67] and Strange et al. [68]. In particular these authors could present the first numerical solution to the coupled radial equations for a single potential well. With this crucial step done, it is possible to derive for any band structure method its spin-polarized relativistic (SPR) version. This has been done for example for the KKR [67,68], the LMTO [57,69], and the ASW [70] methods.

In the case of the SPR-KKR the k-space mode based on the variational principle [71] as well as the multiple scattering mode leading directly to the Green's function [24,67,72] has been generalized accordingly. In particular the later approach (SPR-KKR-GF) was extensively used during the last 10 years. Some reasons for this are that one does not require Bloch symmetry for the investigated system and that one can link it straightforwardly to the Coherent Potential Approximation (CPA) alloy theory [73]. A major drawback of the SPR-KKR is its numerical effort required for complex systems. This problem could be overcome for many situations by the development of a TB-version

[74,75] and the use of real space cluster techniques [76–78]. Nearly all SPR-KKR-GF calculations performed so far were based on a muffin-tin or atomic sphere approximation (ASA) construction of the potential. A full-potential version of it has been worked out by various authors [79–81] and could be implemented recently in a self-consistent way [82,83]. Although in practice sometimes tedious because of the complex wave functions, it is possible to investigate more or less any property of magnetic solids using the SPR-KKR-GF formalism. In particular it is now one of the standard starting points to deal with magnetic dichroism in many kinds of electron spectroscopy as for example in X-ray absorption [84,85], in X-ray fluorescence [86], core-level XPS [87,88], valence band XPS [89], angular resolved valence band UPS [90], magnetic scattering [91], and the Faraday effect in the X-ray regime [92].

Most of the benefits supplied by the SPR-KKR-GF method can also be obtained using a corresponding version of the TB-LMTO-method [93–95]. Thus, it seems for many purposes and situations just a matter of taste which one of the band structure schemes is used. In the following the SPR-KKR-GF is described in some detail and a number of applications is used to demonstrate its great flexibility.

2 Formalism

2.1 Relativistic Density Functional Theory

When dealing with the electronic structure of magnetic solids one usually neglects the influence of orbital magnetism on it. Accordingly, corresponding band structure calculations are in general done on the basis of spin density functional theory (SDFT) as it has been derived among others by von Barth and Hedin [62] in a non-relativistic way. This framework seems still to be acceptable when relativistic effects are included by introducing corresponding corrections terms to the Schrödinger equation. If fully relativistic calculations are performed instead, in principle a corresponding basis should be adopted to deal with many-body effects. The first step in this direction has been done by Rajagopal and Callaway [61], who derived the SDFT starting from a relativistic level. These authors demonstrated in particular that quantum electrodynamics supplies the proper framework for a relativistically consistent density functional theory and derived the corresponding relativistic Kohn-Sham-Dirac equations [96,97]:

$$\left[c\boldsymbol{\alpha} \cdot \left(\frac{\hbar}{\mathrm{i}} \boldsymbol{\nabla} + \frac{e}{c} \boldsymbol{A}_{\mathrm{eff}}(\boldsymbol{r}) \right) + \beta mc^2 + V_{\mathrm{eff}}(\boldsymbol{r}) \right] \Psi_i(\boldsymbol{r}) = \epsilon_i \Psi_i(\boldsymbol{r}) \tag{1}$$

with

$$V_{\mathrm{eff}}(\boldsymbol{r}) = -e \left[A_{\mathrm{ext}}^0(\boldsymbol{r}) + \frac{1}{c} \int d^3 r' \frac{J^0(\boldsymbol{r}')}{|\boldsymbol{r} - \boldsymbol{r}'|} + c \frac{\partial E_{\mathrm{xc}}[J^\mu]}{\partial J^0(\boldsymbol{r})} \right] \tag{2}$$

$$\boldsymbol{A}_{\mathrm{eff}}(\boldsymbol{r}) = -e \left[\boldsymbol{A}_{\mathrm{ext}}(\boldsymbol{r}) + \frac{1}{c} \int d^3 r' \frac{\boldsymbol{J}(\boldsymbol{r}')}{|\boldsymbol{r} - \boldsymbol{r}'|} + c \frac{\partial E_{\mathrm{xc}}[J^\mu]}{\partial \boldsymbol{J}(\boldsymbol{r})} \right] . \tag{3}$$

Here the $\Psi_i(\boldsymbol{r})$ are four-component wave functions (see below) with corresponding single particle energies ϵ_i. The matrices α_i and β are the standard 4×4-Dirac matrices [7]. The effective scalar and vector potentials, V_{eff} and $\boldsymbol{A}_{\mathrm{eff}}$, respectively, contain as a first term the corresponding external contributions. The second terms in Eqs. (2) and (3) are the familiar Hartree potential and the vector potential due to the Breit-interaction, respectively. Finally, the third terms are caused by exchange and correlation with the corresponding exchange and correlation energy $E_{\mathrm{xc}}[J^\mu]$ being a functional of the electronic four-current J^μ. This central quantity that determines all properties of the system is given by:

$$J^0 = -ec \sum_i \Psi_i^\dagger \Psi_i \tag{4}$$

$$J^\mu = -ec \sum_i \Psi_i^\dagger \beta \alpha^\mu \Psi_i , \tag{5}$$

where J^0/c is identical with the familiar electronic charge density ρ, while the other components J^μ give the spatial components of the electronic current density $\boldsymbol{j}$.

Thus, in contrast to non-relativistic SDFT, where the central quantities are the spin densities $n^{\uparrow(\downarrow)}$ or equivalently the particle density n and spin magnetization density $\boldsymbol{m}$, the relativistic formalism leads in a natural way to a current density functional theory (CDFT). Because of the problems connected with this very general scheme an approximate relativistic version of SDFT has been worked out by several authors [61,63,98–100]. The first step in this direction is the Gordon decomposition of the spatial current density into its orbital and spin parts [61,97,101]:

$$\boldsymbol{j}_{\mathrm{orb}} = \frac{1}{2m} \Psi^\dagger \beta \left[\frac{1}{i} \overset{\leftarrow}{\boldsymbol{\nabla}} - \frac{1}{i} \boldsymbol{\nabla} + 2e\boldsymbol{A} \right] \Psi + \frac{1}{2m} \boldsymbol{\nabla} \times \Psi^\dagger \beta \boldsymbol{\sigma} \Psi \tag{6}$$

where $\boldsymbol{\sigma}$ is the vector of 4×4-Pauli matrices [7]. The coupling of the spin part $\boldsymbol{j}_{\mathrm{spin}}$ (the second term in Eq. (6)) to the vector potential $\boldsymbol{A}_{\mathrm{eff}}$ may alternatively be described by introducing the corresponding spin magnetization density

$$\boldsymbol{m} = -\mu_{\mathrm{B}} \sum_i \Psi_i^\dagger \beta \boldsymbol{\sigma} \Psi_i . \tag{7}$$

This leads to the coupling term

$$-\boldsymbol{m} \cdot \boldsymbol{B}_{\mathrm{eff}} , \tag{8}$$

with $\boldsymbol{B}_{\mathrm{eff}}$ the effective magnetic field corresponding to $\boldsymbol{A}_{\mathrm{eff}}$. Thus, ignoring the orbital current density contribution $\boldsymbol{j}_{\mathrm{orb}}$ one arrives at a Kohn-Sham-Dirac equation completely analogous to the non-relativistic SDFT Schrödinger equation [97,101]:

$$\left[\frac{\hbar}{i} c\boldsymbol{\alpha} \cdot \boldsymbol{\nabla} + \beta mc^2 + V_{\mathrm{eff}}(\boldsymbol{r}) + \beta \boldsymbol{\sigma} \cdot \boldsymbol{B}_{\mathrm{eff}}(\boldsymbol{r}) \right] \Psi_i(\boldsymbol{r}) = \epsilon_i \Psi_i(\boldsymbol{r}) \tag{9}$$

with

$$B_{\text{eff}}(r) = B_{\text{ext}}(r) + \frac{\partial E_{\text{xc}}[n, m]}{\partial m(r)} . \tag{10}$$

This approach has been suggested among others by MacDonald and Vosko [63,98], who justified this simplification by introducing a fictitious magnetic field that couples only to the spin degree of freedom as reflected by Eq. (9). This formal justification has been criticized by Xu et al. [102] because describing a relativistic electronic system in terms of the particle density n and spin magnetization density m alone the magnetic interaction part connected with the electronic current density is not Lorentz invariant. This problem could be circumvented by Rajagopal and coworkers [100,102,103] by considering first the problem for the rest frame of an electron – for which j_{orb} vanishes – giving a consistent justification for the use of relativistic SDFT.

The orbital current density contribution to E_{xc} – ignored within SDFT – has first been considered by Vignale and Rasolt on a non-relativistic level [104–107]. As one of the central quantities these authors introduce the paramagnetic orbital current density $j_{\text{orb,p}}$ (see below). Because of the restrictions caused by the demand for gauge invariance this is replaced then by the so-called vorticity:

$$\nu = \nabla \times \frac{j_{\text{orb,p}}(r)}{n(r)} . \tag{11}$$

This step in particular allows to derive a local version of non-relativistic CDFT. A corresponding explicit expression for the corresponding E_{xc} has been given for the first time by Vignale and Rasolt [105]:

$$E_{\text{xc}}[n, \nu] = E_{\text{xc}}[n, 0] + \int dx \left(\frac{9\pi}{4}\right)^{1/3} \frac{1}{24\pi^2 r_{\text{s}}} \left(\frac{\chi_{\text{L}}}{\chi_{\text{L}}^0} - 1\right) |\nu(x)|^2 \tag{12}$$

where $r_{\text{s}} = (\frac{3}{4\pi n})^{1/3}$ and

$$\frac{\chi_{\text{L}}}{\chi_{\text{L}}^0} = 1 + 0.02764 r_{\text{s}} \ln r_{\text{s}} + 0.01407 r_{\text{s}} + O(r_{\text{s}}^2 \ln r_{\text{s}}) \tag{13}$$

is the ratio of the diamagnetic susceptibility for the interacting and non-interacting electron gas. Later, more sophisticated expressions for E_{xc} have been given [108].

The Vignale-Rasolt CDFT-formalism can be obtained as the weakly relativistic limit of the fully relativistic SDFT-Dirac equation (1). This property has been exploited to set up a computational scheme that works in the framework of non-relativistic CDFT and accounts for the spin-orbit coupling at the same time [109]. This hybrid scheme deals with the kinematic part of the problem in a fully relativistic way whereas the exchange-correlation potential terms are treated consistently to first order in $1/c$. In particular, the corresponding modified Dirac equation

$$\left[\frac{\hbar}{i} c\alpha \cdot \nabla + \beta mc^2 + V_{\text{eff}}(r) + \beta\sigma \cdot B_{\text{eff}}(r) + \sum_{\sigma} \beta H_{\text{op},\sigma} P_{\sigma}\right] \Psi_i(r) = \epsilon_i \Psi_i(r) \tag{14}$$

incorporates a term

$$H_{\text{op},\sigma} = -\frac{i\hbar e}{2mc} [A_{\text{xc},\sigma}(r), \nabla]_+ \ , \tag{15}$$

that explicitly represents the coupling of the orbital current and the exchange-correlation vector potential $A_{\text{xc},\sigma}$. Since in the CDFT-formalism of Vignale and Rasolt $A_{\text{xc},\sigma}$ is defined in a spin-dependent way, the spin-projection operator $P_\sigma = \frac{1\pm\beta\sigma_z}{2}$ appears in addition to $H_{\text{op},\sigma}$ in Eq. (14).

Within the above approximate relativistic CDFT scheme the Breit-interaction has been ignored. This radiative correction accounts for the retardation of the Coulomb-interaction and exchange of transversal photons. A more complete version than that included in Eq. (1) is given by the Hamiltonian [2,110]:

$$\mathcal{H}_{\text{Breit}} = \frac{-e^2}{2R}\alpha_1 \cdot \alpha_2 + \frac{e^2}{2R}\left[\alpha_1 \cdot \alpha_2 - \alpha_1 \cdot \hat{R}\alpha_2 \cdot \hat{R}\right] \quad \text{with } R = r_2 - r_1 \ , \tag{16}$$

where the first part is the magnetic Gaunt part and the second one is the retardation term. While inclusion of the Breit-interaction within quantum-chemical calculations for atoms and molecules is nearly standard [2], so far only one model [111] and one fully relativistic [112] calculation have been done in the case of solids. This is quite astonishing, because the Breit-interaction gives rise to the so-called shape anisotropy, that contributes in general to the magneto-crystalline anisotropy energy to the same order of magnitude as the spin-orbit coupling (see below).

An alternative to the CDFT approach is the heuristic suggestion by Brooks and coworkers [113–115] to use a k-space method and to add a so-called orbital polarization (OP) term to the Hamilton matrix. This additional term has been borrowed from atomic theory and is meant to account for Hund's second rule, i.e. to maximize the orbital angular momentum. During the last years, this approach has been applied with remarkable success to d- as well as f-electron systems and has been refined by various authors [116,117]. As it could be shown [118], Brooks' OP-term can be formulated in a way that can be incorporated into the Dirac equation allowing that way for a corresponding extension of band structure methods based on multiple scattering theory [118]. For a d-electron system, i.e. for the case that orbital magnetism is primarily due to an open d-electron shell, Brooks' OP-term takes the form $-B_{m_s}\langle \hat{l}_z\rangle_{m_s} m_l \delta_{l2}$. This term describes a shift in energy for an orbital with quantum numbers $l = 2$, m_l and m_s that is proportional to the average orbital angular momentum $\langle l_z\rangle_{m_s}$ for the m_s-spin subsystem and the so-called Racah parameters B_{m_s} [119] that in turn can be represented by the Coulomb integrals $F^2_{m_s}$ and $F^4_{m_s}$. An operator that corresponds to this energy shift is given by

$$\mathcal{H}^{\text{OP}}_{m_s} = -B^{\text{OP}}_{m_s}(r)\langle \hat{l}_z\rangle_{m_s}\hat{l}_z \delta_{l2} \ , \tag{17}$$

with

$$B^{\text{OP}}_{m_s}(r) = \frac{2}{441} \int [9\frac{r^2_<}{r^3_>} - 5\frac{r^4_<}{r^5_>}] \rho_{\text{d}m_s}(r')4\pi r'^2 \, dr' \ , \tag{18}$$

where ρ_{dm_s} describes the average charge density of a d-electron with spin character m_s. Obviously the operator $\mathcal{H}_{m_s}^{OP}$ has the form expected within CDFT for rotational symmetry [105]. This is emphasized by introducing the vector potential function $A_{m_s}^{OP} = -B_{m_s}^{OP}(r)\langle l_z \rangle$ that leads to the Dirac equation:

$$\left[\frac{\hbar}{i} c\boldsymbol{\alpha} \cdot \boldsymbol{\nabla} + \beta mc^2 + V_{\text{eff}}(\boldsymbol{r}) + \beta\boldsymbol{\sigma} \cdot \boldsymbol{B}_{\text{eff}}(\boldsymbol{r}) + A^{OP}\beta l_z \right] \Psi_i(\boldsymbol{r}) = \epsilon_i \Psi_i(\boldsymbol{r}) . \quad (19)$$

For a further discussion of the connection of this equation with CDFT see below.

In addition to the OP-formalism several alternative schemes have been suggested in the past to account within a relativistic band structure calculation for correlation effects not incorporated within the local approximation to SDFT (LSDA). For example the LDA+U-scheme has been applied to the compound CeSb [56], a system that has a maximum Kerr-rotation angle of 90° [120]. Similar experience has been made for other f-electron systems. Nevertheless, one should point out that by applying the LDA+U-scheme one leaves the framework of DFT. This does not apply to the SIC (self-interaction correction) formalism [96], for which a proper relativistic formulation has been worked out recently [121,122] and applied to magnetic solids [121].

From the above presentation it is obvious that relativistic effects influence the electronic structure in a twofold way. On the one hand side, one has the influence on the electronic kinetics, that is accounted for by working with the Dirac-formalism. On the other hand, relativity influences the electron-electron interaction via the retardation effect, the Breit-interaction and so on, leading to quite pronounced corrections for the exchange and correlation energy E_{xc} compared to its non-relativistic counterpart. This has been studied in detail for the paramagnetic and the spin-polarized case for example by MacDonald and Vosko [63,98], Rajagopal and coworkers [100,102,103] and Engel and coworkers [123]. Until now, however, only very few investigations have been done on the importance of these corrections [19,124–126]. Nevertheless, one may conclude from these few studies that the absolute magnitude of total energies as well as the binding energies of tightly bound core states is affected in a rather appreciable way. However, for properties like the equilibrium lattice parameter or even for magnetic properties no pronounced changes have to be expected. For this reason, the use of parameterizations derived within non-relativistic SDFT seems to be well justified.

2.2 Multiple scattering formalism

Solution of the Single Site Dirac Equation The first step to solve one of the above Dirac-equations for a solid using multiple scattering formalism is to find the solutions to the corresponding Dirac-equation for an isolated potential well. For that purpose Strange et al. [68] investigated the associated Lippmann-Schwinger-equation and derived a set of radial differential equations for the single-site solutions. An alternative scheme has been used by other authors [66,67] dealing with the problem by writing as a first step the single-site Dirac equation in spherical coordinates [7]. For the spin-polarized (SDFT) case this leads to (for simplicity it is assumed in the following that one has $\boldsymbol{B}(\boldsymbol{r}) = B(r)\hat{e}_z$ and atomic Rydberg units will be used throughout):

$$\left[i\gamma_5\sigma_r c \left(\frac{\partial}{\partial r} + \frac{1}{r} - \frac{\beta}{r}\hat{K} \right) + V + \beta\sigma_z B + (\beta - 1)\frac{c^2}{2} - E \right] \psi_\nu = 0 . \tag{20}$$

Here the spin-orbit operator $\hat{K}$ is defined by

$$\hat{K} = \beta(\boldsymbol{\sigma} \cdot \boldsymbol{l} + 1) \tag{21}$$

and the matrices γ_5 and σ_r are given by:

$$\gamma_5 = \begin{pmatrix} 0 & -I_2 \\ -I_2 & 0 \end{pmatrix} \tag{22}$$

and

$$\sigma_r = \hat{\boldsymbol{r}} \cdot \boldsymbol{\sigma} , \tag{23}$$

with $\boldsymbol{\sigma}$ the vector of the 4×4 Pauli matrices [7]. To find solutions to Eq. (20) one makes the ansatz:

$$\psi_\nu = \sum_\Lambda \psi_{\Lambda\nu} , \tag{24}$$

where the partial waves $\psi_{\Lambda\nu}$ are chosen to have the same form as the linearly independent solutions for a spherical symmetric potential:

$$\psi_\Lambda(\boldsymbol{r}, E) = \begin{pmatrix} g_\kappa(r, E) \chi_\Lambda(\hat{\boldsymbol{r}}) \\ i f_\kappa(r, E) \chi_{-\Lambda}(\hat{\boldsymbol{r}}) \end{pmatrix} . \tag{25}$$

Here the large and small components are composed of the radial wave functions $g_\kappa(r, E)$ and $f_\kappa(r, E)$ and the spin-angular functions:

$$\chi_\Lambda(\hat{\boldsymbol{r}}) = \sum_{m_s=\pm 1/2} C(l\tfrac{1}{2}j; \mu - m_s, m_s) Y_l^{\mu-m_s}(\hat{\boldsymbol{r}}) \chi_{m_s} , \tag{26}$$

with the Clebsch-Gordon coefficients $C(l\tfrac{1}{2}j; m_l, m_s)$, the complex spherical harmonics $Y_l^{m_l}$ and the Pauli-spinors χ_{m_s}. The spin-orbit and magnetic quantum numbers κ and μ, respectively, have been combined to $\Lambda = (\kappa, \mu)$ and $-\Lambda = (-\kappa, \mu)$, respectively. The spin-angular functions $\chi_\Lambda(\hat{\boldsymbol{r}})$ are simultaneous eigenfunctions of the operators $\boldsymbol{j}^2$, j_z and $\hat{K}$ with $\boldsymbol{j} = \boldsymbol{l} + \tfrac{1}{2}\boldsymbol{\sigma}$. The corresponding eigenvalues are $j(j + 1)$, μ and $-\kappa$ and are connected by the following relations:

$$\kappa = \begin{cases} -l - 1 & \text{for } j = l + 1/2 \\ +l & \text{for } j = l - 1/2 \end{cases} \tag{27}$$

$$j = |\kappa| - 1/2 \tag{28}$$

$$-j \le \mu \le +j \tag{29}$$

$$\bar{l} = l - S_\kappa , \tag{30}$$

where $S_\kappa = \kappa/|\kappa|$ is the sign of κ and $\bar{l}$ is the orbital angular momentum quantum number belonging to $\chi_{-\Lambda}$.

Inserting the ansatz in Eq. (24) into the single-site Dirac-equation (20) and integrating over the angles leads to the following set of radial differential equations:

$$P'_{\Lambda\nu} = -\frac{\kappa}{r}P_{\Lambda\nu} + \left[\frac{E-V}{c^2} + 1\right]Q_{\Lambda\nu} + \frac{B}{c^2}\sum_{\Lambda'}\langle\chi_{-\Lambda}|\sigma_z|\chi_{-\Lambda'}\rangle Q_{\Lambda'\nu} \qquad (31)$$

$$Q'_{\Lambda\nu} = \frac{\kappa}{r}Q_{\Lambda\nu} - [E-V]P_{\Lambda\nu} + B\sum_{\Lambda'}\langle\chi_\Lambda|\sigma_z|\chi_{\Lambda'}\rangle P_{\Lambda'\nu}, \qquad (32)$$

where the usual notation $P_{\Lambda\nu} = rg_{\Lambda\nu}$ and $Q_{\Lambda\nu} = crf_{\Lambda\nu}$ has been used. The coupling coefficients occurring here are given by:

$$\langle\chi_\Lambda|\sigma_z|\chi_{\Lambda'}\rangle = G(\kappa, \kappa', \mu)\,\delta_{\mu\mu'} \qquad (33)$$

$$= \delta_{\mu\mu'}\begin{cases} -\frac{\mu}{(\kappa+1/2)} & \text{for} \qquad \kappa = \kappa' \\ -\sqrt{1-(\frac{\mu}{\kappa+1/2})^2} & \text{for} \qquad \kappa = -\kappa'-1 \\ 0 & \text{otherwise} \end{cases} \qquad (34)$$

Because of the properties of $G(\kappa, \kappa', \mu)$ only partial waves for the same μ get coupled; i.e. μ is still a good quantum number. In addition, one can see that for the orbital angular momentum quantum numbers l and l' of two coupled partial waves one has the restriction $l - l' = 0, \pm 2, \dots$, i.e. only waves of the same parity are coupled. Nevertheless, this still implies that an infinite number of partial waves are coupled. In practice, however, all coupling terms for which $l - l' = \pm 2$ are ignored. A justification for this restriction has been given by Feder et al. [67] and Cortona et al. [127]. Results of numerical studies furthermore justify this simplification [128,129]. Altogether, this restricts the number of terms in Eqs. (31) and (32) to 2 if $|\mu| < j$. For the case $\mu = j$, there is no coupling at all; i.e. the solutions ψ_ν have pure spin-angular character Λ.

The procedure sketched above to derive, starting from the SDFT-Dirac equation, the corresponding set of coupled radial differential equations can be used straightforwardly for more complex situations. In the case of the formally rather simple Brooks' OP-formalism one has the additional term $\mathcal{H}_{m_s}^{OP}$ (see Eq. (17)) leading to the radial equations:

$$P'_{\Lambda\nu} = -\frac{\kappa}{r}P_{\Lambda\nu} + \left[\frac{E-V}{c^2} + 1\right]Q_{\Lambda\nu} + \frac{B}{c^2}\sum_{\Lambda'}\langle\chi_{-\Lambda}|\sigma_z|\chi_{-\Lambda'}\rangle Q_{\Lambda'\nu} \qquad (35)$$

$$-\frac{1}{c^2}\sum_{\Lambda'}\langle\chi_{-\Lambda}|A^{OP}l_z|\chi_{-\Lambda'}\rangle Q_{\Lambda'}$$

$$Q'_{\Lambda\nu} = \frac{\kappa}{r}Q_{\Lambda\nu} - [E-V]P_{\Lambda\nu} + B\sum_{\Lambda'}\langle\chi_\Lambda|\sigma_z|\chi_{\Lambda'}\rangle P_{\Lambda'\nu} \qquad (36)$$

$$-\sum_{\Lambda'}\langle\chi_\Lambda|A^{OP}l_z|\chi_{\Lambda'}\rangle P_{\Lambda'}.$$

Formally, the term $\mathcal{H}_{m_s}^{\mathrm{OP}}$ can be seen to represent the coupling of the electronic orbital current to a vector potential $\boldsymbol{A}$. A more general form of such a vector potential is encountered for example when one is including the Breit interaction in the Dirac equation. To deal with such a situation, it is most convenient to represent the vectors within the scalar product $\boldsymbol{\alpha} \cdot \boldsymbol{A}$ (see Eq. (1)) using spherical coordinates and to expand the spatial dependency of the components of the vector potential $\boldsymbol{A}$ in terms of complex spherical harmonics:

$$\boldsymbol{\alpha} \cdot \boldsymbol{A} = \sum_m \alpha_m i \sum_{LM} A_{LM}^m(r) Y_L^{-m}(\hat{r}) . \tag{37}$$

When added to the SDFT-Dirac-Hamiltonian this leads to the following radial equations [112]:

$$P'_{\Lambda\nu} = -\frac{\kappa}{r} P_{\Lambda\nu} + \left[\frac{E - V}{c^2} + 1\right] Q_{\Lambda\nu} + \frac{B}{c^2} \sum_{\Lambda'} \langle \chi_{-\Lambda} | \sigma_z | \chi_{-\Lambda'} \rangle Q_{\Lambda'\nu} \tag{38}$$

$$-\frac{1}{c} \sum_{\Lambda'} \sum_{LMm} A_{LM}^m \langle \chi_{-\Lambda} | Y_L^M \sigma_m | \chi_{\Lambda'} \rangle P_{\Lambda'}$$

$$Q'_{\Lambda\nu} = \frac{\kappa}{r} Q_{\Lambda\nu} - [E - V] P_{\Lambda\nu} + B \sum_{\Lambda'} \langle \chi_{\Lambda} | \sigma_z | \chi_{\Lambda'} \rangle P_{\Lambda'\nu} \tag{39}$$

$$-\frac{1}{c} \sum_{\Lambda'} \sum_{LMm} A_{LM}^m \langle \chi_{\Lambda} | Y_L^M \sigma_m | \chi_{-\Lambda'} \rangle Q_{\Lambda'} .$$

The Dirac equation Eq. (14) set up within the framework of CDFT can be treated in a completely analogous way resulting in a similar set of radial differential equations for large and small component wave functions.

Because of the high symmetry of the orbital polarization Hamiltonian $\mathcal{H}_{m_s}^{\mathrm{OP}}$ no coupling between partial waves beyond that caused by the spin-dependent part of the SDFT-Dirac-Hamiltonian is introduced. This holds also if a vector potential term is added to the Hamiltonian that stems from the Breit-interaction or from the CDFT-formalism, as long as one imposes for this rotational symmetry, with the symmetry axes coinciding with the direction of the magnetization.

The problem of deriving a full-potential (FP) version of the KKR-formalism has been discussed in a rather controversial way during the last decades. Now, it is generally accepted that the scheme proposed among others by Dederichs, Zeller and coworkers [130,131] supplies a sound basis for FP-KKR band structure calculations. This implies that in a first step space is subdivided into non-overlapping, space-filling polyhedra usually obtained by means of the Wigner-Seitz-construction. The shape of these Wigner-Seitz cells is represented by the so called shape functions $\Theta_L(r)$ [132] with

$$\Theta_{\mathrm{WS}}(\boldsymbol{r}) = \sum_L \theta_L(r) \mathcal{Y}_L(\hat{r}) , \tag{40}$$

where the step function Θ_{WS} is 1 for $\boldsymbol{r}$ within the cell and 0 otherwise. The functions $\mathcal{Y}_L$ are real spherical harmonics with L standing for (l, m). In addition one defines the radius r_{cr} of the smallest circumscribed sphere, for which $\Theta_{WS}(\boldsymbol{r}) = 0$ for $r > r_{cr}$. Multiplying the potential V of the extended system in Eq. (9) with the function $\Theta_{WS}(\boldsymbol{r})$ centered at an atomic site n leads to the single site problem. To solve the corresponding single site Dirac equation the same ansatz as given in Eqs. (24) and (25) is made. This leads now to the coupled radial Dirac equations:

$$P'_\Lambda = -\frac{\kappa}{r} P_\Lambda + \left[\frac{E}{c^2} + 1\right] Q_\Lambda - \frac{1}{c^2} \sum_{\Lambda'} V^-_{\Lambda\Lambda'} Q_{\Lambda'} \tag{41}$$

$$Q'_\Lambda = \frac{\kappa}{r} Q_\Lambda - E P_\Lambda + \sum_{\Lambda'} V^+_{\Lambda\Lambda'} P_{\Lambda'} \ . \tag{42}$$

Here the underlying Dirac-Hamiltonian has been restricted to the SDFT-form with the corresponding potential matrix elements $V^\pm_{\Lambda\Lambda'}$ defined by

$$V^\pm_{\Lambda\Lambda'}(r) = \langle \chi_{\pm\Lambda} | V_{\text{eff}} \pm \boldsymbol{\sigma} B | \chi_{\pm\Lambda'} \rangle \ . \tag{43}$$

These are straightforwardly evaluated by expanding the potential into real spherical harmonics:

$$V(\boldsymbol{r}) \quad = \sum_L V_L(r) \mathcal{Y}_L(\hat{\boldsymbol{r}}) \tag{44}$$

$$B(\boldsymbol{r}) \quad = \sum_L B_L(r) \mathcal{Y}_L(\hat{\boldsymbol{r}}) \tag{45}$$

with $\boldsymbol{B}(\boldsymbol{r}) = B(\boldsymbol{r})\hat{\boldsymbol{B}}$.

Here it has been assumed that $\boldsymbol{B}(\boldsymbol{r})$ points everywhere along the same direction $\hat{\boldsymbol{B}}$. In the following applications $\hat{\boldsymbol{B}}$ will be oriented along the z-axis. However, these are no necessary restrictions for the formalism; i.e. treatment of other orientations or non-collinear magnetic states can be straightforwardly accounted for.

Compared to a muffin-tin or atomic sphere approximation potential construction inclusion of non-spherical terms in V and B obviously leads to further coupling. In practice, however, the number of coupled partial waves is restricted to $2(l_{\max} + 1)^2$ by fixing an upper limit $l_{\max}$ for the angular momentum expansion of the wave function in Eq. (24). For example, for $l_{\max} = 2$ one may have up to 18 partial waves coupled; i.e., one has to solve up to 36 coupled equations for the functions P_Λ and Q_Λ. However, for a cubic system with $\hat{\boldsymbol{B}} = \hat{\boldsymbol{z}}$ and $l_{\max} = 2$ one has at most 3 partial waves coupled due to the high symmetry of the system.

Single-site t-matrix and Normalization of the Wave Functions Working with one of the above versions of the Dirac equation and using the corresponding coupled radial differential equations a set of $2(l_{\max} + 1)^2$ linearly independent regular solutions Φ_Λ can be created by initializing the outward integration with a selected spin-angular character Λ dominating close to the nucleus; i.e. one demands that:

$$\Phi_\Lambda(\boldsymbol{r}, E) = \sum_{\Lambda'} \Phi_{\Lambda'\Lambda}(\boldsymbol{r}, E) \xrightarrow{r\to 0} \Phi_{\Lambda\Lambda}(\boldsymbol{r}, E) . \tag{46}$$

After having solved all systems of coupled equations for the wave functions Φ_Λ one gets the corresponding single site t-matrix by introducing the auxiliary matrices a and b [133–135]:

$$a_{\Lambda\Lambda'}(E) = -ipr^2[h_\Lambda^-(pr), \Phi_{\Lambda\Lambda'}(\boldsymbol{r}, E)]_r \tag{47}$$
$$b_{\Lambda\Lambda'}(E) = ipr^2[h_\Lambda^+(pr), \Phi_{\Lambda\Lambda'}(\boldsymbol{r}, E)]_r . \tag{48}$$

Here $p = \sqrt{E(1 + E/c^2)}$ is the relativistic momentum [7] and $[\dots]_r$ denotes the relativistic form of the Wronskian [135]:

$$[h_\Lambda^+, \phi_{\Lambda\Lambda'}]_r = h_l^+ c f_{\Lambda\Lambda'} - \frac{p}{1 + E/c^2} S_\kappa h_{\bar{l}}^+ g_{\Lambda\Lambda'} . \tag{49}$$

The functions $h_\Lambda^\pm$ are the relativistic version of the Hankel functions of the first and second kind [7]:

$$h_\Lambda^\pm(\boldsymbol{r}) = \sqrt{\frac{1 + E/c^2}{c^2}} \begin{pmatrix} h_l^\pm(pr)\chi_\Lambda(\hat{r}) \\ \frac{ipcS_\kappa}{E+c^2} h_{\bar{l}}^\pm(pr)\chi_{\bar{\Lambda}}(\hat{r}) \end{pmatrix} . \tag{50}$$

Evaluating all functions in Eqs. (47) – (48) at $r_b = r_{mt}$, r_{ws} or r_{cr}, resp., i.e. the muffin-tin, the Wigner-Seitz or the critical radius depending on whether one is using the muffin-tin, the ASA- or full-potential mode, one finally has [135]:

$$t(E) = \frac{i}{2p}(a(E) - b(E))b^{-1}(E) . \tag{51}$$

By a superposition of the wave functions Φ_Λ according to the boundary conditions

$$Z_\Lambda(\boldsymbol{r}, E) = \sum_{\Lambda'} Z_{\Lambda'\Lambda}(\boldsymbol{r}, E)$$
$$\xrightarrow{r \geq r_b} \sum_{\Lambda'} j_{\Lambda'}(\boldsymbol{r}, E)t(E)_{\Lambda'\Lambda}^{-1} - iph_\Lambda^+(\boldsymbol{r}, E) \tag{52}$$

one gets an alternative set of linearly independent regular solutions Z_Λ to the single site Dirac equation. These functions are normalized in analogy to non-relativistic multiple scattering theory according to the convention of Faulkner and Stocks [10] and allow straightforwardly to set up the electronic Green's function (see below). The additionally needed irregular solutions J_Λ are fixed by the boundary condition

$$J_\Lambda(\boldsymbol{r}, E) \xrightarrow{r \to r_b} j_\Lambda(\boldsymbol{r}, E) \tag{53}$$

and are obtained just by inward integration. The functions j_Λ occurring in Eqs. (52) – (53) are the relativistic counterpart to the spherical Bessel functions defined in analogy to Eq. (50) for $h_\Lambda^\pm$ [7].

Manipulation of the SDFT-Dirac-Hamiltonian Dealing with relativistic effects by adding corresponding corrections to the Schrödinger equation allows one straightforwardly to investigate the role of these corrections individually. For the fully relativistic Dirac formalism sketched above this is obviously not the case. The only way to monitor the importance of all relativistic effects in a consistent way is to manipulate them simultaneously by scaling the speed of light c. To allow in spite of this for a separate investigation of the role of scalar-relativistic effects and the spin-orbit coupling an elimination scheme has been applied recently to the SDFT-Dirac Hamiltonian.

Using this procedure an equation for the large component wave function $\phi(r, E)$ can be introduced that is still exact [137,138]:

$$\left[-\frac{1}{r^2}\frac{\partial}{\partial r}r^2\frac{\partial}{\partial r} + \frac{\hat{l}^2}{r^2} - S_\Lambda T + S_\Lambda B\sigma_z \right.$$
$$\left. +\frac{S'_\Lambda}{S_\Lambda}\left(\frac{d}{dr} - \frac{1}{r} - \frac{\hat{K}-1}{r} \right) \right] \phi(r, E) = 0 . \tag{54}$$

Here the abbreviations

$$T = E - V \tag{55}$$

$$S_\Lambda = \frac{E-V}{c^2} + 1 + \frac{B}{c^2}\langle\chi_{-\Lambda}|\sigma_z|\chi_{-\Lambda}\rangle \tag{56}$$

have been used, where S_Λ would be identical to 1 in the non-relativistic limit which is obtained for $c \to \infty$.

For the wave function $\phi(r, E)$ the ansatz

$$\phi(r, E) = \sum_\Lambda \phi_\Lambda(r, E) = \sum_\Lambda g_\Lambda(r, E)\chi_\Lambda(\hat{r}) , \tag{57}$$

is made in accordance with the adopted Λ-representation. Inserting this ansatz into the wave equation (54) leads to the following second order radial differential equation:

$$P''_\Lambda = \frac{l(l+1)}{r^2} P_\Lambda - S_\Lambda T P_\Lambda + S_\Lambda \sum_{\Lambda'} B_{\Lambda\Lambda'} P_{\Lambda'}$$
$$+ \frac{S'_\Lambda}{S_\Lambda}\left[\frac{d}{dr} - \frac{1}{r} \right] P_\Lambda + \frac{S'_\Lambda}{S_\Lambda}\frac{1}{r}\sum_{\Lambda'} \xi_{\Lambda\Lambda'} P_{\Lambda'} \tag{58}$$

with the spin-orbit coupling operator $\hat{K}$ replaced using the operator

$$\hat{\xi} = \hat{K} - 1 = \boldsymbol{\sigma} \cdot \boldsymbol{l} . \tag{59}$$

Inserting the proper values for the corresponding angular matrix elements

$$\xi_{\Lambda\Lambda'} = \langle\chi_\Lambda|\hat{K} - 1|\chi_{\Lambda'}\rangle = (-\kappa - 1)\delta_{\Lambda\Lambda'} , \tag{60}$$

the exact second order differential equation for the major component is recovered.

Replacing the spin-orbit coupling operator $\hat{K}$ in Eqs. (59) and (60) by its scaled counterpart

$$\hat{K}_x = 1 + x\,\boldsymbol{\sigma}\cdot\boldsymbol{l} \tag{61}$$

with the associated effective spin-orbit quantum number

$$\kappa_x = -1 + x\,(1+\kappa)\,, \tag{62}$$

allows one to scale the strength of the spin-orbit coupling separately. Obviously, setting $x = 1$ nothing changes at all, while for $x = 0$ one gets $\kappa_x = -1$. This is just the value of the spin-orbit quantum number for s-states where there is no spin-orbit coupling. Therefore, replacing κ in Eq. (60) by κ_x switches the spin-orbit coupling off for any partial wave if $x = 0$ and reduces or increases the spin-orbit coupling strength else.

To solve the resulting second order differential equation Eq. (58) for the wave functions $P_\Lambda(r, E)$ the auxiliary function $Q_\Lambda(r, E)$ is introduced by the definition

$$Q_\Lambda = \left[P'_\Lambda + \frac{\kappa_x}{r}\,P_\Lambda \right] \frac{1}{S_\Lambda}\,. \tag{63}$$

This leads after some simple transformations to a coupled set of first order differential equations:

$$P'_\Lambda = -\frac{\kappa_x}{r}\,P_\Lambda + S_\Lambda Q_\Lambda \tag{64}$$

$$Q'_\Lambda = \frac{\kappa_x}{r}\,Q_\Lambda - T P_\Lambda + \sum_{\Lambda'} B_{\Lambda\Lambda'}\,P_{\Lambda'} + \frac{l\,(l+1) - \kappa_x(\kappa_x + 1)}{r^2}\,\frac{1}{S_\Lambda}P_\Lambda\,. \tag{65}$$

Apart from allowing one to manipulate the strength of the spin-orbit coupling, Eq. (58) also permits to modify the form of the spin-orbit coupling operator. For this purpose $\hat{\xi}$ is splited according to

$$\begin{aligned}
\hat{\xi} &= \boldsymbol{\sigma}\cdot\boldsymbol{l} \\
&= \sigma_z l_z + (\sigma_x l_x + \sigma_y l_y) \\
&= \hat{\xi}_{zz} + \hat{\xi}_{xy}
\end{aligned} \tag{66}$$

into two parts. The first term, $\hat{\xi}_{zz}$, gives rise only to a splitting of levels with different quantum numbers m_l. Because no mixing of states with different spin character is introduced that way, m_s is left as a good quantum number. In contrast to this, the second term, $\hat{\xi}_{xy}$, gives rise to a hybridization of different spin states while no obvious splitting of m_l-levels is caused by it. Because the two parts of $\hat{\xi}$ have quite different consequences it is interesting to investigate their effect separately by replacing $\hat{\xi}$ in Eq. (58) either by $\hat{\xi}_{zz}$ or by $\hat{\xi}_{xy}$. The corresponding angular matrix elements to be inserted are

$$\langle \chi_\Lambda | \hat{\xi}_{zz} | \chi_{\Lambda'} \rangle = \delta_{ll'} \delta_{\mu\mu'} \begin{cases} -\mu \sqrt{1 - \left(\frac{2\mu}{2l+1}\right)^2} & \text{for } \kappa \neq \kappa' \\ -S_\kappa \frac{2\mu^2}{2l+1} - \frac{1}{2} & \text{for } \kappa = \kappa' \end{cases} \tag{67}$$

$$\langle \chi_\Lambda | \hat{\xi}_{xy} | \chi_{\Lambda'} \rangle = \delta_{ll'} \delta_{\mu\mu'} \begin{cases} +\mu \sqrt{1 - \left(\frac{2\mu}{2l+1}\right)^2} & \text{for } \kappa \neq \kappa' \\ -\kappa + S_\kappa \frac{2\mu^2}{2l+1} - \frac{1}{2} & \text{for } \kappa = \kappa' . \end{cases} \tag{68}$$

A solution of the resulting second order differential equation for the two different cases can again be achieved by introducing the auxiliary function Q_Λ in Eq. (63). This leads to the following sets of coupled first order differential equations:

$$P'_\Lambda = -\frac{\kappa}{r} P_\Lambda + S_\Lambda Q_\Lambda \tag{69}$$

$$Q'_\Lambda = \frac{\kappa}{r} Q_\Lambda - T P_\Lambda + \sum_{\Lambda'} B_{\Lambda\Lambda'} P_{\Lambda'}$$

$$- \frac{S'_\Lambda}{S_\Lambda} \frac{1}{r} \left[(\kappa+1) P_\Lambda + \sum_{\Lambda'} \xi^\lambda_{\Lambda\Lambda'} P_{\Lambda'} \right] , \tag{70}$$

with $\xi^\lambda_{\Lambda\Lambda'} = \langle \chi_\Lambda | \hat{\xi}_{zz} | \chi_{\Lambda'} \rangle$ or $\xi^\lambda_{\Lambda\Lambda'} = \langle \chi_\Lambda | \hat{\xi}_{xy} | \chi_{\Lambda'} \rangle$, respectively.

The final coupled radial equations obtained for the two manipulation schemes sketched above differ only with respect to the last term in Eqs. (65) and (70), respectively, from the original equation (31) for their small component wave function corresponding to the proper SDFT-Dirac-Hamiltonian. Implementation of the two manipulation schemes therefore requires only minor modifications in the corresponding programs.

However, one has to keep in mind that $Q_\Lambda(r, E)$ defined by Eq. (63) has not the meaning of a small component occurring within the bi-spinor formalism. For this reason the boundary conditions to match the wave functions to solutions outside the sphere boundary have to be specified by $P_\Lambda(r, E)$ alone [57]. To set up the corresponding single site t-matrix $t_{\Lambda\Lambda'}(E)$ used within the KKR-formalism one therefore has to replace the relativistic Wronskian (see Ref. [135]) by its standard form $g_\Lambda(r, E) j'_l(r, E) - g'_\Lambda(r, E) j_l(r, E)$ with $j_l(r, E)$ the spherical Bessel function. Furthermore, one has to note that for the evaluation of the matrix elements of any operator it has to be transformed in such a way that no coupling of large and small component occurs. This applies, for example, to the operator $\boldsymbol{\alpha} \cdot \boldsymbol{A}$ that describe the interaction of electrons with the vector potential $\boldsymbol{A}$. In this case, for example, the $\boldsymbol{\nabla} \cdot \boldsymbol{A}$-form of the matrix elements can be used [42,139].

Green's function The problem of setting up the electronic Green's function $G(\boldsymbol{r}, \boldsymbol{r}' E)$ for a solid on the basis of relativistic multiple scattering theory for arbitrary scalar and vector potentials has been investigated in great detail by Tamura [79]. In analogy to the non-relativistic formalism of Faulkner and Stocks [10] the Green's function $G(r, r', E)$ can be written as:

$$G(r, r', E) = \sum_{\Lambda\Lambda'} Z_\Lambda^n(r, E)\tau_{\Lambda\Lambda'}^{nn'}(E)Z_{\Lambda'}^{n'\times}(r', E)$$
$$- \sum_\Lambda \left[Z_\Lambda^n(r, E)J_\Lambda^{n\times}(r', E)\Theta(r' - r) \right.$$
$$\left. + J_\Lambda^n(r, E)Z_\Lambda^{n\times}(r', E)\Theta(r - r') \right] \delta_{nn'} \tag{71}$$

for r (r') within the cell n (n'). Here the quantity $\tau_{\Lambda\Lambda'}^{nn'}(E)$ is the so-called scattering path operator [140] that represents all multiple scattering events in a many-atom system in a consistent way (see below). The wave functions Z_Λ^n and J_Λ^n are the properly normalized regular and irregular solutions of the corresponding single site problem for site n (see above).

The most important technical point to note is that in Eq. (71) the sign "$\times$" indicates that the wave functions $Z^\times$ and $J^\times$ are the left-hand side regular and irregular solutions of the corresponding modified Dirac equation [79]. Fortunately, for the SDFT-Dirac-Hamiltonian these are obtained from the same radial differential equations as the conventional right-hand side solutions Z_Λ and J_Λ; i.e. from Eqs. (41)-(42) with the potential matrix elements $V_{\Lambda\Lambda'}^\pm$ replaced by $V_{\Lambda'\Lambda}^\pm$. For highly symmetric systems one may have the situation that $V_{\Lambda\Lambda'}^\pm = V_{\Lambda'\Lambda}^\pm$. In this case $Z^\times$ and $J^\times$ are obtained from Z and J by simple complex conjugation and transposition:

$$Z_\Lambda^\times(r, E) = \sum_{\Lambda'}(g_{\Lambda'\Lambda}(r, E)\chi_{\Lambda'}^\dagger(\hat{r}); -if_{\Lambda'\Lambda}(r, E)\chi_{\bar{\Lambda}'}^\dagger) \tag{72}$$

and

$$J_\Lambda^\times(r, E) = \sum_{\Lambda'}(\tilde{g}_{\Lambda'\Lambda}(r, E)\chi_{\Lambda'}^\dagger(\hat{r}); -i\tilde{f}_{\Lambda'\Lambda}(r, E)\chi_{\bar{\Lambda}'}^\dagger) , \tag{73}$$

since left and right hand side solutions are identical with respect to their radial parts. This applies in particular to the single site problem with spherically symmetric potential terms V and B, but also to cubic systems with the magnetization along the z-axis, as investigated here. Fortunately, this is still true if the OP-term is included in the Dirac equation (19) because here the relation (Eq. (11) in [79]) for the vector potential corresponding to the OP-potential term in Eq. (19) holds.

Calculation of the Scattering Path Operator $\tau_{\Lambda\Lambda'}^{nn'}$

The scattering path operator $\tau_{\Lambda\Lambda'}^{nn'}$ – introduced by Gyorffy and Stott [140] – transfers a wave with spin-angular character Λ' coming in at site n' into a wave outgoing from site n with character Λ and with all possible scattering events that may take place in between accounted for. According to this definition it has to fulfill the following self-consistency condition:

$$\underline{\tau}^{nn'} = \underline{t}^n \delta_{nn'} + \underline{t}^n \sum_{k \neq n} \underline{G}^{nk} \underline{\tau}^{kn'} , \qquad (74)$$

where all quantities are energy dependent and the underline denotes matrices with their elements labeled by $\Lambda = (\kappa, \mu)$. Here the single site t-matrix $\underline{t}^n$ is fixed by the solutions to the single-site Dirac equation for site n. Furthermore, $\underline{G}^{nn'}$ is the relativistic real space Green's function or structure constants matrix that represents the propagation of a free electron between sites n and n' [133]. It is related to its non-relativistic counterparts $G_{\mathcal{L}L'}^{nn'} = G_{LL'}^{nn'} \delta_{m_s m_s'}$ by the simple relation [80]:

$$G_{\Lambda\Lambda'}^{nn'} = (1 + E/c^2) \sum_{\mathcal{L}L'} S_{\Lambda\mathcal{L}}^+ G_{\mathcal{L}L'}^{nn'} S_{L'\Lambda'} , \qquad (75)$$

where $\mathcal{L}$ and L stand as usual for the sets (l, m_l, m_s) and (l, m_l), resp., of non-relativistic quantum numbers. The matrix elements of the unitary transformation matrix $\underline{S}$ occurring in Eq. (75) are given by the Clebsch Gordon coefficients $C(l\frac{1}{2}j, \mu - m_s, m_s)$ [7].

For many situations it is often sufficient to consider a system consisting of only a finite number of atoms, as for example within the local interaction zone (LIZ) formalism [78] or EXAFS-theory [76,77]. In that case Eq. (74) can be solved by inverting the corresponding real-space KKR-matrix [77]:

$$\underline{\underline{\tau}} = [\underline{\underline{m}} - \underline{\underline{G}}]^{-1} , \qquad (76)$$

where the double underline indicates super-matrices with the elements being labeled by the site indices of the cluster. The elements themselves are matrices labeled by Λ as for example $(\underline{\underline{G}})^{nn'} = \underline{G}^{nn'}$ with $(\underline{G}^{nn'})_{\Lambda\Lambda'} = G_{\Lambda\Lambda'}^{nn'}$. The matrix $\underline{\underline{m}}$ in Eq. (76) is site-diagonal and has the inverse of the single site t-matrix $\underline{t}^n$ as its diagonal elements; i.e. $(\underline{\underline{m}})^{nn'} = \underline{m}^n \delta_{nn'} = (\underline{t}^n)^{-1} \delta_{nn'}$.

Alternatively one may instert Eq. (74) repeatedly into itself to arrive at the scattering path expansion

$$\underline{\tau}^{nn} = \underline{t}^n + \underline{t}^n \sum_{k \neq n} \underline{G}^{nk} \underline{t}^k \underline{G}^{kn} \underline{t}^n$$

$$+ \underline{t}^n \sum_{k \neq n} \sum_{l \neq k} \underline{G}^{nk} \underline{t}^k \underline{G}^{kl} \underline{t}^l \underline{G}^{ln} \underline{t}^n + ... \qquad (77)$$

for the site-diagonal scattering path operator $\underline{\tau}^{nn}$. Compared to the matrix inversion the scattering path expansion technique is highly efficient. On the other hand, one may encounter convergence problems using it. To avoid these problems several alternative and efficient schemes have been suggested in the literature [141,142] that have not been applied so far for relativistic calculations.

For ordered infinite systems Eq. (74) can be solved exactly by means of Fourier-transformation. For one atom per unit cell, the term $\tau^{nm}_{\Lambda\Lambda'}$ is obtained from the Brillouin-zone integral

$$\tau^{nn'}_{\Lambda\Lambda'}(E) = \frac{1}{\Omega_{\mathrm{BZ}}} \int\limits_{\Omega_{\mathrm{BZ}}} d^3k [\underline{t}^{-1}(E) - \underline{G}(\boldsymbol{k}, E)]^{-1}_{\Lambda\Lambda'} e^{i\boldsymbol{k}(\boldsymbol{R}_n - \boldsymbol{R}_{n'})} . \qquad (78)$$

Here $\boldsymbol{R}_{n(n')}$ denotes the lattice vector for site $n(n')$ and the relativistic $\boldsymbol{k}$-dependent structure constant matrix $\underline{G}(\boldsymbol{k}, E)$ is connected to its non-relativistic counterpart in analogy to Eq. (75) for the real space formulation.

As usual, group theory can be exploited to restrict the integration in Eq. (78) to the irreducible part of the Brillouin-zone, that depends on the orientation of the magnetization [143,144]. For cubic systems, the site-diagonal case $n = n'$ has been dealt with in detail by Hörmandinger and Weinberger [145], while the site-off-diagonal case $n \neq n'$ has been worked out by Zecha and Kornherr [146,147].

To deal with the electronic structure of surfaces within the framework of the spin-polarized relativistic KKR-formalism, the standard layer techniques used for LEED and photoemission investigations [148] have been generalized by several authors [90,149]. As an alternative to this, Szunyogh and coworkers introduced the TB-version of the KKR-method [74,150,75]. To invert the emerging layer-, $\boldsymbol{k}_\parallel$- and Λ-indexed KKR-matrix, that has in principle an infinite number of rows and columns, these authors applied techniques that are also used within the TB-LMTO-Green's function formalism [95]. Finally, the scheme to deal with surfaces and layered systems developed by Dederichs and coworkers [151], that represents the vacuum region by layers of empty atomic sites, has been generalized recently by Huhne and Nonas [152,153].

Treatment of Disordered Alloys One of the appealing features of the multiple scattering formalism described above is that it can be applied straightforwardly to deal with disordered alloys. Within the Coherent Potential Approximation (CPA) [154] alloy theory the configurationally averaged properties of a disordered alloy are represented by a hypothetical ordered CPA-medium, which in turn may be described by a corresponding scattering path operator $\tau^{nn,\mathrm{CPA}}_{\Lambda\Lambda'}$. This operator is determined by the CPA-condition:

$$x_{\mathrm{A}} \underline{\tau}^{nn,\mathrm{A}} + x_{\mathrm{B}} \underline{\tau}^{nn,\mathrm{B}} = \underline{\tau}^{nn,\mathrm{CPA}} , \qquad (79)$$

where the matrices are defined with respect to the index Λ and the binary alloy system has components A and B at relative concentrations x_{A} and x_{B}, respectively. The above equation represents the requirement that the concentration-weighted sum of the component-projected scattering path operators $\underline{\tau}^{nn,\alpha}$ should be identical to that of the CPA-medium; i.e. embedding an A- or a B-atom into the CPA-medium should not cause any additional scattering.

The $\underline{\tau}^{nn,\alpha}$ describes the scattering properties of an α-atom embedded in the CPA-medium, i.e. of a substitutional impurity and is given by the expression

$$\underline{\tau}^{nn,\alpha} = \underline{\tau}^{nn,\mathrm{CPA}} \left[1 + \left(\underline{t}_\alpha^{-1} - \underline{t}_{\mathrm{CPA}}^{-1} \right) \underline{\tau}^{nn,\mathrm{CPA}} \right]^{-1} , \qquad (80)$$

where $\underline{t}_\alpha$ and $\underline{t}_{\mathrm{CPA}}$ are the single site t-matrices of the component α and of the CPA-medium, respectively. The later quantity is connected with the CPA scattering path operator by an equation analogous to Eq. (78) or (76), respectively. To obtain the quantities $\underline{t}_{\mathrm{CPA}}$ and $\underline{\tau}^{nn,\mathrm{CPA}}$, for given concentrations x_α, one must solve Eqs. (79), (80) and (78) or (76), respectively, iteratively.

3 Applications

3.1 Basic Electronic Properties

Dispersion relations The impact of the spin-orbit coupling on the dispersion relations $E_j(\boldsymbol{k})$ of spin polarized relativistic solids have been investigated by several authors in the past [71,138,155]. Corresponding results for fcc-Ni are shown in Fig. 1 for the magnetization $\boldsymbol{M}$ along the [001]-axis and the wavevector $\boldsymbol{k}$ along the [100]-axis [138]. As one notes, spin-orbit coupling gives rise to a lifting of degeneracies (e.g. at A and B in Fig. 1, left) and causes hybridization or mixing of bands (e.g. at C, D, E and F) that simply cross within a non- or scalar relativistic treatment. In addition, one finds that for Bloch states $|\Psi_{j\boldsymbol{k}}\rangle$ the

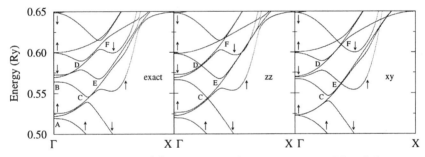

Fig. 1. Dispersion relation $E_j(\boldsymbol{k})$ of fcc-Ni for the magnetization $\boldsymbol{M}$ and the wavevector $\boldsymbol{k}$ along the [001]- and [100]-axis, respectively. The panels show from left to the right results based on the full Dirac equation and those obtained keeping the zz- and xy-terms in Eq. (66).

expectation value $\langle \Psi_{j\boldsymbol{k}} | \sigma_z | \Psi_{j\boldsymbol{k}} \rangle$ is not restricted to ± 1 (see e.g. Ref. [65,71]), i.e. spin is no more a good quantum number. However, remarkable deviations from the values ± 1 occur only in the region where bands cross if spin-orbit coupling is neglected. For this reason it is justified to attach the labels $\downarrow$ and $\uparrow$ to the bands to indicate their dominant spin character for a certain range of $\boldsymbol{k}$.

Keeping only the $\hat{\xi}_{zz}$-part of the spin-orbit interaction the most important consequence is that now all states have pure spin character that cannot change if one goes along a certain band. However, this does not rule out the hybridization of bands induced by $\hat{\xi}_{zz}$. As one can see from the middle panel of Fig. 1 hybridization takes place at E and F. On the other hand, no hybridization is found at C and D, where now bands of different spin character cross. Furthermore one notes that the splitting of the bands, e.g. at A, B, E and F caused by the $\hat{\xi}_{zz}$-part is quite comparable to that due to the full spin-orbit interaction.

Concerning the hybridization, the situation is more or less opposite to the situation for $\hat{\xi}_{zz}$, if the $\hat{\xi}_{xy}$-part is used. The right panel of Fig. 1 demonstrates that there is now a pronounced hybridization of bands of different spin character (C and D) – just as for the full spin-orbit interaction. While hybridization is also present at E and F, it is much less pronounced than for $\hat{\xi}_{zz}$. Surprisingly, the splitting of the bands caused by $\hat{\xi}_{xy}$, while being in general smaller than for $\hat{\xi}_{zz}$, is still quite appreciable. In spite of this, both parts have a rather different importance for many spin-orbit induced properties, as it will be demonstrated below.

The influence of the spin-orbit coupling on the dispersion relation of ordered spin-polarized solids shown in Fig. 1 can also be demonstrated for disordered ones. Instead via the dispersion relation, the band structure is represented in the later case by means of the Bloch-spectral function $A_B(\mathbf{k}, E_F)$ that can be viewed as a $\mathbf{k}$-dependent density of states (DOS) function [156]. Due to the chemical disorder, $A_B(\mathbf{k}, E_F)$ for a given energy is in general spread out in $\mathbf{k}$-space, implying that the wave vector $\mathbf{k}$ is not a good quantum number. For an ordered system, on the other hand, the smearing out in $\mathbf{k}$-space does not occur and $A_B(\mathbf{k}, E_F)$ can be written as a superposition of δ-functions $\delta(E - E_j(\mathbf{k}))$ and the conventional dispersion relations are recovered.

Fig. 2 shows results for spin-projected Bloch-spectral function $A_B(\mathbf{k}, E_F)$ obtained for fcc-$Fe_{0.2}Ni_{0.8}$ for the Fermi energy E_F, the wave vector $\mathbf{k}$ in the (010)-plane and the magnetization $\mathbf{M}$ along the [001]-direction [157]. As one notes, disorder has quite a different impact on the majority and minority spin Bloch-spectral functions. Nevertheless, there is a well-defined Fermi surface for both of them that – due to the specific composition – is very similar to that of pure Ni [26,158]. Comparing the details of the Bloch-spectral function for the two spin sub-systems a hybridization is recognized. This especially holds for the single majority sheet centered at the Γ-point that has a pronounced minority spin admixture. Because the majority spin states primarily carry the electric current [159] and because the spin hybridization leads effectively to a short-cut it should have a great influence on the electric resistivity. In fact, calculations of the isotropic residual resistivity of magnetic alloys using the SPR-KKR-CPA give always higher values than calculations that neglect spin-orbit coupling and make use of the so-called two-current model [160,161].

In addition to the spin hybridization, one notes a small anisotropy of the Bloch spectral function; i.e. it depends on the relative orientation of the wave vector $\mathbf{k}$ and the magnetization $\mathbf{M}$. This applies not only to the location of $A_B(\mathbf{k}, E)$ in $\mathbf{k}$-space, but also for the width of the Bloch spectral function. The former property is also found for the dispersion relation of ordered systems and clearly indicates the lowered symmetry of the system compared to a paramagnetic state. For the situation considered in Figs. 1 and 2 the symmetry is effectively tetragonal instead of being cubic. Changing the orientation of the

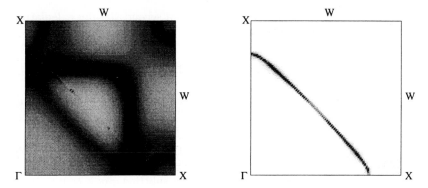

Fig. 2. Gray-scale representation of the Bloch spectral function $A_B(\mathbf{k}, E_F)$ for fcc-$Fe_{0.2}Ni_{0.8}$ for the energy fixed to the Fermi energy E_F and the magnetization $\mathbf{M} \parallel [001]$. The wave vector $\mathbf{k}$ is in the (010)-plane; i.e. the horizontal axis gives the component of $\mathbf{k}$ perpendicular to $\mathbf{M}$, while the vertical axis gives that parallel to $\mathbf{M}$. The white background corresponds to $A_B(\mathbf{k}, E) = 0$, while the black regions represent $A_B(\mathbf{k}, E) \geq$ 50 a.u.; i.e. the cusps of $A_B(\mathbf{k}, E)$ have been cut for a more resolved representation. The left and right parts give the Bloch spectral function decomposed into their minority and majority spin part, respectively.

magnetization $\mathbf{M}$ will in general change the symmetry but also the electronic structure itself. The corresponding changes in the total energy give rise to the magneto-striction [162,163] and the magneto-crystalline anisotropy [164,165]. The conventional approach to evaluate magneto-crystalline anisotropy energies is to apply the force theorem [166], that allows to approximate the energy difference for two orientations of the magnetization as the energy difference of the corresponding single particle energies. This means that the small changes found for the dispersion relation when the magnetization is rotated [155,158] are the microscopic origin of the magneto-crystalline anisotropy. In practice the mentioned energy differences are calculated by integrating the various energy-weighted density of states curves or equivalently the integral DOS curves up to a fixed Fermi energy. On the basis of this procedure Újfalussy et al. performed for multi-layer systems a layer-wise decomposition of the magneto-crystalline anisotropy energy by using corresponding layer projected DOS functions [167]. Of course, an analogous composition can be made for any multi-component system. Using the KKR-formalism an elegant way to perform the above mentioned energy-integral for the integrated DOS function is to make use of Lloyd's formula [168]. This has recently been used for a calculation of the magneto-crystalline anisotropy energy of disordered fcc-Co_xPt_{1-x} alloys by Razee et al. [169] who also worked out corrections to the expression based on the force theorem.

Finally, it should be emphasized that all calculations of the magneto-crystalline anisotropy energy done so far on the basis of the force theorem or alternative schemes [164,165] account only for the spin-orbit coupling as its microscopic source. The Breit interaction, that gives rise to contributions in the same order of magnitude, is usually accounted for only in a second step by calculating the so-called shape anisotropy energy [165,170,171].

Charge and Orbital Current Density Distribution For a paramagnetic solid time reversal symmetry implies that spin-orbit coupling viewed as a perturbation leads for states with quantum numbers (m_l, m_s) to the same changes as for $(-m_l, -m_s)$. As a consequence the spatial symmetry of the charge distribution is not affected and no orbital current is induced. For a spin-polarized solid, on the other hand, this does not hold anymore; i.e. states with quantum numbers (m_l, m_s) and $(-m_l, -m_s)$ are affected by the inclusion of the spin-orbit coupling in a different manner because of the exchange splitting. As a consequence the charge distribution will be rearranged according to the lowered symmetry of the system compared to its paramagnetic state. For a magnetic solid with a cubic lattice and the magnetization along the z-axis, for example, the effective symmetry is only tetragonal. Accordingly, self-consistent full-potential spin polarized relativistic calculations lead to non-cubic terms like ρ_{20} and V_{20}, respectively, if for the charge density ρ and the potential V the conventional expansion into spherical harmonics is used [83] (see Eq. (44)).

A further consequence of the presence of the spin-orbit coupling for a spin-polarized solid is that its orbital angular momentum is no more quenched. This corresponds to the occurrence of a finite paramagnetic orbital current density j_{p} (the adjective paramagnetic can be omitted in the following because external magnetic fields are assumed to be absent; i.e. the physical and paramagnetic current densities are identical).

Within the Green's function formalism used here the current density j_{p} can be obtained from the expression

$$j_{\mathrm{p}} = -\frac{1}{\pi} \operatorname{Trace} \Im \int^{E_F} dE \, \frac{1}{\mathrm{i}} [\overrightarrow{\nabla} - \overleftarrow{\nabla}] G(\boldsymbol{r}, \boldsymbol{r}', E)|_{\boldsymbol{r}=\boldsymbol{r}'} . \tag{81}$$

Corresponding results [83] obtained for the current density j_{p} in bcc-Fe are shown in Fig. 3. Here the direction and magnitude of j_{p} is represented by arrows for the (001)-plane with the z- and magnetization axes pointing upwards. At first sight the current density distribution seems to be rotational symmetric. However, a closer look reveals that it has in fact a lower symmetry. This is demonstrated in the right part of Fig. 3 which gives the radial component of j_{p} within the (001)-plane. This component is about 2-3 orders of magnitude smaller than j_{p} itself and has been scaled by a factor of approximately 350 with respect to the left part of Fig. 3. As one notes, there is only a four-fold symmetry axis along the z-axis. For the paramagnetic state the x- and y-axes as well as the diagonal axes in between would be twofold symmetry axes. Obviously, the corresponding symmetry operation C_2 is missing here because of the ferromagnetic state and the spin-orbit coupling accounted for. However, one can also clearly see from the right part of Fig. 3 that this symmetry operations combined with the time reversal operator T result in proper symmetry operations $(TC_{2\perp})$ for the ferromagnetic state [144].

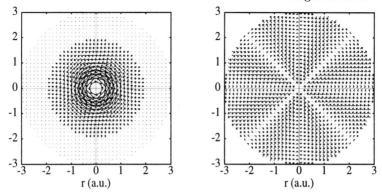

Fig. 3. Orbital current density j_p for bcc-Fe in the (001)-plane (left). The right part gives the corresponding radial component scaled by a factor of around 350 with respect to the left part. For display j_p has been weighted with r^2.

3.2   Orbital Magnetic Moments

With the spin-orbit-induced orbital current density in magnetic solids there is obviously a finite orbital angular momentum density associated. For a rotational symmetric current density distribution, for example, one has circular currents implying the simple relationship [83]

$$\langle \beta l_z(\boldsymbol{r}) \rangle = \frac{1}{2\sqrt{2}} \langle r j_{p,\phi}(\theta, r) \rangle \,, \tag{82}$$

where $j_{p,\phi}$ is the ϕ-component of $\boldsymbol{j}_p$ that gives its magnitude along a closed circular loop.

Connected with $\langle \beta l_z \rangle$ there is of course a corresponding orbital magnetic moment μ_{orb} that can be obtained via Eq. (82) or directly from the conventional expression [24,83]:

$$\mu_{orb} = -\frac{\mu_B}{\pi} \text{Trace} \, \Im \int^{E_F} dE \int d^3r \, \beta l_z G(\boldsymbol{r}, \boldsymbol{r}, E) \,. \tag{83}$$

As Fig. 4 shows, the spin-orbit induced μ_{orb} contributes 5 to 10 % of the total magnetic moments of the elemental ferromagnets Fe, Co and Ni. However, one also notes from this figure that the results obtained on the basis of plain SDFT are much too small compared with experiment in the case of Fe and Co. To cure this problem, that also occurs for f-electron systems, Brooks introduced the OP-formalism [113], that was originally restricted to $\boldsymbol{k}$-space band structure methods. Using the real-space formulation given above, one can see that it effectively leads to a feed-back of the spin-orbit induced orbital current into the potential term of the Dirac equation (see Eq. (19)). Based on the corresponding spin- and orbital polarized relativistic (SOPR) KKR-formalism [118], one finds a strong enhancement of the orbital magnetic moment for Fe and Co leading to

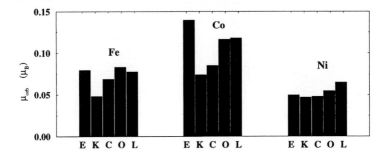

Fig. 4. Orbital magnetic moments for bcc-Fe, fcc-Co and fcc-Ni. The various columns represent from left to right the experimental data (E) [172] and the theoretical data obtained by the plain SPR-KKR- (K), within CDFT (C) [109] as described in the text as well as the SOPR-KKR (O) [118] including the OP-potential term. The last column labeled with L gives results obtained using the LMTO that account for spin-orbit coupling and the OP-term in the variational step [51,173].

a rather satisfying agreement with experiment (see Fig. 4). The spin magnetic moment, on the other hand, is hardly affected by inclusion of the OP-term. Furthermore, calculations done in the full-potential mode [83] clearly demonstrated that the OP-term does not include aspherical potential terms that would be accounted twice in a full-potential calculation, as it was sometimes suspected in the past.

Apart from minor numerical differences, the results obtained with the SOPR-KKR are completely in line with those obtained before using the LMTO-method [51,173,174]. However, the latter approach accounts for spin-orbit coupling and the OP-term only in the variational step, while for the SOPR-KKR these are also included when calculating the wave functions and the corresponding single-side t-matrices. As a consequence the SOPR-KKR can straightforwardly be combined with the CPA to deal with disordered alloys. As an example for an application of the SOPR-KKR-CPA results for μ_{orb} of bcc-Fe_xCo_{1-x} are shown in Fig. 5 [118].

In contrast to the investigations of Söderlind et al. [175] done using the LMTO together with the virtual crystal approximation (VCA) alloy theory [156] the SOPR-KKR-CPA supplies component-resolved results. As one can see in Fig. 5 the enhancement of μ_{orb} for Fe and Co in bcc-Fe_xCo_{1-x} are very similar to that found for the pure metals. Again this enhancement brings the average orbital magnetic moment for the alloy in very satisfying agreement with experiment.

Because the OP-term is very similar in form to the operator representing spin-orbit coupling as a correction or perturbation, one may expect that it will not only affect the spin-orbit induced orbital magnetic moments but also any other quantity caused by spin-orbit coupling. This is in general indeed the case as it could be demonstrated by investigations on the spin-orbit induced band-splittings [176], the orbital contributions to the hyperfine fields [118], the magneto-crystalline anisotropy [177], galvano-magnetic properties [178], the

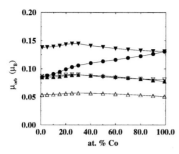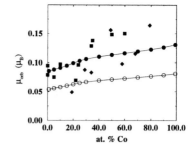

Fig. 5. Orbital magnetic moments in bcc-Fe_xCo_{1-x}. The triangles pointing up- and downwards represent the theoretical moments of Fe and Co, respectively, while the concentration weighted sum is given by circles. Full and open symbols stand for results obtained with and without the OP-term included (SOPR- and SPR-KKR-CPA, resp.). Experimental data [172] for the average magnetic moment (right part) stemming from magnetomechanical and spectroscopic g-factors are given by full squares and diamonds.

magneto-optical Kerr effect [179] and the magnetic dichroism in X-ray absorption [180].

Using the OP-formalism, one obviously leaves the framework of density functional theory and arrives at a heuristic hybrid scheme. From a formal point of view CDFT therefore supplies a much more satisfying basis to deal with orbital magnetism. Results for μ_{orb} of Fe, Co and Ni, that have been obtained using the relativistic version of Vignale and Rasolt's CDFT-formalism, are given in Fig. 4 [109]. Obviously using CDFT instead of plain SDFT leads indeed to an enhancement of μ_{orb} for Fe and Co. Although this effect is found to be too small, one may expect that the remaining deviation from experiment will be reduced with improved parameterizations for the exchange-correlation potentials available.

The basic CDFT-Hamiltonian in Eq. (14) does not rule out the existence of a finite orbital magnetic moment in the non-relativistic limit. With the help of model calculations, it could be demonstrated that this is not the case [109]. Starting a SCF-calculation with a finite spin-orbit induced orbital current density and switching off the spin-orbit coupling during the SCF-cycle the orbital magnetic moment vanished.

Using Vignale and Rasolt's formulation of CDFT [104,105], one has to use the current density in a spin-projected way. This leads to quite large contributions in the nucleus near region stemming from the core states, that essentially cancel if the spin contributions are summed up (see for example Fig. 3). However, for transition metals the corresponding core contributions to the exchange-correlation potential $A_{xc,\sigma}(r)$ has not much overlap with the current density of the valence d-electrons. Because μ_{orb} stems primarily from these, the core contribution to $A_{xc,\sigma}(r)$ can be neglected. For bcc-Fe the corresponding valence band part of the polar component of the spin-dependent exchange-correlation vector potential $A_{xc,\sigma}(r)$ is given in Fig. 6 (left part). Because the OP-term can be manipulated to represent a coupling to the electronic orbital degree of freedom or current,

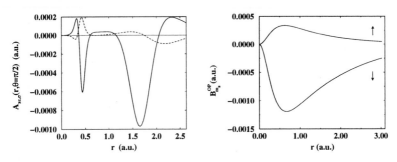

Fig. 6. Left: Valence band part of the polar component of the spin-dependent exchange-correlation vector potential $A_{xc,\sigma}(r,\theta)$ for bcc-Fe (in atomic units). The full and dashed lines give the potential for minority and majority character, respectively, for $\theta = \pi/2$. Right: The OP potential term B_σ^{OP} for bcc-Fe as calculated by the self-consistent FP-OP-SPR-KKR.

respectively, a corresponding vector potential function B_σ^{OP} can be constructed within the OP-formalism [83]. However, one has to keep in mind that the physical picture behind the OP-formalism is quite different from the CDFT as used here. While for the former case one tries to account in an approximate way for intra-atomic correlations, the vector potential occurring within CDFT is due to diamagnetic contributions to the exchange-correlation energy of the electron gas. Accordingly, it is not surprising that the resulting vector potential function (see Fig. 6) for the OP-formalism is quite different from that obtained within CDFT. In spite of this fundamental difference one finds the current density $j_{p,\sigma}$ calculated within the extended OP-formalism to be very similar to that calculated within the framework of CDFT [83], i.e. both differ only with respect to their absolute magnitude but not concerning their radial variation. For this reason, the OP-formalism, that is extremely simple to be implemented, may be used to study the influence of corrections to the exchange-correlation energy due to finite orbital currents as long as no better parameterizations for this have been derived within CDFT.

3.3 Hyperfine Interaction

Quadrupolar Hyperfine Interaction The nuclear quadrupolar hyperfine interaction, that can be investigated experimentally, for example, by means of Mößbauer spectroscopy or NMR, denotes the coupling of the nuclear quadrupole moment Q with the electric field gradient (EFG) stemming from the surrounding electronic charge distribution. Because the nuclear quadrupolar hyperfine interaction reflects the local site-dependent symmetry of the charge and potential distribution in a rather direct way, it provides a unique tool to investigate the consequence of the spin-orbit coupling for the symmetry of a spin-polarized cubic solid.

The electric field gradient, that is only non-zero for a site-symmetry lower than cubic, can be calculated straightforwardly from the Coulomb part of the electronic potential V. Expanding V into spherical harmonics (see Eq. (44)) one has, for example, for the zz-component of the electric field gradient tensor [181]:

$$\Phi_{zz} = eq = 2 \lim_{r \to 0} V_{20}(r) . \tag{84}$$

As an alternative, the EFG can also be calculated from the corresponding non-spherical charge distribution term $\rho_{20}(r)$ within the central Wigner-Seitz cell and a Madelung contribution stemming from the multipoles on the neighboring sites [130,146,181].

For the case that the electric field gradient is caused by the low symmetry of the underlying lattice it was found in the past that it is often sufficient to perform self-consistent calculations for the charge density only in the spherical approximation and to determine the non-spherical charge density $\rho_{lm_l}(r)$ only in the final iteration (spherical approximation). In addition it was found from FP-SPR-KKR calculations on Fe that the spin-orbit induced electric field gradient stems nearly exclusively from the non-spherical charge density within the central cell containing the nucleus while the contribution of the surrounding can be ignored [146]. Using these simplifications together with the ASA-version of the SPR-KKR formalism the electric field gradient tensor component Φ_{zz} can be written as [139]:

$$\Phi_{zz} = \frac{8\pi}{5} \int_0^S \frac{\rho_{20}(r)}{r} \, dr \tag{85}$$

$$= \frac{8\pi}{5} \frac{-1}{\pi} e \, \Im \int dE \sum_{\Lambda\Lambda'} \tau_{\Lambda\Lambda'} \sum_{\Lambda''\Lambda'''} \tag{86}$$

$$\left[B_{\Lambda''\Lambda'''} \int_0^S \frac{g_{\Lambda''\Lambda} \, g_{\Lambda'''\Lambda'}}{r} \, dr + B_{-\Lambda''-\Lambda'''} \int_0^S \frac{f_{\Lambda''\Lambda} \, f_{\Lambda'''\Lambda'}}{r} \, dr \right] .$$

with the angular matrix elements $B_{\Lambda\Lambda'}$ given by

$$B_{\Lambda\Lambda'} = \delta_{\mu\mu'} (-1)^{\mu - 1/2} \frac{\sqrt{(2l+1)(2l'+1)}}{4\pi} C(ll'2;00) \tag{87}$$

$$\sum_{m_s} C_\Lambda^{m_s} C_{\Lambda'}^{m_s} C(ll'2; (\mu - m_s)(-\mu + m_s))$$

and the Clebsch Gordon coefficients $C(l\frac{1}{2}j; (\mu - m_s)m_s)$ represented by the short hand notation $C_\Lambda^{m_s}$.

This approach has been used to study the properties of 5d-transition metal impurities dissolved substitutionally in bcc-Fe [182]. These impurity type calculations have been done in the single-site approximation ignoring the distortion of Fe-atoms in the vicinity of the impurity. The resulting electric field gradient parameter $q = \Phi_{zz}/e$ is shown in Fig. 7 for the whole 5d-series. First of all one notes that q is only around one order of magnitude smaller than that usually found for a lattice-induced EFG [187]. Taking into account that the existing experimental data are scattering quite strongly and that measurements on single crystals with a definite relative orientation of the magnetization and

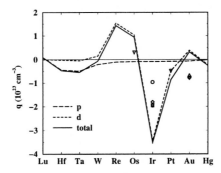

Fig. 7. EFG parameter q of 5d-transition metal-impurities in Fe. Theoretical results for the magnetization direction along the z-axis together with their decomposition into p- and d-electron contributions are given by full, dotted and dashed lines, respectively. Experimental data stem from powder as well as from single crystal measurements [183–186].

the crystal axis were done only in some few cases, agreement of the theoretical results with experiment is quite satisfying. In particular the pronounced dip of the EFG as a function of the ordering number Z for Ir seems to be confirmed by experiment. The variation found for q as a function of Z strongly resembles that obtained earlier for the spin-orbit induced contributions to the magnetic hyperfine field [188], which are predominantly of orbital origin. Earlier, more qualitative investigations [189,190] assumed that this contribution is exclusively due to the d-electrons. However, it turned out that the p-electrons contribute to a similar extent to the magnetic hyperfine field showing only a weak variation with Z [188]. A similar situation is encountered here for the EFG parameter q. As can be seen in Fig. 7 the p-electrons contribute in particular for the early 5d-transition metals, where they exceed the d-electron contribution. Nevertheless, one also notes that the variation of q with Z is primarily due to its d-electron part.

Based on perturbation theory it was expected that the spin-orbit induced EFG should depend quadratically on the spin-orbit coupling strength [184]. Using the manipulation scheme described in section 2.2 this could be verified for the d-electron contribution, while for the p-electrons a pronounced deviation from the quadratic dependency has been found. So far it has been assumed that the spin-orbit induced EFG stems predominantly from the spin-diagonal part ξ_{zz} of the spin-orbit coupling (see Eq. (66)). However, performing corresponding model calculations it was found that the spin-mixing part ξ_{xy} contributes to the same order of magnitude but with opposite sign.

Because the EFG reflects the local symmetry one expects it to change with the orientation of the magnetization; i.e. it should monitor the fact that the spin-orbit induced anisotropy is anisotropic. This could indeed be demonstrated recently by Seewald et al. [46], who determined the EFG of Ir in bcc-Fe for the magnetization pointing along the [001]-, [111]- and [110]-directions. It is interesting to note that this type of anisotropy was too weak to be detected in the case of magneto-optical Kerr effect (MOKE) investigations on fcc-Co [191,192].

Magnetic Hyperfine Interaction The magnetic hyperfine interaction represents the interaction of the nuclear magnetic moment μ_n with the current density j of the surrounding electronic system. In its relativistic form the corresponding operator $\mathcal{H}_{hf}$ is given by [7,193]:

$$\mathcal{H}_{hf} = ec\boldsymbol{\alpha} \cdot \boldsymbol{\mu}_n \times \boldsymbol{r}/r^3 \ . \tag{88}$$

Here only the static part of the hyperfine interaction is considered. Assuming the magnetization and quantization axis to point along the z-axis, only the part $\mathcal{H}_{hf,z} \propto \alpha_z(\boldsymbol{\mu}_n \times \boldsymbol{r})_z$ has to be accounted for.

Within a non-relativistic theoretical description of the hyperfine interaction it is conventional to split the total hyperfine interaction operator into three distinct contributions: the Fermi-contact, the spin-dipolar and the orbital terms. While the first is relevant only for s-electrons, the other two are connected exclusively to non-s-electrons. Starting from a Gordon-decomposition of the electronic current, a corresponding decomposition of the hyperfine interaction operator $\mathcal{H}_{hf}$ in Eq. (88) can also be made within relativistic theory [139,182,194]. For the orbital part one gets for example the expression:

$$\mathcal{H}_{hf,orb} = 2\mu_B\beta\mu_n l \cdot \begin{cases} r^{-3} \text{ for } r > r_n \\ r_n^{-3} \text{ for } r < r_n \end{cases} , \tag{89}$$

where r_n is the nuclear radius. This expression already indicates that for the decomposition of the relativistic hyperfine interaction operator a nucleus of finite size has to be considered [182,194]. Furthermore one has to note that the various parts of $\mathcal{H}_{hf}$ are no more exclusively due to s- or non-s-electrons, respectively.

For spontaneously magnetized solids the central hyperfine interaction parameter is the hyperfine field B_{hf}. This quantity is determined by the expectation value of the static part of the hyperfine interaction operator:

$$B_{hf} = \langle \mathcal{H}_{hf,z} \rangle / \hbar\gamma_n \ , \tag{90}$$

with γ_n the nuclear gyromagnetic ratio. Representing the underlying electronic structure by means of the Green's function formalism $\langle \mathcal{H}_{hf,z} \rangle$ in turn is given by [24]:

$$\langle \mathcal{H}_{hf,z} \rangle = -\frac{1}{\pi} \text{Trace} \ \Im \int dE \int d^3r \, \mathcal{H}_{hf,z}(\boldsymbol{r}) \, G(\boldsymbol{r},\boldsymbol{r},E) \ . \tag{91}$$

Dealing with this expression on a non-relativistic level one gets contributions to the hyperfine field B_{hf} only from the Fermi-contact and spin-dipolar terms because the orbital magnetization density is quenched in the solid [109] (see above). Contributions due to the spin-dipolar term are in general ignored because they arise only from a non-cubic electronic spin density distribution. For these reasons the standard approach to calculate hyperfine fields is to determine just its Fermi-contact contribution stemming from s-electrons. In contrast to this simple but conventional approach, the fully relativistic scheme given above leads to contributions to the hyperfine field from non-s-electrons as well. These are caused by the spin-orbit coupling and are non-negligible even for cubic systems [25].

The left part of Fig. 8 shows the total hyperfine fields of Fe in fcc-Fe_xPd_{1-x} together with a decomposition into contributions stemming from the core, valence and non-s-electrons. The experimental data available for Fe indicate that

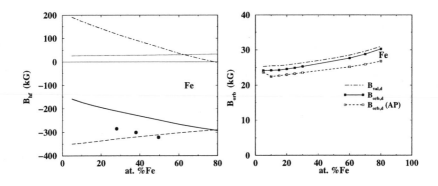

Fig. 8. Left: Hyperfine fields of Fe in fcc-Fe_xPd_{1-x}. In addition to the total field B_{hf} (total) the contributions of the core, valance and non-s-electrons are given separately. Available experimental data have been added. Right: Contributions B_{val} of the d-like valence electrons to the hyperfine fields of Fe in fcc-Fe_xPd_{1-x}. The fields B_{orb} represent the corresponding orbital part. This has also been determined by the approximation due to Abragam and Pryce (AP, see Eq. (3.3)) [195].

the theoretical fields are too small in magnitude. Discrepancies similar to these have been found before for many other systems and have been ascribed to problems in dealing with the core polarization contribution when the spin density functional theory is used on a local-density approximation (LDA) level [20,24]. However, improvements to the LDA, like the generalized gradient approximation (GGA) [196] or the self-interaction correction (SIC) [196,125] did not give much better results. Using the optimized effective potential (OEP), on the other hand, very satisfying results for the hyperfine fields of Fe, Co and Ni could be achieved recently [197].

Within a non-relativistic calculation of the hyperfine fields in Fe_xPd_{1-x} one would get only contributions due to the s-electrons via the Fermi-contact interaction. Within a fully relativistic investigation this part is enhanced by about 10% for Fe [25]. In addition one finds quite appreciable contributions from non-s-electrons. These are induced by the spin-orbit coupling and in general opposite to the normally dominating negative core polarization fields.

The dominating part of the non-s-fields stemming from the valence band electrons is given once more in Fig. 8 (right part). For Fe this stems nearly exclusively from the d-electrons. With the proper relativistic decomposition of the hyperfine interaction in Eq. (88), the origin of these fields can be investigated in a detailed way. The corresponding fields $B_{\mathrm{orb,d}}$ obtained using the relativistic orbital hyperfine interaction operator (see Eq. (89)) have been added to Fig. 8. As one can see these fields differ only slightly from $B_{\mathrm{val,d}}$ implying that the fields coming from d-electrons via the Fermi-contact and spin dipolar interaction are in general negligible. This is also confirmed by an additional and direct calculation of these fields. For this reason it is quite well justified to call the spin-orbit induced hyperfine fields coming from non-s-electrons in a somewhat loose way *orbital* [25].

One of the most important consequences of the spin-orbit coupling for magnetic solids is the presence of a spin-orbit induced orbital electronic current density that gives rise – according to Eq. (89) – to the orbital hyperfine fields but that causes also a corresponding orbital contribution μ_{orb} to the total magnetic moment. Because of their common physical origin one can expect the fields B_{orb} and the moment μ_{orb} to be related via [195]:

$$B_{\mathrm{orb},l} = 2\mu_{\mathrm{B}} \langle r^{-3} \rangle \mu_{\mathrm{orb},l} \quad (l = p, d) .$$

As it can be seen in Fig. 8, this simple approximation works quite well justifying once more the designation *orbital* used above.

3.4   Linear Response

Static Magnetic Susceptibility and Knight Shift Using the Green's function formalism for a description of the underlying electronic structure gives several important advantages when dealing with response quantities. In the case of the static magnetic susceptibility, for example, it is straightforward that way to deal with inhomogeneous systems. This has been demonstrated among others by Terakura et al. [198], who calculated the non-local site-dependent susceptibility χ^{ij} of several paramagnetic transition metals. A corresponding relativistic approach has been worked out by Staunton [199] that has been applied to pure transition metals [200,201] with fcc and hcp structure, respectively. The first step of this approach is to use the first order-approximation to the Dyson equation to represent the Green's function G^B of the investigated system in the presence of an external magnetic field B_{ext} in terms of the Green's function G of the unperturbed system:

$$G^B(\boldsymbol{r},\boldsymbol{r}',E) = G(\boldsymbol{r},\boldsymbol{r}',E) + \tag{92}$$
$$\int d^3r'' G(\boldsymbol{r},\boldsymbol{r}'',E)\mathcal{H}_{\mathrm{pert}}^{\mathrm{spin}}(\boldsymbol{r}'')G(\boldsymbol{r}'',\boldsymbol{r}',E) .$$

Within the framework of SDFT the perturbing Hamiltonian $\mathcal{H}_{\text{pert}}^{\text{spin}}$ is given by [199–201]:

$$\mathcal{H}_{\text{pert}}^{\text{spin}} = \sigma_z[2\mu_B B_{\text{ext}} - (V_{\text{xc}}^{\uparrow} - V_{\text{xc}}^{\downarrow})] . \tag{93}$$

Here the second term represents a feedback of the induced spin-magnetization via the modified exchange and correlation potentials $V^{\uparrow(\downarrow)}$ that gives rise to the Stoner enhancement mechanism [202]. The spin susceptibility χ_{spin} is obtained from Eqs. (92) and (93) by calculating the expectation value of the operator $\beta\sigma_z$ and eliminating the external field B_{ext} [199]:

$$\chi_{\text{spin}} = -\frac{\mu_B}{\pi B_{\text{ext}}} \text{Trace} \, \Im \int dE \int d^3r \, \beta\sigma_z \, G^B(\boldsymbol{r}, \boldsymbol{r}, E) . \tag{94}$$

The procedure sketched here is not restricted to calculations of the spin susceptibility but has a much broader range of application. First of all one has to note that the perturbing Hamiltonian in Eq. (93) represents just the coupling of the external magnetic field to the spin of the electrons. In addition, there is a coupling to the orbital degree of freedom. Within a non-relativistic treatment this gives rise to the diamagnetic Langevin and Landau susceptibilities and the paramagnetic VanVleck susceptibility [203,204]. A straightforward way to obtain the later contribution in a fully relativistic way is to replace the operator $\chi_{\text{pert}}^{\text{spin}}$ in Eq. (93) by:

$$\mathcal{H}_{\text{pert}}^{\text{orb}} = 2\mu_B l_z B_{\text{ext}} \tag{95}$$

and replacing σ_z by l_z in Eq. (94) giving the VanVleck susceptibility χ_{VV} instead of χ_{spin}. Here it has to be noted that the standard non-relativistic treatment of the spin- and orbital magnetic susceptibility do not lead to any spin-orbital cross terms. Within a fully relativistic treatment, on the other hand, the perturbing Hamiltonian $\mathcal{H}_{\text{pert}}^{\text{spin}}$ in Eq. (93) will lead to a non-vanishing orbital susceptibility due to the spin-orbit coupling; i.e. replacing σ_z by l_z in Eq. (94) for χ_{spin} will give an orbital contribution in addition to the pure spin susceptibility χ_{spin}. Analogously, calculating the expectation value of $\beta\sigma_z$ with the orbital perturbation Hamiltonian $\mathcal{H}_{\text{pert}}^{\text{orb}}$ in Eq. (94) will give rise to a spin contribution in addition to the pure orbital VanVleck susceptibility. The remaining contribution to the orbital susceptibility – the Langevin and Landau susceptibilities – can also be treated in a fully relativistic way by a corresponding extension of the non-relativistic theory [203].

A quantity that is closely connected to the susceptibility is the Knight shift that gives the ratio of the induced hyperfine field B_{hf} seen by a nucleus and the external inducing magnetic field. Using the linear response formalism sketched above one has [25,205]:

$$K = -\frac{1}{\pi} \frac{1}{\hbar\gamma_n B_{\text{ext}}} \text{TrIm} \int dE \int d^3r \, \mathcal{H}_{\text{hf},z}(\boldsymbol{r}) \, G(\boldsymbol{r}, \boldsymbol{r}, E) \tag{96}$$

where $\mathcal{H}_{\text{hf},z}$ is the z-part of the relativistic hyperfine interaction Hamiltonian (see Eq. (88)). Making use of the Gordon decomposition of the electronic current density, K can be decomposed into the conventional Fermi-contact, the spin dipolar and the orbital part.

So far, there are only very few theoretical investigations on the Knight shift in transition metal systems that can be found in the literature. Very similar to the situation for the hyperfine field of spontaneously magnetized solids nearly all of these considered only the Fermi contact interaction due to the s-electrons. Using a non-relativistic version of the linear response formalism presented above, the first calculation of all contributions to the Knight shift has been done for the transition metals V, Cr, Nb and Mo [206]. For these metals the VanVleck contribution to the magnetic susceptibility and to the Knight shift was found to be of the same order of magnitude as the various spin contributions and to stem nearly exclusively from the d-electrons. Concerning the magnetic susceptibility similar results have been obtained by Yasui and Shimizu using a non-relativistic [207] as well as fully relativistic [208] approach.

Results obtained using the above relativistic linear response formalism are given in Fig. 9. The left panel of this figure shows the VanVleck susceptibility

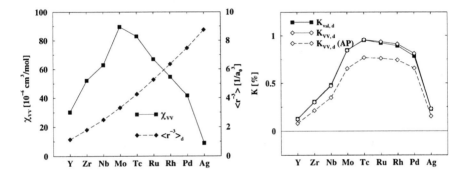

Fig. 9. VanVleck susceptibility χ_{VV} (left), expectation value of r^{-3} (left) and Van-Vleck contribution to the Knight shift K_{VV} (right) for the d-valence band electrons of the pure 4d-transition metal elements. The estimation for K_{VV} based on the approximation proposed by Abragam and Pryce (AP) as well as the total valence contribution K_{val} due to the orbital perturbation term $\mathcal{H}_{pert}^{orb}$ in Eq. (95) has been added (right).

χ_{VV} of the pure 4d-transition metals. As found within earlier studies a maximum is present for χ_{VV} roughly in the middle of the row. This can be explained by using a simplified expression for χ_{VV} [209] and the fact that here the product $n_o \cdot n_u$ of the number of occupied (n_o) and unoccupied (n_u) d-states is at its maximum. From the relationship of the orbital parts of the magnetic moment and the hyperfine field (see Eq. (3.3)) one can expect an analogous relationship for the VanVleck contributions to the susceptibility and Knight shift:

$$K_{VV,l} = 2\mu_B \langle r^{-3} \rangle \chi_{VV,l} \quad (l = p, d) . \tag{97}$$

As one can see in Fig. 9 (left) the expectation value $\langle r^{-3} \rangle_d$ for the 4d-elements increases rapidly along the 4d row. The reason for this is that the corresponding d-like wave function gets more and more localized with increasing atomic number. Combining the results for χ_{VV} and $\langle r^{-3} \rangle_d$ using Eq. (97) leads to an estimate for the VanVleck Knight shift K_{VV} that is shown in Fig. 9 (right). Due to the variation of $\langle r^{-3} \rangle_d$ the maximum in K_{VV} is obviously shifted to the right compared with the χ_{VV} curve. Fig. 9 shows in addition the shift K_{VV} that has been obtained by the full formalism; i.e. Eqs. (89), (92), (95), and (96). As one can see the estimation of K_{VV} using χ_{VV} and $\langle r^{-3} \rangle_d$ reproduces the variation with atomic number quite well. However, in line with previous non-relativistic results [206], the absolute values differ by up to about 20%. This means that estimations based on Eq. (97) are in general less reliable than spin-orbit induced orbital hyperfine fields estimated using Eq. (3.3).

Finally, it should be emphasized that the VanVleck Knight shift K_{VV} given in Fig. 9 stems from the coupling of the external magnetic field to the orbital degree of freedom (see Eq. (95)). Because of the use of the relativistic orbital hyperfine interaction operator (see Eq. (89)) it is by definition of pure orbital nature. If the full hyperfine interaction operator (Eq. (88)) is used instead, the Knight shift denoted K_{val} in Fig. 9 is obtained. The small difference between K_{VV} and K_{val} is of pure spin nature. Within a non-relativistic formalism this cross contribution cannot be accounted for, because it is a consequence of spin-orbit coupling.

This section is somewhat outside the main issue of this review, because it presents results for paramagnetic solids. However, one should notice that the formalism presented above can be applied without modifications to deal with the high-field susceptibility of spontaneously magnetically ordered solids, as well as the Knight shift in such systems. Furthermore, one can apply the linear response formalism also to deal with magneto-crystalline anisotropy or the various kinds of spin-spin coupling constants [168,210].

Transport properties A further interesting application of the linear response formalism is the treatment of galvano-magnetic properties of disordered alloys. Here the Kubo-Greenwood formalism allows one to express the elements of the electrical conductivity tensor $\boldsymbol{\sigma}$ in terms of a current-current correlation function [211,212]:

$$\sigma_{\mu\nu} = \frac{\hbar}{\pi V_{\mathrm{cryst}}} \mathrm{Tr} \left\langle j_\mu \, \Im \, G^+(E_F) \, j_\nu \, \Im \, G^+(E_F) \right\rangle_{\mathrm{conf.}} , \qquad (98)$$

where j_μ is the μth spatial component of the electronic current density operator $\boldsymbol{j} = ec\boldsymbol{\alpha}$. In the following it is assumed that a finite conductivity or resistivity, respectively, of the investigated system stems exclusively from chemical disorder, i.e. contributions caused by lattice imperfections, grain boundaries, phonons, magnons, and so on are ignored. This implies in particular that one is dealing with the residual resistivity and that $\langle \cdots \rangle_{\text{conf.}}$ in Eq. (98) denotes the atomic configuration average for a disordered alloy.

For a paramagnetic cubic solid the conductivity tensor that results from Eq. (98) is diagonal with all elements identical, i.e. the conductivity is isotropic. For a ferromagnetic cubic solid, however, this is not the case and the form of the conductivity tensor depends on the direction of the magnetization reflecting the lowered symmetry of the system. A very general procedure to work out the corresponding symmetry properties of response functions has been developed by Kleiner [213] (for a somewhat alternative approach, see for example Ref. [214]). For example, for a cubic solid with the magnetization along the z-axis one finds that way the form [215]:

$$\rho = \sigma^{-1} = \begin{pmatrix} \rho_\perp & -\rho_H & 0 \\ \rho_H & \rho_\perp & 0 \\ 0 & 0 & \rho_\parallel \end{pmatrix} . \tag{99}$$

Here ρ is the resistivity tensor with $\rho_\perp$ and $\rho_\parallel$ the transverse and longitudinal resistivities, respectively, and ρ_H the spontaneous or anomalous Hall resistivity. In addition one defines the spontaneous magnetoresistance anisotropy (SMA) or anomalous magnetoresistance (AMR) ratio by [215]:

$$\frac{\Delta \rho}{\bar{\rho}} = \left. \frac{\rho_\parallel(B_{\text{ext}}) - \rho_\perp(B_{\text{ext}})}{\bar{\rho}(B_{\text{ext}})} \right|_{B_{\text{ext}} \to 0} , \tag{100}$$

where $\bar{\rho} = \frac{1}{3}(2\rho_\perp + \rho_\parallel)$ is the isotropic resistivity. Here the notation emphasizes that experimentally the SMA is determined by measuring $\rho_\parallel(B_{\text{ext}})$ and $\rho_\perp(B_{\text{ext}})$ as a function of an applied external magnetic field B_{ext} and extrapolating to $B_{\text{ext}} = 0$. The reason for this is that in contrast to investigations on the conventional magnetoresistance $\Delta\rho/\rho = (\rho(B_{\text{ext}}) - \rho(0))/\rho(0)$ of paramagnetic solids, the external magnetic field is used here only to align the magnetization of a sample along a certain direction.

The CPA formalism directly gives access to the configurationally averaged Green's function of disordered alloys. However, when calculating response functions for disordered alloys – as for example the conductivity given in in Eq. (98) – one has to deal with the configurational average of the product of two Green's functions. Within the framework of the KKR-formalism, this problem has first been investigated by Staunton [199] based on the work of Durham et al. [216]. With respect to the conductivity tensor a corresponding expression that is consistent with the single-site CPA-formalism has been worked out in great detail by Butler [212]. This approach, derived originally for the non-relativistic case, can straightforwardly be applied also for the spin polarized relativistic case [44] leading – in contrast to a non-relativistic scheme – in particular to an anisotropic conductivity tensor (see for example Eq. (99)). This makes clear that the galvano-magnetic effects AHR and SMA are all spin-orbit induced phenomena.

As an example of an application of the formalism sketched here the calculated isotropic resistivities $\bar{\rho}$ for the alloy systems $Co_x Pd_{1-x}$ and $Co_x Pt_{1-x}$ are shown in the left part of Fig. 10 [160] together with corresponding experimental data measured at low temperature. As one can see, the agreement between calculated and measured resistivities is very good for Co_xPd_{1-x}. The maximum value of the resistivity in this system ($16\mu\Omega$·cm) as well as the composition for which the maximum occurs (about 20% Co) are well reproduced by the calculations.

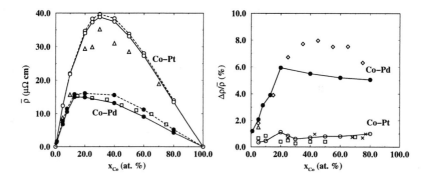

Fig. 10. Left: residual isotropic resistivity $\bar{\rho}$ of disordered Co_xPd_{1-x} (•) and Co_xPt_{1-x} (○) alloys. Full lines: calculated including vertex corrections, broken lines: calculated omitting vertex corrections. Right: calculated spontaneous magnetoresistance anisotropy (SMA) ratio $\Delta\rho/\bar{\rho}$ of Co_xPd_{1-x} (•) and Co_xPt_{1-x} (○) alloys. Experimental data presented by open squares, diamonds, triangles and crosses stem from various sources (see Ref. [160])

Using Butler's approach in dealing with Eq. (98) one accounts for the so-called vertex corrections within the framework of the CPA. For Co_xPd_{1-x} it was found that their contribution increases from about 2% for 5 at.% Co to about 25% for 80 at.% Co.

For the system Co_xPt_{1-x} the calculated resistivities are much higher than for Co_xPd_{1-x}, reaching almost 40 $\mu\Omega$·cm for 30 at% Co. This agrees in a satisfactory way with the experimental maximum of about 35 $\mu\Omega$·cm at that composition. In contrast to Co_xPd_{1-x} the vertex corrections are quite small for Co_xPt_{1-x}, contributing less than 3% to the total conductivity over the entire composition range. Previous investigations on paramagnetic alloy systems [217], lead to the conclusion that the vertex corrections are the more important the lower the d-like DOS at the Fermi level is. For Cu_xPt_{1-x} [217], for example, this applies to the noble metal rich side of this system. For ferromagnetic systems, on the other hand, the vertex corrections seem to be more important, if the d-like DOS at the Fermi level is low at least for one spin subsystem. For this reason, they are more pronounced for Co_xPd_{1-x} compared to Co_xPt_{1-x} and more important on the Co-rich side of both systems.

The anisotropy ratios (SMA) calculated from the transverse and longitudinal resistivities are shown in the right part of Fig. 10 for the two alloy systems Co_xPd_{1-x} and Co_xPt_{1-x}. Experimental values for both systems are included for comparison. Co_xPd_{1-x} shows remarkably high SMA values of more than 6% for concentrations higher than 20 at.% Co [218,219]. The calculations reproduce the increase of the experimental data at low Co concentrations very well. For higher Co concentrations the calculated values are slightly too low. Note that the SMA in Co_xPd_{1-x} is still as large as 1.5% even for very low Co contents [219,220] which was attributed to local orbital moments on the magnetic sites in Ref. [219]. In contrast to Co_xPd_{1-x} the SMA for Co_xPt_{1-x} was found to be below 1% throughout the whole concentration range [221,222]. These findings are perfectly reproduced by the relativistic calculation which reflects the slowly varying SMA in Co_xPt_{1-x}.

It was realized already years ago that the SMA and also the AHR are caused by the spin-orbit coupling. Nevertheless, for the discussion of experimental data corresponding phenomenological descriptions had to be used in the past. These approaches were based on Mott's two-current model that ascribe to each spin subsystem an independent current contribution and introduced a number of model parameters. The SPR-KKR-CPA formalism, on the other hand, does not rely on Mott's two-current model and allows for a parameter-free and quantitative investigation of galvano-magnetic properties. By manipulating the strength of the spin-orbit coupling it was possible in particular to demonstrate explicitly the dependency of the SMA and the AHR on the spin-orbit coupling [223]. In addition it could be shown that even the isotropic resistivity $\bar{\rho}$ can be strongly influenced by the spin-orbit coupling, as it has been expected before [159].

Further insight into mechanisms giving rise to galvano-magnetic effects can be obtained by decomposition of the spin-orbit coupling. To demonstrate this, corresponding results for $\bar{\rho}$ and SMA ratio $\Delta\rho/\bar{\rho}$ are given in Fig. 11. The left

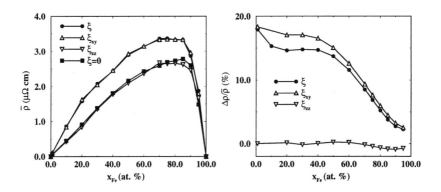

Fig. 11. Isotropic residual resistivity $\bar{\rho}$ (left), and spontaneous magnetoresistance anisotropy ratio $\Delta\rho/\bar{\rho}$ (right) of disordered bcc-Fe_xCo_{1-x} alloys calculated in four different ways. The results obtained using the full spin-orbit coupling – indicated by ξ – are represented by full circles. Open triangles give the results obtained keeping the xy-part ξ_{xy} and zz-part ξ_{zz}, respectively, of the spin-orbit coupling. Full squares give the result with the spin-orbit coupling completely suppressed ($\xi = 0$).

part of this figure shows the isotropic residual resistivity $\bar{\rho}$ of bcc-$Fe_x Co_{1-x}$ obtained from calculations using the full spin-orbit coupling (ξ). As one notes, the variation of $\bar{\rho}$ with composition is strongly asymmetric. This corresponds to the experimental findings [224,225]. The deviation from a parabolic shape can be qualitatively explained by the change of the DOS at the Fermi energy $n(E_F)$ that decreases monotonously with increasing Fe content. Keeping only the spin mixing part ξ_{xy} of the spin-orbit coupling (see Eq. (66)), one finds that $\bar{\rho}$ hardly changes. This already indicates that the spin mixing is the primary source for the relativistic enhancement of $\bar{\rho}$. This spin mixing or hybridization has already be demonstrated by means of the spin-projected Bloch spectral function $A_B(\mathbf{k}, E)$ in Fig. 2 showing for fcc-$Fe_{0.2}Ni_{0.8}$ that there is an appreciable minority spin character admixed to the majority spin states which form a Γ-centered sheet of the Fermi surface and which primarily carry the electronic current [159,157]. Admixture of minority spin character clearly opens for these states a new scattering channel that is very effective because of the high DOS $n^{\downarrow}(E_F)$ at the Fermi energy with minority spin character. As a consequence, the total resistivity has to go up remarkably compared to a calculation based on the two-current model [161]. This interpretation is confirmed by the results obtained by keeping just ξ_{zz}; i.e. suppressing the spin mixing effect of the spin-orbit coupling. In Fig. 11 one can see that this manipulation leads to a strong reduction of the total resistivity throughout the whole range of concentration. To demonstrate that the remaining part ξ_{zz} of the spin-orbit coupling has practically no influence on $\bar{\rho}$, an additional calculation has been carried out with the spin-orbit coupling completely suppressed ($\xi = 0$). The corresponding results nearly completely coincide with the ξ_{zz}-data confirming this expectation. Here one should note that the latter calculational mode ($\xi = 0$) – although technically somewhat different – corresponds essentially to a calculation on the basis of the two-current model, where the electronic structure is calculated in a scalar relativistic way; i.e. with the relativistic corrections Darwin- and mass-velocity-terms taken into account [223].

For the spin-orbit induced SMA ratio the results obtained by the various calculations are given in the right part of Fig. 11. Here one finds that keeping only ξ_{xy} slightly reduces $\Delta\rho/\bar{\rho}$. This means that in contrast to $\bar{\rho}$, ξ_{zz} has some small effect on this quantity. Nevertheless, one finds that keeping ξ_{zz} alone brings $\Delta\rho/\bar{\rho}$ essentially to zero. From this result it can be concluded that the part ξ_{zz} of the spin-orbit coupling can in general be neglected as a source for the SMA compared to ξ_{xy}. Finally, setting $\xi = 0$ of course reduces $\Delta\rho/\bar{\rho}$ exactly to zero [223].

The model calculations performed for the residual resistivity tensor elements of $Fe_x Co_{1-x}$ allow to check the above mentioned phenomenological models for the galvano-magnetic effects. For example, Smit ascribed the occurrence of the SMA to the spin hybridization caused by the spin-orbit coupling [226]. From an analysis of experimental data, on the basis of corresponding expressions for $\Delta\rho/\bar{\rho}$, Jaoul et al. concluded that there should be an additional contribution due to the spin-diagonal part of the spin-orbit coupling [227]. The results presented in Fig. 11 clearly demonstrate that the mechanism discussed by Jaoul et al. can be neglected for the isotropic resistivity $\bar{\rho}$ and has only minor contribution to the SMA in the case of the alloy system bcc-$Fe_x Co_{1-x}$.

Finally, it should mentioned that the expression for the conductivity tensor σ given in Eq. (98) has been generalized recently by Butler et al. [228] to deal with the giant magnetoresistance (GMR) of multilayer systems. A corresponding spin polarized relativistic formulation has been given by Weinberger et al. [229] that includes in particular the influence of the spin-orbit coupling. An extension of Eq. (98) to finite frequencies ω is also straightforward leading to absorptive part of the optical conductivity tensor $\sigma(\omega)$. This has been done recently by Banhart for the visible regime of the light and for paramagnetic alloy systems [230]. For ferromagnetic systems this extension gives directly access to the spin-orbit induced magneto-optical Kerr effect (MOKE) that has been investigated so far exclusively by means of conventional k-space band structure methods as for example the SPR-LMTO [42]. The only exception to this is the work Huhne et al. [231], who developed a very general expression for $\sigma(\omega)$ within the framework of SPR-KKR. This approach should in particular supply a sound basis for an investigation of the oscillations of the Kerr-rotation observed for layered surface systems like Au/Fe/Au [232–234].

3.5 Spectroscopy

Magnetic Circular Dichroism in X-ray Absorption Magneto-optical effects in the visible regime of light are known now for more than 100 years [41,235] and it was realized more than 60 years ago that spin-orbit coupling plays a central role for these [236]. Guided by their experience with the magneto-optical Kerr effect (MOKE) Erskine and Stern [237] suggested that there should be a corresponding magnetic dichroism in X-ray absorption when circularly polarized radiation is used. This magnetic circular X-ray dichroism (MCXD) could be demonstrated for the first time for transition metals by Schütz et al. [43] by measurements at the K-edge of Fe in bcc-Fe in the XANES-region. Later on these authors could also observe the magnetic dichroism in the EXAFS region by investigation on the $L_{2,3}$-edge spectra of Gd in hcp-Gd [238]. Motivated by the MCXD-measurements on bcc-Fe a corresponding fully relativistic description has been developed that is based on the SPR-KKR-formalism [84,85] and that has been applied since then to a great variety of different systems [42]. Recently, this approach was extended to deal with magnetic EXAFS (MEXAFS) by making use of the cluster approximation for the multiple scattering representation of the final states.

Using the SPR-KKR-formalism the X-ray absorption coefficient $\mu^{q\lambda}(\omega)$ is given by [42]:

$$\mu^{q\lambda}(\omega) \propto \Im \sum_{i\,\mathrm{occ}} \left[\sum_{\Lambda\Lambda'} M_{\Lambda i}^{q\lambda*}(E_i + \hbar\omega)\tau_{\Lambda\Lambda'}^{nm}(E_i + \hbar\omega)M_{\Lambda'i}^{q\lambda}(E_i + \hbar\omega) \right.$$

$$\left. + \sum_{\Lambda} I_{\Lambda i}^{q\lambda}(E_i + \hbar\omega) \right] . \tag{101}$$

Here the sum i runs over all involved core states with energy E_i and wave function Φ_i. The electron-photon interaction operator $X_{q\lambda}$, occurring in the matrix elements $M_{\Lambda i}^{q\lambda}$, carries in particular information on the wave vector q of the radiation and on its polarization λ. The last term $I_{\Lambda i}^{q\lambda}$ in Eq. (101) is an atomic-like matrix element [42] and is connected to the term in the Green's function involving the irregular solution to the Dirac equation (see Eq. (71)). Accordingly, it contributes only when working with complex energies.

Although there are also various forms of linear magnetic dichroism, most experimental investigations on the magnetic dichroism in X-ray absorption spectroscopy use circularly polarized radiation because the circular dichroism is most pronounced. To allow for a sound interpretation of the corresponding dichroic signal $\Delta\mu = \mu^+ - \mu^-$, given by the difference in absorption of left and right circularly polarized radiation, a set of so-called sum rules have been derived by several authors [239–242]. The main virtue of these rules is that they should allow one to obtain a reasonable estimate for expectation values $\langle\sigma_z\rangle$ and $\langle l_z\rangle$ of an absorber atom from its energy integrated dichroic signals $\int \Delta\mu(E)dE$. Of course, this is a very appealing property because these quantities are directly proportional to the spin and orbital magnetic moments, μ_{spin} and μ_{orb}, respectively. However, in applying the sum rules one of the main problems is to fix the upper energy integration limit. For that reason it has been suggested to apply the sum rules in their differential form and to discuss the dichroic spectra $\Delta\mu(E)$ directly. For the $L_{2,3}$-edges these differential sum rules are given by [77]:

$$3\left[\Delta\mu_{\mathrm{L}_3} - 2\Delta\mu_{\mathrm{L}_2}\right] = C_{\mathrm{d}}\left(\frac{d}{dE}\langle\sigma_z\rangle_{\mathrm{d}} + 7\frac{d}{dE}\langle T_z\rangle_{\mathrm{d}}\right) \tag{102}$$

$$2\left[\Delta\mu_{\mathrm{L}_3} + \Delta\mu_{\mathrm{L}_2}\right] = C_{\mathrm{d}}\frac{d}{dE}\langle l_z\rangle_{\mathrm{d}} . \tag{103}$$

Here C_{d} is a normalization constant and T_z is the magnetic dipole operator, that often can be ignored. Thus, the basic information to be deduced from the dichroic signal are the spin- and orbital polarization, $\frac{d}{dE}\langle\sigma_z\rangle_{\mathrm{d}}$ and $\frac{d}{dE}\langle l_z\rangle_{\mathrm{d}}$, respectively, of final states with d-character.

The magnetic dichroism of the $L_{2,3}$-edge spectra of Pt in disordered Fe_xPt_{1-x} alloys has been studied experimentally as well as theoretically in great detail in the past [243–247]. Typically for Pt $L_{2,3}$-spectra it was found that the white lines at the L_2- and L_3-edges are quite different because of the influence of the spin-orbit coupling acting on the final states. This finding makes clear that a fully relativistic approach is indispensable to achieve a quantitative description of the $L_{2,3}$-absorption spectra of Pt. This applies in particular if one is dealing with magnetic EXAFS (MEXAFS). To deal with the MCXD in the X-ray absorption of Pt in ordered Fe_3Pt the scattering path operator $\tau_{\Lambda\Lambda'}^{nn}$ entering the expression for the absorption coefficient $\mu^{q\lambda}$ was calculated using the matrix inversion technique for a cluster of 135 atoms in the XANES and 55 atoms in the EXAFS region, respectively, including the central absorber site. The effects of self-energy corrections [141,248] have been accounted for after calculating the spectra.

The top panel of Fig. 12 shows the results of calculations for the L_2-edge EXAFS-spectra of Pt in ordered Fe_3Pt. Corresponding experimental data, obtained by Ahlers and coworkers [249] for an ordered but slightly off-stoichiometric sample, are added. As one can see, the agreement of the theoretical and experimental spectra is quite satisfying, demonstrating in particular that the experimental sample is indeed ordered [250].

The circular dichroic spectrum $\Delta\mu_{L_2}$ for the L_2-edge is shown in the bottom panel of Fig. 12. Again a very satisfying agreement with the corresponding experimental results could be achieved. The results for $\Delta\mu_{L_2}$ clearly demonstrate that the occurrence of magnetic dichroism is by no means restricted to the white line region. Although the amplitude for $\Delta\mu_{L_2}$ is quite small compared to the white line region, it is present throughout the whole EXAFS-range.

As mentioned above, the applicability of the sum rules in their conventional form seems to be somewhat doubtful because of these findings. Nevertheless a clear-cut interpretation of the MEXAFS-spectra can be given making use of the sum rules in their differential form. In the upper part of Fig. 13 a superposition of the theoretical magnetic dichroism spectra $\Delta\mu_{L_2}$ and $\Delta\mu_{L_3}$ according to Eq. (102) is given (here the very small contribution $\frac{d}{dE}\langle T_z\rangle_d$ has been neglected). This is compared to the spin polarization $\frac{d}{dE}\langle\sigma_z\rangle_d$ of the d-states that have been obtained directly from the band structure calculations. In the lower part of Fig. 13 the superposition according to Eq. (103) of the dichroic spectra is compared with the directly calculated orbital polarization $\frac{d}{dE}\langle l_z\rangle_d$ of the d-states. To compare the spectroscopic data with the band structure results the normalization factor C_d in Eq. (102) and (103) has been used as a free scaling parameter using the same value for the upper and lower part of Fig. 13.

The nearly perfect coincidence of the various curves in the upper and lower part of Fig. 13 convincingly demonstrates that the primary information that can be deduced from circular $L_{2,3}$-MEXAFS spectra is the spin and orbital polarization for the final d-like states of the absorber atom. Of course, these are no pure atomic-like properties but concerning their variation with energy they strongly depend on the bonding to their surrounding. For this reason it is quite reasonable to perform a Fourier transform to the MEXAFS-spectra to seek for information on the magnetization distribution around the absorber atom [249]. However, the relationship of the magnetic radial distribution is by no means trivial. Nevertheless, it seems to be worth to investigate this relationship in more

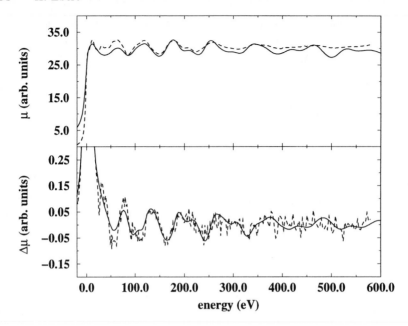

Fig. 12. EXAFS- (top) and MEXAFS-spectra (bottom) at the L_2-edge of Pt in Fe_3Pt. Calculations for the ordered compound (full line), compared against the experimental data for the $Fe_{0.72}Pt_{0.28}$ (dotted line) [249].

detail to be able to deduce further magnetic information from MEXAFS-spectra in a sound and reliable way.

Magnetic Dichroism in the VB-XPS Nowadays, the standard approach to deal with angular-resolved photoemission is to use the one-step model [251] together with the layer-KKR formalism [148]. A corresponding relativistic version for paramagnetic solids of this approach has been developed by Ackermann and Feder [252] and Ginatempo et al. [253] to study spin-orbit induced polarization effects in photoemission. In addition, a spin polarized relativistic and full-potential version has been introduced recently by Feder et al. [90] and Fluchtmann et al. [149] to deal with magnetic dichroic effects in magnetic solids. As an alternative to the above mentioned layer-KKR formalism, one can use the real-space multiple scattering formalism as well [254]. A corresponding expression for the time-reversed LEED-state used to represent the final states of a photo-emission experiment has been worked out first by Durham [255]. For the

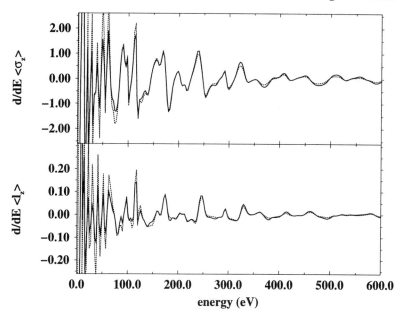

Fig. 13. Top: Spin polarization $\frac{d}{dE}\langle\sigma_z\rangle_\mathrm{d}$ for the d-states of Pt (full line) compared to those derived from the MCXD-spectra using Eq. (102) (dashed line). Bottom: Orbital polarization $\frac{d}{dE}\langle l_z\rangle_\mathrm{d}$ for the d-states of Pt (full line) compared to those derived from the MCXD-spectra using Eq. (103) (dashed line).

spin polarized relativistic case the LEED-state $\phi^{\mathrm{LEED}}_{k,m_s}(r,E)$ is given by [256]:

$$\phi^{\mathrm{LEED}}_{k,m_s}(r_n,E) = 4\pi\sqrt{\frac{E+c^2}{2E+c^2}}\sum_\Lambda i^l\, C^{m_s}_\Lambda\, Y^{\mu-m_s}_l{}^*(\hat{k}) \qquad (104)$$

$$\sum_m e^{ikR_m}\sum_{\Lambda'}\tau^{mn}_{\Lambda\Lambda'}(E)\, Z^n_{\Lambda'}(r_n,E)\ .$$

where k and m_s denote the wave vector and spin character of the photo-electron. With the LEED-state available, it is straightforward to derive expressions for the photo-electron current intensity $I(E,m_s;\omega,q,\lambda)$ for any kind of photoemission experiment making use of Fermi's golden rule [251,256]. For excitation with high energy photons, multiple scattering may be neglected for the final states, leading to the so-called single scatterer approximation [257]. Dealing in addition with angle-integrated spectra leads to a rather simple expression for the photo-current intensity. In the case of VB-XPS one finds in particular [89]:

$$I(E, m_s; \omega, q, \lambda) \propto \sum_{\alpha} x_\alpha \Im \sum_{\substack{\Lambda \Lambda'' \\ \mu = \mu''}} C_\Lambda^{-m_s} C_{\Lambda''}^{-m_s} \tag{105}$$

$$\left\{ \sum_{\Lambda_1 \Lambda_2} \tau_{\Lambda_1 \Lambda_2}^{0\,0,\,\alpha}(E) \left[\sum_{\Lambda'} t_{\Lambda'\Lambda}^{0,\,\alpha}(E') M_{\Lambda'\Lambda_1}^{q\lambda,\,\alpha} \right] \left[\sum_{\Lambda'''} t_{\Lambda'''\Lambda''}^{0,\,\alpha}(E') M_{\Lambda'''\Lambda_2}^{q\lambda,\,\alpha} \right]^* \right.$$

$$\left. - \sum_{\Lambda' \Lambda''' \Lambda_1} t_{\Lambda'\Lambda}^{0,\,\alpha}(E') I_{\Lambda'\Lambda_1\Lambda'''}^{q\lambda,\,\alpha} t_{\Lambda'''\Lambda''}^{0,\,\alpha*}(E') \right\} .$$

Dealing with a paramagnetic solid this expression can be further simplified leading to the familiar result that the VB-XPS intensity is given by the sum over the κ- or angular-momentum j resolved DOS n_κ of the occupied part of the valence band with each contribution weighted by an appropriate cross-section σ_κ [258].

The spin polarized relativistic approach presented above has been applied among others to calculate the VB-XPS spectra of the disordered alloy system $Co_x Pt_{1-x}$ [89]. Theoretical spectra for various concentrations are given in Fig. 14 for unpolarized radiation with $\hbar\omega = 1253.6$ eV. As can be seen, these spectra agree very well with corresponding experimental data reported by Weller and Reim [259].

As mentioned above, for the paramagnetic case the expression for the photo current in Eq. (105) can be simplified to a concentration weighted sum over the products of the κ-resolved partial DOS $n_\alpha^\kappa(E)$ and a corresponding matrix element that smoothly varies with energy [258]. This means that the XPS-spectra map the DOS-curves in a rather direct way. This essentially holds also for the spin-polarized case. As can be seen from the DOS curves of $Co_x Pt_{1-x}$ (see for

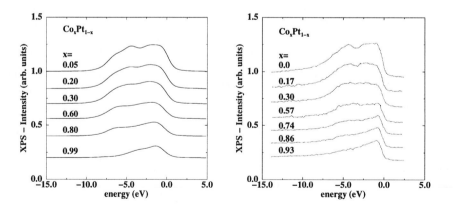

Fig. 14. Theoretical VB-XPS spectra (left) for $Co_x Pt_{1-x}$ for unpolarized radiation and photon energy $\hbar\omega = 1253.6$ eV. The corresponding experimental data (right) have been taken from Weller and Reim [259].

example Ref. [89]) both components retain the gross features of their electronic structure when they are combined to an alloy system: while Co has a narrow and strongly exchange-split d-band complex close to the Fermi level, the Pt d-band is rather broad and only slightly spin-split due to hybridization in the region of the Co d-band. This behavior supplies a very simple explanation for the variation of the spectra in Fig. 14 with composition and allows one to ascribe prominent features as shoulders and peaks to either Co or Pt. Nevertheless, one has to note that these spectra are not just a concentration weighted sum of the spectra for pure Co and pure Pt. One reason for this is that there are non-negligible changes of the component-resolved DOS-curves compared to those of the pure constituents. In addition one can see from Fig. 14 that for the selected photon energy the partial photoemission absorption cross section that depends on the initial state energy E as well as the photon energy $\hbar\omega$ is about a factor of two higher for Pt than for Co.

Performing angular-resolved photoemission experiments for disordered alloys seems to be profitable only in some favorable cases for which disorder does not prevent the existence of a rather well-defined dispersion relation [260]. While this applies for example to Fe_xNi_{1-x} (see Fig. 2) this is surely not the case for Co_xPt_{1-x}. Nevertheless, deeper insight into its electronic structure and in particular its magnetic aspects can be obtained by VB-XPS experiments done using circularly polarized radiation. In the case of Co_xPt_{1-x} [89] but also for Fe_xCo_{1-x} [261] the calculation of the VB-XPS spectra predict a non-negligible circular dichroism even for the angular averaged spectra that should be strong enough to be detected within an experiment (corresponding measurements have not yet been done). The circular dichroism occurs primarily in the region of pronounced spin-polarization and has the same sign throughout that energy region. This behavior differs from that of the angular resolved theoretical spectra obtained by Scheunemann et al. [90] for the normal emission from perpendicular magnetized Ni (001) that show a circular dichroism that change sign with binding energy with the energy integrated dichroism spectrum essentially vanishing [90]. While these specific spectra map states well below the Fermi energy with both spin characters contributing roughly to the same extent, the angular integrated spectra are dominated by the majority spin character [89]. Therefore, shifting the Fermi energy artificially to higher energies should result in a reversal in sign for the dichroism in the VB-XPS spectrum with the energy integrated signal going to zero.

A more pronounced circular dichroism than for the spin-averaged case is found for the spin-polarized case, with corresponding spectra given in Fig. 15. Again the circular dichroism occurs primarily in the energy region showing pronounced spin-polarization. Comparing the circular dichroism signal for the two different spin characters one notes that they are opposite in sign. This behavior has also been found before for the spin-resolved spectra mentioned above and explained as a direct consequence of the spin orbit coupling [90]. Finally it should be mentioned that circular dichroism spectra similar to that shown in Fig. 15 are found if the photon energy $\hbar\omega$ is varied. Setting $\hbar\omega$ to a value for which

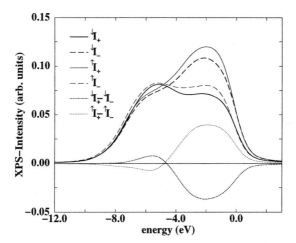

Fig. 15. Theoretical spin-resolved VB-XPS spectra of $Co_{0.6}Pt_{0.4}$ for circularly polarized radiation and photon energy $\hbar\omega = 1253.6$ eV.

one of the components possess a Cooper minimum allows one to get additional component specific information. This could supply very helpful information on the nature of the initial states and the contribution of the various components for the very complex magneto-optical Kerr-rotation spectra of transition metal alloys and compounds [262].

4   Summary

The formal background to deal with the electronic structure of magnetic solids, accounting simultaneously for relativistic effects, has been reviewed. The main emphasize has been laid on a four-component formalism that is based on a Dirac equation set up within the framework of DFT. As recent developments in this field, the inclusion of the Breit-interaction and the use of CDFT has been discussed. Solving the corresponding single site problem allows in principle to set up for any band structure method the necessary four-component spherical basis functions that account on a corresponding level of sophistication simultaneously and on the same footing for magnetic order and relativity. Here, the KKR-method has been used and accordingly several recent developments concerning this – as the incorporation of the OP-formalism, the full-potential formalism or the TB-version – have been discussed or mentioned. The main motivation for using the KKR-method stems from the fact that all features of the underlying Hamiltonian are directly incorporated in the corresponding wave functions, in contrast to schemes that account for corrections to an unperturbed Hamiltonian within a variational procedure. In addition, it directly gives access to the electronic Green's function. This very appealing feature has been exploited within many of the presented applications, as for example in the case of impurity systems lacking Bloch-symmetry.

Independent of the use of the KKR-method, the presented applications demonstrated that the spin-orbit coupling influences the properties of magnetic solids in various, quite different, ways. In some cases one gets just a correction to results that are obtained by a non- or scalar relativistic calculation. Of course, most interesting are pure spin-orbit induced properties that would not occur if the spin-orbit coupling or magnetic order would be absent. Nearly all properties discussed here fall into this category.

Using the full Dirac-formalism to deal with magnetic systems is of course very satisfying from a formal point of view but makes it in general quite cumbersome to give a simple interpretation of the numerical results. For this reason, approximate schemes, as those presented here, are very helpful and important for the discussion of the various spin-orbit induced properties. In particular, they allow to decide which consequence of the spin-orbit coupling is more important: lifting of energetic degeneracies or spin-hybridization. Concerning this question, it was demonstrated that the relevance of this two effects can be quite different for a spin-orbit induced property.

Altogether, one can claim that in general a rather satisfying quantitative agreement of theoretical and experimental data can be achieved when calculating spin-orbit induced properties. Remarkable deviations occur primarily if one is investigating properties closely connected with orbital magnetism. The most promising way to remove these problems in a consistent way, i.e. in particular within the framework of DFT, is to apply CDFT. Because of the delicate situation it might be necessary to perform the corresponding calculations in the full-potential mode and/or to use the optimized effective potential (OEP) scheme in addition. Anyhow, whatever will happen in this and related fields in the future, working with the four-component Dirac-Green's function-formalism will guaranty that one is open and prepared for any new developments.

Acknowledgment

The author would like to thank H. Akai, D. Ahlers, J. Banhart, M. Battocletti, P. H. Dederichs, M. Deng, H. Freyer, E. K. U. Gross, B. L. Gyorffy, T. Huhne, V. Popescu, J. Schwitalla, P. Strange, W. Temmerman, C. Zecha, R. Zeller, and A. Vernes for their fruitful and pleasant collaboration during the last years. The common work presented here benefited a lot from financial support and collaborations within the HCM-networks *Ab-initio (from electronic structure) calculation of complex processes in materials* and *Novel probes for magnetic materials and magnetic phenomena: linear and circular X-ray dichroism*, the TMR-network on *Ab-initio Calculations of Magnetic Properties of Surfaces, Interfaces and Multilayers* of the European Union, the programme *Relativistic Effects in Heavy-Element Chemistry and Physics (REHE)* of the European Science Foundation, the programme *Theorie relativistischer Effekte in der Chemie und Physik schwerer Elemente* funded by the DFG (Deutsche Forschungsgemeinschaft) and the programme *Zirkular polarisierte Synchrotronstrahlung: Dichroismus, Magnetismus und Spinorientierung* run by the German ministry for research and technology (BMBF).

References

1. N. E. Christensen and B. O. Seraphin, Phys. Rev. B **4**, 3321 (1971).
2. P. Pyykkö, Adv. Quantum. Chem. **11**, 353 (1978).
3. Y. Cauchois and I. Manescu, C. R. Acad. Sci. **210**, 172 (1940).
4. N. F. Mott, Proc. Phys. Soc. A **62**, 416 (1949).
5. A. R. Mackintosh and O. K. Andersen, in *Electrons at the Fermi Surface*, edited by M. Springford (Cambridge University Press, Cambridge, 1980), Chap. The Electronic Structure of Transition Metals.
6. L. Tterlikkis, S. D. Mahanti, and T. P. Das, Phys. Rev. **176**, 10 (1968).
7. M. E. Rose, *Relativistic Electron Theory* (Wiley, New York, 1961).
8. F. Rosicky and F. Mark, J. Phys. B: Atom. Molec. Phys. **8**, 2581 (1975).
9. F. Rosicky, P. Weinberger, and F. Mark, J. Phys. B: Atom. Molec. Phys. **9**, 2971 (1976).
10. D. D. Koelling and B. N. Harmon, J. Phys. C: Solid State Phys. **10**, 3107 (1977).
11. H. Gollisch and L. Fritsche, phys. stat. sol. (b) **86**, 145 (1978).
12. T. Takeda, Z. Physik B **32**, 43 (1978).
13. J. H. Wood and A. M. Boring, Phys. Rev. B **18**, 2701 (1978).
14. H. L. Skriver, *The LMTO-method* (Springer, Berlin, 1983).
15. N. J. M. Geipel and B. A. Hess, Chem. Phys. Lett. **273**, 62 (1997).
16. E. van Lenthe E J Baerends and J. G. Snijders, J. Chem. Phys. **101**, 1272 (1994).
17. F. Gesztesy, H. Grosse, and B. Thaller, Ann. Inst. Poincaré **40**, 159 (1984).
18. C. Brouder, M. Alouani, and K. H. Bennemann, Phys. Rev. B **54**, 7334 (1996).
19. D. D. Koelling and A. H. MacDonald, in *Relativistic Effects in Atoms Molecules and Solids*, edited by G. L. Malli (Plenum Press, New York, 1983).
20. S. Blügel, H. Akai, R. Zeller, and P. H. Dederichs, Phys. Rev. B **35**, 3271 (1987).
21. T. Gasche, Ph.D. thesis, University of Uppsala, 1993.
22. J. E. Lee, C. L. Fu, and A. J. Freeman, J. Magn. Magn. Materials **62**, 93 (1986).
23. T. Asada and K. Terakura, J. Phys. F: Met. Phys. **12**, 1387 (1982).
24. H. Ebert, P. Strange, and B. L. Gyorffy, J. Phys. F: Met. Phys. **18**, L135 (1988).
25. H. Ebert and H. Akai, Hyperfine Interactions **78**, 361 (1993).
26. C. S. Wang and J. Callaway, Phys. Rev. B **9**, 4897 (1974).
27. D. K. Misemer, J. Magn. Magn. Materials **72**, 267 (1988).
28. G. Y. Guo and H. Ebert, Phys. Rev. B **51**, 12633 (1995).
29. J. Callaway and C. S. Wang, Phys. Rev. B **7**, 1096 (1973).
30. O. K. Andersen, Phys. Rev. B **12**, 3060 (1975).
31. D. Singh, *Plane Waves, Pseudopotentials and the LAPW Method* (Kluwer Academic, Amsterdam, 1994).
32. A. Delin, P. Ravindran, O. Eriksson, and J. M. Wills, Intern. J. Quantum. Chem. **69**, 349 (1997).
33. T. L. Loucks, Phys. Rev. A **139**, 1333 (1965).
34. P. Soven, Phys. Rev. **137**, A 1706 (1965).
35. Y. Onodera and M. Okazaki, J. Phys. Soc. Japan **21**, 1273 (1966).
36. S. Takada, Progr. Theor. Phys. Suppl. **36**, 224 (1966).
37. T. Takeda, J. Phys. F: Met. Phys. **9**, 815 (1979).

38. U. Fano, Phys. Rev. **184**, 250 (1969).
39. U. Fano, Phys. Rev. **178**, 131 (1969).
40. U. Heinzmann, K. Jost, J. Kessler, and B. Ohnemus, Z. Physik **251**, 354 (1972).
41. J. Kerr, Phil. Mag. **3**, 321 (1877).
42. H. Ebert, Rep. Prog. Phys. **59**, 1665 (1996).
43. G. Schütz *et al.*, Phys. Rev. Letters **58**, 737 (1987).
44. J. Banhart and H. Ebert, Europhys. Lett. **32**, 517 (1995).
45. B. Újfalussy, L. Szunyogh, P. Bruno, and P. Weinberger, Phys. Rev. Letters **77**, 1805 (1996).
46. G. Seewald *et al.*, Phys. Rev. Letters **78**, 1795 (1997).
47. M. Singh, C. S. Wang, and J. Callaway, Phys. Rev. B **11**, 287 (1975).
48. M. S. S. Brooks and P. J. Kelly, Phys. Rev. Letters **51**, 1708 (1983).
49. J. Sticht and J. Kübler, Solid State Commun. **53**, 529 (1985).
50. W. M. Temmerman and P. A. Sterne, J. Phys.: Condensed Matter **2**, 5529 (1990).
51. O. Eriksson *et al.*, Phys. Rev. B **42**, 2707 (1990).
52. S. P. Lim, D. L. Price, and B. R. Cooper, IEEE transactions on magnetics **27**, 3648 (1991).
53. B. I. Min and Y.-R. Jang, J. Phys.: Condensed Matter **3**, 5131 (1991).
54. O. Hjortstam *et al.*, Phys. Rev. B **53**, 9204 (1996).
55. D. S. Wang, R. Q. Wu, L. P. Z. LP, and A. J. Freeman, J. Magn. Magn. Materials **144**, 643 (1995).
56. V. P. Antropov, A. I. Liechtenstein, and B. N. Harmon, J. Magn. Magn. Materials **140-144**, 1161 (1995).
57. H. Ebert, Phys. Rev. B **38**, 9390 (1988).
58. J. M. MacLaren and R. H. Victora, J. Appl. Physics **76**, 6069 (1994).
59. H. Akai, private communication, 1995.
60. A. Ankudinov, Ph.D. thesis, University of Washington, 1985.
61. A. K. Rajagopal and J. Callaway, Phys. Rev. B **7**, 1912 (1973).
62. U. von Barth and L. Hedin, J. Phys. C: Solid State Phys. **5**, 1629 (1972).
63. A. H. MacDonald and S. H. Vosko, J. Phys. C: Solid State Phys. **12**, 2977 (1979).
64. M. V. Ramana and A. K. Rajagopal, Adv. Chem. Phys. **54**, 231 (1983).
65. M. Richter and H. Eschrig, Solid State Commun. **72**, 263 (1989).
66. S. Doniach and C. Sommers, in *Valence Fluctuations in Solids*, edited by L. M. Falicov, W. Hanke, and M. B. Maple (North-Holland, Amsterdam, 1981), p. 349.
67. R. Feder, F. Rosicky, and B. Ackermann, Z. Physik B **52**, 31 (1983).
68. P. Strange, J. B. Staunton, and B. L. Gyorffy, J. Phys. C: Solid State Phys. **17**, 3355 (1984).
69. I. V. Solovyev *et al.*, Soviet Physics - Solid State **31**, 1285 (1989).
70. B. C. H. Krutzen and F. Springelkamp, J. Phys.: Condensed Matter **1**, 8369 (1989).
71. B. Ackermann, R. Feder, and E. Tamura, J. Phys. F: Met. Phys. **14**, L173 (1984).
72. G. Schadler, R. C. Albers, A. M. Boring, and P. Weinberger, Phys. Rev. B **35**, 4324 (1987).
73. H. Ebert, B. Drittler, and H. Akai, J. Magn. Magn. Materials **104-107**, 733 (1992).
74. L. Szunyogh, B. Újfalussy, P. Weinberger, and J. Kollar, Phys. Rev. B **49**, 2721 (1994).
75. R. Zeller *et al.*, Phys. Rev. B **52**, 8807 (1995).
76. H. Ebert, V. Popescu, and D. Ahlers, J. Phys. (Paris) **7**, C2 131 (1997).
77. H. Ebert, V. Popescu, and D. Ahlers, Phys. Rev. B, submitted (1998).

78. S. V. Beiden *et al.*, Phys. Rev. B **57**, 14247 (1998).
79. E. Tamura, Phys. Rev. B **45**, 3271 (1992).
80. X. Wang *et al.*, Phys. Rev. B **46**, 9352 (1992).
81. S. C. Lovatt, B. L. Gyorffy, and G.-Y. Guo, J. Phys.: Condensed Matter **5**, 8005 (1993).
82. T. Huhne, Master's thesis, University of Munich, 1997.
83. T. Huhne *et al.*, Phys. Rev. B **58**, 10236 (1998).
84. H. Ebert, P. Strange, and B. L. Gyorffy, J. Appl. Physics **63**, 3055 (1988).
85. H. Ebert, P. Strange, and B. L. Gyorffy, Z. Physik B **73**, 77 (1988).
86. P. Strange, P. J. Durham, and B. L. Gyorffy, Phys. Rev. Letters **67**, 3590 (1991).
87. E. Tamura, G. D. Waddill, J. G. Tobin, and P. A. Sterne, Phys. Rev. Letters **73**, 1533 (1994).
88. H. Ebert, L. Baumgarten, C. M. Schneider, and J. Kirschner, Phys. Rev. B **44**, 4466 (1991).
89. H. Ebert and J. Schwitalla, Phys. Rev. B **55**, 3100 (1997).
90. T. Scheunemann, S. V. Halilov, J. Henk, and R. Feder, Solid State Commun. **91**, 487 (1994).
91. E. Arola, P. Strange, and B. L. Gyorffy, Phys. Rev. B **55**, 472 (1997).
92. H. J. Gotsis and P. Strange, J. Phys.: Condensed Matter **6**, 1409 (1994).
93. A. B. Shick, V. Drchal, J. Kudrnovský, and P. Weinberger, Phys. Rev. B **54**, 1610 (1996).
94. P. Weinberger, I. Turek, and L. Szunyogh, Intern. J. Quantum. Chem. **63**, 165 (1997).
95. I. Turek *et al.*, *Electronic structure of disordered alloys, surfaces and interfaces* (Kluwer Academic Publ., Boston, 1997).
96. R. M. Dreizler and E. K. U. Gross, *Density Functional Theory* (Springer-Verlag, Heidelberg, 1990).
97. H. Eschrig, *The Fundamentals of Density Functional Theory* (B G Teubner Verlagsgesellschaft, Stuttgart, Leipzig, 1996).
98. A. H. MacDonald, J. Phys. C: Solid State Phys. **16**, 3869 (1983).
99. A. K. Rajagopal, J. Phys. C: Solid State Phys. **11**, L943 (1978).
100. M. V. Ramana and A. K. Rajagopal, J. Phys. C: Solid State Phys. **12**, L845 (1979).
101. H. Eschrig, G. Seifert, and P. Ziesche, Solid State Commun. **56**, 777 (1985).
102. B. X. Xu, A. K. Rajagopal, and M. V. Ramana, J. Phys. C: Solid State Phys. **17**, 1339 (1984).
103. M. V. Ramana and A. K. Rajagopal, J. Phys. C: Solid State Phys. **14**, 4291 (1981).
104. G. Vignale and M. Rasolt, Phys. Rev. Letters **59**, 2360 (1987).
105. G. Vignale and M. Rasolt, Phys. Rev. B **37**, 10685 (1988).
106. G. Vignale and M. Rasolt, Phys. Rev. Letters **62**, 115 (1989).
107. G. Vignale, in *Current Density Functional Theory and Orbital Magnetism*, Vol. 337 of *Nato ASI Series, Series B*, edited by E. K. U. Gross and R. M. Dreizler (Plenum Press, NewYork, 1995), p. 485.
108. P. Skudlarski and G. Vignale, Phys. Rev. B **48**, 8547 (1993).
109. H. Ebert, M. Battocletti, and E. K. U. Gross, Europhys. Lett. **40**, 545 (1997).
110. H. Bethe and E. Salpeter, *Quantum Mechanics of One- and Two-Electron Atoms* (Springer, New York, 1957).
111. H. J. F. Jansen, J. Appl. Physics **64**, 5604 (1988).
112. H. Ebert, unpublished (1995).

113. M. S. S. Brooks, Physica B **130**, 6 (1985).

114. O. Eriksson, B. Johansson, and M. S. S. Brooks, J. Phys.: Condensed Matter **1**, 4005 (1989).

115. O. Eriksson, M. S. S. Brooks, and B. Johansson, Phys. Rev. B **41**, 7311 (1990).

116. L. Severin, M. S. S. Brooks, and B. Johansson, Phys. Rev. Letters **71**, 3214 (1993).

117. A. B. Shick and V. A. Gubanov, Phys. Rev. B **49**, 12860 (1994).

118. H. Ebert and M. Battocletti, Solid State Commun. **98**, 785 (1996).

119. G. Racah, Phys. Rev. **42**, 438 (1942).

120. R. Pittini, J. Schoenes, O. Vogt, and P. Wachter, Phys. Rev. Letters **77**, 944 (1996).

121. W. M. Temmerman, Z. Szotek, and G. A. Gehring, Phys. Rev. Letters **79**, 3970 (1997).

122. J. Forstreuter, L. Steinbeck, M. Richter, and H. Eschrig, Phys. Rev. B **55**, 9415 (1997).

123. E. Engel *et al.*, Phys. Rev. A **58**, 964 (1998).

124. A. H. MacDonald, J. M. Daams, S. H. Vosko, and D. D. Koelling, Phys. Rev. B **23**, 6377 (1981).

125. L. Severin, M. Richter, and L. Steinbeck, Phys. Rev. B **55**, 9211 (1997).

126. R. N. Schmid *et al.*, Adv. Quantum. Chem. in press (1999).

127. P. Cortona, S. Doniach, and C. Sommers, Phys. Rev. A **31**, 2842 (1985).

128. B. Ackermann, Ph.D. thesis, University of Duisburg, 1985.

129. A. C. Jenkins and P. Strange, J. Phys.: Condensed Matter **6**, 3499 (1994).

130. B. Drittler, H. Ebert, R. Zeller, and P. H. Dederichs, Phys. Rev. B **67**, 4573 (1990).

131. P. H. Dederichs, B. Drittler, and R. Zeller, Mat. Res. Soc. Symp. Proc. **253**, 185 (1992).

132. O. K. Andersen and R. G. Wooley, Molecular Physics **26**, 905 (1973).

133. J. S. Faulkner, J. Phys. C: Solid State Phys. **10**, 4661 (1977).

134. J. S. Faulkner, Phys. Rev. B **19**, 6186 (1979).

135. H. Ebert and B. L. Gyorffy, J. Phys. F: Met. Phys. **18**, 451 (1988).

136. J. S. Faulkner and G. M. Stocks, Phys. Rev. B **21**, 3222 (1980).

137. H. Ebert, H. Freyer, A. Vernes, and G.-Y. Guo, Phys. Rev. B **53**, 7721 (1996).

138. H. Ebert, H. Freyer, and M. Deng, Phys. Rev. B **56**, 9454 (1997).

139. H. Ebert, habilitaion thesis, University of München, 1990.

140. B. L. Gyorffy and M. J. Stott, in *Band Structure Spectroscopy of Metals and Alloys*, edited by D. J. Fabian and L. M. Watson (Academic Press, New York, 1973), p. 385.

141. T. Fujikawa, R. Yanagisawa, N. Yiwata, and K. Ohtani, J. Phys. Soc. Japan **66**, 257 (1997).

142. H. Wu and S. Y. Tong, Phys. Rev. B submitted (1998).

143. A. P. Cracknell, J. Phys. C: Solid State Phys. **2**, 1425 (1969).

144. A. P. Cracknell, Phys. Rev. B **1**, 1261 (1970).

145. G. Hörmandinger and P. Weinberger, J. Phys.: Condensed Matter **4**, 2185 (1992).

146. C. Zecha, Master's thesis, University of Munich, 1997.

147. C. Kornherr, Master's thesis, University of Munich, 1997.

148. J. B. Pendry, *Low energy electron diffraction* (Academic Press, London, 1974).

149. M. Fluchtmann, J. Braun, and G. Borstel, Phys. Rev. B **52**, 9564 (1995).

150. L. Szunyogh, B. Újfalussy, and P. Weinberger, Phys. Rev. B **51**, 9552 (1995).

151. R. Zeller, P. Lang, B. Drittler, and P. H. Dederichs, Mat. Res. Soc. Symp. Proc. **253**, 357 (1992).

152. T. Huhne, Ph.D. thesis, University of München, in preparation, 1989.
153. B. Nonas, Ph.D. thesis, RWTH Aachen, in preparation, 1989.
154. P. Soven, Phys. Rev. **156**, 809 (1967).
155. P. Strange, H. Ebert, J. B. Staunton, and B. L. Gyorffy, J. Phys.: Condensed Matter **1**, 2959 (1989).
156. J. S. Faulkner, Prog. Mater. Sci. **27**, 1 (1982).
157. H. Ebert, A. Vernes, and J. Banhart, Solid State Commun. **104**, 243 (1997).
158. H. Eckardt and L. Fritsche, J. Phys. F: Met. Phys. **17**, 925 (1987).
159. I. Mertig, R. Zeller, and P. H. Dederichs, Phys. Rev. B **47**, 16178 (1993).
160. H. Ebert, J. Banhart, and A. Vernes, Phys. Rev. B **54**, 8479 (1996).
161. J. Banhart, H. Ebert, and A. Vernes, Phys. Rev. B **56**, 10165 (1997).
162. A. B. Shick, D. L. Novikov, and A. J. Freeman, Phys. Rev. B **56**, R14259 (1997).
163. A. B. Shick, D. L. Novikov, and A. J. Freeman, J. Appl. Physics **83**, 7258 (1998).
164. X. Wang, D. Wang, R. Wu, and A. J. Freeman, J. Magn. Magn. Materials **159**, 337 (1996).
165. G. H. O. Daalderop, P. J. Kelly, and M. F. H. Schuurmans, Phys. Rev. B **41**, 11919 (1990).
166. P. Strange, J. B. Staunton, and H. Ebert, Europhys. Lett. **9**, 169 (1989).
167. Újfalussy, L. Szunyogh, and P. Weinberger, in *Properties of Complex Inorganic Solids*, edited by A. Gonis, A. Meike, and P. E. A. Turchi (Plenum Press, New York, 1997), p. 181.
168. V. A. Gubanov, A. I. Lichtenstejn, and A. V. Postnikov, *Magnetism and the electronic structure of crystals* (Springer, Berlin, 1992).
169. S. S. A. Razee, J. B. Staunton, and F. J. Pinski, Phys. Rev. B **56**, 8082 (1997).
170. G. Y. Guo, H. Ebert, and W. M. Temmerman, J. Phys.: Condensed Matter **3**, 8205 (1991).
171. G. Y. Guo, H. Ebert, and W. M. Temmerman, J. Magn. Magn. Materials **104-107**, 1772 (1992).
172. M. B. Stearns, in *Magnetic Properties of 3d, 4d and 5d Elements, Alloys and Compounds*, Vol. III/19a of *Landolt-Börnstein, New Series*, edited by K.-H. Hellwege and O. Madelung (Springer, Berlin, 1987).
173. O. Eriksson *et al.*, Phys. Rev. B **41**, 11807 (1990).
174. J. Trygg, B. Johansson, O. Eriksson, and J. M. Wills, Phys. Rev. Letters **75**, 2871 (1995).
175. P. Söderlind *et al.*, Phys. Rev. B **45**, 12911 (1992).
176. M. S. S. Brooks and B. Johansson, in *Spin-orbit influenced spectroscopies of magnetic solids*, Vol. 466 of *Lecture Notes in Physics*, edited by H. Ebert and G. Schütz (Springer, Berlin, 1996), p. 211.
177. G. H. O. Daalderop, P. J. Kelly, and M. F. H. Schuurmans, Phys. Rev. B **44**, 12054 (1991).
178. H. Ebert, A. Vernes, and J. Banhart, Phys. Rev. B submitted (1998).
179. Gasche, M. S. S. Brooks, and B. Johansson, Phys. Rev. B **53**, 296 (1996).
180. H. Ebert, Solid State Commun. **100**, 677 (1996).
181. P. Herzig, Theoret. Chim. Acta. (Berlin) **67**, 323 (1985).
182. M. Battocletti, Ph.D. thesis, University of München, 1997.
183. P. C. Riedi and E. Hagn, Phys. Rev. B **30**, 5680 (1984).
184. M. Aiga and J. Itoh, J. Phys. Soc. Japan **37**, 967 (1974).
185. Eder, E. Hagn, and E. Zech, Phys. Rev. C **32**, 582 (1985).
186. M. Kawakami, H. Enokiya, and T. Okamoto, J. Phys. F: Met. Phys. **15**, 1613 (1985).

187. P. Blaha, K. Schwarz, and P. H. Dederichs, Phys. Rev. B **37**, 2792 (1988).
188. H. Ebert, R. Zeller, B. Drittler, and P. H. Dederichs, J. Appl. Physics **67**, 4576 (1990).
189. G. A. Gehring and H. C. W. L. Williams, J. Phys. F: Met. Phys. **4**, 291 (1974).
190. C. Demangeat, J. Phys. F: Met. Phys. **5**, 169 (1975).
191. G. Y. Guo and H. Ebert, Phys. Rev. B **50**, 10377 (1994).
192. D. Weller *et al.*, Phys. Rev. Letters **72**, 2097 (1994).
193. G. Breit, Phys. Rev. **35**, 1447 (1930).
194. N. C. Pyper, Molecular Physics **64**, 933 (1988).
195. A. Abragam and M. H. L. Pryce, Proc. Roy. Soc. (London) A **205**, 135 (1951).
196. M. Battocletti, H. Ebert, and H. Akai, Phys. Rev. B **53**, 9776 (1996).
197. H. Akai and T. Kotani, Hyperfine Interactions, in press (1998).
198. K. Terakura, N. Hamada, T. Oguchi, and T. Asada, J. Phys. F: Met. Phys. **12**, 1661 (1982).
199. J. B. Staunton, Ph.D. thesis, University of Bristol, 1982.
200. M. Matsumoto, J. B. Staunton, and P. Strange, J. Phys.: Condensed Matter **2**, 8365 (1990).
201. M. Matsumoto, J. B. Staunton, and P. Strange, J. Phys.: Condensed Matter **3**, 1453 (1991).
202. J. F. Janak, Phys. Rev. B **16**, 255 (1977).
203. J. Benkowitsch and Winter, J. Phys. F: Met. Phys. **13**, 991 (1983).
204. J. Benkowitsch and Winter, Physica Scripta **31**, 222 (1985).
205. H. Ebert, Ph.D. thesis, University of München, 1986.
206. H. Ebert, H. Winter, and J. Voitländer, J. Phys. F: Met. Phys. **16**, 1133 (1986).
207. M. Yasui and M. Shimizu, J. Phys. F: Met. Phys. **9**, 1653 (1979).
208. M. Yasui and M. Shimizu, J. Phys. F: Met. Phys. **15**, 2365 (1985).
209. G. C. Carter, L. H. Bennett, and D. J. Kahan, Prog. Mater. Sci. **20**, 153 (1977).
210. I. V. Solovyev, P. H. Dederichs, and I. Mertig, Phys. Rev. B **5**, 13419 (1995).
211. D. A. Greenwood, Proc. Phys. Soc. **71**, 585 (1958).
212. W. H. Butler, Phys. Rev. B **31**, 3260 (1985).
213. W. H. Kleiner, Phys. Rev. **142**, 318 (1966).
214. P. R. Birss, *Selected Topics in Solid State Physics, vol.III* (North-Holland, Amsterdam, 1966).
215. T. R. McGuire and R. I. Potter, IEEE Transactions on Magnetics **11**, 1018 (1975).
216. P. J. Durham, B. L. Gyorffy, and A. J. Pindor, J. Phys. F: Met. Phys. **10**, 661 (1980).
217. J. Banhart, H. Ebert, P. Weinberger, and J. Voitländer, Phys. Rev. B **50**, 2104 (1994).
218. S. U. Jen, Phys. Rev. B **45**, 9819 (1992).
219. S. Senoussi, I. A. Campbell, and A. Fert, Solid State Commun. **21**, 269 (1977).
220. A. Hamzić, S. Senoussi, I. A. Campbell, and A. Fert, J. Phys. F: Met. Phys. **8**, 1947 (1978).
221. S. U. Jen, T. P. Chen, and B. L. Chao, Phys. Rev. B **48**, 12789 (1993).
222. T. R. McGuire, J. A. Aboaf, and E. Klokholm, J. Appl. Physics **55**, 1951 (1984).
223. J. Banhart, A. Vernes, and H. Ebert, Solid State Commun. **98**, 129 (1996).
224. F. P. Beitel and E. M. Pugh, Phys. Rev. **112**, 1516 (1958).
225. P. P. Freitas and L. Berger, Phys. Rev. B **37**, 6079 (1988).
226. J. Smit, Physica **16**, 612 (1951).
227. O. Jaoul, I. A. Campbell, and A. Fert, J. Magn. Magn. Materials **5**, 23 (1977).
228. W. H. Butler, X. G. Zhang, D. M. C. Nicholson, and J. M. MacLaren, J. Magn. Magn. Materials **151**, 354 (1995).

229. P. Weinberger *et al.*, J. Phys.: Condensed Matter **8**, 7677 (1996).

230. J. Banhart, private communication, 1998.

231. T. Huhne and H. Ebert, Solid State Commun. to be submitted (1998).

232. W. Geerts *et al.*, Phys. Rev. B **50**, 12581 (1994).

233. H. Ebert *et al.*, Mat. Res. Soc. Symp. Proc. **475**, 407 (1998).

234. Y. Suzuki, T. Katayama, P. B. S. Yuasa, and E. Tamura, Phys. Rev. Letters **80**, 5200 (1998).

235. M. Faraday, Phil. Trans. Roy. Soc. **136**, 1 (1846).

236. H. R. Hulm, Proc. Roy. Soc. (London) A **135**, 237 (1932).

237. J. L. Erskine and E. A. Stern, Phys. Rev. B **12**, 5016 (1975).

238. G. Schütz *et al.*, Phys. Rev. Letters **62**, 2620 (1989).

239. R. Wienke, G. Schütz, and H. Ebert, J. Appl. Physics **69**, 6147 (1991).

240. B. T. Thole, P. Carra, F. Sette, and G. van der Laan, Phys. Rev. Letters **68**, 1943 (1992).

241. G. Schütz, M. Knülle, and H. Ebert, Physica Scripta **T49**, 302 (1993).

242. P. Carra, B. T. Thole, M. Altarelli, and X. Wang, Phys. Rev. Letters **70**, 694 (1993).

243. H. Maruyama *et al.*, J. Magn. Magn. Materials **104**, 2055 (1992).

244. H. Ebert *et al.*, Int. J. Mod. Phys. B **7**, 750 (1993).

245. H. Ebert and H. Akai, Int. J. Mod. Phys. B **7**, 922 (1993).

246. F. Baudelet *et al.*, J. Phys. (Paris) **7**, C2 441 (1997).

247. S. Stähler *et al.*, J. Appl. Physics **73**, 6063 (1993).

248. J. Mustre de Leon *et al.*, Phys. Rev. B **44**, 4146 (1991).

249. D. O. Ahlers, Ph.D. thesis, Universität Würzburg, 1998.

250. V. Popescu, H. Ebert, and A. C. Jenkins, J. Synchr. Rad. **5**, in press (1998).

251. P. J. Feibelman and D. E. Eastman, Phys. Rev. B **10**, 4932 (1974).

252. B. Ackermann and R. Feder, J. Phys. C: Solid State Phys. **18**, 1093 (1985).

253. B. Ginatempo, P. J. Durham, B. L. Gyorffy, and W. M. Temmerman, Phys. Rev. Letters **54**, 1581 (1985).

254. P. J. Durham, in *The Electronic Structure of Complex Systems*, edited by P. Phariseau and W. M. Temmerman (Plenum Press, New York, 1984), p. 709.

255. P. J. Durham, J. Phys. F: Met. Phys. **11**, 2475 (1981).

256. H. Ebert and G. Y. Guo, J. Magn. Magn. Materials **148**, 174 (1995).

257. H. Winter, P. J. Durham, and G. M. Stocks, J. Phys. F: Met. Phys. **14**, 1047 (1984).

258. P. Marksteiner *et al.*, Phys. Rev. B **34**, 6730 (1986).

259. D. Weller and W. Reim, Appl. Physics A **49**, 599 (1989).

260. M. Schröder *et al.*, Phys. Rev. B **52**, 188 (1995).

261. S. Ostanin and H. Ebert, Phys. Rev. B **58**, 11577 (1998).

262. D. Weller *et al.*, Mater. Res. Soc **313**, 501 (1993).

First Principles Theory of Magneto–Crystalline Anisotropy

O. Eriksson[1] and J. Wills[2]

[1] Department of Physics, Uppsala University, Box 530, S-75112 Uppsala, Sweden
[2] Los Alamos National Laboratory, Los Alamos, NM87545, USA

Abstract. A review of a state of the art, theoretical method for calculating the magneto crystalline anisotropy (MAE) is given. The fundamentals of first principles, total energy calculations, i.e. density functional theory, are described. In addition one of the most common methods for solving the Kohn–Sham equation, the linear muffin–tin orbital method, is described briefly. Technical aspects and difficulties for performing theoretical studies of the MAE are discussed and several examples are given. It is pointed out that the orbital magnetism and the MAE often are intimately connected. The MAE is also argued to be connected to other details of the electronic structure, such as the values of the density of states (DOS) at the Fermi level, the partitioning of the DOS into crystal field components and the hybridization with orbitals of possible ligand atoms.

1 Introduction

Today we are witnessing how modern electronic structure theory has become a very useful theoretical tool in explaining and complementing experiments. Often accurate calculations can reproduce observed magnetic properties such as the occurrence of ferro-, ferri-, antiferro- and paramagnetism. Also, the direction and magnitude of the magnetic moments of bulk, surface and inter-phase systems are normally described well, as is the preferred easy axis of the magnetization direction, something which is called the magneto–crystalline anisotropy-MAE. In addition it has become possible to reproduce the finding that magnetic moments do not always point in the same direction, as is the case in non–collinear magnets and systems with spin spirals [1]. Parallel to this development, theories of micro–magnetism [2] have advanced attempting to describe processes such as magnetization reversal and nucleation when the direction of an external field is changed. These theories describe 'realistic' magnetization phenomena, such as domain formation, domain walls (Bloch and Neel walls), magnetization rotation and curling, etc., which may not always be accessible directly to first principles theorists. However, one important parameter in these theories is the magneto-crystalline anisotropy (MAE). Information of this physical property is often given in terms of the so called anisotropy constants (e.g. K_1 and K_2). Apart from being an important parameter in micro–magnetic models, a deeper insight about the MAE, which shows the directions in the crystal which are easy and hard magnetization directions, is important knowledge in itself. This is not least due to the possibility of designing new materials which are optimal for information storage. Let us illustrate this statement with an example. A small fraction of a magnetic disc with the magnetic moment pointing in one direction may be thought of as representing a 0 in a binary code whereas another section with a reversed moment direction would represent a 1. It becomes important to have the directions well defined, and as a consequence materials with an optimal MAE are searched for. Recently a lot of attention has been focused on orbital magnetism, $\langle L_z \rangle$, of itinerant electron systems, especially since it has been observed that there may be a connection between orbital magnetism and the magneto–crystalline anisotropy (MAE) [3]. As will be argued below, this means that one avenue to find a material with a large MAE is to search for a large orbital moment. At least materials which have a large directional dependence (anisotropy) of the orbital moment are expected to have enhanced MAE [3].

This chapter deals with theoretical aspects of the MAE of transition metals, its relation to important parameters such as the crystal structure, spin and orbital moments, electronic structure and hybridization with possible neighboring atoms. Although several examples of bulk, surfaces and interfaces will be given, this chapter is by far not complete and should not be viewed as an overview of the subject. It is merely intended to illustrate ideas, give understanding and to show ways to think about the MAE. In this chapter we shall thus outline how one can calculate the orbital moment of an itinerant electron system. We also point out that most of the discussion in this chapter deals with elements and compounds where the magnetism is provided by itinerant electrons, electrons which occupy band states that have dispersion (an example of this is bcc Fe). This is in contrast to localized electrons which are atomic like, even in the crystalline environment (an example is hcp Gd, a rare earth element). Calculation of the MAE requires resolving the difference in total energy (which often is of the order of several thousands of Ry) when the magnetization is pointing in two different directions of the solid, with a requirement of accuracy sometimes better than 1 μRy. It is the purpose of this chapter to outline some of the techniques used at present and to give a few examples of the theoretical modeling of magnetism of solids by means of first principles calculations. This chapter begins by reviewing the most important parts of the basis of modern electronic structure theory, namely Density Functional Theory (DFT) [4,5] and the approximations which make it useful for calculations, i.e. the Local Density Approximation (LDA) and the Generalized Gradient Approximation (GGA). The latter two approximations give rise to explicit but approximate expressions of the total energy of a material. At this point we would like to draw attention to the fact that there exist textbooks dealing with this topic; for a more detailed study of DFT, LDA and GGA we refer the reader to Dreitzler and Gross [6], and Eschrig [7]. There are also good review articles on this such as Gunnarsson and Jones [8], and Callaway and March [9]. We will also describe in some detail one of the many methods which exist for calculating the electronic structure and total energy of a solid or a surface, a basis function method based on so–called linear muffin–tin orbitals (LMTO). There are already detailed descriptions of methods for calculating the electronic structure of a solid based on linear muffin–tin orbitals [10–12]. However the method outlined here makes fewer approximations and is more accurate. In achieving this there are obstacles which must be over-come which have not been described before, we will try to point these out and compare with the previous methods [10–12]. We would also like to mention that detailed descriptions of other computational methods, such as the Augmented Plane Wave (APW) method of Slater [13] and the modified, linearized version of it (LAPW) by Andersen [10], can be found in Ref. [14] and [15], respectively. Another computational method which calculates the LDA or GGA total energy is based on the so called Linear Combination of Atomic Orbitals (LCAO), or-bitals which may be recognized from text books on chemistry and physics of molecules [16]. The LCAO method is described in detail in Refs. [17] and [18] and in addition calculations of the electronic structure of a variety of elements [19] and compounds [20] have been compiled in book form.

2   Introductory Remarks on Electronic Structure Theory

All theories for calculating the magnetic properties and the total energy of solids, surfaces and interfaces start out by adopting the Born–Oppenheimer approxi-mation. This approximation simply neglects the movement of the atomic nuclei and the electrons are considered to be moving around in a material where all nu-clei are at fixed positions. The motivation for this is that the electrons are much lighter than the nuclei and thus move much faster. For the materials discussed in this chapter, where the lightest element studied belongs to the 3d series, this approximation is a very good one. One can now focus solely on the electrons, which in itself is a formidable problem. The electrons interact with the positive atomic nuclei and with each other electrons via Coulomb forces. Although the former interaction is by no means simple it can be treated, whereas the latter interaction is impossible to calculate and one must resort to approximations.

Attempts to estimate the electron–electron interaction in solids in order to calculate the electronic dispersion of solids or the total energy of them dates back to the days of the Thomas–Fermi model [21], the Hartree approximation and to the X–α method of Slater [13]. An extension of these ideas culminated in what we today know as Density Functional Theory (DFT) [4,5]. As we will see below this has made it possible to calculate the total energy of for instance a solid, using the electron density, $n(\mathbf{r})$, as the key variable $(n(\mathbf{r})=n^\uparrow(\mathbf{r})+n^\downarrow(\mathbf{r}))$ where $n^\uparrow(\mathbf{r})$ is the spin up electron density and $n^\downarrow(\mathbf{r})$ the spin down electron density). For magnetic systems one has also to consider the magnetization density, $m(\mathbf{r})$ $(m(\mathbf{r})=n^\uparrow(\mathbf{r})-n^\downarrow(\mathbf{r}))$, and examples will be given in this chapter of calculations based on DFT where the total energy and the magnetic properties of a solid are successfully reproduced.

3 Density Functional Theory

3.1 The Hohenberg–Kohn Theorem

Let us start the description of the basic ideas of DFT by considering a non–magnetic system with spin degeneracy[1], the spin polarized case will be discussed later. It is now our purpose to prove the two theorems that are the basis of DFT.

Theorem 1. *The total energy of a system is a unique functional of the ground state electron density.*

To demonstrate this we consider a Hamiltonian, $H = T + V + W$, where T represents the kinetic energy of the system, V the interaction of the electrons with an external potential (normally this is the potential given by the atomic nuclei in the solid) and W the (exact but complex) electron–electron interaction. The solution to this Hamiltonian results in a ground state many body wave function $\Psi(\mathbf{r}_1, \mathbf{r}_2, ...\mathbf{r}_N)$ (for N electrons), and we have

$$H\Psi = E_{\mathrm{gs}}\Psi. \tag{1}$$

The electron density can now be calculated from

$$n(\mathbf{r}) = \int d^3 r_i \Pi_{i=1}^N \mid \Psi(\mathbf{r}_1, \mathbf{r}_2, ...\mathbf{r}_N) \mid^2 \delta(\mathbf{r} - \mathbf{r}_i) \tag{2}$$

and the interaction V is written as $V = \int n(\mathbf{r})v(\mathbf{r})d^3r$, where $v(\mathbf{r})$ is the external potential. We will now demonstrate that two different external potentials $v(\mathbf{r})$ and $v'(\mathbf{r})$ must give rise to different ground state electron densities. To show this we note firstly that for the system with potential $v'(\mathbf{r})$ we have

$$H'\Psi' = E'_{\mathrm{gs}}\Psi'. \tag{3}$$

From the variational principle it follows that

$$E_{\mathrm{gs}} = \langle \Psi \mid H \mid \Psi \rangle < \langle \Psi' \mid H \mid \Psi' \rangle. \tag{4}$$

By adding and subtracting $v'(r)$ on the rhs. of (4) we obtain

$$\langle \Psi' \mid H \mid \Psi' \rangle = \langle \Psi' \mid H' + V - V' \mid \Psi' \rangle \tag{5}$$

$$= E'_{\mathrm{gs}} + \int n'(\mathbf{r})(v(\mathbf{r}) - v'(\mathbf{r}))d^3r.$$

Combining the expressions in (4) and (5) gives,

$$E_{\mathrm{gs}} < E'_{\mathrm{gs}} + \int n'(\mathbf{r})(v(\mathbf{r}) - v'(\mathbf{r}))d^3r. \tag{6}$$

---

[1] A more elaborate discussion of the contents of this chapter may be found in the text book by Dreitzler and Gross [6].

A similar argument starting from the expression

$$E'_{gs} = \langle \Psi' \mid H' \mid \Psi' \rangle < \langle \Psi \mid H' \mid \Psi \rangle, \tag{7}$$

results in

$$E'_{gs} < E_{gs} + \int n(\mathbf{r})(v'(\mathbf{r}) - v(\mathbf{r}))d^3r. \tag{8}$$

We will now show that if we assume that $n'(\mathbf{r}) = n(\mathbf{r})$ an absurd result emerges and hence this assumption must be wrong, i.e. $n'(\mathbf{r}) \neq n(\mathbf{r})$. The absurdity comes from the fact that if (6) and (8) are added and one assumes $n'(\mathbf{r}) = n(\mathbf{r})$ one obtains the equation

$$E_{gs} + E'_{gs} < E'_{gs} + E_{gs}, \tag{9}$$

which is clearly wrong. Hence $n'(\mathbf{r}) \neq n(\mathbf{r})$ and we conclude that knowledge of the electron density, $n(\mathbf{r})$, implies that we know it was calculated from a Hamiltonian with an external potential $v(\mathbf{r})$ (since we just showed that $v(\mathbf{r})$ and $v'(\mathbf{r})$ give rise to different densities $n(\mathbf{r})$ and $n'(\mathbf{r})$). As the kinetic energy, T, and electron–electron interactions, W, are known and specified we conclude that knowledge of the ground state electron density determines the entire Hamiltonian and hence the ground state energy, which proves Theorem 1 (although an explicit and practical form for calculating E_{gs} from $n(\mathbf{r})$ is not clear from the arguments given above). One can thus express a functional relationship between the ground state energy and the corresponding electron density as

$$E[n(\mathbf{r})] = T[n(\mathbf{r})] + V[n(\mathbf{r})] + W[n(\mathbf{r})]. \tag{10}$$

Since the arguments that lead to this relationship do not depend on the form of $v(\mathbf{r})$ (and hence is valid for atoms, molecules and solids) the kinetic energy and electron–electron interaction $T + W \equiv F[n(\mathbf{r})]$ is called a *universal* functional of the electron density.

A second important theorem of DFT is

Theorem 2. *The exact ground state density minimizes $E[n(\mathbf{r})]$.*

This statement partly follows from the fact that the many electron wave function is also specified by the electron density [6], since the ground state density specifies the Hamiltonian and hence also the wave function (of the ground state and of excited states), and to illustrate this dependence we write $\Psi[n(\mathbf{r})]$. To prove Theorem 2 we next note that for a given external potential $v_0(\mathbf{r})$ we can from Theorem 1 write

$$E_{v_0}[n(\mathbf{r})] = < \Psi[n(\mathbf{r})] \mid T + W + V_0 \mid \Psi[n(\mathbf{r})] >, \tag{11}$$

where the subscript v_0 indicates that this is the energy functional for a system with external potential $v_0(\mathbf{r})$. If we now denote the ground state electron density by $n_0(\mathbf{r})$ we can express the ground state as $\Psi[n_0(\mathbf{r})]$. From the variational principle we again obtain

$$\langle \Psi[n_0(\mathbf{r})] \mid T + W + V_0 \mid \Psi[n_0(\mathbf{r})]\rangle <$$
$$\langle \Psi[n(\mathbf{r})] \mid T + W + V_0 \mid \Psi[n(\mathbf{r})]\rangle, \tag{12}$$

which can also be expressed as

$$E_{v_0}[n_0(\mathbf{r})] < E_{v_0}[n(\mathbf{r})], \tag{13}$$

i.e., the ground state density minimizes the energy functional $E[n(\mathbf{r})]$, which is what Theorem 2 states. If we now had an explicit form for $E[n(\mathbf{r})]$ we could go ahead and minimize it with respect to the electron density and in this way calculate the ground state energy. Unfortunately one must resort to approximations to obtain an explicit expression for $E[n(\mathbf{r})]$, due to the complexity provided by the electron–electron interactions. We will outline such approximations in the next section but before we do this we note that the arguments above can also be repeated for spin polarized systems and one may show that the ground state energy is a unique functional of the electron and magnetization density. The proof of this is quite similar to the proof outlined above, and we start out by modifying the Hamiltonian to include an external magnetic field, $\mathbf{B}(\mathbf{r})$, so that we have $H = T + W + U$, where $U = \int v(\mathbf{r})n(\mathbf{r}) - \mathbf{B}(\mathbf{r})\cdot\mathbf{m}(\mathbf{r}) \, d^3r$. Based on the variational principle we may, analogous to the discussion around Eqns.4-6, arrive at

$$E_{\text{gs}} < E_{\text{gs}}' + \int n'(\mathbf{r})(v(\mathbf{r}) - v'(\mathbf{r}))d^3r - \int \mathbf{m}'(\mathbf{r})(\mathbf{B}(\mathbf{r}) - \mathbf{B}'(\mathbf{r}))d^3r \tag{14}$$

and

$$E_{\text{gs}}' < E_{\text{gs}} + \int n(\mathbf{r})(v'(\mathbf{r}) - v(\mathbf{r}))d^3r - \int \mathbf{m}(\mathbf{r})(\mathbf{B}'(\mathbf{r}) - \mathbf{B}(\mathbf{r}))d^3r. \tag{15}$$

If we again assume that $n(\mathbf{r}) = n'(\mathbf{r})$ and $\mathbf{m}(\mathbf{r}) = \mathbf{m}'(\mathbf{r})$ and add (14) and (15) we arrive at the same absurd result as in the discussion of spin degenerate systems, i.e. (9), and we must draw the conclusion that $n(\mathbf{r}) \neq n'(\mathbf{r})$ and $\mathbf{m}(\mathbf{r}) \neq \mathbf{m}'(\mathbf{r})$. Hence we may conclude that for magnetic systems we can write the ground state energy as a unique functional of the electron density and the magnetization density.

3.2   The Kohn–Sham Approach

The theorems described above are also valid for non–interacting electron systems where the part of the Hamiltonian describing electron–electron interactions, W, is absent. In this case electrons which move in the field of an external potential which, we for reasons that will be obvious below, call V_{eff}, are solutions to a one–electron Schrödinger equation,

$$[\frac{-\nabla^2}{2} + V_{\text{eff}}]\psi_i = E_i\psi_i. \tag{16}$$

There is an infinity of solutions to this equation and to specify a special solution the subscript i is introduced. From (16) one can calculate an electron density from the lowest lying one–particle (op) states. If there are N electron states which are solutions to (16) one simply calculates the one-particle (a label introduced to show that there are no electron–electron interactions considered) electron density from

$$n_{\text{op}}(\mathbf{r}) = \sum_{i=1}^{N/2} 2 \mid \psi_i(\mathbf{r}) \mid^2, \tag{17}$$

where the factor 2 comes from spin degeneracy. In this case the energy functional which describes the total energy of the N electrons may be written as,

$$E_{\text{op}}[n_{\text{op}}(\mathbf{r})] \equiv T_{\text{op}}[n_{\text{op}}] + V_{\text{eff}}[n_{\text{op}}] \equiv \sum_{i=1}^{N/2} \langle \psi_i(\mathbf{r}) \mid \frac{-\nabla^2}{2} \mid \psi_i(\mathbf{r}) \rangle$$
$$+ \int n_{\text{op}}(\mathbf{r}) V_{\text{eff}}(\mathbf{r}) d^3 r, \tag{18}$$

and the electron density which minimizes this functional is obtained from the requirement that the energy functional is stationary for small variations of the electron density around the ground state density. This can be written as,

$$0 = \delta E_{\text{op}} = E_{\text{op}}[n_{\text{op}}(\mathbf{r}) + \delta n_{\text{op}}(\mathbf{r})] - E_{\text{op}}[n_{\text{op}}(\mathbf{r})], \tag{19}$$

which may also be written as

$$0 = \delta T_{\text{op}}[n_{\text{op}}] + \int \delta n(\mathbf{r}) V_{\text{eff}}(\mathbf{r}) d^3 r. \tag{20}$$

Carrying out the minimization in (20) leads to (16), and we have demonstrated that independent particles which are the solution to (16) give rise to a density which minimizes the total energy expression of independent particles in (18).The reason for introducing (16) to (20) is mainly that they can be solved, at least approximately, to within a desired accuracy (this is the topic of the next sub-section), but at this state it is unclear if equations (16) to (20) have anything to do with a 'real' interacting system. However, as will be clear in a moment they can, via the Kohn–Sham approach, be used to actually calculate the ground state energy of a 'real' electron system. The basic principle of the Kohn–Sham approach is now to assume that one can find an effective potential, V_{eff}, so that $n_{\text{op}}(\mathbf{r}){=}n(\mathbf{r})$, where $n(\mathbf{r})$ is the electron density of the fully interacting ('real') system. Since we know that the total energy of a system is uniquely determined by the electron density it seems to be an efficient route to obtain the correct electron density from a one–electron like problem, in our ultimate quest of calculating the ground state energy of a solid.

The question now is how to determine V_{eff}, so that $n_{op}(\mathbf{r})$ becomes equal to $n(\mathbf{r})$. To do this we recast the energy functional in (10) as

$$E[n(\mathbf{r})] = T_{op}[n(\mathbf{r})] + \int n(\mathbf{r})v(\mathbf{r})d^3r \tag{21}$$

$$+\frac{1}{2}\int\int e^2\frac{n(\mathbf{r})\cdot n(\mathbf{r}')}{|\mathbf{r}-\mathbf{r}'|}d^3rd^3r' + E_{xc}[n(\mathbf{r})].$$

Since we require that $n_{op}(\mathbf{r})$ should be equal to $n(\mathbf{r})$ we have in the expression above, for simplicity, skipped the subscript op on the electron density of the right hand side. In (21) we have introduced the one–particle kinetic energy functional instead of the true kinetic energy functional of (10) and we have introduced the Hartree electrostatic interaction instead of the true electron–electron interaction. Hence in order to make (21) equal to (10) we must introduce a term that corrects for these replacements, and this is what the exchange and correlation energy, $E_{xc}[n(\mathbf{r})]$, does. Since the first three terms on the right hand side of (21) are possible to calculate numerically we have moved the complexity of the fully interacting system to finding the exchange and correlation functional. So far it has been impossible to find the exact exchange and correlation functional so that (21) holds for all densities and all systems. However, for a uniform electron gas one can calculate $E_{xc}[n(\mathbf{r})]^2$ for all values of the electron density and parameterized forms of $E_{xc}[n(\mathbf{r})]$ as a function of $n(\mathbf{r})$ is available. The local density approximation assumes that these parameterizations work even in cases where the electron gas is not uniform, but varies is space, as it does in a solid, surface or interface. The local density approximation introduces the following expression for the exchange-correlation energy [6],

$$E_{xc}[n(\mathbf{r})] = \int \epsilon_{xc}[n(\mathbf{r})]n(\mathbf{r})d^3r, \tag{22}$$

where $\epsilon_{xc}[n(\mathbf{r})]$ is named the exchange–correlation energy density and in a parameterized form its dependence on $n(\mathbf{r})$ is relatively simple and may, for example, be found in Ref. [6]. We are now armed with an (approximate) expression for the ground state energy functional and in analogy with (19) and (20) we determine the ground state density from this functional by requiring that the functional (equation (21)) is stationary for small variations of the electron density around the ground state density. By doing this we obtain an expression which is quite similar to (20), i.e. we obtain

$$0 = \delta T_{op}[n] + \int \delta n(\mathbf{r})[v(\mathbf{r}) + \int e^2\frac{n(\mathbf{r}')}{|\mathbf{r}-\mathbf{r}'|}d^3r' + \frac{\partial(\epsilon_{xc}[n(\mathbf{r})]n(\mathbf{r}))}{\partial n(\mathbf{r})}]. \tag{23}$$

---

[2] This can be done in the high electron density limit [22] and in the low electron density limit [23]. Interpolation between these two limits gave rise to parameterized forms of the exchange and correlation functional of a uniform electron gas for all values of the density [23]. However, this interpolation is in modern electronic structure calculations replaced by approaches which are based on quantum Monte–Carlo simulations for the intermediate values of the electron gas [24].

We have now achieved our goal in finding the effective potential which (within the approximations and assumptions introduced) ensures that $n_{op}(\mathbf{r})=n(\mathbf{r})$, since when we compare (20) and (23) we can identify V_{eff} as,

$$V_{eff}(\mathbf{r}) = v(\mathbf{r}) + \int e^2 \frac{n(\mathbf{r}')}{|\mathbf{r} - \mathbf{r}'|} d^3r' + \mu_{xc}(n(\mathbf{r})), \tag{24}$$

where

$$\mu_{xc}(n(\mathbf{r})) = \frac{\partial(\epsilon_{xc}[n(\mathbf{r})]n(\mathbf{r}))}{\partial n(\mathbf{r})} = \epsilon_{xc}[n(\mathbf{r})] + n(\mathbf{r})\frac{\partial(\epsilon_{xc}[n(\mathbf{r})])}{\partial n(\mathbf{r})}. \tag{25}$$

All we have to do now is to solve (16) with the effective potential specified by (24). Since the effective potential to be used in (16) depends on the electron density, the property we want to calculate, one has to perform a self–consistent field calculation (SCF) where an initial electron density is more or less guessed and an effective potential is calculated from (24). This potential is then used to solve (16) and a new electron density is calculated from (17), which is then put back into (24). This procedure is repeated until convergence is obtained, i.e. until the density does not change appreciably with successive iterations[3]. Once a self–consistent electron density has been found one can calculate the ground state energy of the Kohn–Sham (LDA) energy functional (via (21)) and hence one of the main goals in electronic structure calculations has been achieved.

Before entering the details of how one might solve the self consistent Kohn–Sham equations, (16) and (17), we note that the entire procedure outlined above can also be made to work for magnetic systems. At the end of the previous section we concluded that for magnetic systems the ground state energy may be written as a unique functional of the electron density and of the magnetization density. An alternative way of expressing this is to state that there is an energy functional which depends both on the majority and the minority spin density (since $n(\mathbf{r}) = n^\uparrow(\mathbf{r}) + n^\downarrow(\mathbf{r})$ and $m(\mathbf{r}) = n^\uparrow(\mathbf{r}) - n^\downarrow(\mathbf{r})$[4]) and we can write $E[n^\uparrow(\mathbf{r}), n^\downarrow(\mathbf{r})]$. We can then make an analogous assumption to the discussion around (21) and obtain a Kohn–Sham scheme for spin polarized systems via,

$$E[n^\uparrow(\mathbf{r}), n^\downarrow(\mathbf{r})] = T_{op}[n^\uparrow(\mathbf{r}), n^\downarrow(\mathbf{r})] + \int n(\mathbf{r})v(\mathbf{r})d^3r \tag{26}$$

$$+ \frac{1}{2} \int \int e^2 \frac{n(\mathbf{r}) \cdot n(\mathbf{r}')}{|\mathbf{r} - \mathbf{r}'|} d^3r d^3r' + E_{xc}[n^\uparrow(\mathbf{r}), n^\downarrow(\mathbf{r})].$$

[3] Normally one is forced to mix the electron density which is the output of (17) with the electron density which is in input for that particular loop in the SCF iterational procedure before one takes this *mixed* density and puts it in (24). The whole procedure of *mixing* is quite complex where many suggestions of how to achieve self consistency with as few iterations as possible have been suggested [25].

[4] We note here that this approach has simplified the situation somewhat since the magnetization density is a scalar property with both magnitude and spin. In this analysis we are assuming that the magnetization is pointing only in one direction, the z–direction, of the system.

In a real solid the preference for occupying one spin channel (to some degree) more than the other is traditionally explained as due to the exchange interaction and the driving force for it is the electron–electron interaction in the Hamiltonian. Hence in the spin polarized Kohn–Sham scheme this necessarily means that the exchange and correlation potential, which is supposed to absorb all complex electron–electron interactions, must depend both on the charge and the spin (magnetization) density. Turning again to studies on the uniform electron density is useful and parameterizations for $E_{xc}[n^\uparrow(\mathbf{r}), n^\downarrow(\mathbf{r})]$, as a function of $n^\uparrow(\mathbf{r})$ and $n^\downarrow(\mathbf{r})$, have been made. We can now proceed quite analogously to the discussion around (16) and (17) and analyze a one particle Hamiltonian with spin up (down) effective potentials,

$$[\frac{-\nabla^2}{2} + V_{\text{eff}}^{\uparrow(\downarrow)}]\psi_i^{\uparrow(\downarrow)} = E_i^{\uparrow(\downarrow)}\psi_i^{\uparrow(\downarrow)}, \tag{27}$$

where the electron density for electrons with a given spin is obtained from,

$$n_{\text{op}}^{\uparrow(\downarrow)}(\mathbf{r}) = \sum_{i=1} |\psi_i^{\uparrow(\downarrow)}(\mathbf{r})|^2 . \tag{28}$$

Repeating the discussion which led to (24), with the only modification that we now require the energy functional to be stationary with regard to both the spin up and the spin down density, leads to effective potentials which are different for the two spin directions due to differences in the exchange and correlation potential,

$$V_{\text{eff}}^{\uparrow(\downarrow)}(\mathbf{r}) = v(\mathbf{r}) + \int e^2 \frac{n(\mathbf{r}')}{|\mathbf{r} - \mathbf{r}'|} d^3r' + \mu_{xc}(n^\uparrow(\mathbf{r}), n^\downarrow(\mathbf{r})). \tag{29}$$

Hence the simplest forms[5] of spin polarized calculations treat the spin up and spin down electrons separately and for every iteration in the self–consistent loop one solves a Kohn–Sham equation for both spin directions. The spin up and spin down densities are then calculated by occupying the N lowest (spin up or spin down) eigenvalues of the separate two Kohn–Sham equations. Since for a given $V_{\text{eff}}^\uparrow(\mathbf{r})$ which may be different from $V_{\text{eff}}^\downarrow(\mathbf{r})$ there may be more spin up states, $E_i^\uparrow$ than spin down states, $E_i^\downarrow$, which have an energy lower than the highest occupied state (the Fermi level-E_F) it is clear how spin polarization might occur. With a self consistent spin and magnetization density the magnetic moment is calculated as $\int m(\mathbf{r}) d^3r$ (in Bohr magneton units) and the total energy may be calculated from (26).

4   Solving the Kohn–Sham Equations: Bulk

We have shown in the previous subsection that the equations to be solved for calculating the magnetic properties of a solid from first principles are (27)–(29). Before entering the details of one of the most used methods, the Linear

---

[5] We will discuss below, in connection to the section about orbital moments and relativity, complications which makes this scheme somewhat more involved.

Muffin–Tin Orbital (LMTO) method, we note that due to the symmetry of bulk materials a number of simplifications evolve. This discussion can also be found in, for instance, the textbook by Ashcroft and Mermin [26].

First of all one normally assumes in a bulk material that the potential which enters (27) is periodic, i.e. $V_{\text{eff}}^{\uparrow(\downarrow)}(\mathbf{r}) = V_{\text{eff}}^{\uparrow(\downarrow)}(\mathbf{r} + \mathbf{R})$, where $\mathbf{R}$ is a translation vector (a Bravais lattice vector) of the solid. This periodic boundary condition leads to Bloch's theorem [26] which states that as an effect of the periodicity of the bulk material the one–electron wave function must obey the following condition,

$$\psi_{i,\mathbf{k}}^{\uparrow(\downarrow)}(\mathbf{r} + \mathbf{R}) = e^{i\mathbf{k}\cdot\mathbf{R}}\psi_{i,\mathbf{k}}^{\uparrow(\downarrow)}(\mathbf{r}), \tag{30}$$

and we note that a vector $\mathbf{k}$ has been introduced. This is a vector of reciprocal space[6] and due to the translation symmetry one has only to consider $\mathbf{k}$–vectors which lie inside the first Brillouin zone when looking for solutions to (27)–(29) [26]. In addition one can solve the (27)–(29) for each $\mathbf{k}$–vector being separate and independent of the others. However, the dependence of the one–electron wave function on $\mathbf{k}$ makes the calculation of the one–electron density somewhat more complex since we have to include a sum over all possible $\mathbf{k}$–vectors, and (28) is in a crystal replaced by

$$n_{\text{op}}^{\uparrow(\downarrow)}(\mathbf{r}) = \sum_{i}\sum_{\mathbf{k}} \mid \psi_{i,\mathbf{k}}^{\uparrow(\downarrow)}(\mathbf{r}) \mid^2 . \tag{31}$$

In a similar way one often needs to sum all the possible Kohn–Sham eigenvalues E_i (equation (16)) to be used for calculating the total energy, (26). This is needed since one often writes, $T_{\text{op}} = E_{\text{sum}} - \int v_{\text{eff}}(\mathbf{r})n(\mathbf{r})d^3r$). The sum of the eigenvalues, E_{sum}, is often referred to as the eigenvalue sum and it is calculated from

$$E_{\text{sum}} = \sum_{i}\sum_{\mathbf{k}} E_{i,\mathbf{k}}. \tag{32}$$

In principle all $\mathbf{k}$–vectors inside the first Brillouin zone should be considered in the sums above, but since this number is enormous one would like to replace the sum with an integral. However, if one does not have an analytic dependence of, for instance, E_i of $\mathbf{k}$, we must find ways to approximate (32). In order to do this it is useful to introduce the concept of density of states, DOS. A derivation of the DOS may be found in most textbooks on solid state physics and it is not repeated here. Instead we quote the result, that the DOS can be calculated from,

$$D(E) = \sum_{i}\frac{1}{8\pi^3}\int_{BZ} \delta(E - E_{i\mathbf{k}})d^3k. \tag{33}$$

---

[6] Reciprocal space is spanned by the vectors $\mathbf{G}_i$, defined as $\mathbf{G}_i \cdot \mathbf{R}_j = 2\pi\delta_{ij}$, where V is the volume of the primitive cell of the Bravais lattice.

With this definition of the DOS one can calculate the eigenvalue sum from

$$E_{\text{sum}} = \int_{-\infty}^{E_F} ED(E)dE. \tag{34}$$

4.1   Different Types of k-Space Integration

Having defined the DOS in (33) and the eigenvalue sum, in (34) we are now ready to discuss different ways to approximate the integral over the Brillouin zone (BZ), which is necessary in numerical methods where one does not have an analytical expression of $E_{i\mathbf{k}}$. First we write the eigenvalue sum as,

$$E_{\text{sum}} = \sum_i \int_{-\infty}^{\infty} Ef(E)\frac{1}{8\pi^3}\int_{BZ} \delta(E - E_{i\mathbf{k}})d^3kdE, \tag{35}$$

where $f(E)$ is a step function which attains the value one below the Fermi energy and zero above. E_{sum} is numerically very sensitive to the $\mathbf{k}$–point sampling and the choice of numerical method to perform the Brillouin–zone integration, BZI. We will discuss three different BZI schemes [27]. In all three cases a uniform mesh of $\mathbf{k}$–points is used, distributed as to fulfill the symmetry of the space–group. The BZI can in all three cases be written as weighted sums over the bands, i, and the discrete set of sampled $\mathbf{k}$–points, k_j, with weight functions, w_{ji}. In the so–called linear tetrahedron method [28], LTM, the uniform mesh is divided into corner–sharing tetrahedra. A linear interpolation of the eigenvalues is performed between the $\mathbf{k}$–points belonging to one tetrahedron, resulting in a weight function w_{ji}. The step function is used directly by using E_F as an upper limit in the energy integration. A modification of the linear tetrahedron method, MTM, was suggested by Blöchl et al. [29] In the MTM the linear weights, w_{ji}, are corrected by, $dw_{ij}=\sum_T \frac{1}{40}D_T(E_F)\sum_l^4(E_{il}-E_{ij})$, where T is an index for the tetrahedra and l is an index for the $\mathbf{k}$–points at the corners of the tetrahedron T. $D_T(E_F)$ is the contribution to the density of states from the tetrahedron T at the Fermi level and E_{ij} is the ith eigenvalue of the jth $\mathbf{k}$-point belonging to the tetrahedron T. The MTM corrects for the curvature of the energy band to leading order. Another way to perform the BZI is to use a Gaussian broadening method, GBM, which convolute each discrete eigenvalue with a Gaussian function of width W. This method and the related Fermi-Dirac broadening method are very popular in total energy methods since they lead to a fast and stable convergence of the charge and spin densities. The GBM can be seen as a truncation of a complete series expansion of a δ–function in terms of Hermite polynomials, H_n, with a Gaussian weight function [30]. Then the step function, $f(E)$, can be written as

$$f(E) = f_0(E, W) + \sum_{n=1}^{\infty} A_n H_{2n-1}(\frac{E - E_F}{W})e^{-(\frac{E-E_F}{W})^2}, \tag{36}$$

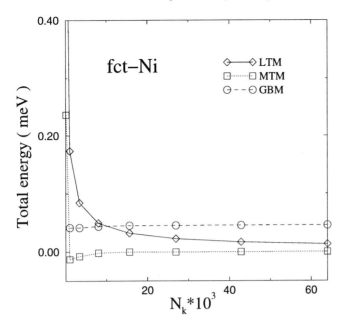

Fig. 1. Calculated total energy of fct Ni as a function of the number of **k**–points use for sampling the BZ. Three different sampling techniques were used, LTM, MTM and GBM (see text).

where $f_0(E,W) = \frac{1}{2}(1 - erf(\frac{E-E_F}{W}))$ (erf stands for error function) and A_n are coefficients which may be calculated analytically. In practical calculations one has to truncate the sum in the equation above and the resulting "step" function is called $f_N(E;W)$, if N terms are kept in the sum. If the function which is "smeared" by $f_N(E;W)$ (for instance the DOS times the energy – see equation (34)) can be represented by a polynomial of $2N$ there is no error involved in the truncation [30].

An example of how the different BZ integration methods work is show in Fig. 1 where the total energy of fct Ni (c/a ratio of 0.945) is presented as a function of the number of **k**–points in the irreducible wedge of the BZ [31]. The calculations were based on a full–potential linear muffin–tin orbital method, described in the next section. Note from Fig. 1 that the LTM converges much slower than the MTM and GBM. Notice also that the GBM does not converge to the same value as the other two methods which seem to converge to the same value. This is due to the fact that in the GBM the sum over Hermite polynomials is truncated, resulting in an approximate step function. However, the error in the energy differences is much smaller [31,32].

4.2   The FP–LMTO Method

We have outlined above a number of concepts which appear due to the trans-lational symmetry of a solid, i.e. the Kohn–Sham equation must be solved for a number of k–vectors which, for a given cycle in the self consistent loop, may be treated as independent of each other. In addition one has to find ways to approximate the k–space summation of the electron states and we have given examples of how one may do this. We are now ready to tackle the toughest part of the problem, namely to solve (27). One approach is to expand the (unknown) one–electron wave function in a set of (known) basis functions as,

$$\psi_{i,\mathbf{k}}(\mathbf{r}) = \sum_{l}^{l_{\mathrm{max}}} c_{lik} \chi_{l\mathbf{k}}(\mathbf{r}). \tag{37}$$

The sum in the equation above is truncated after sufficiently many basis functions have been included and the coefficients c_{lik} are, via the Rayleigh–Ritz principle [16], determined from the following secular equation,

$$\sum_{l}^{l_{\mathrm{max}}} [H_{ll'} - E_{i\mathbf{k}} O_{ll'}] c_{lik} = 0, \tag{38}$$

where

$$H_{ll'} = \int_{U_c} \chi_{l\mathbf{k}}(\mathbf{r}) [\frac{-\nabla^2}{2} + V_{\mathrm{eff}}^{\uparrow(\downarrow)}] \chi_{l'\mathbf{k}}(\mathbf{r}) d^3 r \equiv \int_{U_c} \chi_{l\mathbf{k}}(\mathbf{r}) h_{\mathrm{eff}} \chi_{l'\mathbf{k}}(\mathbf{r}) d^3 r \tag{39}$$

and

$$O_{ll'} = \int_{U_c} \chi_{l\mathbf{k}}(\mathbf{r}) \chi_{l'\mathbf{k}}(\mathbf{r}) d^3 r, \tag{40}$$

where the integral is over the unit cell (U_c). Once $H_{ll'}$ and $O_{ll'}$ have been evaluated the eigenvalues, $E_{i\mathbf{k}}$ ($i = 1 - l_{\mathrm{max}}$), are determined by [16]

$$\det |H_{ll'} - E_{i\mathbf{k}} O_{ll'}| = 0, \tag{41}$$

a standard numerical problem, which may be solved by existing software.

4.3   Defining the LMTO Basis Functions

The difficulty is now to choose a basis set which is flexible and converges fast, i.e. as few basis functions as possible are needed to represent with sufficient accuracy a given eigenfunction, $\psi_{i,\mathbf{k}}(\mathbf{r})$. One efficient basis set is comprised of so called linear muffin–tin orbitals [10], which may be used in a full–potential mode, as described by Wills [33] or with the use of the atomic sphere approximation [10]. Both these methods are reviewed in other chapters of this book[34,35], hence we give only some of the details here.

Let us note here that the FP–LMTO method is defined using a base geometry, which is the usual construction of muffin–tin spheres centered around the atoms and a region outside these spheres, called the interstitial region.

In the muffin–tins, the basis functions, electron density, and potential are expanded in spherical waves, whereas in the interstitial, the basis functions, electron density, and potential are expanded in Fourier series. The calculation of the Hamiltonian and overlap matrix elements thus involves replacing the integrals in (39) and (40) with two parts, one coming from the muffin–tins and one coming from the interstitial. To be specific we break up integrals of $f(\mathbf{r})$ as $\int_{U_c} f(\mathbf{r})d^3r = \int_{MT} f(\mathbf{r})d^3r + \int_{Int} f(\mathbf{r})d^3r$.

The basis functions used in the LMTO method are best described as augmented (or rather, inside the muffin–tin spheres, replaced) Bessel functions. These functions are described mathematically as,

$$K_\ell(\kappa, r) \equiv -\kappa^{\ell+1} \begin{cases} n_\ell(\kappa r) - ij_\ell(\kappa r)\kappa^2 < 0 \\ n_\ell(\kappa r)\kappa^2 > 0 \end{cases} \tag{42}$$

$$K_L(\kappa, \mathbf{r}) \equiv K_\ell(\kappa, \mathbf{r})\mathcal{Y}_L(\hat{\mathbf{r}}) \tag{43}$$

$$\mathcal{J}_L(\kappa, \mathbf{r}) \equiv \mathcal{J}_\ell(\kappa, \mathbf{r})\mathcal{Y}_L(\hat{\mathbf{r}}) \tag{44}$$

$$\mathcal{J}_\ell(\kappa, r) \equiv j_\ell(\kappa r)/\kappa^\ell \tag{45}$$

In the equation above we have used,

$$\mathcal{Y}_{\ell m}(\hat{\mathbf{r}}) \equiv i^\ell Y_{\ell m}(\hat{\mathbf{r}}) \tag{46}$$

$$C_{\ell m}(\hat{\mathbf{r}}) \equiv \sqrt{4\pi\frac{2}{\ell}+1}\, Y_{\ell m}(\hat{\mathbf{r}}) \tag{47}$$

$$\mathcal{C}_{\ell m}(\hat{\mathbf{r}}) \equiv i^\ell C_{\ell m}(\hat{\mathbf{r}}), \tag{48}$$

where Y is a spherical harmonic.

4.4   The Muffin–Tin Matrix Element

The evaluation of matrix elements over the muffin–tin region requires a mathematical description of the basis function used in this part of the crystal. In case the effective potential in (24) is a constant (a purely hypothetical situation) the eigenfunctions can be written as plane waves or as a linear combination (a so called Bloch sum) of spherical Hankel or Neumann functions. To be specific the expression is (for a mono–atomic solid)

$$\chi_L(\mathbf{k}, \mathbf{r}) = \sum_{\mathbf{R}} e^{i\mathbf{k}\cdot\mathbf{R}} K_\ell(\kappa, |\mathbf{r} - \mathbf{R}|)\mathcal{Y}_{\ell m}(\mathbf{r} - \mathbf{R}), \tag{49}$$

where $\mathcal{K}$ is a basis function defined in (42). Since part of the crystal has an effective potential which is rather constant, and does not vary much (in the interstitial) it is meaningful to use the above quoted functions as basis functions in the interstitial region. However, before describing how these functions are used to calculate (39) and (40) let us first deal with the muffin–tin region. We first note that (49)

$$\chi_L(\mathbf{k}, \mathbf{r}) = \sum_R e^{i\mathbf{k}\cdot\mathbf{R}} \sum_{L'} \mathcal{Y}_{\mathbf{L'}}(\hat{\mathbf{r}}) \, [\mathcal{K}_{\ell'}(\kappa, s_\tau)\delta(R, 0)\delta(L, L') \tag{50}$$
$$+\mathcal{J}_{L'}(\kappa, s)B_{L',L}(\kappa, \mathbf{R}),]$$

can be rewritten inside a muffin–tin sphere at $\mathbf{R}=0$ as a multi–pole expansion which yields,

$$\chi_L(\mathbf{k}, \mathbf{r}) = \sum_{L'} \mathcal{Y}_{\mathbf{L'}}(\hat{\mathbf{r}})(\mathcal{K}_{\ell'}(\kappa, s)\delta(L', L) + \mathcal{J}_{L'}(\kappa, s)B_{\mathbf{L'},\mathbf{L}}(\kappa, \mathbf{k})). \tag{51}$$

This function is continuous and differentiable at the muffin–tin spheres. Since the effective potential around the nuclei, inside the muffin–tins, is more or less spherically symmetric and atomic like an efficient choice of basis set is to replace the Bessel and Neumann functions in (50) with numerical functions that are solutions to a Kohn–Sham like differential equation. However, the numerical functions are only determined inside the region where the effective potential is close to being spherical, i.e. inside the muffin–tins. They are calculated from $\Psi = a\psi + b\dot{\psi}$, where $(h_{\text{eff}}^{\text{spherical}} - \epsilon_\nu)\psi = 0$ and $\dot{\psi}$ is the energy derivative of ψ. The constants a and b are determined to ensure continuity and differentiability. Hence the Bessel and Neumann functions are replaced, in a continuous and differential way, with numerical functions. The basis function to be used in the muffin–tin region is then,

$$\chi_L(\mathbf{k}, \mathbf{r}) = \sum_{L'} \mathcal{Y}_{\mathbf{L'}}(\hat{\mathbf{r}})(\Phi_{\ell'}(\kappa, s_\tau)\delta(L', L) + \Phi_{\mathbf{L'}}(\kappa, s)B_{\mathbf{L'},\mathbf{L}}(\kappa, \mathbf{k})). \tag{52}$$

With this basis function, matrix elements of the Hamiltonian and overlap can be evaluated over the muffin–tin region. Added to each such matrix elements is then the contribution from the interstitial region, which is described next.

4.5   The Interstitial Matrix Element

The basis function in the interstitial region is described by (49). However, since it has been found to be efficient to represent the effective potential and charge density as Fourier series, it becomes computationally efficient to make use of a Fourier series also for the basis function. This is possible since a basis function to be used in a crystalline environments can be written as a periodic function multiplied with a plane wave, and one can of course express the periodic function in a Fourier series. Hence one makes use of,

$$\chi_L(\mathbf{k}, \mathbf{r}) = \sum_G \chi_G e^{i(\mathbf{k}+\mathbf{G})\cdot\mathbf{r}}, \tag{53}$$

in the interstitial region from which the interstitial contribution to the Hamiltonian and overlap matrix elements are evaluated. Since the potential in the interstitial region is represented as a Fourier series the evaluation of the matrix element of the Hamiltonian (and the overlap) becomes relatively simple, involving a sum of Fourier components. In practice it becomes faster to perform this integral via a combination of inverse Fourier transforms and Fourier transforms, which is not described here.

5   Magneto–Crystalline Anisotropy of Selected Materials

5.1   General Remarks

In the remaining part of this chapter we will give examples of how the theories outlined above can be used to calculate the magneto–crystalline anisotropy, as well as spin and orbital moments. However, before entering the details of this section which describes the magneto–crystalline anisotropy (MAE) as well as spin and orbital moments of selected materials, it should be mentioned that van Vleck pointed out, already in 1937 [36], that the spin–orbit interaction is responsible for the coupling of the spin to the lattice, which then gives rise to a magnetic anisotropy in a magnetic solid. Hence, much of the physical understanding of the mechanisms behind the MAE of 3d elements have been known for a long time. Actual calculations of the MAE have been thwarted because efficient methods for solving the Kohn–Sham equations, in combination with fast computers, are only recently available. Pioneering and important work in the field of MAE was also made by Brooks [37] and Kondorsii and Straub [38] where technical details of, for instance, problems with BZI were analyzed. We are interested here in the electronic contribution to the MAE, provided by the coupling of spin and orbital angular momenta, via the spin–orbit coupling. There is also another contribution to the MAE which is called the shape anisotropy (described in the following section). However, in the theoretical description of the MAE below we shall focus on the contribution given by the spin–orbit coupling. Hence, when comparing theory and experiment we have, unless otherwise explicitly stated, removed the shape anisotropy from the experimental data.

5.2   Shape Anisotropy

The shape anisotropy results from the magnetic dipole interaction. To calculate it we can consider a lattice of magnetic dipoles with magnetic moment μ_i. The energy of the dipole interaction can be written as [2] $E_{\mathrm{dipole}} = -\frac{1}{2}\sum_i \mu_i \cdot \mathbf{h}_i$, where $\mathbf{h}_i$ is the field intensity at lattice point i due to all the other dipoles. This interaction can be written as [39,40],

$$E_{\mathrm{dipole}} = \frac{1}{c^2}\sum_{i\neq i'}\left[\frac{\mu_i \cdot \mu_{i'}}{|\mathbf{R}_i - \mathbf{R}_{i'}|^3} - 3\frac{(\mu_i \cdot \{\mathbf{R}_i - \mathbf{R}_{i'}\})(\mu_{i'} \cdot \{\mathbf{R}_i - \mathbf{R}_{i'}\})}{|\mathbf{R}_i - \mathbf{R}_{i'}|^5}\right], \tag{54}$$

where $\mathbf{R}_i$ is a vector describing the lattice point i. Due to the long range nature of the magnetic dipole interaction the sum in (55) depends on the shape of the sample, and this leads to the shape anisotropy. For thin films, for instance, it always favors in–plane moments.

5.3   Orbital Moments and Orbital Polarization

An important part of magnetism, especially when considering the MAE, is the orbital contribution to the magnetic moment. Just as the spin moment can be written as the expectation value of s_z using the wavefunctions in (37) one can calculate the orbital moment from [41],

$$L_\sigma = \sum_{\mathbf{k}i} \left\langle \psi_{i,\mathbf{k}}^\sigma(\mathbf{r}) \left| l_z \right| \psi_{i,\mathbf{k}}^\sigma(\mathbf{r}) \right\rangle . \tag{55}$$

The expectation value in (55) is non–zero only if the relativistic spin–orbit coupling is included in the calculation and if the spin degeneracy is lifted. Via the spin–orbit interaction there is now a coupling between spin and orbital angular momentum and it is this interaction which is responsible for the electronic contribution to the MAE. Hence the orbital moment and the electronic contribution to the MAE have in common the fact that they both are a consequence of the spin–orbit interaction and below we will show that one can sometimes write an explicit relationship between the MAE and the orbital moment. From the arguments above it is clear that any theoretical method which attempts to calculate the MAE or orbital moment must consider the spin–orbit interaction. This can either be done by solving the Dirac equation, by use of perturbation theory, or to diagonalise a fully relativistic Hamiltonian using basis functions from a non–relativistic Hamiltonian [10,42] (or more precisely scalar relativistic Hamiltonian, since so–called scalar relativistic corrections such as mass velocity and Darwin shifts [23] are normally included in most band structure methods anyway). Using perturbation theory is more approximate than the other two methods, which give almost identical results[7]. The latter finding in not too surprising since the same relativistic Hamiltonian is considered, only the form of the basis functions is different. Diagonalising a relativistic Hamiltonian with scalar–relativistic basis functions has the advantage of being relatively easy and scalar–relativistic codes may, without too much effort, in this way be modified to include also the spin–orbit interaction. This can either be done in a straight forward way [10,42], as will be described below, or by using the so–called second variation approach [45]. The basic principle in these two implementations is however the same, and we describe only the first version here. If the effective

---

[7] This was found for instance when comparing orbital moments of Fe, Co and Ni calculated using these two different methods, but also by inspection of the DOS of Pu [44], a very heavy element where the spin–orbit interaction is approximately 1 eV.

Hamiltonian in (39) is augmented to include also the relativistic spin–orbit interaction, $h_{\text{eff}}^{\text{relativistic}} = h_{\text{eff}}^{\text{scalar}} + \xi \mathbf{l} \cdot \mathbf{s}$, (where ξ represents the strength of the spin–orbit interaction) the secular equation in (38) is modified as follows,

$$\sum_l^{l_{\max}} \sum_\sigma [H_{l\sigma l'\sigma'} - E_{i\mathbf{k}} O_{ll'} \delta_{\sigma\sigma'}] c_{l\sigma i\mathbf{k}} = 0. \tag{56}$$

In this equation matrix elements of the form

$$\langle \chi_{l\sigma\mathbf{k}}(\mathbf{r})\sigma \, | \xi \mathbf{l} \cdot \mathbf{s} | \, \sigma' \chi_{l'\sigma'\mathbf{k}}(\mathbf{r}) \rangle$$

$$= \langle \chi_{l\sigma\mathbf{k}}(\mathbf{r})\sigma \, | \xi \, (l_x s_x + l_y s_y + l_z s_z) | \, \sigma' \chi_{l'\sigma'\mathbf{k}}(\mathbf{r}) \rangle$$
$$= \left\langle \chi_{l\sigma\mathbf{k}}(\mathbf{r})\sigma \, \Big| \xi \left(\tfrac{l_+ s_- + l_- s_+}{2} + l_z s_z \right) \Big| \, \sigma' \chi_{l'\sigma'\mathbf{k}}(\mathbf{r}) \right\rangle, \tag{57}$$

must be evaluated. Since the basis functions involve spherical harmonic functions, spinors,σ, and a radial component, these matrix elements are rather straight forward to calculate. Diagonalisation of (56) now yields eigenvalues which are not pure spin states. Instead, the wave function of a given eigenvalue is written as a linear combination of states with different l and σ character.

A practical complication of (57) is that the Hamiltonian matrix is doubled in size, making the numerical diagonalisation procedure slower. A further complication of including both the spin–orbit coupling and a lifting of spin degeneracy is that the symmetry (of the crystal) is reduced. Consider for example a cubic crystal, such as bcc Fe. Normally there are 48 point group operations which leaves the lattice invariant. However, if magnetism is considered and if the magnetization is coupled to the lattice via the spin–orbit interaction, the so–called double group [46] has to be considered, and the effect of the time reversal operator has to be taken into consideration. If the magnetization of our cubic lattice is pointing along the crystallographic z–axis we can, for instance, not perform a 90 degree rotation around the x- or y–axis, since the magnetization direction then is rotated out from the z–axis and the system is not invariant. However, a 90 and 180 degree rotation around the z–axis is allowed. Also, a 180 degree rotation around the x–axis flips the magnetization direction from the z–direction to the z–direction and in itself such a symmetry operation is not allowed. However, the product of this rotation and the time reversal operator (which changes the sign of the magnetization) is an allowed operation and this element of the double group does leave the crystal invariant. Our example of a cubic lattice when the magnetization is pointing along the z–direction has 16 allowed symmetry operations. Hence, the lattice becomes tetragonal in symmetry, an effect which we will discuss below in connection with calculations of magneto–striction. At present we simply note that the reduced symmetry simply means that the irreducible part of the BZ is larger, and that more $\mathbf{k}$–points need to be sampled in order to have a well converged total energy.

Since the spin–polarization (spin pairing) is treated via a spin–dependent effective potential, such as (29), and the relativistic spin–orbit interaction is treated via (57), the physical mechanisms behind Hund's first and third rules, respectively, are (at least approximately) included. One can at this stage ask the question why total energy calculations based on different approximations of DFT ignore Hund's second rule? The answer to this question is of course that an exact formalism for including effects responsible for Hund's second rule is lacking at present. However, it has been pointed out [47] that an approximate way to incorporate Hund's second rule in total energy calculations, using DFT, is to add to the LDA (or GGA) energy functional a term,

$$E_{\mathrm{OP}} = -\frac{1}{2} \sum_{\sigma} E_{\sigma}^3 L_{\sigma}^2. \tag{58}$$

In this equation E_{σ}^3 is the so called Racah parameter (note that this parameter is normally denoted B for d–electrons and E^3 for f–electrons), which may be calculated from Slater integrals; F^2, F^4 and F^6 [48,49]. The correction in (58) is normally referred to as the orbital polarization (OP) correction. This form for including electron–electron interactions comes from the finding that a vector model, involving interactions of the form $l_i \cdot l_j$ between electron pairs [50,51], can be used to calculate the lowest energy multiplet. In fact an explicit expression for electron–electron interactions having a term $l_i \cdot l_j$, is quoted both by van Vleck [50] and Ballhausen [51]. This term in the electron–electron interaction was also used by Norman et al. [52] to give a correction term which is similar to the form in (58). Summing the interaction, $l_i \cdot l_j$, over electron pairs as well as replacing the average of this interaction with the sum over average interactions, $l_{zi}l_{zj}$, gives rise to an expression for orbital correlations given by (58) [47]. The approximation of replacing $l_i \cdot l_j$ with the average is in the spirit of replacing the spin pairing energy, $\sum_{i \neq j} s_i \cdot s_j$, with a Stoner expression $\sum_i s_{zi} \sum_j s_{zj} = S_z^2$. Since this form of the energy is absent in the LDA or GGA energy functionals one must add the energy of (58) to these total energy functionals. The interactions which are responsible for Hund's second rule, which in physical language are a reflection of that states with different angular momentum have different angular shape and hence a different Coulomb interaction, are now in an approximate way included in the energy functional.

To illustrate the importance of (58) we show in Fig. 2 the energy correction corresponding to (58) for the 4f electrons of the lanthanide atoms (or ions). Specifically Fig. 2 shows equation (58) neglecting the effect of E^3, hence illustrating only the angular behavior of the correction. The correction in (58) is compared to the exact values [53], which are calculated from the energy difference between the lowest atomic multiplet of the f^n configurations and those corresponding to the Grand Barry Centre (an average of the multiplets). This energy difference involves of course also spin–pairing energy and spin—orbit interaction, but shown in Fig. 2 is only the part which depends on the orbital angular momentum. To get a feeling for how important the correction in (58) is, we note that for a rare–earth system the value of E^3 is ~5 mRy which means that the correction in (58) can be as large as ~100 mRy. From Fig. 2 we note that the agreement between the approximate form of the orbital part of the electron–

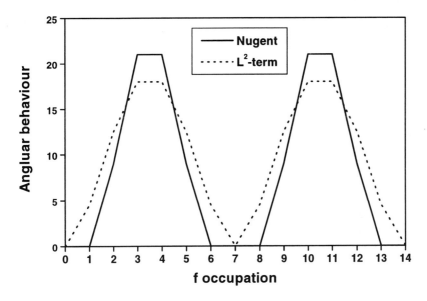

Fig. 2. Angular behavior of the electron–electron interactions of the f–elements, as given by Nugent [53] and from equation (58).

electron interaction, given in (58), and the exact values given by Nugent [53] is rather good. The largest disagreement is found for an f–occupation close to 1, 6, 8, and 13, since the exact values should be zero here (there is for instance no spin- and orbital pairing energy of a single electron), whereas the approximate correction does yield a non–zero correction. This is analogous to the fact that LSDA gives a spin–pairing energy of a single electron system.

The derivation of (58) was made having nearly localized electron systems in mind [47] and for other systems other corrections should be tried. We have previously mentioned the corrections of Norman [52], but there is also a third form suggested by Severin and co–workers [54]. We will not describe all the proposed corrections in detail here, but instead quote some results of calculations using (58), in order to illustrate the importance of these types of corrections, not only for f–electron systems but also for d–electron systems and in particular for orbital magnetism. In Figs. 3 and 4 we show the calculated spin and orbital moments [55], respectively, of Fe, Co and Ni, including alloys between these elements. The calculations were using a standard LSDA calculation (with spin–orbit interaction), as well as the correction in (58). In the figures experimental values are also shown, and we note that an improved agreement between theory and experiment is consistently found for the orbital moment, when the OP correction is used. This is encouraging as well as reasonable, since the correction is supposed to include an interaction which is present in an interacting

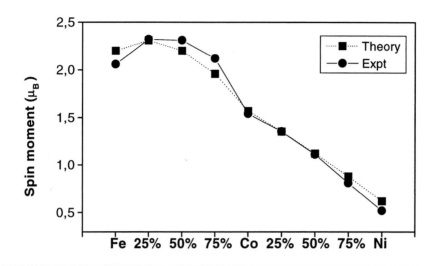

Fig. 3. Calculated and experimental values of the spin moments of the 3d elements, Fe, Co and Ni, as well as for alloys between them.

electron system. In addition we also observe that the orbital moments show a rather irregular behavior and as discussed by Söderlind [55] this is due to band filling effects. Since the size of the orbital moment is to a large degree caused by a redistribution of electron states around the Fermi level, [56,57] it generally scales to some degree with the value of the DOS at E_{F}. Hence a large DOS at E_{F} normally results in a larger orbital moment and the irregular behavior of the orbital moment is to some degree a reflection of an irregular behavior of the DOS at E_{F}, for these alloys. In addition the crystal symmetry is important since hcp Co is found to have the largest orbital moment. It can be argued from perturbation theory that for a cubic material the influence of the spin–orbit coupling strength, ξ, enters as ξ^4. For non–cubic materials the dependence is stronger, and this generally results in larger orbital moments.

6   MAE of 3d Elements

6.1   Fe, Co and Ni

The first example of calculated MAE values which we would like to discuss is at first glance the simplest, namely the MAE of the ferromagnets Fe, Co and Ni. However, due to that the MAE is extremely small for these elements, at least for bcc Fe and fcc Ni (of order μeV) the task of calculating this from first principles

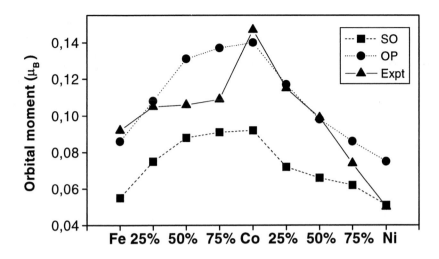

Fig. 4. Calculated and experimental values of the orbital moments of the 3d elements, Fe, Co and Ni, as well as for alloys between them. Calculations with (OP) and without (SO) the orbital polarization term are shown.

is enormously difficult, where especially a high accuracy of the computational method is needed as well as a very dense sampling of the BZI. For this reason different values of the MAE of bcc Fe, hcp Co and fcc Ni have been reported in the literature. Some of the most recently calculated [58] values of the MAE are listed in Table 1.

Table 1. MAE of Fe, Co and Ni in different crystal structures. All units in μeV. The MAE is calculated as the difference between the axis: 001-111 for the bcc Fe, 001-100 for the hcp Co, 001-111 for the fcc Co and finally 001-111 for the fcc Ni.

MAE	bcc Fe	hcp Co	fcc Co	fcc Ni
ASA-force theorem [40]	-0.5	16	-	-0.5
ASA-total energy [60]	-2.6		2.4	1.0
FP-total energy [58]	-0.5 (-1.8)	-29 (-110)	0.5 (2.2)	-0.5 (-0.5)
experiment	-1.4	-65	1.8 [59]	2.7

In the Table 1 we quote results obtained using the atomic sphere approximation (ASA) where spherical potentials in overlapping spheres are used to replace the actual effective potential of the lattice (in equation (29)) in combination with

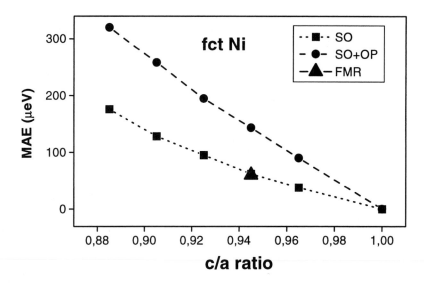

Fig. 5. Calculated and experimental values of the MAE of bcc Fe (001 vs. 111 direction), hcp Co (0001 vs. 100 direction), fcc Co (001 vs. 111 direction) and fcc Ni.(001 vs. 111 direction). Calculations with and without the orbital polarization term are shown.

a minimal basis set. This approximation is estimated to give an error in each calculated eigenvalue of a few mRy's [10]. The full potential (FP) calculations make no approximation to the shape of the charge– or spin density as well as the shape of the effective potential (equation (29)) and as described above makes use of a double or sometimes even a triple basis set. One may note from Table 1 that a variety of signs and sizes of the MAE may be obtained for Fe, Co and Ni, depending on which approximation one uses. The calculations of Ref. [40] seem to be well converged with respect to the number of **k**–points used, as do the calculations of Ref. [60]. Since both these methods use the ASA the disagreement in the MAE of bcc Fe and fcc Ni must be caused by the use of the so called force theorem[8] in Ref. [40]. This finding is consistent with the analysis of Ref. [62] where it was argued that for cubic systems the applicability of the force theorem may be limited. In the Table 1 we have quoted a value of 16 μeV for hcp Co, using ASA and the force theorem. This actually corresponds to the wrong easy axis. However, this value was obtained using an spdf basis set, whereas an spd

---

[8] The force theorem [61] states that the total energy difference between two different magnetization directions may simply be calculated from the difference in the eigenvalue sum of the the two magnetization directions, using the same effective potential for the two directions. This approximation is correct to first order changes in the charge and magnetization density.

basis set gave a value of -29 μeV, i.e. the correct easy magnetization direction. One can possibly draw the conclusion from this finding that the MAE of hcp Co is hard to converge with a minimal basis set. Unfortunately no total energy calculation using the ASA [60] was reported for hcp Co so that one can not examine if the poor convergence in the number of basis functions is connected to the use of the force theorem or not. The total energy, full–potential calculations did report on convergence tests with respect to the number of **k**–points, the width of the Gaussians used in the GBM for the BZI and the basis functions used (however, mostly for Ni). In fact the convergence of the double basis set (which is a standard size of the basis set of FP–LMTO calculations) was tested by a calculation which adopted a triple basis set (i.e. three s basis functions, three p basis functions and three d basis functions, each connecting to an envelope function with a unique kinetic energy). Since the full–potential calculations have the most flexible basis set as well as a more exact effective potential (29) these results must be considered as the most accurate of the ones in Table 1[9]. If we now examine the MAE values of Table 1, using the full–potential method, we find that the correct easy axis is found for all elements except fcc Ni. We also note that including the OP influences the size of the MAE quite strongly, as it did for the orbital moments in Fig. 4. In order to show this dependence more clearly we compare in Fig. 5 different calculated MAE values (using the FP–LMTO method) with experimental data. Note that for bcc Fe and fcc Co including the OP correction improves the agreement with experiment, whereas for hcp Co the experimental value is between the two different theoretical values.[10] For fcc Ni the OP correction has a very small effect on the MAE and the 001 axis is calculated to be the easy axis, in disagreement with observations. As we shall see below, where more examples are given, first principles calculations normally reproduce the correct easy axis, even if the size of the MAE may deviate somewhat. The fact that first principles calculations, based on the LSDA, are incapable of providing the correct easy axis for fcc Ni is probably connected to deficiencies in the form of the exchange and correlation potential in LSDA. From Table 1 and Fig. 5 we also note that the effect of crystal symmetry has a large influence on the size and direction of the MAE, since Co in the fcc phase has a much smaller value of the MAE compared to the value of the hcp phase. This can be understood from perturbation theory where, as stated, it is known that due to the high symmetry of the cubic crystal structure the influence of the spin–orbit coupling strength (ξ) on the MAE enters as ξ^4. Since the value of ξ is rather small for Fe, Co and Ni, it follows that in cubic phases the magnitude of the MAE is tiny. For lower symmetries of the crystal a stronger interaction of ξ on the MAE is present, resulting in larger MAE values.

---

[9] Although this statement was debated in Ref. [60] we maintain that these arguments are not valid and the tests made to compare the two methods [60] are of poor quality.

[10] Recent work by P. James [63] show that a better converged calculation of hcp Co yields a MAE value of ~ 20 μeV when the OP correction is omitted, with the 001 axis as the easy direction.

6.2   Effects of Straining the Crystal Structure

In this section we discuss theoretical results of the MAE and spin and orbital moments of tetragonally strained Fe (bct), Co (fct) and Ni (fct). Many studies of orbital magnetism and MAE are devoted to Fe, Co and Ni as over–layers on a substrate with a small in-plane lattice mismatch. This causes a strain in the over-layer material (Fe, Co or Ni) for thicknesses sometimes up to 50 atomic layers. As a rule of thumb one can estimate this strain by equating the in–plane lattice parameter to that of the substrate, then the out–of–plane lattice parameter of the over–layer is adjusted so that the volume/unit cell of the over–layer is constant and the same as for the elemental form of the over–layer material. The effect of strain on the MAE can, as we shall see in a moment, be large and since many experimental studies are devoted to studying this situation it is important to have a theoretical understanding of how crystallographic strain and MAE are connected.

In this section we have chosen one primary example for which there are both theoretical and experimental data available; Ni on Cu (001) [64]. A few recent LEED studies of Co and Ni films on Cu(001) [65,66] have shown that these films can be grown epitaxially in a face centered tetragonal (fct) structure which deviates with a few percent from that of the fcc structure. In addition these films have been shown to have an out–of–plane magnetization for film thicknesses *above* ~7 atomic mono–layers [67] and an in–plane magnetization for thinner thicknesses [68]. This is a rather unusual behavior since in most systems the shape anisotropy will be the dominating term in the thicker limit and it always favors the in-plane magnetization. In many epitaxially grown magnetic films it is known that the magnetic anisotropy energy, MAE, (per atom) of the system can be expressed by the empirical formula, $E = E_{\mathrm{v}} + \frac{2E_{\mathrm{s}}}{n_d}$, where n_d is the number of atomic layers in the film and E_{v} and E_{s} are the so–called volume and surface (interface) contributions to the magnetic anisotropy energy, respectively. Normally the surface contribution (E_{s}) is much larger than the volume contribution (E_{v}) so that for thin film thicknesses the surface dominates and determines the easy axis magnetization. Normally the volume contribution is assumed to be a result of the material in an undistorted structure (e.g. fcc) and hence is assumed to be small. For instance the volume contribution of fcc Ni is small compared to the surface contribution. Hence only two terms need be considered, the surface contribution to the MAE and the shape anisotropy. With increasing film thickness the surface contribution becomes less important since it scales with the inverse of the film thickness, so that eventually the shape contribution dominates. Since this contribution always favors an in–plane magnetization, a change of magnetization direction as a function of increasing film thickness will be from out–of–plane to in–plane, in contrast to the experimental result for Ni on Cu. It was speculated [67] that the unusual behavior of the Ni/Cu system is driven by a large and positive volume contribution, E_{v}, to the magnetic anisotropy energy in the Ni film which occurs due to the tetragonal distortion of Ni grown on a Cu(001) substrate, in combination with a negative surface contribution, E_{s}. We

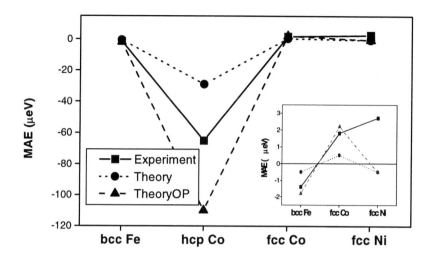

Fig. 6. Calculated values of the MAE (001 vs. 110 direction) as a function of crystallographic strain of fct Ni. Calculations with and without the orbital polarization term are shown. Also shown is the experimental value of Ref. [67].

will demonstrate in this section that this indeed is the case and that a breaking of the crystal symmetry enhances the value of the MAE dramatically.

Using the full–potential LMTO method, E_v for Ni in a face centered tetragonal (fct) structure was calculated [64]. E_v is here defined as the difference in total energy, between the [110] and [001] magnetization directions, per atom. In Fig. 6 E_v is plotted as a function of tetragonal strain, or equivalently c/a-ratio, of the fcc lattice (in reality it is, due to the strain, an fct lattice), assuming a constant in–plane parameter (i.e. this distortion is not volume conserving). (The in–plane lattice parameter, is designated 3.58 Å, which is the same as the value measured by LEED for thin Ni films on Cu [65]. The experimental lattice parameter of bulk fcc Ni is 3.52 Å). At c/a=1.0 the fct crystal has a volume which is larger than the volume of bulk fcc Ni. For comparison, both the calculated E_v with and without orbital polarization is shown. Note the large increase of E_v due to the orbital polarization. It is found that the [001] direction is the easy direction for all c/a <1.0. Further, we notice that E_v is linear in c/a, to first order, especially when c/a is close to 1.0. This indicates that our distortions in the range c/a ˜0.88 to 1.0 are in the magneto–elastic regime. Notice also that in the calculations with only the spin–orbit interaction included, this linear behavior is not as pronounced. In addition it may be observed that due to the breaking of the cubic symmetry the value of the MAE becomes very large. In

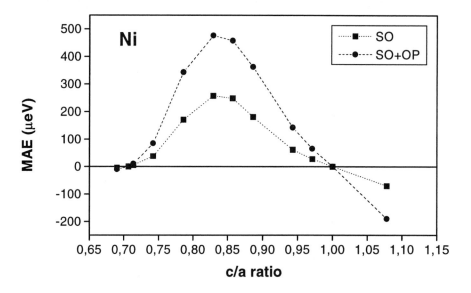

Fig. 7. Calculated values of the MAE (001 vs. 110 direction) of Ni, for strains along the Bains path. Calculations with and without the orbital polarization term are shown.

the figure we have also displayed one experimental value obtained from FMR experiments [64]. In Fig. 6 we see that, for the experimental value of the c/a ratio (c/a=0.945, corresponding to the tetragonal distortion in Ni films on Cu(001)), the theoretical calculation gives $E_v = 140$ μeV. This is in reasonable agreement with the measured value, E_v=60 μeV. The most important thing to note is that theory and experiment agree on the sign of E_v and that they both show that it is substantial in tetragonal Ni. As previously discussed by Schulz and Baberschke [67], the large and positive E_v term is larger than the shape anisotropy for the same Ni film. This leads to a change in magnetization direction for film thicknesses larger than a critical thickness of ˜7 mono–layers (i.e. E_v dominates over the negative surface term, E_s, and the shape anisotropy), which is consistent with observations.

Since Fig. 6 demonstrates that a strain of the fcc crystal structure influences the value of the MAE very strongly it is of interest to investigate the MAE over large regions of strain. The so–called Bains path, which connects the bcc and fcc structures via a tetragonal strain, is especially useful here since one may calculate the MAE along the Bains path and observe how it is modified with respect to changes in the strain. For this reason we show in Fig. 7 calculated values of E_v along the Bains path. The calculations were done in such a way that the volume is the same for all different c/a ratios. For c/a close to fcc or bcc one observes an almost linear behavior as expected from magneto-elastic arguments. Since both

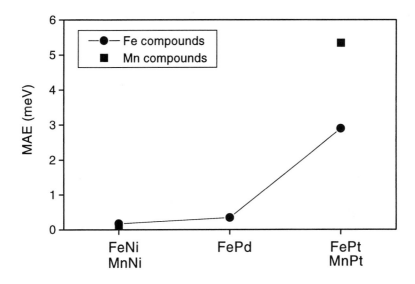

Fig. 8. Calculated values of the MAE (001 vs. 110 direction) of fct Ni as a function of calculated values of the orbital moment anisotropy. The behavior of (59) is shown as a straight line.

fcc and bcc are cubic structures, and therefore (for symmetry reasons) can only have magnetic anisotropy constants of 4th order and higher, E_v must be small for c/a close to 1.0 and $\frac{1}{\sqrt{2}}$. Therefore the E_v curve must deviate from the linear behavior at some intermediate c/a ratio. In Fig. 7 it is seen that this happens when c/a is between 0.8 and 0.9. Further we notice that E_v is positive in the interval $\frac{1}{\sqrt{2}}<$c/a <1 and negative for c/a >1.0 and $<\frac{1}{\sqrt{2}}$. It seems likely that this information can serve as a prediction of the volume contribution to the MAE, of pseudomorphically grown tetragonal Ni films on any substrate (if possible to fabricate). Due to the elasticity the volume of fct Ni will never deviate much from the volume of fcc Ni and the curve in Fig. 7 should resemble the experimental reality.

6.3   The Correlation between MAE and Orbital Moment

An alternative to measuring the MAE directly (which could be difficult in certain cases) may be found from a simple relationship between the orbital moment anisotropy (OMA), which is defined as the difference in the orbital moment when the magnetization is pointing in two different directions and the total energy difference between these two directions, i.e. the MAE. Bruno showed [3], using perturbation theory and assuming that the exchange–splitting is larger than the band width, that the following relationship holds between MAE and OMA,

$$MAE = \frac{\xi}{4\mu_B}OMA, \qquad (59)$$

where ξ is the spin–orbit parameter. Hence instead of measuring the MAE of a system one could measure the orbital moment when the magnetization is pointing in one direction and subtract the measured orbital moment of a different magnetization direction. Forcing the magnetization to lie in a specific direction may be obtained by an applied external field. Hence, if it is judged that the OMA is easier to measure than the MAE one could simply take the measured OMA and use (59) to estimate the MAE. In addition this relationship may be used to analyze the behavior of measured MAE values of compounds, interfaces and so on. Since the relationship between OMA and MAE relies on different assumptions it is of interest to investigate how well this relationship holds for different systems. We present one example here, Ni on Cu, and show that the relationship between MAE and OMA actually is linear, which is what (59) suggests. To illustrate this we plot in Fig. 8 the MAE of fct Ni (the same calculation as described in the section above) as a function of the difference in orbital moment (OMA) when the magnetization is pointing in the 001 and 110 direction. Since the calculation of fct Ni was made for several values of the c/a ratio, several different values of both the MAE and OMA may be calculated and it is these data which are shown in the figure. The relationship between MAE and OMA are also shown in Fig. 8, using a value of 105 meV for ξ. Overall, the relationship between the MAE and OMA is roughly linear although for certain c/a ratios there are larger disagreements. This probably implies that the assumption of an exchange splitting which is larger that the band width is starting to break down.

7   Selected Compounds

There can be quite strong modifications of the MAE as well as spin and orbital moments when a 3d element is situated in a compound. A good understanding of the mechanisms which modify the MAE in one direction or the other is important since one then can start tuning material parameters in order to optimize the magnetic properties. We will outline here some observed trends and explain their origins. We illustrate this with some recent results by Ravindran et al. [69] on the FeX and MnX compounds (X=Ni, Pd and Pt).

7.1   FeX and MnX Compounds (X=Ni, Pd or Pt)

The MnX and FeX compounds crystallize in a layered tetragonal structure with two atoms per primitive unit cell. Experimentally the MAE has been measured for a large number of compounds of this type, some which are compiled in Ref. [69]. We will by no means report on all the details of the theoretical work of

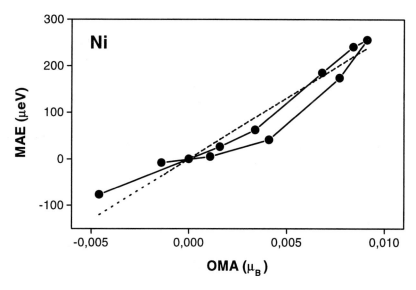

Fig. 9. Calculated values of the MAE (001 vs 100) of FeX and MnX compounds (X=Ni, Pd or Pt).

Ref. [69] but merely quote the results which are of largest importance for the discussion here. In Fig. 9 we show the calculated MAE of iso–structural FeNi, FePd and FePt as well as of MnNi and MnPt. This figure illustrates that for inter–metallic compounds the influence of the spin–orbit coupling of the ligand atom strongly influences the MAE. This finding was also observed for multi layer systems [70]. In particular one observes that a larger spin–orbit coupling of the ligand atom results in a larger value of the MAE. The reason for this is that the magnetism of the 3d atoms will induce a magnetic moment also in the ligand atoms, even if these atoms are non–magnetic as pure elements.

In Fig. 10 the effect of the exchange–splitting of the 3d atom is illustrated and we observe that a larger exchange–splitting results in a larger MAE. Since the MAE is a result of spin–orbit coupling in combination with spin polarization the results in Figs. 9 and 10 are maybe not too surprising but it is not entirely obvious that there should be a rather smooth relationship between the MAE and exchange–splitting, as Fig. 10 shows. All details of the calculated results for all studied FeX and MnX compounds are reported in Ref. [69] and are not given here. However, we note that theory obtains the correct easy axis of magnetization for all these compounds [69], although the absolute values sometimes deviate from experiments with as much as 50%.

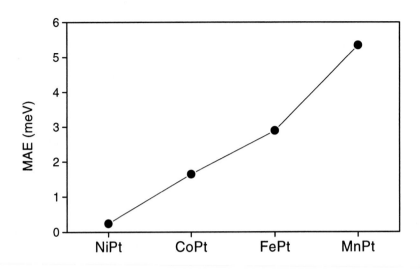

Fig. 10. Calculated values of the MAE (001 vs 100) of TPt (T=Ni, Pd or Pt).

8   Surface and Interface Magnetism

When an atom is close to an interface or a surface it will experience a different environment compared to its elemental environment and as a consequence, the wave function overlap between atoms, the hybridization and the direct hopping will be different. In short, the electronic structure will be modified. In this section we will illustrate how this rearrangement of the electronic structure produces magnetic properties such as spin and orbital moments and in addition how it modifies the MAE. These effects are possible to measure on an almost atomic level. For instance, the surface contribution to the MAE and the spin and orbital moments have been measured [66,67,71,72]. In addition to describing some arcetypal features of surface and interface magnetism we will illustrate with a few examples how, from an electronic structure point of view, one can use the knowledge of these modifications in order to influence the magnetic properties in a desired direction.

8.1   Spin and Orbital Moments of Selected Surfaces and Interfaces

The example to be discussed in this section is that of Co grown on Cu (001) [71]. This system is chosen since it is known experimentally that thin films of Co grow in a well characterized way (fcc) for up to 50 atomic layers, which enables measurement of the thickness dependence of the spin and orbital moments. In the experiments of Tisher et al. [71]. X–ray dichroism was combined with the sum rules [73], which connect x-ray dichroism to the size of magnetic moments. The advantage with this experimental technique is that it gives atom specific information about the magnetism. By studying the thickness dependence of the ratio between orbital and spin moment one can calculate the interface, bulk and surface contribution to this ratio. The so obtained ratio between orbital and spin moments are in Table 2 compared to theoretical values [71]. Note that both for the interface and the surface the ratio between orbital and spin moments is increased over the bulk value. Theory is in good agreement with this finding, and the observed trend with the largest ratio between orbital and spin moments for the interface and smallest ratio for bulk, is reproduced by theory. Table 2 also lists the individual spin and orbital moments for the bulk, interface and surface atomic layers. As may be observed both the spin and orbital moments are enhanced for atoms which have a chemical surrounding which differs from bulk. However, the enhancement is larger for the orbital moment than it is for the spin moment. The enhancement of the spin moment may simply be understood from the fact that atoms which are close to the surface or close to the Co/Cu interface have a reduced number of nearest neighboring atoms. From a simple tight binding model, where the band width scales as the square root of the coordination number, the reduced bandwidth and enhanced spin moments at a surface or even an interface (if the interaction–hybridization between the atoms is weak over the interface) may be understood.

Table 2. Spin and orbital moments for 1 ML of Co on Cu (001), bulk fcc Co and Co surface (001).

quantity	1 ML Co/Cu (001)	Co (fcc, bulk)	Co (fcc, surface)
$M_L(\mu_B)$	0.261	0.134	0.234
$M_S(\mu_B)$	1.850	1.724	1.921
M_L/M_S (theo.)	0.141	0.078	0.122
M_L/M_S (expt.)	0.195	0.078	0.113

In Fig. 11 we display as an example the DOS of bulk fcc Co and the DOS projected on a Co atom for a mono–layer of Co on Cu (001). The narrowing of the DOS at the surface is clear from this figure. Hence, since reduced bandwidths almost always (but curiously not always) produce enhanced spin moments, the trends of the spin moments in Table 2 may be understood. Before entering the discussion about orbital moments we remark here first on the fact that even though the DOS at, for instance, a surface may become substantially more narrow compared to bulk, details of the filling of the bands as well as the hybridization with the underlying substrate may in rare cases produce reduced moments. An example of this is Ni on Cu (001) [74]. Let us now turn to the enhancement of the orbital moment at the interface and at the surface. For the surface it has been pointed out that the crystal field is different and that hence

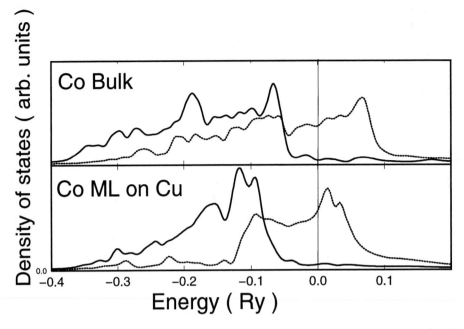

Fig. 11. Calculated density of states of bulk (hcp) Co and a Co monolayer on Cu (001). The Fermi level is marked by a vertical line.

the crystal field quenching is reduced at the surface [75]. This tends to enhance the orbital moments. Moreover, since the bands become more narrow at the surface the value of the DOS at E_F is enhanced compared to bulk producing larger orbital moments. Finally we observe that orbital moments appear only when one theoretically includes the relativistic spin–orbit interaction in a spin polarized calculation. A spin degenerate calculation, even if it includes the spin–orbit interaction, will yield a zero orbital moment. Thus, a spin moment which is reduced, approaching zero, produces an orbital moment which also is reduced and approaches zero. Conversely, increased spin moments, such as the ones for the surface and interface in Table 2, generally produce larger orbital moments. We have thus identified three mechanisms which for atoms at, or close to, the surface result in enlarged orbital moments. For the interface atoms all the above applies, and using these arguments one may also expect enhancements, at least if the interaction (hybridization) with the atoms across the interface is sufficiently small. Previously we discussed the simple relationship between MAE and OMA. Since the size of the orbital moments often is larger for surfaces and interfaces it is likely that the orbital moment anisotropy (OMA) also is enhanced, from which it follows that the value of the MAE is larger than it would be in the bulk. In addition to this the symmetry is lowered and the spin–orbit interaction influences the MAE more strongly (than ξ^4) which also produces larger values of the MAE at surfaces and interfaces.

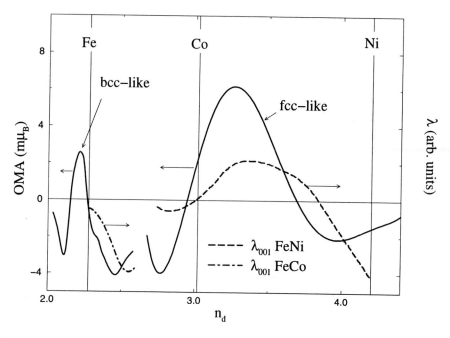

Fig. 12. Calculated OMA (full line, see text) and experimental magneto–striction coefficient of FeCo (bcc, dashed–dotted line) and FeNi (fcc, dashed line) alloys.

Analysis of the MAE of different types of over-layers, for instance of Co on Au(111) [76], Fe on Au(001) [77], and Co on Cu(001) [78]gives additional valuable information. For the case of Co on Au(111) [76] it was demonstrated that enhancement in the MAE could be understood from details in the DOS projected on the Co site, and in particular a large DOS at E_F with the 'correct' symmetry explained the large values of the MAE for certain thicknesses of an Au capping layer [76]. This way of explaining variations of the MAE was also reported by Pick and Dreyssé [79] who argued that a large MAE is expected when a large value of the DOS (of XY and YZ character) is situated on E_F. Another important finding was also reported in Ref. [79], where the MAE of free standing mono–layers of Fe, Co and Ni was shown to have a minimum number of nodes as a function of band–filling.

9   Magneto–Striction

For strains close to the cubic phase, bcc and fcc, the change in the MAE is linearly dependent on the strain, ϵ ($\epsilon=c/a-1$). Using arguments from magneto–elastic theory, the MAE of small tetragonal distortions can be used to calculate the magneto–striction coefficient, λ_{001}, as illustrated in Ref. [80] for fcc Ni. A similar scheme has been used by Wu et al. [81].The total energy can be written as the sum of an elastic and a magneto–elastic energy (E_{el} and E_{me}) which are assumed to be quadratic respectively linear in small distortions, $E_{el} = C'\epsilon^2$ and $E_{me} = \alpha B \epsilon$. Here ϵ is the volume conserving tetragonal distortion ($\epsilon \tilde{} 2(c/a-1)/3$), α is a constant which takes the value 1 for magnetization parallel to the tetragonal axis and $-\frac{1}{2}$ for a perpendicular direction, and B and C are the magneto–elastic and tetragonal elastic coefficients, respectively. The magneto–striction coefficient, λ_{001}, is defined as the equilibrium distortion, i.e. where the total energy, $E_{el} + E_{me}$, has its minimum, for the case with the magnetization along the [001] direction. By differentiation of the total energy with respect to ϵ this turns out to be $\lambda = -\frac{B}{2C}$. The uniaxial MAE is, for small distortions, related to the magneto–elastic energy as, $E_{MAE} = \frac{3}{2}B\epsilon = 3\epsilon C\lambda$. Hence one may connect the λ and MAE values and compare their trends directly [32]. Hence, in order to study trends in the MAE as a function of, for instance, alloy concentration one may, as an alternative, study experimental trends in λ or the trend in the OMA, and this is something we will do next [32]. Before entering the details of such a comparison we comment that a rigid band approximation works well for studying the trends in the MAE and OMA of alloys involving Fe, Co and Ni [32]. In Fig. 12 we compare the calculated OMA of bcc Fe and fcc Co to the corresponding experimental magneto–striction coefficients, λ. All analyses [32] indicate that it is sensible to compare these two properties, at least their trends. Note that the theoretical calculations reproduce the experimental trends very well. For the FeCo alloy (bcc) theory reproduces the drop in λ with increasing Co concentration and the dip at a Co concentration corresponding to ˜2.5 spin down d–electrons. The trends in λ of the FeNi (fcc) alloys is somewhat more intricate where λ changes sign two times as a function of alloy concentration. However, theory and experiment agree rather well and we conclude that magneto–striction is a property which, at least concerning the trend, is accessible for state of the art, total energy calculations [32].

10   Summary

To summarize, we have outlined some of the most central aspects of first principles calculations of the magneto–crystalline anisotropy, and the spin and orbital moments of elements, compounds, interface and surface systems. One of the more commonly used methods for solving the Kohn–Sham equations, the full–potential linear muffin–tin orbital method is also reviewed, briefly. We give several examples of the accuracy of theory for reproducing the MAE, the spin and orbital moments, and the magneto–striction coefficients. A more detailed discussion of these effects can however be found in Ref. [63]. The most important results reviewed here is that the anisotropy of the orbital moment very often is connected to the MAE and the orbital moment may be influenced by a 'suitable' tuning of the electronic structure, via geometry, ligand atoms, crystal field splitting and so on. A theoretical engineering of suitable magnetic properties, such as magnitude of the moment and its directional behavior is now starting to become feasible. By good knowledge of how the electronic structure, and especially the

DOS projected on a specific atom type, is modified for different crystallographic environments, it is possible to predict materials with novel magnetic properties.

Acknowledgment

The support from the Swedish Natural Science Foundation is acknowledged. The support from the TMR is also acknowledged. Many of the results presented here are the result of a collaboration with O.Hjortstam, P.James, B. Johansson, L.Nordström, P.Ravindran, P.Söderlind and J.Trygg. We would like to thank the listed persons for this fruitful collaboration.

References

1. L.Nordström and D.Singh, Phys. Rev. Lett. **76**, 4420 (1996).
2. A.Aharoni, *Introduction to the Theory of Ferromagnetism*, (Oxford Science Publications, Oxford, 1998).
3. P.Bruno, Phys. Rev. B **39**, 865 (1989).
4. P.Hohenberg and W.Kohn, Phys. Rev. B**136**, 864 (1964).
5. W.Kohn and L.Sham, Phys. Rev. A **140**, 1133 (1965).
6. R.M.Dreitzler and E.K.U.Gross *Density Functional theory* (springer, berlin, 1990).
7. H.Eschrig, *The Fundamentals of Density Functional Theory* (Stutgart;Teubner, 1996).
8. R.Jones and O.Gunnarsson, Rev. Mod. Phys. **61**, 689 (1989).
9. J.Callaway and N.M.March, Rev. Mod. Phys. **38**, 689 (1989).
10. O.K.Andersen, Phys. Rev. B **12**, 3060 (1975).
11. H.L.Skriver, *The LMTO Method* (Springer verlag, Berlin, 1984).
12. I.Turek, V.Drchal, J.Kudrnovsky, M.Sob and P.Weinberger, *Electronic Structure of Disordered Alloys, Surfaces and Interfaces* (Kluwer Academic Publishers, Boston, 1997).
13. J.C.Slater, *The Self–Sonsistent Field of Molecules and Solids* (McGraw-Hill, New York, 1974).
14. T. L. Loucks, *Augmented Plane Wave Method* (W. A. Benjamin, Inc., New York, 1967).
15. D.J.Singh, *Planewaves, Pseudopotentials and the LAPW Method* (Kluwer Academic Publishers, Boston, 1994).
16. P.W.Atkins, *Molecular Quantum Mechanics*, Oxford University Press, Oxford, (1983).
17. H.Eschrig, *LCAO Method and the Electronic Structure of Extended Systems*, Springer-Verlag, Berlin (1989).
18. W.Harrison, *Electronic Structure and the Properties of Solids,* W.H.Freeman and Company, San Fransisco, 1980.
19. V.L.Morruzi, J.F.Janak, and A.R.Williams, *Calculated Electronic Properties of Metals* (Pergmon,New York,1978)
20. V.L.Morruzzis and C.B.Sommers, *Calculated Electronic Properties of Ordered Alloys, A Handbook* (World Scientific, Singapore, 1995).
21. L.H.Thomas, Proc.Cambridge Phil. Soc. **23**, 542 (1927); E.Fermi, Rend. Accad. Naz.Linzei **6**, 602 (1927).
22. Gell-Mann and K.A.Brueckner, Phys. Rev. **106**, 369 (1957).

23. E.P.Wiegner, Phys. Rev. 46, 1002, (1934).
24. D.M.Ceperly and B.J.Alder, Phys. Rev. Lett. 45, 566 (1980)
25. G.P. Srivastava, J.Phys. A 17, L317 (1984); O.Eriksson, J.Trygg, O.Hjortstam, B.Johansson and J.M.Wills, Surface Science 382, 93 (1997).
26. N.W.Ashcroft and N.D.Mermin, *Solid State Physics* (Holt-Saunders Japan LTD.,1976)
27. O.Hjortstam (unpublished).
28. O.Jepsen and O.K.Andersen, Solid State Commun. 9, 1763 (1971).
29. P.E.Blöchl, O.Jepsen, and O.K.Andersen, Phys. Rev. B 49, 16223 (1994).
30. M.Methfessel and T.Paxton, Phys. Rev. B 40, 3616 (1989).
31. O.Hjortstam, L.Nordström, B.Johansson, J.M.Wills, P.James and O.Eriksson (to be published).
32. P.James, L.Nordstrom, B.Johansson and O.Eriksson (to be published).
33. J.M.Wills (unpublished).
34. O.K.Andersen, First chapter in this book.
35. J.M.Wills, O.Eriksson, M.Alouani, and D.L. Price, Chapter 4 in this book.
36. J.H.van Vleck, Phys. Rev. B 52, 1178 (1937).
37. H.Brooks, Phys. Rev. B 58, 909 (1940).
38. E.I.Kondorski and E.Straub, Zh. Eksp. Teor. Fiz. 63, 356 (1972).
39. B.R.A.Nijboer and F.W.De Wette, Physica (Amsterdam) 23, 309 (1957); F.S.Ham and B.Segall, Phys. Rev. B 124, 1786 (1961); B.R.A.Nijboer and F.W.De Wette, Physica (Amsterdam) 24, 422 (1958).
40. G.H.O.Daalderop, P.J.Kelly and M.F.H.Schuurmans, Phys. Rev. B 41, 11919 (1990).
41. M.S.S.Brooks and P.J.Kelly, Phys. Rev. Lett. 51, 1708 (1983).
42. D.D.Koelling and B.N.Harmon, J.Phys. C 10, 3107 (1977).
43. M.E.Rose, *Relativistic Electron Theory*, (John Wiley and Sons. New York, 1961).
44. O.Eriksson, Thesis (Uppsala University).
45. C.Li, A.J.Freeman, H.J.F.Jansen, and C.L.Fu, Phys. Rev. B 42, 869 (1990).
46. J.C.Bradley and A.P.Cracknell, *The Mathematical Theory of Symmetry in Solids*, Clarendon Press, Oxford, 1972.
47. O.Eriksson, M.S.S.Brooks and B.Johansson, Phys. Rev.B 41, 7311(1990).
48. G.Racah, Phys. Rev. 61, 186 (1949).
49. J.K.Jorgensen, *Electrons in Atoms and Molecules* (Academic Press, London, 1962).
50. J.H.van Vleck, Phys. Rev. 45, 405 (1934).
51. C.J.Ballhausen, *Crystal Field Theory*, McGraw-Hill, New York, 1962.
52. M.Norman, Phys. Rev. Lett. 64, 1162 (1990).
53. L.J.Nuget, J.Inorg. Nucl. Chem. 32, 3485 (1970).
54. L.Severin, M.S.S.Brooks, and B.Johansson, Phys. Rev. Lett. 71, 3214 (1993).
55. P.Söderlind, O.Eriksson, R.C.Albers., A.M.Boring, and B.Johansson, Phys. Rev. B 45, 12911 (1992).
56. O.Eriksson, B.Johansson, R.C.Albers, A.M.Boring, M.S.S.Brooks, Phys. Rev. B 42, 2707 (1990).
57. H.Ebert et al.,J.Appl. Phys. 67, 4576 (1990).
58. J.Trygg, B.Johansson, O.Eriksson and J.M.Wills, Phys. Rev. Lett.75, 2871 (1995).
59. T.Suzuki (unpublished).
60. S.V.Halilov, A.Ya.Perlov,P.M.Oppeneer,.A.N.Yaresko and V.N.Antonov, Phys. Rev. B 57, 9557 (1998).
61. O.K.Andersen, H.L.Skriver, H.Nohl, and B.Johansson, Pure and Appl. Chem. 52, 93 (1980).

62. X.Wang, D.-S.Wang, R.Wu and A.J.Freeman, J.M.M.M. **159**, 337 (1996).
63. P.James, Thesis (in print).
64. O.Hjortstam, K.Baberschke, J.M.Wills, B.Johansson, and O.Eriksson, Phys. Rev. B **55**, 15026 (1997).
65. S.Muller, B.Schulz, G.Kostka, M-Fahrle, K.Heinz, and K.Baberschke, Surf. Sci. **364**, 235 (1996).
66. O.Heckmann, H.Magnan, P.le Fevre, D.Chanderis, and J.J.Rehr, Surf. Sci. **312**, 62 (1994).
67. B.Schulz, K.-Baberschke, Phys. Rev. B **50**, 13467 (1994); K.Baberschke, Appl. Phys. A **62**, 417 (1996).
68. U.Gradman, Ann. Phys. (N.Y.) **17**, 91 (1966).
69. P.Ravindran, P.James, J.M.Wills and O.Eriksson (unpublished).
70. G.H.O.Daalderop, P.J.Kelly, M.F.H.Schuurmans, Phys. rev. B **42**, 7270 (1990).
71. M.Tischer, O.Hjortstam, D.Arvanitis, J.Hunter Dunn, F.May, K.Baberschke, J.Trygg, J.M.Wills, B.Johansson and O.Eriksson, Phys. Rev. Lett. **75**, 1602 (1995).
72. enhancement of ni spin moment at surface
73. B.T.Tole, et al. Phys. Rev. Lett. **68**, 1943 (1992); P.Carra et al. Phys. Rev. Lett. **70**, 694 (1993).
74. O.Hjortstam, J.Trygg, J.M.Wills, B.Johansson and O.Eriksson, Phys. Rev B **53**, 9204 (1996).
75. O.Eriksson, et al. Solid State Commun.**78**, 801 (1991); O.Eriksson, et al. Phys. Rev. B **45**, 2868 (1992).
76. B.Ujfalussy, L.Szunyogh, P.Bruno, and P.Weinberger, Phys. Rev. Lett. **77**, 1805 (1996).
77. L.Szunyogh, B.Ujfalussy and P.Weinberger, Phys., Rev. B **51**, 9552 (1995).
78. L.Szunyogh, B.Ujfalussy,U.Pustogowa, and P.Weinberger, Phys. Rev. B **57**, 8838 (1998).
79. S.Pick and H.Dreysse, Phys. Rev. B **46**, 5802 (1992).
80. P.James et al (unpublished)
81. R.Wu et al. (unpublished).

On the Implementation of the Self-Interaction Corrected Local Spin Density Approximation for d- and f-Electron Systems

W. M. Temmerman[1], A. Svane[2], Z. Szotek[1], H. Winter[3], and S. V. Beiden[4]

[1] Daresbury Laboratory, Daresbury, Warrington, WA4 4AD, UK
[2] Institute of Physics and Astronomy, University of Aarhus,
    DK-8000 Aarhus C, Denmark
[3] Forschungszentrum Karlsruhe, INFP, Postfach 3640, Karlsruhe, Germany
[4] Department of Physics, University of Sheffield, Sheffield, UK
    and Department of Physics, University of West Virginia,
    Morgantown, West Virginia 26506-6315, USA

Abstract. The ab-initio self-interaction corrected (SIC) local-spin-density (LSD) approximation is elaborated upon, with emphasis on the ability to describe localization phenomena in solids. Two methods for minimizing the SIC–LSD total energy functional are considered, one using an unified Hamiltonian for all electron states, thus having the advantages of Bloch's theorem, the other one employing an iterative scheme in real space. Moreover, an extension of the formalism to the relativistic case is discussed. Results for NiO, cerium and cerium compounds are presented. For NiO a significant charge transfer gap is produced, in contrast to the near vanishing band gap seen in the LSD approximation. Also, the magnetic moment is larger in the SIC–LSD approach than in the LSD approach. For the cerium compounds, the intricate isostructural phase transitions in elemental cerium and cerium pnictides may be accurately described. A sizeable orbital moment for elemental cerium metal is obtained which, upon lattice expansion, is seen to reach the atomic limit.

1 Introduction

Density Functional Theory (DFT)[1] is a very powerful tool for performing *ab initio* electronic structure calculations for complex systems. It provides an exact mapping of a many–body electron problem which occurs in solids onto a one–electron problem. Instead of considering, for N interacting electrons in an external potential $V_{ext}(\mathbf{r})$, the 3N-dimensional Schrödinger equation for the wavefunction $\Psi(\mathbf{r}_1, \mathbf{r}_2, \mathbf{r}_3, ..., \mathbf{r}_N)$, DFT expresses this many–body problem in terms of the electronic density distribution $n(\mathbf{r})$ and a universal exchange and correlation functional of the density, $E_{xc}[n]$. The task of solving the many–body problem is then reduced to finding sufficiently accurate expressions for $E_{xc}[n]$ and then solving the relevant one–electron Schrödinger equation with an effective potential of which the exchange–correlation potential is a prominent part. Generalizing DFT to the spin-polarised case, the spin-density functional theory (SDFT) allows to study magnetic systems, where the magnetization density is the order parameter of the theory, and appears explicitly in the exchange–correlation energy functional, i.e. , $E_{xc}[n, \mathbf{m}]$.

The local-spin-density (LSD) approximation to SDFT provides a simple and rather successful scheme[2]. This is owing to a simple and practical approximation for $E_{xc}[n, \mathbf{m}]$, where the exchange and correlation energy of the electrons in the solid is expressed in terms of the exchange–correlation energy per particle, $\varepsilon_{xc}(n, \mathbf{m})$, of a homogeneous electron gas of homogeneous density n and magnetization $\mathbf{m}$. Specifically, in each point in space the electrons present are assumed to contribute an exchange–correlation energy given as if they were in a homogeneous electron gas at the local density $n(\mathbf{r})$ and $\mathbf{m}(\mathbf{r})$:

$$E_{xc}^{\mathrm{LSD}}[n, \mathbf{m}] \equiv \int n(\mathbf{r})\varepsilon_{xc}(n(\mathbf{r}), \mathbf{m}(\mathbf{r}))d\mathbf{r},$$

The simple function $\varepsilon_{xc}(n, \mathbf{m})$ is known with great precision, and hence allows for an accurate determination of the ground state energies and charge densities of any system.

The LSD is a highly accurate approximation for systems where the electrons are delocalized and travel 'fast' through the solid. However, when the 'static' electron-electron interactions become so strong that some electrons get localized on atomic sites in the solid, the LSD, as well as its gradient corrections, fail to describe the correct groundstate. This obviously means that the $E_{xc}[n, \mathbf{m}]$ can not anymore be adequately represented by the LSD. For when the electron slows down upon localization it starts responding to a different potential than the effective LSD potential.

One of the most prominent examples where electron correlations are too strong to be properly treated within LSD is the high-temperature superconductors, for which LSD fails to produce the antiferromagnetic and semiconducting ground states of the generic La_2CuO_4 and $YBa_2Cu_3O_6$ compounds [3]. Other examples are the $3d$ transition metal oxides MnO, FeO, CoO and NiO, which are Mott insulators characterized by localized d-electrons [4,5]. These materials have antiferromagnetic order and large energy gaps although the metal d-shell is incompletely filled. The LSD approximation does reveal the magnetic ordering but with somewhat too small magnetic moments and vanishing (for FeO and CoO) or very small (for MnO and NiO) gaps [6]. In addition, the persistence of the magnetic moments above the Néel temperature is difficult to explain in the Slater-Stoner picture of magnetism inherent in the LSD band picture. The failure of LSD in producing the correct gaps may be traced to the fact that the LSD eigenenergies do not have built in the large on-site Coulomb repulsion, which characterizes the separation between occupied and unoccupied states [5,7,8]. LSD also fails to give a physically correct picture of trivalent cerium compounds, which are characterized by each Ce atom having a localized f−electron. Cerium is the first element of the Periodic Table to accomodate an f−electron. This $4f$−electron is peculiar in being spatially localized with a radial extent much smaller than that of the $5s$ and $5p$ semicore states, yet having an energy in the region of the valence $6s$ and $5d$ electrons. Cerium metal has been widely studied over the years, both experimentally and theoretically, mostly due to its famous isostructural $\gamma \rightarrow \alpha$ phase transition, occuring at a pressure of ~ 8 kbar.

This phase transition is believed to occur as a consequence of a change in bonding properties of the f-electrons. In the (high–volume) γ phase the f-electrons behave as non–bonding localized moments with a Curie–Weiss type susceptibility, while the (low–volume) α-phase is characterized by the f-electrons taking more active part in the cohesion, either by hybridizing into the Bloch states, as envisaged in the Mott transition model [9], or by forming a complicated Kondo lattice groundstate [10]. In an LSD calculation the total energy minimum is found in the region somewhat below the α phase volume. Upon expansion of the lattice, the $\gamma \to \alpha$ transition is signalled by the onset of magnetization around the observed lattice parameter for the γ phase [11]. The calculated pressure is much too negative, however, since the f-electrons still contribute significantly to the cohesion in the spin-polarized phase. The loss of cohesion upon localization has been successfully described within the LSD in cases where the f-shell is half filled, as in americium [12].

The concept of an electron being 'localized' is not particularly well-defined. In a solid a periodic array of deep core states is equally well described by a set of atom centered Wannier states or by k-dependent Bloch functions. A unitary transformation connects the two representations, but the physical picture of a core state is that of a localized wavefunction. The observation of multiplet effects in photoemission experiments on $3d$ monoxides and rare-earth systems seems to suggest that the local description is most appropriate for such systems. On the other hand the great success of modern solid state theory in explaining most conventional metals, semiconductors and insulators leaves little doubt that the Bloch picture of normal valence electrons is the more fruitful one.

The Hubbard model [13] has often been used to describe materials where localization phenomena occur. The substantial Coulomb correlations on particular atoms, say on a Ni d^8 ion in NiO, induce an orbital polarization [5], which may well be described already in a mean-field treatment of the Hubbard model, as has been shown in the LDA+U approach [8,14]. However, the 'localizing' Hubbard term is usually too large to facilitate an accurate description of delicate localization-delocalization transitions, as for example observed in rare-earth and actinide systems.

Another way of taking into account the 'slowing down' of an electron upon localization is to consider orbital–dependent functionals, $E_{xc}[\{\psi_\alpha\}]$, where $\{\psi_\alpha\}$ is a set of one–electron orbitals with the orbital index α. These functionals are a possible evolution on the LSD and can be considered as a basis of the orbital–dependent DFT. They can be treated as the starting point for new approximations to DFT. The self–interaction–corrected LSD (SIC-LSD) is one such approximation [15]. It is based upon the observation that localized orbitals in the LSD give rise to an error due to a spurious self–interaction contained in the effective one–electron potential. This error increases the more localized the orbital becomes. For a 'fast' electron, however, the self-interaction is negligible or zero. The SIC-LSD energy functional is constructed from the LSD functional by explicit subtraction of the self-interaction term from the E_{xc}^{LSD} term. As we will demonstrate in this paper the SIC-LSD functional allows for an improved treatment of the electron correlations, and is more adequate than the LSD for systems where both localized and delocalized electrons are present.

Solids where both localized and delocalized electrons are present give rise to some of the most facinating phenomema in solid state physics. They include the heavy fermion compounds and their complicated phase diagrams at low temperatures, the still unexplained phenomenon of the high T_c superconductivity, the observation of collossal–magneto resistance, with its potentially wide ranging technological applications. The theoretical investigations have usually been confined to either studies of generic model Hamiltonians which allow for a very accurate treatment of the electron–electron interactions, or materials specific LSD calculations with a not fully satisfactory treatment of the 'static' electron–electron correlations.

As already mentioned, the basic assumption in the SIC–LSD is that 'localized' electron states experience a different potential from that of the normal delocalized valence electrons. One may argue for such a differentiation by thinking in terms of Wigner delay times of electrons on a particular atom [16]. If an electron resides on a given atom for a long time the local electronic structure of the atom accomodates to the presence of the added electron, while a fast electron has no influence on the effective structure of that atom. This picture is implemented by assuming a fast ('delocalized') electron to experience an effective potential as given by the LSD approximation, while a slow ('localized') electron experiences the LSD potential corrected for the self-interaction of the electron in question. Taking the d^8 configuration of a Ni ion in NiO as an example, one of the 8 d-electrons residing on a particular atomic site experiences Coulomb interactions with 7 other d-electrons. On the other hand, a conduction electron injected into a perfect NiO crystal will be made out of excited states, which ride on top of an array of Ni d^8 ions, and therefore Coulomb interacts with all the 8 d-electrons. That the LSD is adequate for describing the 'fast' electron is validated by the great success of this approach in describing the bonding properties of conventional metals. This picture is still very idealized, since the transition from localized to delocalized must be a gradual one. The added electron in NiO still may show some localized characteristics, and the ground state electrons do sometimes show **k**-dependent dispersions. However, one may be able to distinguish phases of solids being either predominantly 'localized' or predominantly 'delocalized', which will be the subject of the present paper.

In Section II the general formalism of the SIC–LSD approximation will be discussed, and in Sections III and IV two implementations of the approach into *ab-initio* linear-muffin-tin orbitals (LMTO)[17] band structure codes will be outlined. Section V is devoted to a relativistic extension of the formalism, while in Section VI we concentrate on applications. Section VII concludes the paper.

2   The SIC Formalism

In the LSD all electrons feel the same mean-field single particle potential, V^{LSD}, and the same magnetic field, $\mathbf{B}_{e\!f\!f}$, which depends on the total charge density, n, and magnetization density, $\mathbf{m}$. The LSD energy functional, E^{LSD}, has the form:

$$E^{\text{LSD}} = E_{\text{kin}} + U[n] + \int d\mathbf{r} \, V_{\text{ext}}(\mathbf{r}) n(\mathbf{r}) + E_{xc}^{\text{LSD}}$$
$$+ \int d\mathbf{r} \, \mathbf{m}(\mathbf{r}) \cdot \mathbf{B}_{\text{ext}}(\mathbf{r}) \tag{1}$$

with

$$
\begin{aligned}
E_{\text{kin}} &= \sum_\alpha \langle \psi_\alpha | \hat{T} | \psi_\alpha \rangle \\
U[n] &= \frac{1}{2} \int dr V_H(\mathbf{r}) n(\mathbf{r}) \\
V_H(\mathbf{r}) &= \int d\mathbf{r}' \, \frac{2n(\mathbf{r}')}{|\mathbf{r} - \mathbf{r}'|} \\
V_{\text{ext}}(\mathbf{r}) &= -\sum_{\mathbf{T}, \tau} \frac{2Z_\tau}{|\mathbf{T} + \tau - \mathbf{r}|} \\
E_{xc}^{\text{LSD}}[n, \mathbf{m}] &= \int d\mathbf{r} \, n(\mathbf{r}) \, \varepsilon_{xc}(n(\mathbf{r}), \mathbf{m}(\mathbf{r})).
\end{aligned}
$$

Here, $\hat{T}$ is a kinetic energy operator, and the ψ_α's are the one-electron wavefunctions which in the nonrelativistic as well as the scalar relativistic case are two-component spinors. $\mathbf{B}_{\text{ext}}(\mathbf{r})$ is the external magnetic field, Z_τ is the atomic number of site τ, $\mathbf{T}$ is a lattice translation vector of the Bravais lattice, and τ is the postion vector of site τ in the unit cell. n and $\mathbf{m}$ may be expressed in terms of the ψ_α's in the usual way

$$
\begin{aligned}
n(\mathbf{r}) &= \sum_\alpha n_\alpha(\mathbf{r}) = \sum_\alpha \psi_\alpha^*(\mathbf{r}) \psi_\alpha(\mathbf{r}) \\
\mathbf{m}(\mathbf{r}) &= \sum_\alpha \mathbf{m}_\alpha(\mathbf{r}) = \sum_\alpha \psi_\alpha^*(\mathbf{r}) \sigma \psi_\alpha(\mathbf{r}),
\end{aligned}
\tag{2}
$$

with σ being the spin operator.

Minimizing the above energy functional by taking the functional derivative of E^{LSD} with respect to ψ_α^*, leads to the single particle wave equation of the form

$$h^{\text{LSD}}(\mathbf{r})\psi_\alpha(\mathbf{r}) = (T + V^{\text{LSD}}(\mathbf{r}) + \sigma \cdot \mathbf{B}^{\text{LSD}}(\mathbf{r})) \, \psi_\alpha(\mathbf{r}) = \varepsilon_\alpha \, \psi_\alpha(\mathbf{r}), \tag{3}$$

with

$$
\begin{aligned}
V^{\text{LSD}}(\mathbf{r}) &= V_H(\mathbf{r}) + V_{xc}^{\text{LSD}}(\mathbf{r}) \\
V_{xc}^{\text{LSD}}(\mathbf{r}) &= \frac{\delta E_{xc}^{\text{LSD}}[n, \mathbf{m}]}{\delta n(\mathbf{r})} \\
\mathbf{B}^{\text{LSD}}(\mathbf{r}) &= \mathbf{B}_{\text{ext}}(\mathbf{r}) + \mathbf{B}_{xc}^{\text{LSD}}(\mathbf{r}) \\
\mathbf{B}_{xc}^{\text{LSD}}(\mathbf{r}) &= \frac{\delta E_{xc}^{\text{LSD}}[n, \mathbf{m}]}{\delta \mathbf{m}(\mathbf{r})}.
\end{aligned}
\tag{4}
$$

As discussed in the introduction the LSD successfully describes many solid state properties, but suffers from a deficiency caused by the spurious self-interaction (SI). Namely, such contributions to E^{LSD} as $U[n]$, E_{xc}^{LSD}, and $\mathbf{B}_{xc}^{\mathrm{LSD}}$ contain spurious self-interactions of the single particle charges, n_α, and magnetic moments, $\mathbf{m}_\alpha$, which do not cancel. The exact SDFT energy functional does not contain any self-interaction, and also in some existing approximations, e.g., the Hartree-Fock theory, the SI terms cancel out. Of course, due to its great merits, it does not seem justified to discard the LSD altogether in cases where this inherent self-interaction matters. Instead, in such cases, it is sufficient to augment LSD with terms removing this deficiency. The resulting approach, the self-interaction corrected LSD (SIC-LSD) formalism,[15] is defined by the following functional

$$E^{\mathrm{SIC\text{-}LSD}} = E^{\mathrm{LSD}} + E^{\mathrm{SIC}}, \tag{5}$$

with

$$E^{\mathrm{SIC}} = -\sum_\alpha e_\alpha[n_\alpha, \mathbf{m}_\alpha] = -\sum_\alpha \left(U[n_\alpha] + E_{xc}^{\mathrm{LSD}}[n_\alpha, \mathbf{m}_\alpha] \right)$$

$$U[n_\alpha] = \frac{1}{2} \int d\mathbf{r}\, n_\alpha(\mathbf{r}) V_{H,\alpha}(\mathbf{r})$$

$$V_{H,\alpha} = \int d\mathbf{r}' \frac{2n_\alpha(\mathbf{r}')}{|\mathbf{r} - \mathbf{r}'|}.$$

$$E_{xc}^{\mathrm{LSD}}[n_\alpha, \mathbf{m}_\alpha] = \int d\mathbf{r}\, n_\alpha(\mathbf{r}) \varepsilon_{xc}(n_\alpha(\mathbf{r}), \mathbf{m}_\alpha(\mathbf{r}))$$

The corresponding single-particle wave equation, obtained by taking the functional derivative of $E^{\mathrm{SIC\text{-}LSD}}$ with respect to ψ_α^*, reads

$$\left(h^{\mathrm{LSD}}(\mathbf{r}) + w_\alpha^{\mathrm{SIC}}(\mathbf{r}) \right) \psi_\alpha(\mathbf{r}) = \sum_{\alpha'} \lambda_{\alpha,\alpha'} \psi_{\alpha'}(\mathbf{r}), \tag{6}$$

with h^{LSD} given in Eq. (3) and

$$w_\alpha^{\mathrm{SIC}} = v_\alpha^{\mathrm{SIC}} + \boldsymbol{\sigma} \cdot \mathbf{b}_\alpha^{\mathrm{SIC}}$$

$$v_\alpha^{\mathrm{SIC}} = -\left(V_{H,\alpha}(\mathbf{r}) + V_{xc,\alpha}^{\mathrm{LSD}}(\mathbf{r}) \right)$$

$$v_{xc,\alpha}^{\mathrm{LSD}}(\mathbf{r}) = \frac{\delta E_{xc}^{\mathrm{LSD}}[n_\alpha, \mathbf{m}_\alpha]}{\delta n_\alpha(\mathbf{r})}$$

$$\mathbf{b}_\alpha^{\mathrm{SIC}}(\mathbf{r}) = -\frac{\delta E_{xc}^{\mathrm{LSD}}[n_\alpha, \mathbf{m}_\alpha]}{\delta \mathbf{m}_\alpha(\mathbf{r})}.$$

Note that the problem of finding the single-particle states is now complicated by the fact that each state sees a different potential, w_α^{SIC}. Instead of determining the energy eigenvalues, it is therefore necessary to evaluate the Lagrange multipliers matrix, λ, to ensure orthonormality within the set of states ψ_α. Furthermore, we should remark that in contrast to E^{LSD}, $E^{\mathrm{SIC\text{-}LSD}}$ is not invariant with respect to unitary transformations in the space of the states ψ_α: Suppose we perform a unitary transformation among the occupied orbitals specified by the matrix $U = 1 + i(dS)$, with $(dS)_{\alpha_1,\alpha_2} = \varepsilon$, $(dS)_{\alpha_2,\alpha_1} = \varepsilon^*$. Then the states ψ_{α_1}, and ψ_{α_2} change to first order in ε by $i\varepsilon\psi_{\alpha_2}$ and $i\varepsilon^*\psi_{\alpha_1}$, respectively. Using the definition of w_α^{SIC} (Eq. (6)), we find the first order change of E^{SIC} to be: $dE^{SIC} = i\varepsilon \langle \psi_{\alpha_1}|w_{\alpha_1}^{SIC} - w_{\alpha_2}^{SIC}|\psi_{\alpha_2}\rangle + c.c.$. If we intend to study the ground state properties we should therefore choose the states ψ_α so that they minimize E^{SIC} with respect to all unitary transformations, that is to say, they fulfil the so-called localization criterion

$$\langle \psi_{\alpha_1}|w_{\alpha_1}^{SIC} - w_{\alpha_2}^{SIC}|\psi_{\alpha_2}\rangle = 0. \tag{7}$$

It appears that the SIC-potentials w_α^{SIC} are significant for those states whose charges are localized in space. Thus the importance of SI corrections for atoms, molecules and core electrons in solids seems obvious. However, our main interest concentrates on the valence electrons in periodic solids. If extended Bloch states are used to describe these electrons, the corresponding w_α^{SIC}'s turn out to be negligible. If, on the other hand, spatially localized states are used the w_α^{SIC} may be an attractive potential contribution of considerable size. In the remainder of this and the following section we will describe one method of solving the SIC-LSD energy minimization problem of Eq. (6), while section IV describes another yet equivalent procedure.

In what follows we assume that the electronic system in question may be described by Wannier states Φ, each centered around some atom τ in some unit cell, displaced from the central unit cell by the lattice translation vector $\mathbf{T}$. For some of these states, defined by ϕ_α with $\alpha = (m, \mathbf{T})$, the SI-corrections are assumed to be negligible, while for the others, defined by ψ_β with $\beta = (n, \mathbf{T})$, the SI-corrections have to be taken into account explicitly. Indices n and m thus enumerates the Wannier states within the unit cell. For a periodic solid we assume a periodic repetition of localized states, i. e.,

$$\psi_{n,\mathbf{T}}(\mathbf{r}) = \psi_{n,\mathbf{T}=0}(\mathbf{r} - \mathbf{T}) = \psi_n(\mathbf{r} - \mathbf{T}). \tag{8}$$

$$\phi_{m,\mathbf{T}}(\mathbf{r}) = \phi_{m,\mathbf{T}=0}(\mathbf{r} - \mathbf{T}) = \phi_m(\mathbf{r} - \mathbf{T}). \tag{9}$$

The same symmetry applies to the SIC-potential of state $\psi_{n,\mathbf{T}}$:

$$w_{n,\mathbf{T}}^{SIC}(\mathbf{r}) = w_{n,\mathbf{T}=0}^{SIC}(\mathbf{r} - \mathbf{T}) = w_n^{SIC}(\mathbf{r} - \mathbf{T}), \tag{10}$$

and the Lagrange multipliers matrix as

$$\lambda_{n,\mathbf{T};n',\mathbf{T}'} = \lambda_{n;n'}(\mathbf{T} - \mathbf{T}'). \tag{11}$$

Considering the states ψ_n and ϕ_m as the elements of the column vector Φ, and their complex conjugates as the elements of the row vector $\Phi^\dagger$, Eq. (6) may be rewritten in matrix form as

$$(H^{LSD}(\mathbf{r}) + W^{SIC}(\mathbf{r} - \mathbf{T})) \Phi(\mathbf{r} - \mathbf{T}) = \sum_{\mathbf{T}'}\lambda(\mathbf{T} - \mathbf{T}') \Phi(\mathbf{r} - \mathbf{T}'). \tag{12}$$

Here, H^{LSD} and W^{SIC} are diagonal matrices, and the subblock of W^{SIC} acting on the states ϕ_m is zero. Since the λ's are chosen to make the states Φ orthonormal, the Lagrange multipliers may be expressed as the following matrix elements of the SIC-Hamiltonian

$$\lambda(\mathbf{T} - \mathbf{T}') = \sum_{\mathbf{T}_1} \int_{\Omega_{unit}} d\mathbf{r} \Phi^\dagger(\mathbf{r} + \mathbf{T} - \mathbf{T}' + \mathbf{T}_1)(H^{\mathrm{LSD}}(\mathbf{r})$$
$$+ W^{\mathrm{SIC}}(\mathbf{r} + \mathbf{T}_1))\Phi(\mathbf{r} + \mathbf{T}_1). \tag{13}$$

When deriving Eq. (13) we have made use of Eqs. (8) to (10). To construct the Bloch state vector, $\Psi_{\mathbf{k}}$, we form the following lattice sum

$$\Psi_{\mathbf{k}}(\mathbf{r}) = \sum_{\mathbf{T}} M^{-1}(\mathbf{k}) \exp(-i\mathbf{k}\mathbf{T}) \, \Phi(\mathbf{r} + \mathbf{T}), \tag{14}$$

while the inverse transformation is

$$\Phi(\mathbf{r} + \mathbf{T}) = \frac{1}{\Omega_{BZ}} \int d\mathbf{k} \, M(\mathbf{k}) \exp(i\mathbf{k}\mathbf{T}) \, \Psi_{\mathbf{k}}(\mathbf{r}). \tag{15}$$

Here, the $\mathbf{k}$-integration is over the Brillouin zone (BZ) whose volume is Ω_{BZ}. The unitary matrix M will be elaborated upon at a later stage. By acting from the left with the operator

$$A = \sum_{\mathbf{T}} M^{-1}(\mathbf{k}) \exp(-i\mathbf{k}\mathbf{T}),$$

on the wave equation for $\Phi(\mathbf{r} + \mathbf{T})$, Eq. (12), one obtains the following wave equation for the Bloch state vector $\Psi_{\mathbf{k}}(\mathbf{r})$ with the wave vector $\mathbf{k}$:

$$(H^{\mathrm{LSD}}(\mathbf{r}) + V^{\mathrm{SIC}}_{\mathbf{k}}(\mathbf{r})) \, \Psi_{\mathbf{k}}(\mathbf{r}) = \lambda_{\mathbf{k}} \, \Psi_{\mathbf{k}}(\mathbf{r}). \tag{16}$$

Here $V^{\mathrm{SIC}}_{\mathbf{k}}$ is a diagonal matrix whose elements, $v^{\mathrm{SIC}}_{\mathbf{k},\nu}$, are given by

$$v^{\mathrm{SIC}}_{\mathbf{k},\nu}(\mathbf{r})\Psi_{\mathbf{k},\nu}(\mathbf{r}) = \sum_{n}\sum_{\mathbf{T}}(M^{-1}(\mathbf{k}))_{\nu,n} \exp(i\mathbf{k}\mathbf{T})w^{\mathrm{SIC}}_{n}(\mathbf{r} + \mathbf{T})$$
$$\psi_n(\mathbf{r} + \mathbf{T}), \tag{17}$$

where the subscripts n and ν refer to bands. The Lagrange multipliers matrix, $\lambda_{\mathbf{k}}$, is defined as

$$\lambda_{\mathbf{k}} = \sum_{\mathbf{T}} \exp(i\mathbf{k}\mathbf{T}) \, M^{-1}(\mathbf{k}) \, \lambda(\mathbf{T}) \, M(\mathbf{k}). \tag{18}$$

Note that V^{SIC} has lattice translational symmetry: $V^{\mathrm{SIC}}_{\mathbf{k}}(\mathbf{r} + \mathbf{T}) = V^{\mathrm{SIC}}_{\mathbf{k}}(\mathbf{r})$, due to the symmetry of the Bloch states: $\Psi_{\mathbf{k},\nu}(\mathbf{r}+\mathbf{T}) = \exp(i\mathbf{k}\mathbf{T})\Psi_{\mathbf{k},\nu}(\mathbf{r})$. In practical applications $v^{\mathrm{SIC}}_{\mathbf{k},\nu}$ is chosen to be nonzero only for the bands which are expected to be extremely narrow, i.e. having well localized Wannier states. We shall call the states corresponding to such bands the SI corrected Bloch states, and label them by c, while the rest of states shall be referred to as the non-SI corrected Bloch states, and labelled by nc.

It is straightforward to write E^{LSD} in the usual way in terms of the Bloch states, $\Psi_{\mathbf{k}}$, while E^{SIC} is expressed through the Wannier states $\psi_n(\mathbf{r} - \mathbf{T})$ as:

$$E^{SIC} = -\sum_{n,\mathbf{T}} e_n[n_n, \mathbf{m}_n]. \tag{19}$$

Due to the translational symmetry of the localized states, and making use of Eqs. (8) to (10), the localization criterion, Eq. (7), takes the following form

$$\sum_{\mathbf{T}} \int_{\Omega_{unit}} d\mathbf{r}\, \psi_{n_1}^*(\mathbf{r}+\mathbf{T})\,(w_{n_1}(\mathbf{r}+\mathbf{T}) - w_{n_2}(\mathbf{r}+\mathbf{T}+\mathbf{T}_1-\mathbf{T}_2))$$
$$\psi_{n_2}(\mathbf{r}+\mathbf{T}+\mathbf{T}_1-\mathbf{T}_2) = 0. \tag{20}$$

This relation is valid for any pair of states ψ_{n_1} and ψ_{n_2}, and arbitrary translation lattice vectors $\mathbf{T}_1, \mathbf{T}_2$. Here the integration is over the volume of the central unit cell. By inserting this relation into Eq. (13) it is easy to see that the subblock of the Lagrange multipliers matrix λ, built by the SIC-corrected localized states ψ_n, is hermitian

$$\lambda_{n1,n2}(\mathbf{T}_1 - \mathbf{T}_2) = \lambda_{n2,n1}^*(\mathbf{T}_2 - \mathbf{T}_1). \tag{21}$$

Finally, it remains to define the matrix $M(\mathbf{k})$ introduced in Eq. (14). The requirement is that this matrix, via Eq. (15), generates well localized states, ψ_n, that satisfy the wave equation (12), and for which the SI corrections are significant. The following expression turned out to meet such requirements

$$M(\mathbf{k}) = U\, M^{(1)}(\mathbf{k}) \tag{22}$$

with

$$(M^{(1)}(\mathbf{k}))_{\nu,\nu'} = N_\nu(\mathbf{k}) \int_\Omega d\mathbf{r}\, \Psi_{\mathbf{k},\nu'}^*(\mathbf{r})\, \Psi_{\mathbf{k}=0,\nu}(\mathbf{r}) \tag{23}$$

and

$$N_\nu(\mathbf{k}) = (\sum_{\nu'} |\int_\Omega d\mathbf{r} \Psi_{\mathbf{k},\nu'}^*(\mathbf{r})\Psi_{\mathbf{k}=0,\nu}(\mathbf{r})|^2)^{-1/2}. \tag{24}$$

Here the integration is over the volume Ω, obtained by applying all symmetry operations of the crystalline point group to the unit cell. The matrix U is determined numerically so that the localized states ψ_n are orthonormal and fulfil the localization criterion (Eq. (20)).

3   The Unified Hamiltonian Approach

In practical applications we have to iterate Eqs. (4), (6) and (16) to (18) to self-consistency. In particular, the Lagrange multipliers matrix, $\lambda_{\mathbf{k}}$, has to be evaluated to ensure the orthonormality of the Bloch states, $\Psi_{\mathbf{k}}$. An alternative method, developed in relation to the Hartree-Fock theory, consists in solving the eigenvalue problem for a hermitian operator, $h_{u,\mathbf{k}}$, the so-called unified Hamiltonian

$$h_{u,\mathbf{k}} |\Psi_{\mathbf{k},\nu}\rangle = \varepsilon_{\mathbf{k},\nu} |\Psi_{\mathbf{k},\nu}\rangle. \tag{25}$$

To define $h_{u,\mathbf{k}}$ appropriate for the SIC-LSD method[18] we introduce the projection operators, $P_{\mathbf{k}}$, onto the subspaces of the SI corrected Bloch states, $\Psi_{\mathbf{k},c}$, and the projection operators, $Q_{\mathbf{k}}$, onto the subspace of the non-SI corrected Bloch states, $\Psi_{\mathbf{k},nc}$, namely

$$P_{\mathbf{k}} = \sum_c P_{\mathbf{k},c} = \sum_c |\Psi_{\mathbf{k},c}\rangle \langle\Psi_{\mathbf{k},c}|$$

$$Q_{\mathbf{k}} = \sum_{nc} Q_{\mathbf{k},nc} = \sum_{nc} |\Psi_{\mathbf{k},nc}\rangle \langle\Psi_{\mathbf{k},nc}|. \tag{26}$$

Subsequently, $h_{u,\mathbf{k}}$ is defined as

$$h_{u,\mathbf{k}} = h^{\mathrm{LSD}} - \sum_{c,c';(c\neq c')} P_{\mathbf{k},c}\, h^{\mathrm{LSD}}\, P_{\mathbf{k},c'} + \sum_c P_{\mathbf{k},c}\, v_{\mathbf{k},c}^{\mathrm{SIC}}\, P_{\mathbf{k},c}$$

$$+ Q_{\mathbf{k}} \sum_c v_{\mathbf{k},c}^{\mathrm{SIC}}\, P_{\mathbf{k},c} + \sum_c P_{\mathbf{k},c}\, v_{\mathbf{k},c}^{SIC\dagger}\, Q_{\mathbf{k}} \tag{27}$$

or alternatively as

$$h_{u,\mathbf{k}} = h^{\mathrm{LSD}} + \sum_c v_{\mathbf{k},c}^{\mathrm{SIC}}\, P_{\mathbf{k},c}$$

$$- \sum_{c,c';(c\neq c')} P_{\mathbf{k},c}\, (h^{\mathrm{LSD}} + v_{\mathbf{k},c'}^{\mathrm{SIC}})\, P_{\mathbf{k},c'}$$

$$+ \sum_c P_{\mathbf{k},c}\, v_{\mathbf{k},c}^{SIC\dagger}\, Q_{\mathbf{k}}. \tag{28}$$

To fulfil the hermiticity of $h_{u,\mathbf{k}}$, the transformation matrix, M, and the localized states, ψ_n, have to be chosen so that the following quantity, named A_c, is real

$$A_c = \sum_{\mathbf{T},n} \int_{\Omega_{unit}} d\mathbf{r}\, \Psi_{\mathbf{k},c}^*(\mathbf{r})\, M_{c,n}^{-1}(\mathbf{k}) \exp(-i\mathbf{k}\mathbf{T}) w_n(\mathbf{r}+\mathbf{T})$$

$$\psi_n(\mathbf{r}+\mathbf{T}). \tag{29}$$

To show that Eqs. (25) and (16) are equivalent let the Hamiltonian h_{uk}, as defined by Eq. (28), act on the SI corrected Bloch state, $\Psi_{\mathbf{k},c}$. This gives

$$(h^{\mathrm{LSD}} + v_{\mathbf{k},c}^{\mathrm{SIC}}) |\Psi_{\mathbf{k},c}\rangle = \varepsilon_{\mathbf{k},c} |\Psi_{\mathbf{k},c}\rangle$$

$$+ \sum_{c';(c'\neq c)} \langle\Psi_{\mathbf{k},c'}|h^{\mathrm{LSD}} + v_{\mathbf{k},c}^{\mathrm{SIC}}|\Psi_{\mathbf{k},c}\rangle |\Psi_{\mathbf{k},c'}\rangle. \tag{30}$$

Similarly, for the non-SI corrected state $\Psi_{\mathbf{k},nc}$ we get

$$h^{LSD} |\Psi_{\mathbf{k},nc}\rangle = \varepsilon_{\mathbf{k},nc} |\Psi_{\mathbf{k},nc}\rangle$$
$$- \sum_c \langle \Psi_{\mathbf{k},c}|v_{\mathbf{k},c}^{SIC\dagger}|\Psi_{\mathbf{k},nc}\rangle |\Psi_{\mathbf{k},c}\rangle. \qquad (31)$$

When interpreting the matrix elements on the right-hand side of Eqs. (30) and (31) as the elements of the Lagrange multipliers matrix, $\lambda_{\mathbf{k}}$, then Eqs. (30) and (31) are, indeed, equivalent to Eq. (16). The eigenvalues $\varepsilon_{\mathbf{k},\nu}$ are the diagonal elements of λ.

The transformation from real space to k-space, as defined by Eq. (14) and applied to Eq. (7), leads to the following localization criterion for the Bloch states

$$\int_{\Omega_{unit}} d\mathbf{r}\Psi_{\mathbf{k},\nu_1}^* \left(v_{\mathbf{k},\nu_1}^{SIC*}(\mathbf{r}) - v_{\mathbf{k},\nu_2}^{SIC}(\mathbf{r})\right) \Psi_{\mathbf{k},\nu_2}(\mathbf{r}) = 0. \qquad (32)$$

Inspecting the r.h.s. of Eq. (30), containing the Lagrange multipliers of the subspace of the SI corrected Bloch states, one can see that the above equation guarantees the hermiticity of the corresponding subblock of $\lambda_{\mathbf{k}}$. Summarizing, the unified Hamiltonian approach is seen to yield both the orthogonalized SI corrected and non-SI corrected Bloch states, as well as the Lagrange multipliers matrix, all just from one matrix diagonalization per k-point.

The unified Hamiltonian scheme may readily be incorporated into the linear muffin-tin orbital (LMTO) band structure method,[17,35] and in what follows we shall describe the necessary steps in some detail.

The lattice Fourier transforms of the muffin tin orbitals (MTO's), $\chi_{\sigma,I,\mathbf{k}}(\mathbf{r})$, are used as a set of one-particle basis states. The label I stands for the angular momentum quantum numbers l, m and the site τ corresponding to the head of the MTO, while σ denotes the spin component. The space coordinate is decomposed as $\mathbf{r} = \rho + \tau$, with ρ restricted to the central unit cell. The Bloch states have the form

$$\Psi_{\sigma,\mathbf{k},\nu}(\rho,\tau) = \sum_I A_{\sigma,I;\mathbf{k}}^{(\nu)} \chi_{\sigma,I,\mathbf{k}}(\rho,\tau). \qquad (33)$$

The wave vector coefficients, $A_{\sigma,I;\mathbf{k}}^{(\nu)}$, and the one particle energies, $\varepsilon_{\sigma,\mathbf{k},\nu}$, are obtained by solving the following eigenvalue problem

$$\sum_{I'} \langle \chi_{\sigma,I,\mathbf{k}}|h_{u\mathbf{k}}|\chi_{\sigma,I',\mathbf{k}}\rangle\rangle A_{\sigma,I';\mathbf{k}}^{(\nu)} =$$
$$\varepsilon_{\sigma,\mathbf{k},\nu} \sum_{I'} \langle \chi_{\sigma,I,\mathbf{k}}|\chi_{\sigma,I',\mathbf{k}}\rangle A_{\sigma,I';\mathbf{k}}^{(\nu)}. \qquad (34)$$

Making use of Eq. (27) leads to the following form of the matrix elements of the unified Hamiltonian

$$
\begin{aligned}
\langle \chi_{\sigma,I,\mathbf{k}} | h_{u\mathbf{k}} | \chi_{\sigma,I',\mathbf{k}} \rangle &= \langle \chi_{\sigma,I,\mathbf{k}} | h^{\mathrm{LSD}} | \chi_{\sigma,I',\mathbf{k}} \rangle \\
&- \sum_{c,c',c \neq c'} \sum_\nu \langle \chi_{\sigma,I,\mathbf{k}} | \Psi_{\sigma,\mathbf{k},c} \rangle \langle \Psi_{\sigma,\mathbf{k},c} | \Psi_{\sigma,\mathbf{k},\nu}^{\mathrm{LSD}} \rangle \\
&\varepsilon_{\sigma,\mathbf{k},\nu}^{\mathrm{LSD}} \langle \Psi_{\sigma,\mathbf{k},\nu}^{\mathrm{LSD}} | \Psi_{\sigma,\mathbf{k},c'} \rangle \langle \Psi_{\sigma,\mathbf{k},c'} | \chi_{\sigma,I',\mathbf{k}} \rangle \\
&+ \sum_c \langle \chi_{\sigma,I,\mathbf{k}} | \Psi_{\sigma,\mathbf{k},c} \rangle \langle \Psi_{\sigma,\mathbf{k},c} | v_{\mathbf{k},c}^{\mathrm{SIC}} | \Psi_{\sigma,\mathbf{k},c} \rangle \langle \Psi_{\sigma,\mathbf{k},c} | \chi_{\sigma,I',\mathbf{k}} \rangle \\
&+ (\sum_{nc,c} \langle \chi_{\sigma,I,\mathbf{k}} | \Psi_{\sigma,\mathbf{k},nc} \rangle \langle \Psi_{\sigma,\mathbf{k},nc} | v_{\mathbf{k},c}^{\mathrm{SIC}} | \Psi_{\sigma,\mathbf{k},c} \rangle \\
&\langle \Psi_{\sigma,\mathbf{k},c} | \chi_{\sigma,I',\mathbf{k}} \rangle + \mathrm{c.c.}).
\end{aligned}
\tag{35}
$$

In the second term on the right-hand side of the above equation, we have inserted a complete set of the LSD Bloch states, $\Psi_{\sigma,\mathbf{k},\nu}^{\mathrm{LSD}}$, with the corresponding eigenvalues $\varepsilon_{\sigma,\mathbf{k},\nu}^{\mathrm{LSD}}$. Next, instead of using the MTO's, we can express all quantities in terms of the normalized single-site solutions, ϕ, and their energy derivatives, $\dot{\phi}$, at fixed energies $\varepsilon_{\sigma,l,\tau}$, and in the spherical potential V^{LSD}, and the magnetic field $\mathbf{B}^{\mathrm{LSD}}$. In the representation of these states $\Phi_{\sigma,I,1}(\boldsymbol{\rho}) = \phi_{\sigma,I}(\boldsymbol{\rho})$, and $\Phi_{\sigma,I,2}(\boldsymbol{\rho}) = \dot{\phi}_{\sigma,I}(\boldsymbol{\rho})$, the Bloch states read as

$$
\Psi_{\sigma,\mathbf{k},\nu}(\boldsymbol{\rho},\tau) = \sum_{I,i} a_{\sigma,I,i;\mathbf{k}}^{(\nu)} \Phi_{\sigma,I,i}(\boldsymbol{\rho}).
\tag{36}
$$

The relation between the coefficients $a_{\sigma,I,i;\mathbf{k}}^{(\nu)}$ on the one hand and the coefficients $A_{\sigma,I;\mathbf{k}}^{(\nu)}$ on the other, is provided by the standard LMTO formalism.[35] Similarly, the MTO's may be written in the form

$$
\chi_{\sigma,I,\mathbf{k}}(\boldsymbol{\rho},\tau') = \sum_{I',i} \Gamma_{\sigma,I,I',i;\mathbf{k}} \Phi_{\sigma,I',i}(\boldsymbol{\rho}).
\tag{37}
$$

Inserting Eq. (36) into Eq. (15) we obtain the corresponding representation of the localized states, ψ_n

$$
\psi_n(\mathbf{r}+\mathbf{T}) = \psi_n(\boldsymbol{\rho},\tau;\mathbf{T}) = \sum_{l,m,i} C_{\sigma,I,i;\mathbf{T}}^{(n)} \Phi_{\sigma,I,i}(\boldsymbol{\rho})
\tag{38}
$$

with

$$
C_{\sigma,I,i;\mathbf{T}}^{(n)} = \frac{1}{\Omega_{BZ}} \sum_\nu \int d\mathbf{k} \, M_{n,\nu}(\mathbf{k}) \exp(i\mathbf{k}\mathbf{T}) a_{\sigma,I,i;\mathbf{k}}^{(\nu)}.
$$

Making use of Eqs. (36) to (38) and (17), we can calculate the matrix elements of the unified Hamiltonian, $h_{u,\mathbf{k}}$, with

$$\langle \chi_{\sigma,I,\mathbf{k}} | \Psi_{\sigma,\mathbf{k},\nu} \rangle = \sum_{I',i} \Gamma_{\sigma,I,I',i;\mathbf{k}} \, a^{(\nu)}_{\sigma,I',i} \, g_{I',i}$$

$$\langle \Psi_{\sigma,\mathbf{k},\nu} | v^{SIC}_{\mathbf{k},c} | \Psi_{\sigma,\mathbf{k},c} \rangle = \sum_{n,\mathbf{T}} M^{-1}_{c,n}(\mathbf{k}) \exp(-i\mathbf{kT}) \sum_{I,i,i'} a^{(\nu)*}_{\sigma,I,i;\mathbf{k}} C^{(n)}_{\sigma,I,i',\mathbf{T}}$$

$$\times \int d\rho \Phi^*_{\sigma,I,i}(\boldsymbol{\rho}) w^{SIC,av}_n(\rho,\tau;\mathbf{T}) \Phi_{\sigma,I,i'}(\boldsymbol{\rho}). \qquad (39)$$

Here $g_{I,i} = \langle \Phi_{I,i} | \Phi_{I,i} \rangle$, and we have approximated the SIC potential w^{SIC}_n by its spherical average, $w^{SIC,av}_n$, over the ASA sphere at site τ in the cell $\mathbf{T}$. Then it is also straightforward to express the energy functionals E^{LSD} and E^{SIC} in terms of quantities defined above.

As can be seen from Eq. (35), the SIC-LSD implementation is much more complicated than the LSD, because the Hamiltonian matrix elements depend explicitly on the Bloch states $\Psi_{\sigma,\mathbf{k},\nu}$. Consequently, one has to iterate the appropriate set of equations until the self-consistency for the one-particle potential, V^{LSD}, is achieved, and in addition, the unified Hamiltonian becomes consistent with its solutions. In practice, it turns out, that the self-consistency with respect to all necessary quantities, can be reached within a moderate number of iterations, if the Hamiltonian, $h^{(i)}_u$, for the ith iteration is evaluated using the Bloch states of the previous iteration. In contrast, the transformation matrix, M, the localized states, ψ_n, and the SIC potentials, $v^{SIC}_{\mathbf{k},\nu}$, are derived from the eigenstates of $h^{(i)}_u$. In the first iteration one starts with LSD Bloch states. The LSD potential, V^{LSD}, is obtained by mixing the total charge densities of previous iterations in the usual way. In practice, the evolution of the total energy, E_{tot}, is a suitable measure of the progress made in approaching overall self-consistency: If E_{tot} is converged to within $10^{-4} - 10^{-5}$ Ry, then also all other quantities turn out to be sufficiently well converged.

In contrast to the LSD, the SIC-LSD formalism gives a great deal of variational freedom to the DFT. At the beginning of each calculation one has to decide on the number and the character of the localized states ψ_n, e.g., their orbital moment- and spin-quantum numbers, and the sites to which they belong. This choice depends on the nature of the system in question and is guided by physical intuition. In general, it is not unique. We have the freedom to try different, but reasonable, 'configurations'. After carrying each of them to self-consistency the resulting total energies, E_{tot}, are compared, and the configuration corresponding to the lowest total energy is considered to provide the best description of the physical situation, as described within the SIC-LSD scheme. Of course, there always exists a possibilty that the LSD will provide the minimum of E_{tot}, meaning that the SI corrections are of no importance for the system in question.

4   The Steepest Descent Approach

An alternative way to solve the SIC-LSD equations, Eq. (6), is by the steepest descent method,[20] whereby the $E^{\text{SIC-LSD}}$ functional is minimized iteratively. If at some point approximative solutions $\tilde{\psi}_\alpha$ are given, the energy may be further minimized by adding a correction proportional to the gradient of $E^{\text{SIC-LSD}}$. Specifically,

$$\tilde{\psi}_\alpha \to \tilde{\psi}_\alpha + \delta\psi_\alpha, \tag{40}$$

with

$$\delta\psi_\alpha(\mathbf{r}) = -x\,\hat{Q}\,\frac{\delta E^{\text{SIC-LSD}}}{\delta\psi_\alpha^*(\mathbf{r})}, \tag{41}$$

where $\hat{Q}$ projects onto the space orthogonal to the occupied states:

$$\hat{Q} = 1 - \sum_\alpha^{occ.} |\tilde{\psi}_\alpha\rangle\langle\tilde{\psi}_\alpha|. \tag{42}$$

The gradient in Eq. (41) is given by the SIC Hamiltonian in Eq. (6), so that

$$\delta\psi_\alpha = -x\left[(h^{\text{LSD}} + w_\alpha^{\text{SIC}})\tilde{\psi}_\alpha - \sum_\beta \lambda_{\alpha,\beta}\,\tilde{\psi}_\beta\right]. \tag{43}$$

The localization criterion, Eq. (7), may be fulfilled by adding to $\tilde{\psi}_\alpha$ a correction term given by

$$\delta\psi_\alpha = -y\sum_\beta r_{\alpha,\beta}\,\tilde{\psi}_\beta \tag{44}$$

$$r_{\alpha,\beta} = \lambda_{\alpha,\beta} - \lambda_{\beta,\alpha}^*, \tag{45}$$

while a third type of correction term is needed to keep the orbitals ψ_α orthonormal:

$$\delta\psi_\alpha = -\frac{1}{2}\sum_\beta t_{\alpha,\beta}\,\tilde{\psi}_\beta, \tag{46}$$

with

$$t_{\alpha,\beta} = \langle\tilde{\psi}_\alpha|\tilde{\psi}_\beta\rangle - \delta_{\alpha,\beta}. \tag{47}$$

The steps (43), (44) and (46) may be interchanged and the parameters x and y chosen according to the problem at hand. At convergence all steps vanish, and it is seen that a vanishing step in Eq. (43) corresponds to Eq. (6), a vanishing step (44) corresponds to Eq. (7), and a vanishing step (46) corresponds to orthonormality, $t_{\alpha,\beta} = 0$.

When the one-electron wavefunctions are expanded in a basis set, χ_i, as

$$\psi_\alpha = \sum_i a_i^\alpha \, \chi_i \, , \tag{48}$$

where i is a composite index labelling the degrees of freedom of the basis, the steps (43), (44) and (46) are turned into matrix operations on the vector a of expansion coefficients. In our actual implementation, the χ_i's are conveniently chosen as TB-LMTO functions.[21] in which case i labels atomic sites and angular momentum and spin quantum numbers, as discussed in connection with Eq. (33). The steps then read

$$\delta a^\alpha = -x \left(\mathbf{O}^{-1} - \Pi \right) \cdot \left(\mathbf{H} + \mathbf{V}^\alpha \right) \cdot a^\alpha \, , \tag{49}$$

$$\delta a^\alpha = -y \sum_\beta r_{\alpha,\beta} \, a^\beta \, , \tag{50}$$

$$\delta a^\alpha = \frac{1}{2} \left(\mathbf{O}^{-1} - \Pi \right) \cdot \mathbf{O} \cdot a^\alpha \, , \tag{51}$$

respectively. Here, $\mathbf{O}$, $\mathbf{H}$ and $\mathbf{V}^\alpha$ are the overlap, the Hamiltonian, and the SIC-potential matrices:

$$\mathbf{O}_{ij} = \langle \chi_i | \chi_j \rangle, \tag{52}$$

$$\mathbf{H}_{ij} = \langle \chi_i | h^{\mathrm{LSD}} | \chi_j \rangle, \tag{53}$$

$$\mathbf{V}_{ij}^\alpha = \langle \chi_i | w_\alpha^{\mathrm{SIC}} | \chi_j \rangle. \tag{54}$$

and Π is given by

$$\Pi_{ij} = \sum_\beta^{occ.} \delta_{\sigma_\alpha \sigma_\beta} \cdot a_i^{\beta *} \, a_j^\beta \, , \tag{55}$$

where the δ-function ensures that only projection on states with the same spin as ψ_α is considered. In a periodic lattice, $\mathbf{O}$, $\mathbf{H}$, and Π are translational invariant. For Π this follows from the translational invariance (8), which for the expansion coefficients means that (setting $\alpha \equiv (n, \mathbf{T})$, with $\mathbf{T}$ a lattice translation):

$$a^{n,\mathbf{T}}_{(\boldsymbol{\tau}+\mathbf{R})L} = a^{n,\mathbf{0}}_{(\boldsymbol{\tau}+\mathbf{R}-\mathbf{T})L} \, . \tag{56}$$

Here L enumerates the angular momentum characteristics of the TB-LMTO function. Only the SIC potential matrix $\mathbf{V}^\alpha$ is not translational invariant, and requires special attention. As in the unified Hamiltonian approach the translational invariance is exploited by switching to $\mathbf{k}$-space: We assume $\mathbf{R}$ running over a finite cluster (M unit cells with periodic boundary clusters, where M must be so large as to ensure that the SIC states vanish outside this region). Let $\mathbf{k}$ enumerate the corresponding reciprocal space, i. e., the $M \times M$ matrix $U_{\mathbf{kR}} \equiv \frac{1}{\sqrt{M}} e^{i\mathbf{k}\cdot\mathbf{R}}$ is unitary: $\mathbf{UU}^+ = \mathbf{U}^+\mathbf{U} = \mathbf{1}$. Then Eqs. (49)-(51) hold for the Fourier transformed quantities as well, except for the product $\mathbf{V}^\alpha a^\alpha$, which must first be evaluated in real space and Fourier transformed afterwards. Specifically, let

$$b^{n,\mathbf{T}}_{(\boldsymbol{\tau}+\mathbf{R})L} = \langle \chi_{(\boldsymbol{\tau}+\mathbf{R})L} | V^{\mathrm{SIC}}_{n,\mathbf{T}} | \psi_{n,\mathbf{T}} \rangle . \tag{57}$$

Then, with

$$b^n_{\boldsymbol{\tau} L}(\mathbf{k}) = \frac{1}{\sqrt{M}} \sum_{\mathbf{R}} e^{i\mathbf{k}\cdot(\mathbf{R}-\mathbf{T})}\, b^{n,\mathbf{T}}_{(\boldsymbol{\tau}+\mathbf{R})L} , \tag{58}$$

we get

$$\delta a^n(\mathbf{k}) = -x\left(\mathbf{O}^{-1}(\mathbf{k}) - \Pi(\mathbf{k})\right) \cdot \left[\mathbf{H}(\mathbf{k}) \cdot a^n(\mathbf{k}) + b^n(\mathbf{k})\right] , \tag{59}$$

$$\delta a^n(\mathbf{k}) = -y \sum_{n'} r_{n,n'}(\mathbf{k})\, a^{n'}(\mathbf{k}) , \tag{60}$$

and

$$\delta a^\nu(\mathbf{k}) = \frac{1}{2}\left(\mathbf{O}^{-1} - \Pi\right) \cdot \mathbf{O} \cdot a^\nu \tag{61}$$

for the gradient-step, the unitarian-mixing step and the orthonormalizing step, respectively. The r-matrix in $\mathbf{k}$-space in (60) is simply given by:

$$r_{n,n'}(\mathbf{k}) = \sum_{\boldsymbol{\tau} L} [a^{n'*}_{\boldsymbol{\tau} L}(\mathbf{k}) \cdot b^n_{\boldsymbol{\tau} L}(\mathbf{k}) - b^{n'*}_{\boldsymbol{\tau} L}(\mathbf{k}) \cdot a^n_{\boldsymbol{\tau} L}(\mathbf{k})] . \tag{62}$$

5   The Relativistic Extension

For many applications it is desirable to take account of all relativistic effects, including the spin-orbit coupling. This is not only important when we are concerned with systems containing heavy atoms, but also if we, e.g., intend to get reasonable results for properties depending on orbital moments and their coupling to the spins of the electrons. Furthermore, the localized states resulting from the SI corrections are especially sensitive to the relativistic effects. Therefore, we generalize the SIC-LSD formalism, described in the previous section, to a fully relativistic spin-polarized case. The steps involved are in close analogy to the derivation of the relativistic LSD-LMTO method by Ebert [22].

The LSD wave equation of the scalar relativistic theory, where the two spin channels, as well as the orbital- and spin-moments are decoupled, is now replaced by the Dirac equation.[23] The relativistic generalization of Eq. (3) reads as

$$h^{\mathrm{LSD\text{-}REL}} = (\hat{T}^{\mathrm{REL}} + V^{\mathrm{LSD}} + \beta\boldsymbol{\sigma}\cdot\mathbf{B}^{\mathrm{LSD}}(\mathbf{r}))\psi_\alpha(\mathbf{r}) = \varepsilon_\alpha \psi_\alpha(\mathbf{r}) \tag{63}$$

with

$$\hat{T}^{\mathrm{REL}} = \boldsymbol{\alpha}\,\frac{1}{i}\nabla + \frac{1}{2}(\beta - I). \tag{64}$$

Here α, β are the Dirac matrices, while I is the unit matrix. The wave functions ψ_α, specifically their kets, are now four-component column vectors, wheras their bras are the row-vectors with complex conjugate elements.

The equations (1) to (32) remain valid in the relativistic case provided one replaces the operators $\boldsymbol{\sigma}$ and $\hat{T}$ with the matrices $\beta\boldsymbol{\sigma}$ and $\hat{T}^{\text{REL}}$, respectively [23]. The other operators should be interpreted as diagonal matrices. The main complication arises from the nature of the single-site wave functions of the Hamiltonian $h^{\text{LSD-REL}}$. They are to a good approximation simultaneously the eigenfunctions of the orbital moment, l, and the z-component, μ, of the total angular momentum, $\mathbf{j}$. In the case of $|\mu| < l + 1/2$ there exist two independent solutions for these quantum numbers $(is = 1, 2)$, and each solution is the sum of two components $(ic=1,2)$, having total angular momentum $j_{ic=1} = l + 1/2$ and $j_{ic=2} = l - 1/2$, respectively. In the case of $|\mu| = l + 1/2$, on the other hand, there is only one solution which in addition is an eigenfunction of the total angular momentum $j = l + 1/2$. The relativistic generalization of the single-site wave functions, Φ_{σ,I,i_1}, of the scalar relativistic theory, as introduced in Eq. (36), become

$$\Phi^{(\kappa)}_{I,is,i}(\boldsymbol{\rho}) = \Phi^{(\kappa)}_{I,1,is,i}(\boldsymbol{\rho}) + (1 - \delta_{|\mu|,l+1/2})\, \Phi^{(\kappa)}_{I,2,is,i}(\boldsymbol{\rho}) \tag{65}$$

with

$$\Phi^{(\kappa)}_{I,ic,is,i}(\boldsymbol{\rho}) = R^{(\kappa)}_{I,ic,is,i}(\rho)\mathbf{Y}_{j_{ic},\mu,l_{ic,\kappa}}(\boldsymbol{\rho}). \tag{66}$$

Here $R^{(\kappa)}_{I,ic,is,i}$ is the radial part of the ic component of the single-site solution is, and the index I comprises the indices l, μ, as well as, the site-index τ. The $\kappa = 1$ corresponds to the large component and $\kappa = 2$ to the small component of the bispinor $\Phi_{I,ic,is,i}$. Then $l_{ic,\kappa}$ takes the values $l_{1,1} = l_{2,2} = l$, $l_{1,2} = l + 1$, and $l_{2,1} = l - 1$. The spinor functions $\mathbf{Y}_{j,\mu,l}$, are combinations of the spherical harmonics, $Y_{l,m}$, the spin functions, χ_s, and the Clebsch-Gordon-coefficients of the well known form

$$\mathbf{Y}_{j,\mu,l}(\boldsymbol{\rho}) = \sum_s\, <j,\mu|l,\mu - s, 1/2, s\rangle Y_{l,\mu-s}(\boldsymbol{\rho})\chi_s.$$

The matching of these single-site functions to the interstitial LMTO envelope functions, and the construction of the relativistic MTO's has been described by Ebert[22] and will not be repeated here. With the information given above, it is easy to see that Eqs. (33) to (39) apply to the relativistic case as well, provided the subscript σ is omitted, and the index $I(l, m, \tau)$ gets replaced by the indices $(I(l, \mu, \tau), is)$. Furthermore, an evaluation of the occuring quantities implies matrix- and vector-operations in spinor space. Since the dimension of the eigenvalue problem is doubled in comparison to the scalar relativistic case, it is obvious that fully relativistic calculations will be considerably more expensive. In addition, the computation of the matrix elements, especially those introduced by the SIC terms, is rather tedious and requires careful programming. As an illustration, we conclude this section by displaying the relativistic version of Eq. (39), namely

$$\langle \Psi_{\mathbf{k},\nu} | v_{\mathbf{k},c}^{\mathrm{SIC}} | \Psi_{\mathbf{k},c} \rangle = \sum_{n,j} M_{c,n}^{-1}(\mathbf{k}) \exp(-i\mathbf{k}\mathbf{T})$$

$$\sum_{I,i,i'} \sum_{\kappa,is,is'} a_{I,is,i;\mathbf{k}}^{(\nu)*} C_{I,is',i'}^{(n)} \langle \Phi_{I,is,i}^{(\kappa)} | w_n^{SIC,av} | \Phi_{I,is',i'}^{(\kappa)} \rangle_{\mathbf{T}} \tag{67}$$

with

$$\langle\langle \Phi_{I,is,i}^{\kappa} | w_n^{SIC,av} | \Phi_{I,is',i'}^{\kappa} \rangle_{\mathbf{T}} = \int \rho^2 d\rho \, v_n^{SIC,av}(\rho,\tau;\mathbf{T}) \sum_{ic} R_{I,ic,is,i}^{(\kappa)}(\rho)$$

$$R_{I,ic,is',i'}^{(\kappa)}(\rho) + \sum_{ic,ic'} Q(j_{ic}, j_{ic'}, \mu, l) \int \rho^2 d\rho \, b_n^{SIC,av}(\rho,\tau;\mathbf{T}) R_{I,ic,is,i}^{(1)}(\rho)$$

$$R_{I,ic',is',i'}^{(1)}(\rho) - \sum_{ic} Q(j_{ic}, j_{ic}, \mu, l_{2,ic}) \int \rho^2 d\rho \, b_n^{SIC,av}(\rho,\tau;\mathbf{T}) R_{I,ic,is,i}^{(2)}(\rho)$$

$$R_{I,ic,is',i'}^{(2)}(\rho)) \tag{68}$$

and

$$Q(j_1, j_2, \mu, l) = \sum_{s=(-1/2,1/2)} 2s < j_1\mu|l, \mu - s, 1/2, s\rangle$$

$$\langle j_2\mu|l, \mu - s, 1/2, s\rangle. \tag{69}$$

6   Applications

In this section we concentrate on results obtained within SIC-LSD for NiO, cerium metal and cerium monopnictides. For all these systems LSD fails in describing the correct physics. In NiO, as mentioned earlier, LSD underestimates the magnetic moment and leads to a vanishing band gap due to an inadequate treatment of the on-site Coulomb repulsion among $3d$ electrons. In solids containing cerium it is the f electrons that are not correctly represented within LSD, and the attention here is turned to a variety of structural and magnetic phase transitions, beyond the reach of LSD.

6.1   NiO

The calculated key parameters of NiO as given by LSD, LDA+U, SIC-LSD and Hartree-Fock theory are quoted in Table 1 and displayed in Fig. 1. Applying SIC-LSD to NiO is seen to lead to a substantial band gap of 3.15 eV, at the experimental volume, which compares reasonably well with the experimental

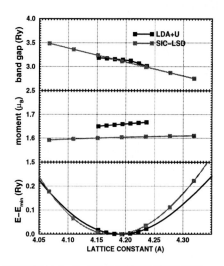

Fig. 1. The band gap, magnetic moment and the total energy of NiO in the antiferromagnetic unit cell as a function of the lattice constant, as calculated by LDA+U and SIC-LSD.

value of about 4.2 eV [27]. Also the magnetic moment is improved with respect to the LSD result, and the lattice parameter is nearly spot on. The experimental values of the magnetic moment fall between 1.66 and 1.90 μ_B [8]. That SIC-LSD provides better treatment for the $3d$ electrons in this system is also supported by comparison with the results of LDA+U calculations [24], performed within the spin polarised generalisation of the method. The latter is yet another scheme that takes the on-site Coulomb repulsion, U, explicitly into account, however, U is often treated as a parameter and chosen such that some quantity agrees with its experimental value. In Fig. 1 we show the results of both schemes for the band gap, spin magnetic moment, and total energies, as functions of the lattice constant. The numerical values calculated by LDA+U and SIC-LSD agree well both between themselves and with the experimental data. It is interesting to note that according to both the LDA+U and SIC-LSD calculations the band gap in NiO increases with increasing pressure. The band gap was found to behave similarly in LSD. The value of the band gap is sensitive to the degree of charge transfer from Ni to O. If this charge transfer increases, the band gap decreases, because the unoccupied Ni d states move down in energy, while the occupied O p states move up in energy. We conclude that the SIC is the mechanism for producing a sizeable band gap in NiO, but the actual value of the gap reflects the charge transfer in the system.

Table 1. Parameters characterizing the electronic structure of crystalline NiO and its structural stability, as calculated by LSD, LDA+U (with a $\bar{U}$=6.2 eV) and SIC-LSD [24]. The Hartree-Fock (H-F) values and experimental data for NiO were taken from Towler et al. [25].

Quantity/Method	LSD	LDA+U	SIC-LSD	H-F	Experiment
lattice constant (Å)	4.08	4.19	4.18	4.26	4.17
band gap (eV)	0.5	3.0	3.15	14.2	4.2
$B = (C_{11} + 2C_{12})/3$ (GPa)	230, 236^a	182	220	214	145, 205, 189

a : Ref. [26] .

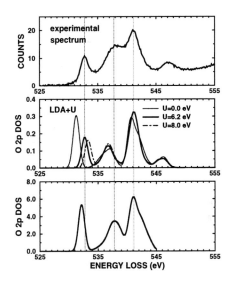

Fig. 2. Comparison of the experimental electron energy loss oxygen K edge spectrum of NiO with the density of empty oxygen 2p states of NiO, as calculated by LSD ($\bar{U} = 0$), LDA+U ($\bar{U} = 6.2$ and 8 eV), and SIC-LSD.

In Table 1 we compare the SIC-LSD results for the band gap, lattice parameter and bulk modulus with the experimental results and those from other calculations within LDA+U, LSD and Hartree-Fock method. Note that, unlike in case of the SIC-LSD, the results for LSD and LDA+U refer to the full potential implementation of the LMTO method. The effect of the full potential can be seen in the LDA+U total energy curve in Fig. 1, at higher values of the lattice parameter, as compared with the one due to SIC-LSD. This, however, has not affected the minima of the respective curves, as can be seen from Table 1.

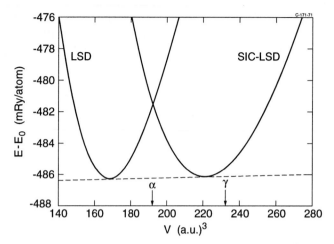

Fig. 3. Cohesive energy of Ce (in mRy/atom) as a function of atomic volume (in a_0^3/atom). The curve marked 'SIC-LSD' corresponds to the calculation with one localized f-electron per Ce atom, while the curve marked 'LSD' corresponds to itinerant f-electrons. The common tangent marks the phase transition.

A further demonstration of the advantages of SIC-LSD over LSD is presented in Fig. 2, where we compare the K edge EELS spectra of NiO with the calculations performed within LSD, LDA+U and SIC-LSD. The structure of the experimental EELS spectrum shown in Fig. 2 is dominated by the the dipole {filled 1s}→{empty 2p} transitions. Figure 2 shows that the unoccupied O $2p$ density of states (DOS) calculated using LSD does not agree well with the experimental EELS spectrum. The peaks in the EELS spectrum are associated with the hybridization between oxygen $2p$ and nickel $3d$, $4s$ and $4p$ states, respectively, and our analysis shows that the latter two peaks are practically unaffected by Hubbard correlations in the $3d$ shell. The separation between the two main peaks, namely, the O $2p$-Ni $3d$ and O $2p$-Ni $4p$ peaks, as seen in the LSD DOS curve, is approximately 2 eV larger than the separation between the same peaks in the experimental spectrum, and the spectral weight of the low-energy peak in the DOS calculated using LSD is far too high. Performing the LDA+U calculation for the O $2p$ DOS, with U=6.2 eV, leads to a significant improvement over LSD. Considering SIC-LSD result, that unlike LDA+U does not contain any adjustable parameters, the agreement with experiment is rather good. The SIC-LSD O $2p$ DOS shows slightly reduced separation between the two main peaks in the spectrum, and thus improves the agreement with the experimental EELS spectra, as compared to LSD.

6.2   Cerium

As mentioned in the introduction, LSD was not able to explain the $\gamma\rightarrow\alpha$ phase transition in cerium although it gave a magnetic solution for the lattice parameter close to the experimental value of γ−Ce. The SIC-LSD, however, provides

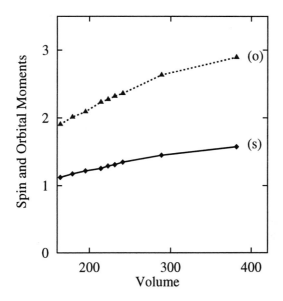

Fig. 4. Spin (s) (in μ_B) and orbital (o) moments as a function of volume (in (a.u.)^3) within the relativistic SIC-LSD approximation in γ–Ce.

a unified description of this transition with the total quenching of the magnetic moment and volume collapse of 24 %. The calculated transition pressure of −1 kbar compared favourably with the value of −7 kbar, extrapolated from the experimental phase diagram to T=0K, at which the calculations were performed. The volume collapse associated with the transition is shown in Fig. 3, where the calculated total energy of Ce as a function of volume is shown [28–30]. The two different curves correspond, respectively, to the α–phase (curve marked LSD; SIC–LSD reduces to the LSD for delocalized electrons) and γ–phase (curve marked SIC–LSD). The LSD minimum of the total energy, corresponding to the α–phase, is located in the non–magnetic region at V = 168 (a.u.)^3, while the spin magnetic moment is m = $1.32\mu_B$ at the SIC–LSD total energy minimum, corresponding to the γ–phase. Within these *ab initio* calculations, the $\gamma \rightarrow \alpha$ transition can be viewed as the transition between the phase with fully localized f electrons and the phase, where the f electrons are fully delocalized.

Performing the fully relativistic SIC-LSD calculations allows one to study also the orbital moment of the γ–phase of Ce [31]. In Fig. 4 the orbital moment can be seen to extrapolate to the atomic value of 3 at large volumes. This is the localization of one f electron within the relativistic SIC-LSD theory that suffices to describe the substantial spin and orbital moments in γ–Ce. The relativistic SIC-LSD scheme allows for each of the 14 possible f–states in Ce to become localized. As seen in Table 2, one finds that these 14 possible solutions have quite different orbital moments and correspond to different total energies. The 14 solutions have the block structure of 2+6+6. The ground state solution has the characteristics of a good crystal field-like state: sizeable spin and orbital moments which are anti-parallel aligned. The other solutions are substantially different from what one would expect from crystal field considerations. We find 6 solutions with energies 9.4 mRy higher than the lowest energy solution. These solutions are characterized by an orbital moment which is over 4 times smaller than the one of the ground state, but a spin moment substantially unaltered. Finally we have 6 solutions, situated 6.3 to 7.9 mRy above the minimum, whose orbital moment is essentially zero and whose spin moment is smaller than that of the other solutions. The total energy differences in Table 2 could be interpreted as excitation energies to populate each of the 14 localized states, corresponding to different spatial symmetry. We note that the solutions fulfill time-reversal symmetry: 7 spin ups and 7 spin downs, nearly all pairwise equal. The small deviations from exact spin up and down symmetry should be considered as the measure of the accuracy of our calculations. In Table 2 we also present the contributions to the spin and orbital moments due to the SIC localized states only, M_s^{loc} and L_z^{loc}, respectively. Regarding the total orbital moment, one can see that nearly the whole contribution comes from the localized state, whilst in case of the total spin moment 74% comes from the localized state, and the delocalized states contribute the remaining 26%.

Table 2. Calculated total energy differences with respect to the ground state (in mRy), orbital and magnetic (in μ_B) moments of the γ−phase of Ce. Also quoted are the local orbital and magnetic moments of the localized state.

M_s	M_s^{loc}	L_z	L_z^{loc}	ΔE
1.310	0.960	−2.290	−2.240	0.0
−1.309	−0.966	2.317	2.305	0.0
1.259	0.923	−0.053	0.034	6.3
1.275	0.935	−0.060	0.028	6.7
−1.257	−0.896	0.054	−0.032	6.9
−1.266	−0.913	0.021	−0.131	7.0
1.271	0.913	−0.024	0.128	7.9
−1.271	−0.912	0.032	−0.120	7.9
1.293	0.986	−0.536	−0.432	9.4
−1.293	−0.985	0.536	0.432	9.4
1.293	0.985	−0.532	−0.429	9.4
−1.293	−0.985	0.532	0.429	9.4
1.293	0.985	−0.510	−0.408	9.4
−1.293	−0.985	0.535	0.432	9.4

The solution with the lowest total energy is consistent with all three Hund's Rules, and the spin and orbital moments are anti-parallel aligned. This gives one confidence that the relativistic SIC–LSD scheme forms a bridge between the atomic and band pictures, and enables one to obtain a good description of both the localized and itinerant properties of a rare earth metal. The SIC-LSD relativistic band theory preserves all characteristic features of the $\gamma \to \alpha$ transition in Ce, and explains it as a transition from a localized state with maximum spin and orbital moments to a delocalized state without spin and orbital moments.

Table 3. Calculated transition pressures for the electronic and structural phase transitions in the cerium pnictides. Also quoted are the specific volumes on the two sides of the transition [32] . The notation (d) and (l) refers to calculations with delocalized or localized Ce f−electrons, i.e. tetravalent or trivalent Ce atoms. B2* denotes a slightly distorted B2 structure (see Ref. [39] for discussion).

compound	transition	P_t (kbar)		V_h (a_0^3)		V_l (a_0^3)	
		theo.	expt.	theo.	expt.	theo	expt.
CeN	B1(d) → B2(d)	620	-	148	-	141	-
CeP	B1(l) → B1(d)	71	90^a,55^b	325	308^a	297	2 98^a
CeP	B1(d) → B2(d)	113	150(40)^a	288	285^a	246	247^a
CeAs	B1(l) → B2(d)	114	140(20)^c	332	315^c	265	274^c
CeSb	B1(l) → B2*(l)	70	85(25)^d	400	398^d	353	354^d
CeSb	B2*(l) → B2*(d)	252	-	311	-	295	-
CeBi	B1(l) → B2*(l)	88	90(40)^e	427	399^e	376	360^e
CeBi	B2*(l) → B2*(d)	370	-	317	-	304	-

a: Ref. [33]. b: Ref. [34].
c: Ref. [35]. d: Ref. [36] .
e: Ref. [37] .

6.3   Cerium Monopnictides

The cerium monopnictides, CeN, CeP, CeAs, CeSb and CeBi, undergo a variety of structural and magnetic phase transitions under pressure. They have been thoroughly studied experimentally, and here we discuss a comparison of those studies with the SIC-LSD calculations. A detailed summary of our SIC-LSD calculations is given in Table 3, where we present the respective transition pressures and volumes associated with a variety of transitions in comparison with the experimental data. The SIC-LSD calculations involved the total energy calculations as a function of volume for CeN, CeP, CeAs, CeSb and CeBi for B1 and B2 structural phases, in the ferromagnetic (F) arrangement of Ce moments, and with the f−electron treated as either delocalized (LSD) or localized (SIC-LSD) [38,39,30].

As can be seen in Table 3, the calculations faithfully reproduce the pressure behaviour of the Ce pnictides. In particular, we find that only for CeN the f electron is delocalized at ambient conditions, while the other cerium pnictides are characterized by localized f electrons, in accordance with experimental observation. For CeP two phase transitions are observed, first the delocalization transition and subsequently the structural transition from B1 to B2 structure. The two transitions merge into a single one for CeAs, namely a transition from a B1 structure with the f electron localized to a B2 structure with the f electron delocalized. For the two remaining systems, CeSb and CeBi, the structural transition occurs first and at higher pressures does the f electron become delocalized. This trend may be understood in terms of increasing f electron localization with increasing nuclear charge of the ligand. Moreover, for CeSb and CeBi the SIC-LSD predicts a second isostructural B2→B2 transition to occur at higher pressures, where the f−electrons are delocalizing. It seems necessary to perform measurements for these systems at pressures above 250 kbar to clarify whether such transitions occur in reality.

The magnetic properties of cerium monopnictides, with the exception of CeN, are rather complicated with several antiferromagnetically ordered phases, including a Devil's staircase, and phase transitions as a function of pressure, temperature and applied magnetic field [34,40–43]. Therefore, we have for CeP performed a series of calculations exploring the energetics of various conceivable magnetic orderings. Apart from the ferromagnetic structure, LSD (delocalized f electrons) and SIC-LSD (localized f electrons) calculations were done in the AF1 and AF2 structures. In the AF1 structure the cerium moments are ferromagnetically ordered within (100) planes, which then are antiferromagnetically stacked in the (100) direction. Similarly, in the AF2 structure the cerium moments are ferromagnetically ordered within (111) planes, which are antiferromagnetically stacked in the (111) direction. In accordance with experiment an AF1 groundstate with the localized f electron [39] was obtained. The equilibrium volumes of the three magnetic structures were found to be virtually identical. CeP was found to be semimetallic in the minimum energy position in both the AF1 and AF2 structures, but at negative pressures (expanded volume), a semimetal-semiconductor transition took place. This semimetallic behaviour was found to originate from small hole pockets around the Γ point in the center of the Brillouin zone and compensated by electron pockets around the M point on the Brillouin zone boundary. These pockets are mostly confined to the basal plane of the Brillouin zone. These results are in agreement with de Haas-van Alphen [44,45] and photoemission [46] findings regarding the location of the electron and hole pockets.

7 Conclusions

We have described the SIC-LSD formalism that provides a mechanism for treating the static Coulomb correlations within *ab initio* band theory. We have demonstrated that it can treat both localized and delocalized electrons on equal footing, and owing to that is capable to study systems and properties for which LSD fails. What SIC-LSD does is to assign an energy contribution, the self-interaction correction, for an electron state to localize. Whether a system is localized or delocalized is then a result of a fight between this explicit 'localization energy' and the band formation energy. Both LSD and SIC-LSD are local minima of the

DFT energy functional. Like LSD, SIC-LSD is still a one-electron theory and both work only in extreme situations. Namely, when $U/t \ll 1$, with t being a typical hopping integral, LSD provides the correct physics, while for $U/t \gg 1$, SIC-LSD gives the valid description of the physical situation. Most importantly, as discussed in the present paper, SIC-LSD provides a correct description of pressure induced transitions from predominantly localized to predominantly delocalized groundstates, and is in general much more appropriate approach for systems with strong Coulomb correlation.

Acknowledgements This work has benefited from collaborations within, and has been partially funded by, the Training and Mobility Network on "Electronic structure calculations of materials properties and processes for industry and basic sciences" (Contract FMRX-CT98-0178).

References

1. P. Hohenberg and W. Kohn, Phys. Rev. **136**, B864 (1964): W. Kohn and L. J. Sham, Phys. Rev. A **140**, 1133 (1965).
2. R. O. Jones and O. Gunnarsson, Rev. Mod. Phys. **61**, 689 (1989).
3. W. Pickett, Rev. Mod. Phys. **61**, 433 (1989).
4. N. F. Mott, "Metal-Insulator Transitions" (Taylor and Francis, London, 1974).
5. B. Brandow, Adv. Phys. **26**, 651 (1977); J. Alloys and Compounds, **181**, 377 (1992).
6. K. Terakura, A. R. Williams, T. Oguchi and J. Kübler, Phys. Rev. Lett. **52**, 1830 (1984) ; Phys. Rev. **B30**, 4734 (1984).
7. J. Zaanen, G. A. Sawatzky and J. W. Allen, Phys. Rev. Lett. **55**, 418 (1985).
8. V. I. Anisimov, J. Zaanen and O. K. Andersen, Phys. Rev. B **44**, 943 (1991).
9. B. Johansson, Phil. Mag. **30**, 469 (1974).
10. J. W. Allen and R. M. Martin, Phys. Rev. Lett. **49**, 1106 (1982); J. W. Allen and L. Z. Liu, Phys. Rev. B **46**, 5047 (1992).
11. D. Glötzel, J. Phys. F **8**, L163 (1978); D. Glötzel and R. Podloucky, Physica **102B**, 348 (1980).
12. H. L. Skriver, O. K. Andersen and B. Johansson, Phys. Rev. Lett. **44**, 1230 (1980).
13. J. Hubbard, Proc. R. Soc. London A**276**, 238 (1963); A**277**, 237 (1964); A**281** 401 (1964)
14. V. I. Anisimov, F. Aryasetiawan and A. I. Liechtenstein, J. Phys.: Condens. Matter, **9**, 767 (1997).
15. J. P. Perdew and A. Zunger, Phys. Rev. B **23**, 5048 (1981); A. Svane, Phys. Rev. **B51**, 7924 (1995).
16. J. Taylor, "Scattering Theory", (Wiley, New York, 1972).
17. O. K. Andersen, Phys. Rev. B **12**, 3060 (1975);
18. J. G. Harrison, R. A. Heaton and C. C. Lin, J. Phys. B **16**, 2079 (1983).
19. H. L. Skriver, "The LMTO Method" (Springer Verlag, Berlin, 1984).
20. A. Svane, Phys. Rev. **B53**, 4275 (1996).
21. O. K. Andersen, O. Jepsen and O. Glötzel, "Canonical description of the band structures of metals", *in* Proc. of Int. School of Physics, Course LXXXIX, Varenna, 1985, ed. by F. Bassani, F. Fumi and M. P. Tosi (North– Holland, Amsterdam, 1985), p. 59.

312   W.M. Temmerman et al.

22. H. Ebert, Phys. Rev. B**38**, 9390 (1988).
23. For a discussion of relativistic DFT, see P. Strange, *Relativistic Quantum Mechanics*, (Cambridge, 1998).
24. S.L. Dudarev, G.A. Botton, S.Y. Savrasov, Z. Szotek, W.M. Temmerman, and A.P. Sutton, Phys. Stat. Sol. (a) **166**, 429 (1998).
25. M. D. Towler, N. L. Allan, N. M. Harrison, V. R. Saunders, W. C. Mackrodt and E. Apra, Phys. Rev. B**50**, 5041 (1994).
26. T. Sasaki, Phys. Rev. **B54**, R9581 (1996).
27. G. A. Sawatzky and J. W. Allen, Phys. Rev. Lett. **53**, 2339 (1984).
28. Z. Szotek, W. M. Temmerman and H. Winter, Phys. Rev. Lett. **72**, 1244 (1994).
29. A. Svane, Phys. Rev. Lett. **72**, 1248 (1994).
30. W.M. Temmerman, A. Svane, Z. Szotek and H. Winter, in "Electronic Density Functional Theory: Recent Progress and New Directions", Eds. J.F. Dobson, G. Vignale and M.P. Das, Plenum Press, New York, 1998.
31. S.V. Beiden, W.M. Temmerman, Z. Szotek and G.A. Gehring, Phys. Rev. Lett. **79**, 3970 (1997).
32. The errorbars in the quoted experimental transition pressures are the present authors estimates based on the hysteresis loop observed in the experimental PV curves. The quoted volume changes are relative to the equilibrium volume and taken at the average transition pressure.
33. I. Vedel, A. M. Redon, J. Rossat-Mignod, O. Vogt and J. M. Leger, J. Phys. C **20**, 3439, (1987).
34. N. Mori, Y. Okayama, H. Takahashi, Y. Haga and T. Suzuki, Physica **B186-188**, 444 (1993).
35. A. Werner, H. D. Hochheimer, R. L. Meng and E. Bucher, Physics Lett. **97A**, 207, (1983).
36. J. M. Leger, D. Ravot and J. Rossat-Mignod, J. Phys. C **17**, 4935, (1984).
37. J. M. Leger, K. Oki, J. Rossat-Mignod and O. Vogt, J. de Physique **46**, 889, (1985).
38. A. Svane, Z. Szotek, W.M. Temmerman and H. Winter, Solid State Commun. **102**, 473 (1997).
39. A. Svane, Z. Szotek, W.M. Temmerman, J. Lægsgaard and H. Winter, J. Phys. Condens. Matter **10**, 5309 (1998).
40. F. Hulliger, M. Landolt, H. R. Ott and R. Schmelczer, J. Low Temp. Phys. **20**, 269 (1975).
41. Y. Okayama, Y. Ohara, S. Mituda, H. Takahashi, H. Yoshizawa, T. Osakabe, M. Kohgi, Y. Haga, T. Suzuki and N. Mori, Physica **B186-188**, 531 (1993).
42. T. Chattopadhyay, Science **264**, 226 (1994).
43. M. Kohgi, T. Osakabe,K. Kakurai,T. Suzuki, Y. Haga, and T. Kasuya, Phys. Rev. B**49**, 7068 (1994).
44. Y. Haga, A. Uesawa, T. Terashima, S. Uji, H. Aoki, Y. S. Kwon, and T. Suzuki, Physica **B206-207**, 792 (1995).
45. T. Terashima, S. Uji, H. Aoki, W. Joss, Y. Haga, A. Uesawa, and T. Suzuki, Phys. Rev. B **55**, 4197 (1997).
46. H. Kumigashira, S.-H. Yang, T. Yokoya, A. Chainani, T. Takahashi, A. Uesawa and T. Suzuki, Phys. Rev. B **55**, R3355 (1997).

Ab Initio Theory of the Interlayer Exchange Coupling

J. Kudrnovský[1,2], V. Drchal[1,2], I. Turek[3,2], P. Bruno[4], P. Dederichs[5], and P. Weinberger[2]

[1] Institute of Physics, Academy of Sciences of the Czech Republic, CZ-182 21 Praha 8, Czech Republic
[2] Center for Computational Materials Science, Technical University, A-1060 Vienna, Austria
[3] Institute of Physics of Materials, Academy of Sciences of the Czech Republic, CZ-616 62 Brno, Czech Republic
[4] Max-Planck Institut für Mikrostrukturphysik, D-06120 Halle, Germany
[5] Institut für Festkörperforschung, Forschungszentrum Jülich, D-52425 Jülich, Germany

Abstract. Ab initio formulations of the interlayer exchange coupling (IEC) between two, in general non-collinearly aligned magnetic slabs embedded in a non-magnetic spacer are reviewed whereby both the spacer and the magnetic slabs as well as their interfaces may be either ideal or random. These formulations are based on the spin-polarized surface Green function technique within the tight-binding linear muffin-tin orbital method, the Lloyd formulation of the IEC, and the coherent potential approximation using the vertex-cancellation theorem. We also present an effective method for the study of the temperature dependence of the IEC. The periods, amplitudes, and phases are studied in terms of discrete Fourier transformations, the asymptotic behavior of the IEC is briefly discussed within the stationary-phase method. Numerical results illustrating the theory are presented.

1 Introduction

Oscillatory interlayer exchange coupling (IEC) has been found in a number of ferromagnetic/non-magnetic multilayer systems and is in some cases accompanied by an oscillatory magnetoresistance. The physical origin of such oscillations is attributed to quantum interferences due to spin-dependent confinement of the electrons in the spacer. The periods of the oscillations with respect to the spacer thickness can be correlated to the spacer Fermi surface, a relation frequently used in experimental studies. A number of models have been proposed to explain this phenomenon and we refer the reader to excellent recent reviews on the subject [1–3].

The situation is much less satisfactory if the amplitudes and/or phases are concerned. They both depend sensitively on the details of the Fermi surface, and, from the experimental point of view, on the quality of the multilayers. Typically, samples include various amounts of disorder at interfaces as well as in the bulk (e.g., surface roughness, intermixing, impurities, grain boundaries, etc.) which can influence the amplitudes and the phases significantly. From the theoretical standpoint of view it is important to keep in mind that the IEC is an oscillatory phenomenon for which, strictly speaking, amplitudes and/or phases are defined only in the asymptotic limit. Experimental data, however, are usually only available for the first few oscillations which are sufficient to extract periods, but not amplitudes and phases, in particular for the so-called long-period oscillations. The presence of impurities not only complicates the theoretical studies but also can provide a valuable insight into the effects controlling the IEC. In particular, substitutional alloying can provide a valuable informations concerning the topology of alloy Fermi surfaces. Alloying has also another, more subtle effect, namely it influences both amplitudes and phases and it can even introduce an extra damping of the oscillation amplitude (an exponential damping in addition to the usual $1/N^2$ decay, where N is the spacer thickness) if $\mathbf{k}_\parallel$-resolved electron states in the neighborhood of so-called callipers (extremal vectors of the Fermi surface) are influenced by disorder. Finally, we mention that a special case of alloying is intermixing of magnetic and spacer atoms at interfaces which can significantly influence coupling amplitudes and which occurs frequently during sample preparation in actual experiments.

It is thus obvious that the study of the effect of alloying on the periods, amplitudes, and phases of the IEC is an important issue which, however, is not properly reflected in the available literature. Conventional bandstructure methods are of limited use for such studies although in particular cases, when combined with the virtual-crystal-type approximations (VCA), they may be justified, e.g., for VCr or CrMn alloy spacers studied recently [4]. However, the complete neglect of alloy disorder makes a reliable determination of the coupling amplitudes or phases and, to some extent, even of the coupling periods, uncertain even in such favorable cases.

In addition, reliable conclusions and verifications of experimental measurements can only be based on a parameter-free theory. In order to determine the IEC one typically estimates the energy difference between the ferromagnetic (F) and antiferromagnetic (AF) alignment of a system consisting of two magnetic slabs separated by a non-magnetic spacer. Using total energy differences (evaluated with the local density approximation to the density functional theory) represents an extremely difficult task as the tiny exchange energies have to be subtracted from the background of huge total energies. Even if one employs very fast and accurate linear methods and computational tricks, the spacer thickness for which the calculated IEC values are reliable, is limited to about 20 layers [5,6]. On the other hand, for thin spacers this is the most accurate approach. One can alternatively employ asymptotic theories which are, strictly speaking, valid in the opposite regime, namely, for large spacer and magnetic slab thicknesses. The idea is to determine reflection (transmission) coefficients for an isolated interface between magnetic and spacer metals and the extremal vectors of the spacer Fermi surface. The former quantities then determine the coupling amplitudes and phases while the latter quantities their periods. In this case the calculations can be performed by using conventional bandstructure methods and, in addition, they will provide a deep insight into the physical nature of the IEC [7]. Note, however, that neither of the above techniques can be extended to treat disorder nor can they be used to interpolate between two limits, namely, the case of thin spacers (preasymptotic region) and of thick spacers (asymptotic limit). For this a theory is needed which can bridge both the preasymptotic and the asymptotic region within a unified framework: IEC values for a large set of spacer thicknesses (say, for 1-100 atomic layers) can be analyzed in terms of discrete Fourier transformation in order to reliably determine not only periods, but also coupling amplitudes and phases. In addition, one can sample various subsets in order to analyze both the preasymptotic and the asymptotic regime as well as long-period oscillations.

The basic idea is to determine the IEC directly by employing the so-called magnetic force theorem [8,9] for rotations in spin space rather then shifting atoms as in the conventional force theorem [10]. We can thus use the same potentials for both the F and AF (or, in general, rotated) alignments of the magnetic slabs (the frozen-potential approximation) and consider only the single-particle (Kohn-Sham) energies.

This allows a direct formulation of interlayer exchange coupling based on an application of the Lloyd formula [11] in order to evaluate the difference between the grand canonical potentials of the F and AF alignment. The first calculations of that type were performed by Dederichs's group in Jülich [12]. The method used in the present paper extends the above approach in three relevant aspects: (i) a reformulation within the framework of a surface Green function technique by which linear scaling of the numerical effort with respect to the number of layers [13,14] is achieved; (ii) a proof of the so-called vertex-cancellation theorem [15] in order to study the influence of alloy disorder on the properties of the IEC, and (iii) an efficient method for a fast and accurate evaluation of integrals involving the Fermi-Dirac distribution function in order to study effects of finite temperature [16,17]. In the present paper we will review these particular techniques that were developed in the past few years and subsequently applied to a number of cases including alloy disorder [18–22]. In addition, we have studied systematically the effect of non-magnetic cap-layers [23,24] on the periods, the amplitudes, and the phases of the oscillations of the IEC.

2   Formalism

In this section we derive an expression for the IEC for in general non-collinearly aligned magnetic slabs embedded in a non-magnetic spacer.

2.1   Geometry of the System

The system considered consists of a stack of layers, namely, from the left to the right: (i) a semi-infinite (nonmagnetic) substrate, (ii) a left ferromagnetic slab of thickness M (in monolayers, MLs), (iii) a nonmagnetic spacer of thickness N, (iv) a right ferromagnetic slab of thickness M', and (v) a semi-infinite (nonmagnetic) substrate. The thickness of the ferromagnetic slabs may extend to infinity. Eventually, one of the semi-infinite substrates may be substituted by a finite nonmagnetic cap of thickness P interfacing semi-infinite vacuum. In general, the various parts of the system can consist of different metals, including disordered substitutional alloys. We assume that the spin orientation of the right magnetic slab is rotated by an angle θ with respect to that of the left magnetic slab. In particular, the cases $\theta = 0$ and $\theta = \pi$ correspond to the ferromagnetic and antiferromagnetic alignments of magnetic moments of two subsystems, respectively.

2.2   Electronic Structure of the System

The electronic structure of the multilayer is described by means of the tight-binding linear-muffin tin orbital (TB-LMTO) method [25]. In particular we employ the all-electron scalar-relativistic version as generalized to the case of random alloys, their surfaces and interfaces [26,27]. The key quantity of the formalism, the physical Green function $G(z)$, is expressed via the auxiliary Green function $g^\alpha(z)$ in the screened tight-binding LMTO representation α as

$$G(z) = \lambda^\alpha(z) + \mu^\alpha(z)\, g^\alpha(z)\, \mu^\alpha(z)\,, \tag{1}$$

where

$$g^\alpha(z) = (P^\alpha(z) - S^\alpha)^{-1}\,. \tag{2}$$

Here S^α is a matrix of screened structure constants $S^\alpha_{\mathbf{R}L,\mathbf{R}'L'}$, and $P^\alpha(z)$ is a site-diagonal matrix of potential functions $P^{\alpha,\sigma}_{\mathbf{R}L}(z)$. The potential functions are diagonal with respect to the angular momentum index $L = (\ell m)$ and the spin index $\sigma = \uparrow, \downarrow$ while the structure constants are spin-independent. The potential functions can be expressed via the so-called potential parameters C, Δ, and γ in the following manner

$$P^\alpha(z) = \frac{z - C}{\Delta + (\gamma - \alpha)(z - C)}\,, \tag{3}$$

where for matters of simplicity all indices are dropped. Similarly, the quantities λ^α and μ^α in (1) can be expressed as

$$\lambda^\alpha(z) = \frac{\gamma - \alpha}{\Delta + (\gamma - \alpha)(z - C)}\,, \qquad \mu^\alpha(z) = \frac{\sqrt{\Delta}}{\Delta + (\gamma - \alpha)(z - C)}\,. \tag{4}$$

As only the screened representation will be used the superscript α is omitted in the following.

A separate problem is the determination of potential functions $P(z)$ for a given layered structure. Here we only mention that by employing the magnetic force theorem we can use the same potential functions for the ferromagnetic and rotated (or, antiferromagnetic) alignments. For random systems treated within the so-called coherent potential approximation (CPA) the potential function $P(z)$ is substituted by its coherent potential counterpart, $\mathcal{P}(z)$, whereby the formal structure of the Green function (2) remains the unchanged. The methods of determination of (coherent) potential functions for collinear alignments of magnetic moments in the present context can be found elsewhere [27,26].

2.3 Definition of the IEC

The exchange coupling energy $\mathcal{E}_x$, evaluated in the framework of the magnetic force theorem, is defined as the difference of the grand canonical potential Ω_λ between the ferromagnetic ($\lambda = F$) and antiferromagnetic ($\lambda = AF$) alignments of two subsystems, i.e. $\mathcal{E}_x = \Omega_{AF} - \Omega_F$. More generally, the quantity of the physical interest is the difference of the grand canonical potentials between a rotated ($\theta \neq 0$) and the ferromagnetic ($\theta = 0$) alignment of the two magnetic slabs, namely, $\mathcal{E}_x(\theta) \equiv \delta\Omega(\theta) = \Omega(\theta) - \Omega(0)$.

The grand canonical potential Ω of a system is defined by

$$\Omega(T,\mu) = -\int_{-\infty}^{\infty} f(E,T,\mu)\, N(E)\, \mathrm{d}E \,, \tag{5}$$

where $N(E)$ is the integrated valence density of states, $f(E,T,\mu)$ is the Fermi-Dirac distribution function at the temperature T and the chemical potential μ of electrons. It should be noted that at zero temperature the chemical potential coincides with the Fermi energy E_F of the system. The integrated valence density of states is then given by

$$N(E) = -\frac{1}{\pi}\,\mathrm{Im}\int_{-\infty}^{E} \mathrm{Tr}\, G(E' + i0)\, \mathrm{d}E' \,, \tag{6}$$

where Tr means the trace over lattice sites $\mathbf{R}$, angular momentum indices $L = (\ell m)$ and spin indices σ. Using (3,4), the following identities can be verified

$$\frac{d}{dz}\lambda(z) = -\lambda^2(z) \,, \qquad \frac{d}{dz}P(z) = \mu^2(z) \,. \tag{7}$$

Together with formula (94), we find

$$\frac{d}{dz}\Big[\mathrm{Tr}\ln\lambda(z) + \mathrm{Tr}\ln g(z)\Big] = -\mathrm{Tr}\, G(z) \,. \tag{8}$$

The grandcanonical potential (5) is then expressed as

$$\Omega(T,\mu) = -\frac{1}{\pi}\,\mathrm{Im}\int_{-\infty}^{\infty} f(E,T,\mu)\,\mathrm{Tr}\ln\lambda(E + i0)\,\mathrm{d}E$$
$$-\frac{1}{\pi}\,\mathrm{Im}\int_{-\infty}^{\infty} f(E,T,\mu)\,\mathrm{Tr}\ln g(E + i0)\,\mathrm{d}E \,. \tag{9}$$

The formula in (9) is the expression for the grandcanonical potential within the TB-LMTO method [28] and for finite temperatures.

The rotated magnetic configuration is characterized by the set of rotation angles $\Theta = \{\theta_{\mathbf{R}}\}$ for all the lattice sites. In the reference (F) state all the angles $\theta_{\mathbf{R}} = 0$ while in the rotated state $\theta_{\mathbf{R}} = \theta$ in the rotated magnetic layer and $\theta_{\mathbf{R}} = 0$ for all other lattice sites. The quantities $\lambda(\Theta, z)$ and $g(\Theta, z)$ for the rotated system are given by

$$\lambda(\Theta, z) = \mathsf{U}(\Theta)\lambda(0, z)\mathsf{U}^\dagger(\Theta) \,, \quad g(\Theta, z) = [\mathsf{U}(\Theta)P(0, z)\mathsf{U}^\dagger(\Theta) - S]^{-1} \,. \tag{10}$$

Here $[U(\Theta)]_{RR'} = \delta_{RR'}U(\theta_R)$ is the rotation matrix for spin $1/2$ particles defined in terms of the single-site matrices $U(\theta_R)$ [29]

$$U(\theta) = \begin{pmatrix} c & s \\ -s & c \end{pmatrix}, \tag{11}$$

where $c = \cos(\theta/2)$, $s = \sin(\theta/2)$, $U(\theta)\,U^\dagger(\theta) = U^\dagger(\theta)\,U(\theta) = 1$, and $\det U(\theta) = \det U^\dagger(\theta) = 1$. We note that in the rotated magnetic configuration $P(\Theta, z) = U(\Theta)P(0,z)U^\dagger(\Theta)$ is generally a non-diagonal matrix with respect to the spin indices σ, σ'.

The first term in (9) is independent of θ because $\lambda(z)$ is site (and layer-) diagonal, it therefore does not contribute to the exchange energy $\mathcal{E}_x(\theta)$, i.e., it is sufficient to consider the second part only,

$$\Omega(\theta, T, \mu) = -\frac{1}{\pi}\,\mathrm{Im}\int_{-\infty}^{\infty} f(E, T, \mu)\,\mathrm{Tr}\ln g(\theta, E + i0)\,dE\,. \tag{12}$$

It should be noted that the above expression is valid only in the absence of spin-orbit coupling.

The magnetic force theorem used here for the evaluation of the IEC was used also in related problems, e.g., for the evaluation of the exchange energies of two impurities embedded in a nonmagnetic host [8] and then extended to the case of Heisenberg exchange parameters between two sites in a magnetic material[9]. In the latter case the magnetic force theorem is valid only for the infinitesimal rotations while in the former case it is valid also for $\theta = \pi$ [30].

2.4   Configurational Averaging

Keeping in mind applications to random systems, one is interested in the configurational average of the expression in (12), namely,

$$\langle\Omega\rangle = -\frac{1}{\pi}\,\mathrm{Im}\int_{-\infty}^{\infty} f(E, T, \mu)\,\langle\mathrm{Tr}\ln g(E + i0)\rangle\,dE\,, \tag{13}$$

where $\langle\ldots\rangle$ denotes a configurational average. Difficulties here arise from the fact that the configurational average of the logarithm $\langle\ln g(z)\rangle$ can differ significantly from the logarithm of the configuration average $\ln\langle g(z)\rangle$. The difference $X \equiv \langle\ln g\rangle - \ln\langle g\rangle$, the so-called vertex correction, is difficult to calculate and usually cannot be neglected. Fortunately, this problem can be circumvented by using the vertex cancellation theorem [15], which states that the contributions from the vertex correction for the F and AF configurations cancel each other exactly, namely $\mathrm{Tr}\,X_{AF} - \mathrm{Tr}\,X_F = 0$, such that to first order with respect to the angle between the magnetizations in the two ferromagnetic layers vertex corrections can be omitted. In other words, the evaluation of (13) simplifies to

$$\langle\Omega\rangle = -\frac{1}{\pi}\,\mathrm{Im}\int_{-\infty}^{\infty} f(E, T, \mu)\,\mathrm{Tr}\ln\langle g(E + i0)\rangle\,dE\,,$$
$$= -\frac{1}{\pi}\,\mathrm{Im}\int_{C} f(z, T, \mu)\,\mathrm{Tr}\ln\langle g(z)\rangle\,dz\,. \tag{14}$$

We have also substituted the energy integral by integration over a contour in the complex energy plane z. The possibility to neglect vertex corrections can conveniently be used in calculations of the interlayer exchange coupling as explicit numerical calculations have shown that it remains valid to a good accuracy even for an angle as large as π [15]. In this respect it is very similar to the force theorem [10]. It is important to note that such an extension is only applicable to the evaluation of exchange energies of magnetic systems interacting via a non-magnetic host. An evaluation of exchange energies in ferromagnetic systems such as parameters of a classical Heisenberg model, was claimed to be limited to infinitesimal rotations only [9]. The use of the vertex-cancellation theorem allows to reduce the computational time in first-principles calculations by almost two orders of magnitude, so that the computational effort for disordered systems is comparable to that for a pure system [15]. We refer the reader to Appendix A for more details concerning the derivation and applicability of the vertex-cancellation theorem. The last remark concerns the fact that the expression for the change in the grandcanonical potential within the magnetic force theorem also includes the classical magnetostatic dipole-dipole interaction energy (DDIE). The DDIE decays with a spacer thickness much faster than the IEC and its contribution can be thus neglected for thicker spacer anyhow. In addition, first-principles fully-relativistic calculations of the IEC [32] have demonstrated that this term has a negligible influence even for a rather thin spacer amounting just to a few layers. Consequently, the DDIE term will be neglected in the following.

2.5   Lloyd Formula

We need to evaluate the difference of configurationally averaged grandcanonical potentials in the rotated and FM configurations. This can be done conveniently with the help of the well-known Lloyd formula [11] applied to layered systems. We formally split the system into two non-interacting fragments, namely a left fragment $\mathcal{L}$, which consists of the left substrate and the left magnetic slab, and a right fragment $\mathcal{R}$, which comprises the rest of the system, i.e., the spacer, the right magnetic slab, and the right substrate (or, eventually, the cap layer interfacing the vacuum). Fragments are described by the unperturbed Green function $\langle g_0(z) \rangle$. In the next step we couple two fragments together with help of a localized potential V which is simply the interlayer screened structure constant. This procedure has a number of advantages as compared to a conventional way of embedding two finite magnetic layers into the infinite (bulk) host spacer [12]: (i) the perturbation V is independent of the thicknesses of magnetic layers; (ii) complicated sample geometries can be treated, including semi-infinite magnetic layers; and (iii) a powerful and efficient method exists for the evaluation of the Green function of fragments, namely the surface Green function technique in the principal-layer formulation [26,27].

Keeping in mind the vertex cancellation theorem, one gets for a difference in the configurationally averaged grandcanonical potential (14), the expression

$$\langle \delta \Omega \rangle = -\frac{1}{\pi} \, \mathrm{Im} \int_C f(z, T, \mu) \, \mathrm{Tr} \ln \left(1 - V \langle g_0(z) \rangle \right) \, dz \,, \tag{15}$$

where $\langle g_0(z) \rangle$ is the configurationally averaged Green function of the decoupled non-interacting fragments $\mathcal{L}$ and $\mathcal{R}$ defined above. For the sake of simplicity, we will denote from here on the configurationally averaged quantities by an overbar, e.g., $\langle g_0(z) \rangle \equiv \bar{g}_0(z)$. The concept of principal layers (PL) [33] as used within the TB-LMTO method leads to a block tridiagonal form of the structure constants and of the inverse Green function. If we apply this tridiagonality to (15), we get for V and $\langle g_0(z) \rangle$ the following expressions by using a supermatrix notation with respect to nearest-neighbor PLs resolved in the wave-vector $\mathbf{k}_{\parallel}$,

$$V(\mathbf{k}_{\parallel}) = \begin{pmatrix} 0 & S_{10}(\mathbf{k}_{\parallel}) \\ S_{01}(\mathbf{k}_{\parallel}) & 0 \end{pmatrix}, \quad \bar{g}_0(\mathbf{k}_{\parallel}, z) = \begin{pmatrix} \bar{\mathcal{G}}_{\mathcal{L}}(\mathbf{k}_{\parallel}, z) & 0 \\ 0 & \bar{\mathcal{G}}_{\mathcal{R}}(\mathbf{k}_{\parallel}, z) \end{pmatrix}, \tag{16}$$

where $S_{10}(\mathbf{k}_{\parallel}) = \left[S_{01}(\mathbf{k}_{\parallel}) \right]^{\dagger}$. Combining (15) and (16) one gets

$$\delta \, \mathrm{Tr} \ln \bar{\mathcal{G}}(z) = -\frac{1}{N_{\parallel}} \sum_{\mathbf{k}_{\parallel}} \mathrm{tr} \ln \left[1 - \bar{\Gamma}_{\mathcal{L}}(\mathbf{k}_{\parallel}, z) \, \bar{\mathcal{G}}_{\mathcal{R}}(\mathbf{k}_{\parallel}, z) \right], \tag{17}$$

$$\bar{\Gamma}_{\mathcal{L}}(\mathbf{k}_{\parallel}, z) \quad = S_{10}(\mathbf{k}_{\parallel}) \, \bar{\mathcal{G}}_{\mathcal{L}}(\mathbf{k}_{\parallel}, z) \, S_{01}(\mathbf{k}_{\parallel}).$$

Here the quantity $\bar{\Gamma}_{\mathcal{L}}(\mathbf{k}_{\parallel}, z)$ has the meaning of an effective embedding potential, and the quantities $\bar{\mathcal{G}}_{\mathcal{L}}$ and $\bar{\mathcal{G}}_{\mathcal{R}}$ are the configurationally averaged surface Green functions (SGF) [33] of the magnetic subsystems $\mathcal{L}$ and $\mathcal{R}$, respectively. By definition, the surface Green function $\bar{\mathcal{G}}_{\mathcal{S}}$ ($\mathcal{S} = \mathcal{L}, \mathcal{R}$) is the top PL projection of the Green function of the corresponding semi-infinite system $\mathcal{S}$. Its determination in the case of random systems was extensively discussed in the literature, see [34–36,26]. The summation in (17) extends over the surface Brillouin zone (SBZ) corresponding to the underlying two-dimensional translational symmetry [37], and $N_{\parallel}$ is the number of sites in a layer.

2.6   The IEC for a General Angle θ

Let us now turn to the evaluation of the energy difference between arbitrary alignments. Consider the following quantity,

$$\mathrm{tr} \ln Z = \mathrm{tr} \ln \left(1 - A_0 \, B \right) - \mathrm{tr} \ln \left(1 - A_0 \, B_0 \right), \tag{18}$$

where the matrices A_0 and B_0 are related to the ferromagnetic alignment and thus are diagonal in spin space

$$A_0 = \begin{pmatrix} A_0^{\uparrow} & 0 \\ 0 & A_0^{\downarrow} \end{pmatrix}, \quad B_0 = \begin{pmatrix} B_0^{\uparrow} & 0 \\ 0 & B_0^{\downarrow} \end{pmatrix}. \tag{19}$$

The particular form of the subblocks A_0^σ and B_0^σ ($\sigma = \uparrow, \downarrow$) is given by

$$A_0^\sigma = S_{10}(\mathbf{k}_\|) \, \bar{\mathcal{G}}_{\mathcal{L}}^\sigma(\mathbf{k}_\|, z) \, S_{01}(\mathbf{k}_\|), \qquad B_0^\sigma = \bar{\mathcal{G}}_{\mathcal{R}}^\sigma(\mathbf{k}_\|, z). \tag{20}$$

The matrix B refers to an alignment in which the orientations of the magnetization in two magnetic slabs are rotated uniformly by a relative angle θ,

$$B = U(\theta) \, B_0 \, U^\dagger(\theta), \tag{21}$$

where $U(\theta)$ is the rotation matrix (11). The quantity $1 - A_0 \, B$ in (18) can therefore be written as

$$1 - A_0 \, B = \Big(U(\theta) - A_0 U(\theta) B_0 \Big) U^\dagger(\theta), \tag{22}$$

where, as follows from (19) and (11),

$$U(\theta) - A_0 \, U(\theta) \, B_0 = \begin{pmatrix} c\,(1 - A_0^\uparrow B_0^\uparrow) & s\,(1 - A_0^\uparrow B_0^\downarrow) \\ -s\,(1 - A_0^\downarrow B_0^\uparrow) & c\,(1 - A_0^\downarrow B_0^\downarrow) \end{pmatrix}. \tag{23}$$

Using now the identity $\operatorname{tr} \ln X = \ln \det X$, which is valid for any non-singular matrix X, and the identity

$$\det \begin{pmatrix} A & B \\ C & D \end{pmatrix} = \det A \cdot \det D \cdot \det(1 - A^{-1} B D^{-1} C), \tag{24}$$

which in turn is valid, if the matrices A and D are non-singular, it is straightforward to prove that

$$\operatorname{tr} \ln Z = \operatorname{tr}_L \ln \left(1 - \frac{1 - \cos(\theta)}{2} M \right), \tag{25}$$

where

$$M = 1 - (1 - A_0^\uparrow B_0^\uparrow)^{-1} (1 - A_0^\uparrow B_0^\downarrow)(1 - A_0^\downarrow B_0^\downarrow)^{-1} (1 - A_0^\downarrow B_0^\uparrow). \tag{26}$$

It should be noted that in (18) tr denotes the trace over angular momenta and spin, while in (24) tr_L denotes the trace over orbital momenta only. The final expression for $\mathcal{E}_x(\theta)$ is thus given by

$$\mathcal{E}_x(\theta) = \frac{1}{\pi N_\|} \sum_{\mathbf{k}_\|} \operatorname{Im} \int_C f(z, T, \mu) \times$$

$$\operatorname{tr}_L \ln \left(1 - \frac{1 - \cos(\theta)}{2} M(\mathbf{k}_\|, z) \right) dz, \tag{27}$$

in which the energy integral is expressed in terms of a contour integral which will be discussed in detail later.

It is interesting to note that the expression (26) for $M(\mathbf{k}_\parallel, z)$ can be rearranged in the following form [19]

$$M = -\left(1 - S_{10} \, \bar{\mathcal{G}}_\mathcal{L}^\uparrow \, S_{01} \, \bar{\mathcal{G}}_\mathcal{R}^\uparrow\right)^{-1} S_{10} \left(\bar{\mathcal{G}}_\mathcal{L}^\uparrow - \bar{\mathcal{G}}_\mathcal{L}^\downarrow\right) \times$$
$$\left(1 - S_{01} \, \bar{\mathcal{G}}_\mathcal{R}^\downarrow \, S_{10} \, \bar{\mathcal{G}}_\mathcal{L}^\downarrow\right)^{-1} S_{01} \left(\bar{\mathcal{G}}_\mathcal{R}^\uparrow - \bar{\mathcal{G}}_\mathcal{R}^\downarrow\right) . \tag{28}$$

It explicitly factorizes the 'spin-asymmetry' of the problem and it is directly related to RKKY-like theories [1]. This result [19] is formally equivalent to the results of the spin current approach [39] as formulated within a Green function formalism based on an empirical single orbital tight-binding model [40]. A matrix version developed in the framework of a semiempirical tight-binding model has appeared recently [2].

For completeness we also give the result for the common case of the antiferromagnetic alignment $(\theta = \pi)$:

$$\mathcal{E}_x \equiv \mathcal{E}_x(\pi) = \frac{1}{\pi N_\parallel} \sum_{\mathbf{k}_\parallel} \mathrm{Im} \int_C f(z, T, \mu) \, \mathrm{tr}_L \ln \mathcal{M}(\mathbf{k}_\parallel, z) \, dz , \tag{29}$$

where $\mathcal{M}$ is a product of four terms,

$$\mathcal{M} = (1 - A_0^\uparrow B_0^\uparrow)^{-1} (1 - A_0^\uparrow B_0^\downarrow) (1 - A_0^\downarrow B_0^\downarrow)^{-1} (1 - A_0^\downarrow B_0^\uparrow) . \tag{30}$$

2.7   The Torque and Infinitesimal Rotations

The differential change in the grand canonical potential $\delta\Omega(\theta)$ with respect to a differential relative angle θ, $-\partial \, \delta\Omega(\theta)/\partial\,\theta$, is usually called the torque. The torque can easily be obtained by differentiating (27) with respect to the angle θ. By definition one gets therefore

$$T(\theta) = -\frac{\partial \mathcal{E}_x(\theta)}{\partial \theta} \quad \text{or} \quad \mathcal{E}_x(\theta) = -\int_0^\theta T(\theta') \, d\theta' , \tag{31}$$

whereby $T(\theta)$ follows immediately from (27)

$$T(\theta) = \frac{\sin(\theta)}{2\pi N_\parallel} \sum_{\mathbf{k}_\parallel} \mathrm{Im} \int_C f(z, T, \mu) \times$$
$$\mathrm{tr}_L \left[M(\mathbf{k}_\parallel, z) \left(1 - \frac{1}{2} [1 - \cos(\theta)] M(\mathbf{k}_\parallel, z) \right)^{-1} \right] dz . \tag{32}$$

By formally expanding the logarithm in (27) in powers of $1 - \cos(\theta)$, one can cast the expression for $\mathcal{E}_x(\theta)$ into the form

$$\mathcal{E}_x(\theta) = B_1 \left[1 - \cos(\theta) \right] + \frac{1}{2} B_2 \left[1 - \cos(\theta) \right]^2 + \ldots , \tag{33}$$

where B_1 and B_2 are the so-called bilinear and the (intrinsic) biquadratic exchange coupling coefficients, respectively,

$$B_1 = \frac{1}{2\pi N_\parallel} \sum_{\mathbf{k}_\parallel} \operatorname{Im} \int_C f(z,T,\mu) \operatorname{tr}_L \mathbf{M}(\mathbf{k}_\parallel, z) \, dz \,, \tag{34}$$

$$B_2 = -\frac{1}{4\pi N_\parallel} \sum_{\mathbf{k}_\parallel} \operatorname{Im} \int_C f(z,T,\mu) \operatorname{tr}_L [\mathbf{M}(\mathbf{k}_\parallel, z)]^2 \, dz \,.$$

It may be, however, more convenient to fit the exact expression (27) into the form (33) by employing calculated values for $\theta = \pi/2$ and $\theta = \pi$ [41]. We obtain

$$B_1 = \frac{\mathcal{E}_x(\pi) + 2\mathcal{E}_x(\pi/2)}{2} \,, \qquad B_2 = \frac{\mathcal{E}_x(\pi) - 2\mathcal{E}_x(\pi/2)}{2} \,. \tag{35}$$

Of particular interest is the expansion of $\mathcal{E}_x(\theta)$ for a small θ, i.e., when $1 - \cos(\theta)$ is a small parameter (the method of infinitesimal rotations (MIR)). This approach becomes particularly relevant in the case when the spacer is a magnetic metal or for complicated geometries, e.g., for so-called periodic multilayers.

2.8   The IEC as Interface–Interface Interaction

We will now discuss briefly an alternative approach of a direct evaluation of the IEC as a difference in the interface-interface interaction energies rather then its indirect determination in terms of the energy of a single interface (13-16). We decouple the system into three fragments, a left, central, and right fragment, $\mathcal{L}, \mathcal{C}$, and $\mathcal{R}$, respectively. The left and the right fragment are formed by corresponding substrates with magnetic slabs whereby the central slab comprises the spacer. Both approaches are physically equivalent because it is irrelevant how the system is divided into an unperturbed part and a perturbation. Note, however, that the interface-interface formulation is more general as it could be used for a determination of interaction energies of two generally different interfaces.

The derivation proceeds in two steps and employs partitioning technique with respect to the trace of the logarithm of the Green function. First, the subsystems $\mathcal{L}$ and $\mathcal{R}$ are downfolded which leads to an effective problem of two localized perturbations in the subsystem $\mathcal{C}$. The second step, a two-potential formula applied to the fragment $\mathcal{C}$ separates directly the interface-interface contribution. The result has formally the same structure as the previous one (17,26), but the subblocks A_0^σ and B_0^σ ($\sigma = \uparrow, \downarrow$) are now of the following form

$$A_0^\sigma = \bar{g}_{N1}(\mathbf{k}_\parallel, z) \, \bar{\tau}_1^\sigma(\mathbf{k}_\parallel, z) \, \bar{g}_{1N}(\mathbf{k}_\parallel, z) \,, \qquad B_0^\sigma = \bar{\tau}_N^\sigma(\mathbf{k}_\parallel, z) \,. \tag{36}$$

The τ-matrices $\bar{\tau}_i$ ($i = 1, N$) corresponding to "multiple scattering" at individual interfaces $\mathcal{L}/\mathcal{C}$, ($i = 1$) and $\mathcal{C}/\mathcal{R}$, ($i = N$) are expressed as

$$\bar{\tau}_i^\sigma(\mathbf{k}_\parallel, z) = \bar{\Gamma}_i^\sigma(\mathbf{k}_\parallel, z) \left[1 - \bar{g}_{ii}(\mathbf{k}_\parallel, z) \, \bar{\Gamma}_i^\sigma(\mathbf{k}_\parallel, z) \right]^{-1} \,, \tag{37}$$

where the effective embedding potentials $\bar{\Gamma}_i^\sigma(\mathbf{k}_\|, z)$ of the left and right interfaces $(i = 1, N)$, respectively, are defined as

$$\bar{\Gamma}_1^\sigma(\mathbf{k}_\|, z) = S_{10}(\mathbf{k}_\|)\,\bar{\mathcal{G}}_{\mathcal{L}}^\sigma(\mathbf{k}_\|, z)\,S_{01}(\mathbf{k}_\|)\,, \tag{38}$$
$$\bar{\Gamma}_N^\sigma(\mathbf{k}_\|, z) = S_{01}(\mathbf{k}_\|)\,\bar{\mathcal{G}}_{\mathcal{R}}^\sigma(\mathbf{k}_\|, z)\,S_{10}(\mathbf{k}_\|)\,.$$

Here, $\bar{\mathcal{G}}_{\mathcal{S}}^\sigma$ $(\mathcal{S} = \mathcal{L}, \mathcal{R})$ are the configurationally averaged SGFs of the left and the right semi-infinite regions, respectively. Details of the derivation can be found in Appendices B and C. The coupling between the two magnetic subsystems is due to the layer off-diagonal projections $\bar{g}_{1N}(\mathbf{k}_\|, z)$ and $\bar{g}_{N1}(\mathbf{k}_\|, z)$ of the Green function (GF) of the finite spacer consisting of N layers. The oscillatory behavior of interlayer coupling is then governed by the oscillatory behavior of these quasi one-dimensional spacer Green functions, a formulation which is very much in the spirit of a simplified RKKY approach [1]. An efficient method of evaluation of the corner-blocks of the Green function, $\bar{g}_{ij}(\mathbf{k}_\|, z)$, $(i, j = 1, N)$, is described in Appendix D [42,36].

2.9   Relation to the KKR Method

We shall discuss now the relation of the present approach (29,30,36) to the method employed in [12] and based on the Korringa-Kohn-Rostoker (KKR) Green function technique. Let us note first the deep internal connection between the KKR and the TB-LMTO-GF approach (see [26,43] for more details). The model in (12) consists of an infinite ideal non-magnetic spacer as a reference system and of two magnetic slabs representing localized perturbations. For simplicity we start from the case of two magnetic monolayers in an infinite spacer. The result

$$A_0^\sigma = \mathcal{G}_{N1}^b(z)(\mathbf{k}_\|, z)\,t_1^\sigma(\mathbf{k}_\|, z)\,\mathcal{G}_{1N}^b(\mathbf{k}_\|, z)\,, \qquad B_0^\sigma = t_N^\sigma(\mathbf{k}_\|, z) \tag{39}$$

is formally the same with the exception that the τ-matrices entering (36) are now substituted by the single-site t-matrices t_i which describe the scattering of electrons from two magnetic monolayers at $i = 1, N$ embedded in an infinite non-random bulk spacer and separated by $N - 2$ spacer layers:

$$t_{\lambda;i}^\sigma(\mathbf{k}_\|, z) = \Delta P_{\lambda;i}^\sigma(z)\left[1 + \mathcal{G}^b(\mathbf{k}_\|, z)\,\Delta P_{\lambda;i}^\sigma(z)\right]^{-1}\,. \tag{40}$$

The strength of the scattering potential, $\Delta P_{\lambda;i}^\sigma(z)$, is given by the difference of the potential functions for the magnetic monolayer $P_{\lambda;i}^\sigma(z)$ and for the non-magnetic spacer $P(z)$, while $\mathcal{G}^b(\mathbf{k}_\|, z)$ is the layer diagonal block of the GF of the bulk spacer. The layer off-diagonal blocks of the bulk spacer GF, $\mathcal{G}_{1N}^b(z)$ and $\mathcal{G}_{N1}^b(z)$, are given by

$$\mathcal{G}_{1N}^b(\mathbf{k}_\|, z) = \left[\mathcal{G}^s(\mathbf{k}_\|, z)\,S_{01}(\mathbf{k}_\|)\right]^{N-1}\mathcal{G}^b(\mathbf{k}_\|, z)\,, \tag{41}$$

and similarly for $\mathcal{G}_{N1}^b(z)$. Here, $\mathcal{G}^s(\mathbf{k}_\|, z)$ is the corresponding SGF of an ideal semi-infinite non-magnetic bulk spacer [33]. It should be noted that also the layer-resolved bulk Green function $\mathcal{G}^b(\mathbf{k}_\parallel, z)$ can be expressed in terms of the SGFs (see, e.g., [34]). Since (41) is exact, there is no need to perform an additional $k_\perp$-integration [12]. It is easy to show that the result is formally identical to the case of two impurities in a simple tight-binding linear chain model with nearest neighbor hopping.

A generalization to the case of magnetic slabs containing a finite number M of magnetic layers is formally straightforward [12]. The t-matrices $t^\sigma_{\lambda;i}(z)$ are then supermatrices with respect to angular momentum and layer indices and the numerical effort to evaluate (40) increases with the third power of M as contrasted with the results of the present approach (17,36) which depend only linearly on M.

2.10   Influence of External Periodicity

Until now it was assumed implicitly that we have a simple "parent" lattice [37]. The periods of the coupling oscillations are closely related to the Fermi surface geometry [1,2] of the bulk spacer. A different translational symmetry (complex lattices) or stacking sequence within layers will thus tend for sufficiently thick spacers to a different kind of bulk periodicity and hence to new periods. For example, an alternating stacking of fcc(001)-layers Cu and ordered c(2 × 2)-CuAu layers tends to an ordered fcc-Cu$_3$Au alloy with a Fermi surface topology different from that of fcc-Cu spacer. For a discussion of "superlattice" formation in magnetic multilayers see also [38]. We will discuss in the following in some detail two possibilities, namely superstructures in the spacer and in the magnetic slabs.

We start with the former case by assuming the same geometry as discussed in Sec. 2.1 but now the spacer slab consists of two non-magnetic metals A and B with respective thickness n_A and n_B periodically alternating. Typically, the spacer layer starts with the layer $A(B)$ and ends with the layer $B(A)$, but the termination of the spacer slab with the same layers is also possible (and interesting [22]). The particular case of $n_A = n_B = 1$ corresponding to an (001)-stacking of an ordered fcc-CuAu alloy was already treated on a first-principles level [22]. The more general case, $(n_A, n_B > 1)$, which corresponds to artificially grown superstructures, was treated only within a simple one-band model [44]. In both cases, new periodicities (in comparison with the spacers consisting from pure A or B metals) arise with an increasing number of repetitions. Alternatively, one can consider a superstructure within a given spacer layer, or combination of both, e.g., the above mentioned example of the ordered fcc-Cu$_3$Au alloy spacer. The similar situation can be encountered also in the magnetic slabs. In particular the case of a c(2 × 2)-CoFe periodicity within the magnetic layers separated by a fcc-Cu(001) spacer [20] leads to the rather surprising appearance of new periods. These new periods can be now correlated to critical points of the spacer Fermi surface folded down to the Brillouin zone corresponding to a c(2×2)-superlattice [20]. A correlated gradual appearance of new periods and the order in statistically disordered layers is a clear indication of their relation to a different bulk periodicity [20,22].

A special case of alternating layers of A and B metals is when one of metals is magnetic and the other is nonmagnetic, all of which sandwiched between two substrates. This is the case of a periodic multilayer.

The generalization of the present formalism to above discussed cases is rather straightforward. In the case of a superlattice within a layer it is just sufficient to substitute matrices appearing in (27,29) by the corresponding supermatrices, e.g., by (2×2)-supermatrices in the case of a $c(2 \times 2)$-superlattice. The key quantity, the surface Green functions $\bar{\mathcal{G}}^{\sigma}_{\mathcal{L},\mathcal{R}}$ (20), can be easily evaluated also in this case (see for details [26]). The generalization of the formalism to the case of alternating layers from A and B metals is as well simple because the surface Green function is constructed in an epitaxial manner, i.e., layer by layer, and it is therefore immaterial if the stacking of layers consists of the same or a different material. In the limit of a periodic multilayer we should just keep in mind that a proper repeating unit consists now from four layers, namely $S-M-S-M$, where the symbols S and M refer to the spacer and magnetic layers, respectively. This is necessary to calculate the F and AF configurations needed for the evaluation of the IEC. We note that the present formalism allows to evaluate efficiently and reliably the IEC for thick spacers (one hundred layers and more) which is important for realistic studies of so-called superlattice spacers and of periodic multilayers.

2.11   Temperature-Dependence of the IEC

We conclude this section by reviewing a recently developed technique for an efficient evaluation of the temperature dependence of the IEC [16]. The main cause for the temperature dependence of the IEC is connected with thermal excitations of electron-hole pairs across the Fermi level as described by the Fermi-Dirac function. It turns out that other mechanisms (as for example electron-phonon and electron-magnon interactions) are less important. We rewrite (29) into the following form

$$\mathcal{E}_x(T) = \operatorname{Im} I(T), \quad I(T) = \int_C f(z, T, \mu) \Psi(z) \, \mathrm{d}z, \tag{42}$$

where

$$\Psi(z) = \frac{1}{\pi N_{\|}} \sum_{\mathbf{k}_{\|}} \operatorname{tr}_L \ln \mathcal{M}(\mathbf{k}_{\|}, z), \tag{43}$$

with the energy integration performed over a contour C along the real axis and closed by a large semicircle in the upper half of the complex energy plane.

The integral in (42) can be recast into a more suitable form using the analytic properties of $\Psi(z)$, namely, (i) $\Psi(z)$ is holomorphic in the upper half of the complex halfplane, and (ii) $z\Psi(z) \to 0$ for $z \to \infty$, $\operatorname{Im} z > 0$. Let us define a new function $\Phi(y) = -i\Psi(E_F + iy)$ of a real variable y, $y \geq 0$. Then at $T = 0$ K,

$$I(0) = \int_0^{+\infty} \Phi(y) \, \mathrm{d}y, \tag{44}$$

while at $T > 0$ K,

$$I(T) = 2\pi k_B T \sum_{k=1}^{\infty} \Phi(y_k) \,, \tag{45}$$

where k_B is the Boltzmann constant and the y_k are Matsubara energies, $y_k = \pi k_B T(2k - 1)$. In the limit $T \to 0$, $I(T) \to I(0)$ continuously.

We have verified that the function $\Phi(y)$ can be represented accurately as a sum of a few complex exponentials of the form

$$\Phi(y) = \sum_{j=1}^{M} A_j \exp(p_j y) \,, \tag{46}$$

where the A_j are complex amplitudes and the p_j are complex wave numbers. An efficient method of finding the parameters A_j and p_j is described elsewhere [16]. The evaluation of $I(T)$ is then straightforward:

$$I(T) = -2\pi k_B T \sum_{j=1}^{M} \frac{A_j}{\exp\left(\pi k_B T p_j\right) - \exp\left(-\pi k_B T p_j\right)} \,, \tag{47}$$

which for $T = 0$ K gives

$$I(0) = -\sum_{j=1}^{M} \frac{A_j}{p_j} \,. \tag{48}$$

The efficiency of the present approach allows to perform calculations with a large number of $\mathbf{k}_{\parallel}$-points in the irreducible part of the surface Brillouin zone (ISBZ) in order to obtain accurate and reliable results. Note also that such calculations have to be done only once and then the evaluation of the IEC for any reasonable temperature is an easy task.

The effect of finite temperatures on the IEC can be evaluated also analytically. The analytical approach assumes the limit of large spacer thickness, for which all the oscillatory contributions to the energy integral cancel out with exception of those at the Fermi energy. The energy integral is then evaluated by a standard saddle-point method [1]. The general functional form of the temperature-dependence of the interlayer exchange coupling $\mathcal{E}_x(T)$ in the limit of a single period is then given by

$$\mathcal{E}_x(T) = \mathcal{E}_x(0)\, t(N, T) \,, \quad t(N, T) = \frac{cNT}{\sinh(cNT)} \,. \tag{49}$$

Here, N denotes the spacer thickness in monolayers, and c is a constant which depends on the spacer Fermi surface. The term $\mathcal{E}_x(0)$ exhibits a standard N^{-2}-dependence [1], while the scaling factor $t(N, T)$ depends on the product N and T. In the preasymptotic regime (small spacer thickness) the functional form of

$t(N,T)$ differs from that of (49), particularly in the case of a complete, but relatively weak confinement due to the rapid variation of the phase of the integrand which enters the expression for the IEC [45]. The present numerical technique is free of the above discussed limitations and can be used to check conclusions of model theories.

3   Numerical Results and Discussion

3.1   Details of Calculations

Special care has to be devoted to the energy and the Brillouin zone integrations. For a finite temperature we determine the parameters of the complex exponentials in (46) through an evaluation of $\Phi(y)$ at 40 Matsubara energies corresponding to $T = 25$ K. We have verified that the results depend weakly on the actual value of the parameter T. For $T = 0$ K we have tested two energy contours C, namely a semicircle between the bottom of the band (E_{min}) and E_F, or, alternatively, a line contour $E_F + i\varepsilon$, $\varepsilon \in (0, \infty)$, using a Gaussian quadrature. The results were very similar in both cases. Using a line contour avoids possible problems connected with the phase of a complex logarithm. Typically a total of 10-15 energy points was used. A large number of $\mathbf{k}_\parallel$-points in the ISBZ is needed only for energy points close to the real axis, whereby generally a greater number is needed for lower temperatures and thicker spacers. The number of $\mathbf{k}_\parallel$-points can significantly be reduced for energies well off the real axis. In particular, for the first energy point on the contour close to the Fermi energy we typically use 5000-10000 $\mathbf{k}_\parallel$-points in the ISBZ, while for the next 3-4 energy points the number of $\mathbf{k}_\parallel$-points is reduced by a factor two for each other point, and about 50-100 $\mathbf{k}_\parallel$-points are taken for all remaining energy points on the contour. The thickness of the spacer, for which well converged results are obtained, is about 100 spacer layers.

3.2   Analysis of the Results

The calculated results, namely $\mathcal{E}_x(\theta, N)$, where N specifies the spacer thickness, can be analyzed in terms of a discrete Fourier transformation

$$F(\theta, q) = \frac{1}{p} \sum_{N=N_{min}}^{N_{max}} N^2 \mathcal{E}_x(\theta, N) \exp(iqN), \tag{50}$$

where $p = N_{max} - N_{min} + 1$ is the number of values used in the Fourier analysis, and N_{min} is chosen in order to eliminate the effect of very thin spacers, or, to analyze intentionally either the preasymptotic or the asymptotic region. Typically p is about 40. The background oscillations thus occurring [14] are due to the discreteness of the Fourier transformation. The background oscillations can be smoothened using the procedure described in [46], namely by multiplying $N^2 \mathcal{E}_x(\theta, N)$ by $C \sin(\pi N/p)/(\pi N/p)$, where C is a normalization factor. The periods of oscillations Λ_α (in monolayers) are then identified with the positions q_α of pronounced peaks of $|F(q_\alpha)|$ as $\Lambda_\alpha = 2\pi/q_\alpha$, the amplitudes of oscillations A_α are estimated from $A_\alpha = (2/p)|F(q_\alpha)|$, and their phases from $\phi_\alpha = \pi/2 - \mathrm{Arg}F(q_\alpha)$, $(\alpha = 1, 2, \dots)$. This analysis can be extended to more complicated cases, namely when the IEC is a function of two variables, e.g., as a function of the spacer and cap thicknesses N and P, respectively. A two-dimensional discrete Fourier transformation

$$F_2(\theta, q_N, q_P) = \sum_{N=N_1}^{N_2} \sum_{P=P_1}^{P_2} (N + P)^2 \, \mathcal{E}_2(\theta, N, P) \, e^{i(q_N N + q_P P)} \qquad (51)$$

is a suitable tool to analyze the quantity $\mathcal{E}_2(\theta, N, P)$, where the prefactor $(N + P)^2$ is consistent with the asymptotic behavior [23,47] for large spacer and cap thickness. Strictly speaking, this is quite an obvious choice for the case when the spacer and cap are formed by the same material, but it can be used also when the spacer and the cap correspond to different materials (for more details, see [23]). In (51) we have introduced the quantity

$$\mathcal{E}_2(\theta, N, P) = \mathcal{E}_x(\theta, N, P) - \mathcal{E}_0(\theta, N), \quad \mathcal{E}_0(\theta, N) = \lim_{P \to \infty} \mathcal{E}_x(\theta, N, P) \qquad (52)$$

in order to remove a trivial peak in the absolute value of $F_2(\theta, q_N, q_P)$ at $q_N = q_P = 0$. A similar two-dimensional discrete Fourier transformation is also useful in the study of the IEC with respect to the thicknesses of the spacer and the magnetic slabs. We note that if one of variables, e.g., the spacer thickness N is fixed, it is possible to analyze the calculated IEC values again with the help of (50).

An alternative of calculating the Fourier transform (50) consists in subdividing the $\mathbf{k}_\|$-integral in (27) into areas around the critical $\mathbf{k}_\|$-vectors (callipers) related to the different oscillation periods [30,49]. In the asymptotic limit each subarea gives then rise to a single oscillation period, while in the preasymptotic regime the resulting division into different periods is only qualitatively valid. In a sense this method bridges the present method of discrete Fourier transformations and the purely asymptotic treatment of calculating only the behavior of the critical $\mathbf{k}_\|$-vectors (see Section 3.3).

3.3 Asymptotic Expansion

Model studies [1,2] indicate that in the asymptotic region, i.e., for large spacer and magnetic layer thickness, and for a random sample, the general form of the spacer-thickness dependence of the IEC is given by

$$\mathcal{E}_x = \mathrm{Im} \sum_\alpha \frac{Z_\alpha}{N^2} \exp(iQ_\alpha N). \qquad (53)$$

Here the sum runs over all possible periods α, the quantities Z_α and Q_α are the complex amplitudes and complex stationary points (callipers), respectively, defined in the following manner

$$Z_\alpha = A_\alpha \exp(i\Phi_\alpha), \qquad Q_\alpha = q_\alpha + i\lambda_\alpha. \qquad (54)$$

The quantities A_α and Φ_α are the amplitudes and phases of coupling oscillations, $p_\alpha = 2\pi/q_\alpha$ their periods, and the quantity λ_α characterizes the damping of oscillations due to the effect of alloying in the sample determined at the Fermi energy. In the limit of non-random samples, $\lambda_\alpha = 0$.

The parameters in (53) can be extracted from a detailed knowledge of the spacer Fermi surface [7]. We briefly sketch a numerical way of determining of the parameters of this asymptotic expansion which requires the knowledge of the integrand of (29) for a set of $\mathbf{k}_\parallel$-points in the neighborhood of the stationary points $\mathbf{k}_\parallel^{(\alpha)}$.

The expression (29) for IEC at $T = 0$ K can be rewritten as

$$\mathcal{E}_x = \frac{1}{N_\parallel} \operatorname{Im} \sum_{\mathbf{k}_\parallel} Y(\mathbf{k}_\parallel), \quad Y(\mathbf{k}_\parallel) = \frac{1}{\pi} \int_C f(z,0) \operatorname{tr}_L \ln \mathcal{M}(\mathbf{k}_\parallel, z) \, dz. \qquad (55)$$

The integration with respect to the energy variable is performed numerically. The function $Y(\mathbf{k}_\parallel)$ for large N decreases as $O(1/N)$ and behaves like

$$Y(\mathbf{k}_\parallel) = \frac{g(\mathbf{k}_\parallel)}{N} \exp(iN\phi(\mathbf{k}_\parallel)), \qquad (56)$$

where the pre-exponential factor $g(\mathbf{k}_\parallel)$ is a smooth function of $\mathbf{k}_\parallel$ and the phase $\phi(\mathbf{k}_\parallel)$ has one, or more stationary points in the SBZ that correspond to callipers of the spacer Fermi surface such that $\nabla_{\mathbf{k}_\parallel} \phi(\mathbf{k}_\parallel) = 0$. The integral over the SBZ in (55) can be evaluated using the stationary-phase method. The contribution of a stationary point $\mathbf{k}_\parallel^{(\alpha)} \equiv (k_x^{(\alpha)}, k_y^{(\alpha)})$ is found in the following way: the integration limits are extended to infinity, and the phase function $\phi(\mathbf{k}_\parallel)$ is approximated by a quadratic function of $\mathbf{k}_\parallel \equiv (k_x, k_y)$ in the vicinity of the stationary point,

$$\phi(\mathbf{k}_\parallel) = \phi(\mathbf{k}_\parallel^{(\alpha)}) + \sum_{i,j=x,y} Q_{ij}(k_i - k_i^{(\alpha)})(k_j - k_j^{(\alpha)})$$
$$= \sum_{i,j=x,y} Q_{ij} k_i k_j + \sum_{i=x,y} P_i k_i + \phi(\mathbf{k}_\parallel^{(\alpha)}). \qquad (57)$$

The expansion coefficients Q_{ij}, P_i, and $\phi(\mathbf{k}_\parallel^{(\alpha)})$ are determined by a least-square fit to values of $\phi(\mathbf{k}_\parallel)$ calculated in the vicinity of $\mathbf{k}_\parallel^{(\alpha)}$. This procedure allows to eliminate numerical inaccuracies with respect to both the values of Q_{ij} and the position of the stationary point $\mathbf{k}_\parallel^{(\alpha)}$, and it is applicable even for disordered surfaces. By inserting (56) and (57) into (55) we find

$$\mathcal{E}_x \approx \frac{1}{\pi N V_{\text{SBZ}}} \operatorname{Im} \left\{ g(\mathbf{k}_\parallel^{(\alpha)}) \times \right.$$

$$\left. \iint_D \exp\left[iN\left(\phi(\mathbf{k}_\parallel^{(\alpha)}) + \sum_{i,j=x,y} Q_{ij}(k_i - k_i^{(\alpha)})(k_j - k_j^{(\alpha)}) \right) \right] dk_x \, dk_y \right\}$$

$$= \frac{\pi}{N^2 V_{\text{SBZ}}} \operatorname{Im} \left\{ \frac{g(\mathbf{k}_\parallel^{(\alpha)})}{\sqrt{-\det|Q|}} \exp\left[iN\phi(\mathbf{k}_\parallel^{(\alpha)}) \right] \right\}, \tag{58}$$

where the two-dimensional integration region D extends to infinity, and V_{SBZ} denotes the volume of the SBZ. The second line in (58) is obtained by diagonalizing the quadratic form in the exponent (57) and by evaluating the resulting one-dimensional Gaussian-like integrals. The identification of the parameters is now straightforward, namely

$$Z_\alpha = \frac{\pi}{V_{\text{SBZ}}} \frac{g(\mathbf{k}_\parallel^{(\alpha)})}{\sqrt{-\det|Q|}}, \quad Q_\alpha = \phi(\mathbf{k}_\parallel^{(\alpha)}). \tag{59}$$

3.4   Free-Electron Limit

The numerical efficiency of the present formalism offers an interesting possibility of testing model theories [1]. The simplest of such models is the free-electron model, because of a spherical Fermi surface with a single critical vector at $\mathbf{k}_\parallel = 0$ and a trivial correspondence between the value of the oscillation period and the band-filling. The free-electron model can be easily simulated by the present formalism by replacing the true metallic potentials by flat potentials (the empty-sphere model). For a suitable choice of the lattice constant and the position of the Fermi energy it is irrelevant what lattice and layer stacking is used, e.g., the fcc(001)-stack is the simplest choice. Such a model is free of the limitations usually adopted [1], e.g., the assumption of large spacer and magnetic slabs thicknesses, or the approximate evaluation of the energy integral for the case of finite temperatures.

3.5   Numerical Illustrations

In Fig. 1 $N^2 \mathcal{E}_x(N)$ is displayed as a function of the spacer thickness N for two semi-infinite Co(001) subsystems sandwiching an fcc-Cu spacer. The corresponding discrete Fourier transformation in Fig. 2 shows a pronounced short-period oscillations of 2.53 monolayers (MLs) while the long-period oscillations are suppressed in this geometry [13,14,49]. The results are insensitive to the choice of the lower and upper index in the summation in (50) provided the preasymptotic region is excluded [14].

For a large enough N the IEC can be approximated by the asymptotic form in (53). The amplitude, phase, and the wave-vector entering this expression can

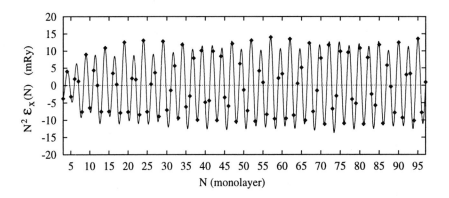

Fig. 1. Exchange coupling $N^2\mathcal{E}_x(N)$ at $T = 0$ K as a function of the spacer thickness N for two semi-infinite fcc Co(001) subsystems sandwiching a Cu spacer. Diamonds refer to the calculated values, the full line (back Fourier transform) serves as a guide to the eye

be determined from the calculated $\mathcal{E}_x(N)$ in the manner as described in Sect. 3.2 and the asymptotic result (53) was compared with the calculated results for a large set of systems including both ideal and alloyed semi-infinite fcc(001) magnetic subsystems sandwiching a Cu-spacer: overall good agreement was found [21]. An example of the complex amplitude for this case is presented in Fig. 3 illustrating the insensitivity of the phase to elements which form the magnetic layers. It is seen that phases corresponding to Co, $Fe_{50}Ni_{50}$, and $Fe_{1/3}Ni_{1/3}Co_{1/3}$ which have the same average electron numbers $N_{el}=9$ are nearly the same [21].

The IEC depends on the temperature T via a factor $x/\sinh(x)$, $x = cNT$, where T is the temperature and N the spacer thickness. This remarkable result of model theories [1] was verified by calculations such as illustrated in Fig. 4. The IEC depends in an oscillatory manner not only on the spacer thickness N but as well on the thickness P of a covering cap. The oscillations are around a biased value which corresponds to coupling for a given spacer thickness assuming a semi-infinite cap. This phenomenon is illustrated in Fig. 5 in terms of discrete Fourier transformations with respect to the spacer and the cap thickness (see Sec. 3.2) for a sample consisting of a semi-infinite fcc-Cu(001) substrate, left and right magnetic layers each five MLs thick, a spacer with varying thickness N, and a Cu-cap of varying thickness P. Fig. 5 shows: (i) long-period oscillations (missing in Fig. 2) in addition to the short-period ones, and (ii) oscillations with respect to the cap thickness which are exactly the same as for the spacer because both are controlled by the same Fermi surface, namely that of fcc-Cu. The more complicated case of different spacer and cap materials is discussed in [23,24].

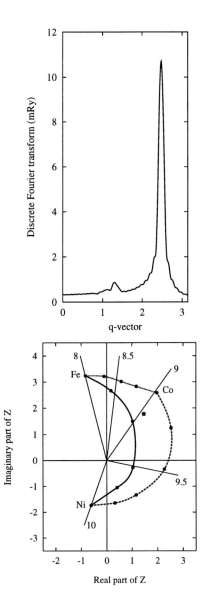

Fig. 2. Absolute value of the discrete Fourier transformation of $N^2\mathcal{E}_x(N)$ for a finite set of spacer layers (N=20–80) corresponding to two semi-infinite fcc Co(001) subsystems sandwiching a Cu spacer. The temperature is $T = 0$ K

Fig. 3. Complex amplitude $Z^{1/2} = A^{1/2}\,e^{i\varPhi/2}$, where A and $\varPhi$ are the oscillation amplitude and phase, respectively, for a semi-infinite fcc(001) subsystems formed by Fe, Co, Ni, their binary alloys (bullets), and the ternary alloy $Fe_{1/3}Co_{1/3}Ni_{1/3}$ (square) sandwiching a Cu spacer. The units are $(mRy)^{1/2}$. The dotted, dashed, and full lines connect various alloys and serve as a guide to the eye. The rays starting at the origin show approximately the phase corresponding to the indicated average number of valence electrons

Ordering in the spacer [22] or in the magnetic layers [20] can induce new periods due to the formation of two-dimensional sublattices. The situation is particularly interesting for a $c(2 \times 2)$-ordering in magnetic layers sandwiching an ideal Cu-spacer [20]. As illustrated in Fig. 6 for full ordering two new periods with complementary periods and phases are formed in addition to a conventional short-period due to a fcc-Cu spacer [20]. These new periods vanish in the completely disordered case.

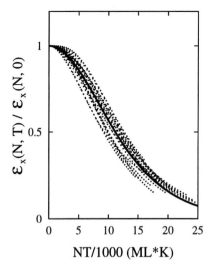

Fig. 4. $\mathcal{E}_x(N,T)/\mathcal{E}_x(N,T=0)$ plotted as a function of $\zeta = NT$ for a trilayer consisting of semi-infinite fcc Co(001)-slabs sandwiching a Cu spacer. The thick line refers to $x/\sinh(x)$, $x = cNT$ with $c = 0.000195$ obtained by a least square-fit to the computed data

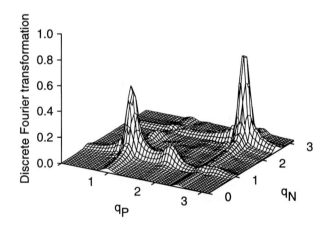

Fig. 5. Absolute values of the discrete two-dimensional Fourier transformation of $(N+P)^2 \mathcal{E}_2(N,P)$ with respect to the spacer and the cap thickness in the case of two magnetic slabs each five monolayers thick with a Cu-substrate, a Cu-spacer, and a Cu-cap. For a definition of $\mathcal{E}_2(N,P)$ see the text

Finally, the effect of disorder in the spacer [19] is illustrated in Fig. 7. Alloying of Cu with Ni decreases the number of average valence electrons and leads to a contraction of the alloy Fermi surface, and in turn to a reduction of the coupling oscillations. The opposite behavior has to be expected for alloying of Cu with

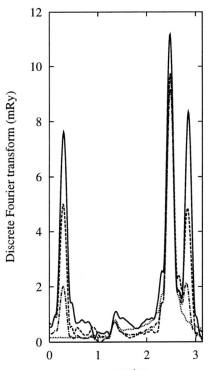

Fig. 6. Absolute values of the discrete Fourier transformation of $N^2 \mathcal{E}_x(N)$ for two semi-infinite fcc $Co_{50}Fe_{50}(001)$ subsystems sandwiching a Cu spacer with different kinds of chemical order in magnetic layers: (a) $S = 1$ (complete $c(2 \times 2)$-order, full line), (b) $S = 0.8$ (dashed line), (c) $S = 0.5$ (dashed-dotted line), and (d) $S = 0.0$ (disordered case, dotted line). The temperature is $T = 0$ K

Zn, whereas only a small concentration dependence of the periods for the CuAu case is seen. The amplitudes of the oscillations are generally reduced by alloying, and in the case of CuZn spacer they are even

exponentially damped. The different behavior of the amplitudes can be related to differently large disorder in the neighborhood of relevant extremal points of the alloy Fermi surfaces.

3.6   Some Published Applications

We briefly review applications of the formalism developed in previous sections to specific problems. Additional details concerning formalism and not discussed here in details, e.g., the expansion of the IEC expression in terms of the small parameter $1 - \cos(\theta)$ or the details concerning the numerical verification of the vertex-cancellation theorem, can be found in [14,15], respectively. The influence of surface roughness (fluctuating spacer thickness and diffusion at the interface between spacer and magnetic layers) on the oscillation amplitudes was studied in [18]. The effect of alloying in the spacer [19] on the oscillation periods and their amplitudes, and in magnetic layers [21] on the oscillation amplitudes and phases was also studied in detail for the trilayer system Co/Cu/Co(001). Ordering in disordered spacers [22] and/or magnetic [20] layers lead to a formation of new periods not present in ideal spacers. Oscillations of the IEC can originate not only due to the spacer but also from adlayers or cap layers. We refer the reader interested in this problem to a recent detailed study [23,24]. Finally, the study

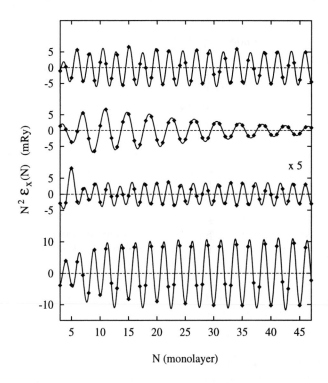

Fig. 7. Exchange coupling $N^2\mathcal{E}_x(N)$ at $T = 0$ K as a function of the spacer thickness N for two semi-infinite fcc Co(001) subsystems sandwiching a spacer of (from bottom to top) Cu, $Cu_{75}Ni_{25}$ (multiplied by a factor 5), $Cu_{50}Zn_{50}$, and $Cu_{50}Au_{50}$. Diamonds refer to the calculated values, the full line (back Fourier transform) serves as a guide to the eye

of the temperature dependence of the IEC and of the combined effect of the temperature and disorder is subject of very recent papers [16,17], respectively.

4   Conclusions

We have derived closed expressions for the exchange coupling between two magnetic subsystems separated by a non-magnetic spacer with a relative angle θ between the orientations of the magnetizations in the magnetic slabs. The derivation is based on a surface Green function formalism. The numerical effort scales linearly with the thickness of both the spacer and the magnetic slabs. The formulation allows also for an efficient evaluation of the temperature dependence of the coupling amplitudes. Numerical examples were chosen to illustrate the theoretical aspects rather than to give a comprehensive overview of results obtained by the present formalism or by related methods.

We wish now briefly to mention some unsolved problems. The following list is neither complete nor are the problems listed according to their importance: (i) The oscillatory dependence of the IEC on the thickness of the magnetic slabs was not yet systematically investigated on an ab initio level. Existing calculations [14,48,49] were performed for too thin magnetic slabs to relate occurring oscillations to extremal points of spin-polarized Fermi surfaces; (ii) The problem of biquadratic and higher order terms also did not receive a proper attention on an ab initio level. A relevant problem is a systematic study of situations for which the non-collinear (biquadratic) coupling can dominate. Obviously, it can happen most probably for the spacer thicknesses for which the IEC values are close to the transition between the F and AF couplings [41]. In addition, it remains to be seen whether a theoretical description of biquadratic coupling has to be based on a fully relativistic spin-polarized level; (iii) The study of superstructures in the spacer and/or in the magnetic slabs (see Sec. 2.10) offers a possibility of a deeper insight into the physical nature of the IEC because of new periods, which are connected with the extremal vectors of the spacer material in a more sophisticated manner than in the canonical cases of Cu or Cr spacers; (iv) The study of oscillatory behavior of exchange interaction across magnetic spacers is of great interest. One possibility here is to employ the method of infinitesimal rotations [9,14]; (v) The study of exchange coupling through the semiconducting or, more generally, through an insulating spacer where one expects exponential rather than N^{-2}-decay has remained limited until now to model studies [1]; (vi) The study of alloying in the spacer, magnetic layers and at interfaces has to be extended to new interesting systems. It offers a straightforward method to obtain valuable informations concerning alloy Fermi surfaces, in particular for the case of alloyed spacers; and, finally (vii) The study of the IEC through spacers with complex Fermi surfaces, in particular through the transition metal spacers.

Acknowledgements This work is a part of activities of the Center for Computational Material Science sponsored by the Academy of Sciences of the Czech Republic. Financial support for this work was provided by the Grant Agency of the Czech Republic (Project No. 202/97/0598), the Grant Agency of the Academy Sciences of the Czech Republic (Project A1010829), the Center for the Computational Materials Science in Vienna (GZ 45.442 and GZ 45.420), and the TMR Network 'Interface Magnetism' of the European Commission (Contract No. EMRX-CT96-0089).

5   Vertex Cancellation Theorem

We present here a general discussion of exchange interactions in the presence of substitutional disorder. The results given here are used in the present paper to study interlayer exchange interactions, but they are also applicable for studying exchange interactions within a ferromagnet, exchange stiffnesses, spin-wave energies, etc. The principal result is the "vertex cancellation theorem" of Bruno *et al.* [15]. In here we give an alternative, more general, derivation of this result.

Let $\hat{u} \equiv \{\hat{u}_R\}$ be a particular configuration of the local moments, where $\hat{u}_R$ is a unit vector pointing in the direction of the R-th local moment. We are interested in the variation of the thermodynamic grandcanonical potential

$$\Omega_{\hat{u}} = -\frac{1}{\pi} \operatorname{Im} \int_{-\infty}^{+\infty} f(E,T) \operatorname{Tr} \langle \ln g_{\hat{u}}(E + i0^+) \rangle \, dE \tag{60}$$

with respect to $\hat{u}$. The Green function $g_{\hat{u}}(z)$ for a particular alloy configuration is defined from the potential function $P_{\hat{u}}(z)$ corresponding to $\hat{u}$ as

$$g_{\hat{u}}(z) = (P_{\hat{u}}(z) - S)^{-1} \ . \tag{61}$$

An immediate consequence of (61) is a trivial commutator relation to be used below, namely

$$[P_{\hat{u}}(z); g_{\hat{u}}(z)]_- = [S; g_{\hat{u}}(z)]_- \ , \tag{62}$$

where $[A; B]_- \equiv AB - BA$. The configuration averaged Green function $\langle g_{\hat{u}}(z) \rangle \equiv \bar{g}_{\hat{u}}(z)$ is usually formulated in terms of the coherent potential function $\mathcal{P}_{\hat{u}}(z)$ as

$$\bar{g}_{\hat{u}}(z) = (\mathcal{P}_{\hat{u}}(z) - S)^{-1} \ , \tag{63}$$

which leads to a relation analogous to (62),

$$[\mathcal{P}_{\hat{u}}(z); \bar{g}_{\hat{u}}(z)]_- = [S; \bar{g}_{\hat{u}}(z)]_- \ . \tag{64}$$

In general, the averaging in (60) cannot be reduced to $\ln \bar{g}_{\hat{u}}(z)$ and an evaluation of the so-called vertex corrections is necessary. We shall show, however, that the variation of (60) due to an infinitesimal change of $\hat{u}$ takes a simple form.

Let us consider the variation of the potential functions $P_{\hat{u}}(z)$ in more detail. To each lattice site R we associate a non-random vector $\Theta_R \equiv \theta_R \hat{n}_R$, where $\hat{n}_R$ refers to the axis of rotation and θ_R to rotation angle by which the reference orientation $\hat{u}_{0,R}$ is transformed into $\hat{u}_R$. The transformed potential functions are therefore given by the following similarity transformation

$$P_{\hat{u}}(z) = U_\Theta \, P_{\hat{u}_0}(z) \, U_\Theta^{-1} \ , \tag{65}$$

where the rotation matrix U_Θ in (65) is defined as

$$(U_\Theta)_{RLs,R'L's'} = \delta_{R,R'} \, \delta_{L,L'} \times$$
$$\left[\cos\left(\frac{\theta_R}{2}\right) \mathbf{1} - i \sin\left(\frac{\theta_R}{2}\right) \hat{n}_R \cdot \boldsymbol{\sigma} \right]_{s,s'} . \tag{66}$$

The symbol $\boldsymbol{\sigma}$ in (66) denotes the vector of the standard 2×2 Pauli matrices and $\mathbf{1}$ is the 2×2 unit matrix. The first-order change of $P_{\hat{u},R}(z)$ caused by an additional infinitesimal rotation $\delta \mathbf{v}_R$ is then expressed as

$$\delta P_{\hat{u}}(z) = [\delta K; P_{\hat{u}}(z)]_- \ , \tag{67}$$

where the matrix elements of the operator $\delta K = U_{\delta \mathbf{v}} - 1$ are explicitly given by

$$(\delta K)_{RLs,R'L's'} = \delta_{R,R'}\,\delta_{L,L'}\,\frac{(-i)}{2}\,[\boldsymbol{\sigma}\cdot\delta\mathbf{v}_R]_{s,s'}\,. \tag{68}$$

The introduced infinitesimal rotation vectors $\delta\mathbf{v}_R$ satisfy $U_{\delta\mathbf{v}}\,U_\Theta = U_{\Theta+\delta\Theta}$ whereas, in general, $U_{\delta\Theta}\,U_\Theta \neq U_{\Theta+\delta\Theta}$. Let us note that δK is a non-random site-diagonal operator.

The first-order variation of $\mathrm{Tr}\,\langle \ln g_{\hat{u}}(z)\rangle$ can be now formulated using (94, 67) as

$$\delta\mathrm{Tr}\,\langle \ln g_{\hat{u}}(z)\rangle = -\,\mathrm{Tr}\,\big\langle g_{\hat{u}}(z)\,[\delta K; P_{\hat{u}}(z)]_-\big\rangle\,, \tag{69}$$

which can be rewritten by applying the permutation invariance of the trace and (62, 64) as

$$\begin{aligned}
\delta\mathrm{Tr}\,\langle \ln g_{\hat{u}}(z)\rangle &= -\,\mathrm{Tr}\,\big\{\delta K\,\langle[P_{\hat{u}}(z); g_{\hat{u}}(z)]_-\rangle\big\}\\
&= -\,\mathrm{Tr}\,\big\{\delta K\,\langle[S; g_{\hat{u}}(z)]_-\rangle\big\}\\
&= -\,\mathrm{Tr}\,\big\{\delta K\,[S; \overline{g}_{\hat{u}}(z)]_-\big\}\\
&= -\,\mathrm{Tr}\,\big\{\delta K\,[\mathcal{P}_{\hat{u}}(z); \overline{g}_{\hat{u}}(z)]_-\big\}\,.
\end{aligned} \tag{70}$$

By using the permutation invariance of the trace once again, (70) can be given the final form

$$\delta\mathrm{Tr}\,\langle \ln g_{\hat{u}}(z)\rangle = -\,\mathrm{Tr}\,\big\{\overline{g}_{\hat{u}}(z)\,[\delta K; \mathcal{P}_{\hat{u}}(z)]_-\big\}\,. \tag{71}$$

Let us note that (71) was derived in a formally exact alloy theory, but is valid in the CPA as well. Within the CPA, the result (71) has an obvious interpretation: the r.h.s. describes the variation of $\mathrm{Tr}\ln\overline{g}_{\hat{u}}(z)$ induced by performing on the site-diagonal coherent potential functions $\mathcal{P}_{\hat{u},R}(z)$ the same rotations (68) as applied to the potential functions $P_{\hat{u},R}(z)$; note however, that this is not equal to the infinitesimal change of the true self-consistent CPA coherent potential function.

Thus, the torque acting on the moment at site $\mathbf{R}$ due to the exchange interactions is given by

$$\begin{aligned}
\Gamma_{\hat{u},R} \equiv -\frac{\delta\Omega_{\hat{u}}}{\delta\mathbf{v}_R} = -\frac{1}{\pi}\int_{-\infty}^{+\infty} f(E,T)\times\\
\mathrm{Im}\,\mathrm{Tr}\Big\{\overline{g}_{\hat{u}}(E+i0^+)\,\frac{(-i)}{2}\,[\Pi_R\boldsymbol{\sigma}\,; \mathcal{P}_{\hat{u}}(E+i0^+)]_-\Big\}\,\mathrm{d}E\,,
\end{aligned} \tag{72}$$

where Π_R is a projector on site $\mathbf{R}$. This exact result constitutes the "vertex cancellation theorem" for the torque. Its usefulness arises from the fact that the "vertex corrections" have been eliminated.

In order to compute the difference of thermodynamic grandcanonical potential between two local moment configurations $\hat{u}_1$ and $\hat{u}_2$ in the CPA, we use a theorem due to Ducastelle [31], which states that the thermodynamic grandcanonical potential, considered as a functional $\tilde{\Omega}[\mathcal{P},P]$ of the independent variables $\mathcal{P}$ and P, is stationary with respect to $\mathcal{P}$ when the latter satisfies the CPA self-consistency condition. This means that a first-order error in $\mathcal{P}_{\hat{u}}$ gives only a second-order error in $\Omega_{\hat{u}}$. Let us approximate $\mathcal{P}_{\hat{u}}(z)$ by

$$\mathcal{P}_{\hat{u}}(z) \approx \mathcal{P}'_{\hat{u}}(z) \equiv U_\Theta \, \mathcal{P}_{\hat{u}_0}(z) \, U_\Theta^{-1} \, , \tag{73}$$

i.e., we assume that $\mathcal{P}_{\hat{u}}(z)$ is transformed like $P_{\hat{u}}(z)$ under a rotation of the local moment direction. This can be expected to be a good approximation, provided the condition

$$m_R \left| \frac{\mathrm{d}\Theta_R}{\mathrm{d}R} \right| \ll k_F \, q_R \tag{74}$$

is satisfied, where q_R and m_R are respectively the charge and spin moment at site R. We then get

$$\overline{g}_{\hat{u}}(z) \approx \overline{g}'_{\hat{u}}(z) \equiv \left(\mathcal{P}'_{\hat{u}}(z) - S \right)^{-1} . \tag{75}$$

Replacing $\mathcal{P}_{\hat{u}}$ by $\mathcal{P}'_{\hat{u}}$ and $\overline{g}_{\hat{u}}$ by $\overline{g}'_{\hat{u}}$ in (71), we obtain

$$\delta \mathrm{Tr} \, \langle \ln g_{\hat{u}}(z) \rangle \approx \delta \mathrm{Tr} \ln \overline{g}'_{\hat{u}}(z) \, , \tag{76}$$

and integrating over the angles, we get

$$\Omega_{\hat{u}_1} - \Omega_{\hat{u}_2} \approx - \frac{1}{\pi} \int_{-\infty}^{+\infty} f(E, T) \times$$
$$\mathrm{Im} \, \mathrm{Tr} \left[\ln \overline{g}'_{\hat{u}_1}(E + \mathrm{i}0^+) - \ln \overline{g}'_{\hat{u}_2}(E + \mathrm{i}0^+) \right] \mathrm{d}E \, , \tag{77}$$

which constitutes the "vertex cancellation theorem" for exchange energies. Note that we have derived here a form of the "vertex cancellation theorem" within the CPA since this is the scheme which is used in practical calculations; however, one can prove that the same result holds if one takes the exact solution to the configuration averaging problem.

In the case of interlayer coupling, the condition (74) is satisfied even for large rotation angles, because $\mathrm{d}\Theta_R/\mathrm{d}R$ differs from zero only in a region where m_R is negligible. This was confirmed by explicit numerical calculations in [15].

6   The Interface–Interface Part of the Grandcanonical Potential

In this Appendix we derive the basic relations for an evaluation of the IEC within the interface-interface interaction formulation.

The subsystems $\mathcal{L}$ and $\mathcal{R}$ can be downfolded using the formula (88)

$$\mathrm{Tr} \ln (P - S) = \mathrm{Tr}_{\mathcal{L}} \ln \left[P - S \right] + \mathrm{Tr}_{\mathcal{R}} \ln \left[P - S \right]$$
$$+ \mathrm{Tr}_{\mathcal{C}} \ln \left[(P - S)_{\mathcal{CC}} - (P - S)_{\mathcal{CL}} \frac{\mathcal{L}}{P - S} (P - S)_{\mathcal{LC}} \right.$$
$$\left. - (P - S)_{\mathcal{CR}} \frac{\mathcal{R}}{P - S} (P - S)_{\mathcal{RC}} \right] . \tag{78}$$

The first two terms are independent of the rotation angle θ and, consequently, they do not contribute to the exchange energy $\mathcal{E}_x(\theta)$. We are thus left with a quantity which is limited to the subspace $\mathcal{C}$ only. It is now easy to identify the individual terms in (78). The potential function blocks between different subspaces such as $P_{\mathcal{L}\mathcal{C}}$ or $P_{\mathcal{C}\mathcal{R}}$ are zero because the potential function P is site-diagonal. The blocks of S between neighboring subspaces do not vanish, but the non-zero subblocks connect only neighboring principal layers. The important part of the $\mathrm{Tr}\ln(P-S)$ is then reduced to

$$\mathrm{Tr}_{\mathcal{C}}\ln(P-S)_{\mathcal{C}\mathcal{C}} + \mathrm{Tr}_{\mathcal{C}}\ln\left[1 - \frac{\mathcal{C}}{P-S}S_{10}\mathcal{G}_{\mathcal{L}}S_{01} - \frac{\mathcal{C}}{P-S}S_{01}\mathcal{G}_{\mathcal{R}}S_{10}\right]. \quad (79)$$

The first term is independent of θ and thus does not contribute to the exchange energy. The second term can be simplified using the two-potential formula (93). We identify $G^{(0)} = \mathcal{C}/(P-S)$, $v_1 = S_{10}\mathcal{G}_{\mathcal{L}}S_{01}$, and $v_2 = S_{01}\mathcal{G}_{\mathcal{R}}S_{10}$. The t-matrices are then identical with the τ-matrices, and the potentials v_1 and v_2 are equal to the embedding potentials Γ_1 and Γ_2. In this way we find the expression for the grandcanonical potential

$$\Omega(\theta, T, \mu) = \Omega_0(T, \mu)$$
$$-\frac{1}{\pi}\,\mathrm{Im}\int_{-\infty}^{\infty} f(E, T, \mu)\,\mathrm{Tr}_1\ln\left[1 - g_{1N}(z)\tau_N(z)g_{N1}(z)\tau_1(z)\right]dz, \quad (80)$$

where $\Omega_0(T, \mu)$ contains all the terms independent of θ and the Tr_1 applies only to the layer 1, i.e., the first spacer layer. If the system is invariant with respect to translations in the planes of atoms, or, if such a symmetry is restored by configuration averaging, (80) can be written as

$$\Omega(\theta, T, \mu) = \Omega_0(T, \mu) - \frac{1}{\pi}\,\mathrm{Im}\int_{-\infty}^{\infty} f(E, T, \mu) \times$$
$$\sum_{\mathbf{k}_{||}} \mathrm{tr}\ln\left[1 - g_{1N}(\mathbf{k}_{||}, z)\tau_N(\mathbf{k}_{||}, z)g_{N1}(\mathbf{k}_{||}, z)\tau_1(\mathbf{k}_{||}, z)\right]dz, \quad (81)$$

where tr means the trace over angular momentum indices $L = (\ell m)$ and the spin index σ.

7 Useful Mathematical Tools

Theoretical developments and many calculations are facilitated by the partitioning technique and the two-potential formula applied to the Green function and its logarithm.

Let P and Q denote projection operators onto the complementary subspaces (i.e. $P+Q=1$). We denote the projections of matrices as $PAP = A_{PP}, PAQ = A_{PQ}$, etc., and P/A means the inversion of A_{PP} in the subspace referring to projector P. In most applications, $A = z - H$ or $A = P(z) - S$ and $G(z) = A^{-1}$.

The projections of the inverse A^{-1} to the matrix A are given by [50]

$$(A^{-1})_{PP} = \frac{P}{A_{PP} - A_{PQ}\frac{Q}{A}A_{QP}}, \tag{82}$$

$$(A^{-1})_{QQ} = \frac{Q}{A_{QQ} - A_{QP}\frac{P}{A}A_{PQ}}, \tag{83}$$

$$(A^{-1})_{PQ} = -\frac{P}{A}A_{PQ}(A^{-1})_{QQ} = -(A^{-1})_{PP}A_{PQ}\frac{Q}{A}, \tag{84}$$

$$(A^{-1})_{QP} = -\frac{Q}{A}A_{QP}(A^{-1})_{PP} = -(A^{-1})_{QQ}A_{QP}\frac{P}{A}. \tag{85}$$

It is sometimes easier to invert the full matrix A than its blocks. In such a case the inverse partitioning is useful

$$\frac{P}{A_{PP}} = (A^{-1})_{PP}\frac{P}{P - A_{PQ}(A^{-1})_{QP}} = \frac{P}{P - (A^{-1})_{PQ}A_{QP}}(A^{-1})_{PP}. \tag{86}$$

This can be used to calculate the surface Green function of a semi-infinite system from the Green function of the infinite system.

Partitioning technique also allows to simplify calculations involving Tr ln of a matrix. The basic relation is

$$\mathrm{Tr}\,\ln A = \ln \det A. \tag{87}$$

It then follows $\mathrm{Tr}\,\ln AB = \mathrm{Tr}\,\ln A + \mathrm{Tr}\,\ln B$, $\mathrm{Tr}\,\ln 1 = 0$, $\mathrm{Tr}\,\ln (A^{-1}) = -\mathrm{Tr}\,\ln A$, and $\mathrm{Tr}\,\ln [(A - B)^{-1}] = -\mathrm{Tr}\,\ln A - \mathrm{Tr}\,\ln [1 - A^{-1}B]$. The Tr ln A can then be partitioned as

$$\mathrm{Tr}\,\ln A = \mathrm{Tr}_P \ln [PAP] + \mathrm{Tr}_Q \ln [QAQ - QA\frac{P}{A}AQ]. \tag{88}$$

To prove (88), let us multiply the matrix A by $L = 1 - A_{QP}(P/A)$ from left and by $R = 1 - (P/A)A_{PQ}$ from right. The result is $LAR = A_{PP} + A_{QQ} - A_{QP}(P/A)A_{PQ}$. Now using (87), and the fact that $\det [L] = \det [R] = 1$ we find (88). In a special, but important case, when $A_{PP} = P$ and $A_{QQ} = Q$ it holds

$$\mathrm{Tr}\,\ln A = \mathrm{Tr}_{P+Q} \ln [P + Q + A_{PQ} + A_{QP}]$$
$$= \mathrm{Tr}_P \ln [P - A_{PQ}A_{QP}] = \mathrm{Tr}_Q \ln [Q - A_{QP}A_{PQ}]. \tag{89}$$

The Green function of a system described by the Hamiltonian $H = H_0 + v_1 + v_2$, where H_0 is the unperturbed part, and $v_i (i = 1, 2)$ are perturbing potentials, is given by $G = G^{(0)} + G^{(0)}TG^{(0)}$, where $G = (z - H)^{-1}$, $G^{(0)} = (z - H_0)^{-1}$, and $T = V(1 - G^{(0)}V)^{-1}$, where $V = v_1 + v_2$. The full T-matrix T can be expressed in terms of the t-matrices, $t_i = v_i(1 - G^{(0)}v_i)^{-1}, (i = 1, 2)$ and of the unperturbed resolvent $G^{(0)}$ by the the two-potential formula

$$T = t_1 [1 - G^{(0)}t_2G^{(0)}t_1]^{-1} (1 + G^{(0)}t_2) + t_2 [1 - G^{(0)}t_1G^{(0)}t_2]^{-1} \times$$
$$(1 + G^{(0)}t_1). \tag{90}$$

It is derived in the following way. Because

$$(1 - A)[1 - (1 - A)^{-1}AB(1 - B)^{-1}](1 - B) = 1 - A - B, \qquad (91)$$

it holds

$$
\begin{aligned}
\mathrm{Tr}\ln[1 - A - B] &= \mathrm{Tr}\ln[1 - A] + \mathrm{Tr}\ln[1 - B] \\
&+ \mathrm{Tr}\ln[1 - (1 - A)^{-1}AB(1 - B)^{-1}].
\end{aligned}
\qquad (92)
$$

By inserting $A = G^{(0)}v_1$ and $B = G^{(0)}v_2$ into (92) one obtains (90). The two-potential formula for the Tr ln of the full Green function

$$
\begin{aligned}
\mathrm{Tr}\ln G &= \mathrm{Tr}\ln G^{(0)}[1 - VG^{(0)}]^{-1} \\
&= \mathrm{Tr}\ln G^{(0)} - \mathrm{Tr}\ln[1 - G^{(0)}v_1 - G^{(0)}v_2] \\
&= \mathrm{Tr}\ln G^{(0)} - \mathrm{Tr}\ln[1 - G^{(0)}v_1] - \mathrm{Tr}\ln[1 - G^{(0)}v_2] \\
&- \mathrm{Tr}\ln[1 - G^{(0)}t_1 G^{(0)}t_2]
\end{aligned}
\qquad (93)
$$

follows directly from (92).

If the matrix A is a function of a variable z (complex in the general case), the derivative with respect to z is given by

$$\frac{d}{dz}\mathrm{Tr}\ln[A(z)] = \mathrm{Tr}\left[\frac{d}{dz}A(z)\,A^{-1}(z)\right], \qquad (94)$$

provided that the matrix $A(z)$ is nonsingular. This identity is used to derive the expression of the grandcanonical potential Ω in terms of the auxiliary Green function (12) within the TB-LMTO.

The identity in (87) is valid up to an integer multiple of $2\pi i$. Neglecting this fact can lead to serious errors. There is no panacea for this kind of difficulties, but in some situations they can be avoided, for example by choosing the integration contour parallel to the imaginary axis, but this is not always possible. In some cases the incremental procedure for calculating the ln det, $\ln f(z_{k+1}) = \ln f(z_k) + \ln[f(z_{k+1})/f(z_k)]$ in the spirit of an analytical continuation can be helpful, provided that the change of phase between two consecutive points z_k is less than 2π. To insure this, one has to choose a sufficiently small grid in z.

8   Inversion of Block-Tridiagonal Matrices

We wish to compute $g = A^{-1}$ for a block-tridiagonal A. The matrix A is divided into $N \times N$ square subblocks of the same dimension m, from which non-zero are only $A_{k,k}$, $A_{k-1,k}$, and $A_{k,k-1}$. The diagonal blocks are a sum of two terms: hermitean matrix and a symmetric complex matrix. They are always non-singular. The off-diagonal blocks under the diagonal are equal to hermitean conjugate of the corresponding blocks above the diagonal ($A_{k,k-1} = A_{k-1,k}^+$). The methods based on repeated use of partitioning are particularly efficient if only diagonal blocks, or four so-called 'corner' blocks $(g_{1,1}, g_{N,N}, g_{1,N}, g_{N,1})$ are needed like in the interlayer exchange coupling calculations.

First, four sequences of auxiliary matrices are calculated

$$
\begin{aligned}
X_{N-k} &= A_{N-k,N-k+1}(A_{N-k+1,N-k+1} - X_{N-k+1})^{-1}A_{N-k+1,N-k}\,, \\
X_N &= 0, \quad (k = 1,\ldots N-1)\,, \\
Y_{k+1} &= A_{k+1,k}(A_{k,k} - Y_k)^{-1}A_{k,k+1}\,, \quad Y_1 = 0, \quad (k = 2,\ldots N) \\
Z_k &= -(A_{k,k} - X_k)^{-1}A_{k,k-1}\,, \quad (k = 2,\ldots N) \\
W_k &= -(A_{k,k} - Y_k)^{-1}A_{k,k+1}\,, \quad (k = 1,\ldots N-1)\,,
\end{aligned}
\tag{95}
$$

that are used to compute the diagonal and off-diagonal blocks of g

$$
\begin{aligned}
g_{k,k} &= (A_{k,k} - X_k - Y_k)^{-1}\,, \\
g_{i,j} &= Z_i\, g_{i-1,j} \quad \text{for} \quad i > j\,, \\
g_{i,j} &= W_i\, g_{i+1,j} \quad \text{for} \quad i < j\,.
\end{aligned}
\tag{96}
$$

It can be proved that the numerical effort to evaluate the corner blocks scales as $O(Nm^3)$. The details, particularly the tests of efficiency can be found in [42].

References

1. P. Bruno, Phys. Rev. B **52**, 411 (1995).
2. J. Mathon, M. Villeret, A. Umerski, R.B. Muniz, J. d'Albuquerque e Castro, and D.M. Edwards, Phys. Rev. B **56**, 11797 (1997).
3. See theoretical and experimental review articles in IBM J. Res. Development **42**, No. 1 (1998).
4. M. van Schilfgaarde, F. Herman, S.S.P. Parkin, and J. Kudrnovský, Phys. Rev. Lett. **74**, 4063 (1995).
5. M. van Schilfgaarde and F. Herman, Phys. Rev. Lett. **71**, 1923 (1993).
6. S. Mirbt, H.L. Skriver, M. Aldén, and B. Johansson, Solid State Commun. **88**, 331 (1993).
7. M.D. Stiles, Phys. Rev. B **48**, 7238 (1993).
8. A. Oswald, R. Zeller, P.J. Braspenning, and P.H. Dederichs, J. Phys. F: Met. Phys. **15**, 193 (1985).
9. A.I. Liechtenstein, M.I. Katsnelson, V.P. Antropov, and V.A. Gubanov, J. Magn. Magn. Mater. **67**, 65 (1987).
10. A.R. Mackintosh and O.K. Andersen, in *Electrons at the Fermi Surface*, Ch. 5.3., ed. M. Springford (Cambridge University Press, Cambridge, England, 1980).
11. P. Lloyd and P.V. Smith, Adv. Phys. **21**, 69 (1972).
12. P. Lang, L. Nordström, R. Zeller, and P.H. Dederichs, Phys. Rev. Lett. **71**, 1927 (1993).
13. J. Kudrnovský, V. Drchal, I. Turek, and P. Weinberger, Phys. Rev. B **50**, 16105 (1994).
14. V. Drchal, J. Kudrnovský, I. Turek, and P. Weinberger, Phys. Rev. B **53**, 15036 (1996).
15. P. Bruno, J. Kudrnovský, V. Drchal, and I. Turek, Phys. Rev. Lett. **76**, 4254 (1996).

16. V. Drchal, J. Kudrnovský, P. Bruno, and P. Weinberger, Phil. Mag. B **78**, 571 (1998).
17. V. Drchal, J. Kudrnovský, P. Bruno, P.H. Dederichs, and P. Weinberger, (to be submitted).
18. J. Kudrnovský, V. Drchal, I. Turek, M. Šob, and P. Weinberger, Phys. Rev. B **53**, 5125 (1996).
19. J. Kudrnovský, V. Drchal, P. Bruno, I. Turek, and P. Weinberger, Phys. Rev. B **54**, R3738 (1996).
20. J. Kudrnovský, V. Drchal, C. Blass, I. Turek, and P. Weinberger, Phys. Rev. Lett. **76**, 3834 (1996).
21. J. Kudrnovský, V. Drchal, R. Coehoorn, M. Šob, and P. Weinberger, Phys. Rev. Lett. **78**, 358 (1997).
22. P. Bruno, J. Kudrnovský, V. Drchal, and I. Turek, J. Magn. Magn. Mater. **165**, 128 (1997).
23. J. Kudrnovský, V. Drchal, P. Bruno, I. Turek, and P. Weinberger, Phys. Rev. B **56**, 8919 (1997).
24. J. Kudrnovský, V. Drchal, P. Bruno, R. Coehoorn, J.J. de Vries, K. Wildberger, P.H. Dederichs, and P. Weinberger, MRS Symposium Proceedings, eds. J. Tolbin et al., Vol. **475**, 575 (1997).
25. O.K. Andersen and O. Jepsen, Phys. Rev. Lett. **53**, 2571 (1984).
26. I. Turek, V. Drchal, J. Kudrnovský, M. Šob, and P. Weinberger, *Electronic Structure of Disordered Alloys, Surfaces and Interfaces* (Kluwer, Boston-London-Dordrecht, 1997)
27. See I. Turek, J. Kudrnovský, and V. Drchal, these Proceedings.
28. V. Drchal, J. Kudrnovský, L. Udvardi, P. Weinberger, and A. Pasturel, Phys. Rev. B **45**, 14328 (1992).
29. See, for example, A. Messiah, *Quantum Mechanics*, vol. II, (North-Holland, Amsterdam, 1969), Appendix C.
30. P. Lang, L. Nördstrom, K. Wildberger, R. Zeller, and P.H. Dederichs, Phys. Rev. B **53**, 9092 (1996).
31. F. Ducastelle, J. Phys. C: Solid State Phys. **8**, 3297 (1975).
32. L. Szunyogh, B. Újfalussy, P. Weinberger, and C. Sommers, Phys. Rev. B **54**, 6430 (1996).
33. B. Wenzien, J. Kudrnovský, V. Drchal, and M. Šob, J. Phys.: Condens. Matter **1**, 9893 (1989).
34. J. Kudrnovský, I. Turek, V. Drchal, P. Weinberger, N.E. Christensen, and S.K. Bose, Phys. Rev. B **46**, 4222 (1992).
35. J. Kudrnovský, I. Turek, V. Drchal, P. Weinberger, S.K. Bose, and A. Pasturel, Phys. Rev. B **47**, 16525 (1993).
36. V. Drchal, J. Kudrnovský, and I. Turek, Comp. Phys. Commun. **97**, 111 (1996).
37. P. Weinberger, Phil. Mag. B **75**, 509 (1997).
38. J. Zabloudil, C. Uiberacker, U. Pustogowa, B. Blaas, L. Szunyogh, C. Sommers, and P. Weinberger, Phys. Rev. B **57**, 7804 (1998).
39. J.C. Slonczewski, Phys. Rev. B **39**, 6995 (1989).
40. D.M. Edwards, A.M. Robinson, and J. Mathon, J. Mag. Mag. Mat. **140-144**, 517 (1995).
41. C. Blaas, P. Weinberger, L. Szunyogh, J. Kudrnovský, V. Drchal, P.M. Levy, and C. Sommers (submitted to J. Phys. I France).
42. E.M. Godfrin, J. Phys.: Condens. Matter **3**, 7843 (1991).
43. P. Weinberger, I. Turek, and L. Szunyogh, Int. J. Quant. Chem. **63**, 165 (1997).

44. M.S. Ferreira, J. Phys. Condens. Matter **9**, 6665 (1997).
45. J. d'Albuquerque e Castro, J. Mathon, M. Villeret, and A. Umerski, Phys. Rev. B **53**, R13306 (1996).
46. C. Lanczos, *Applied Analysis*, (Dover, New York, 1988), p. 219.
47. P. Bruno, J. Magn. Magn. Mater. **164**, 27 (1996).
48. S. Krompiewski, F. Süss, and U. Krey, Europhys. Lett. **26**, 303 (1994).
49. L. Nördstrom, P. Lang, R. Zeller, and P.H. Dederichs, Phys. Rev. B **50**, 13058 (1994).
50. P.O. Löwdin, J. Chem. Phys. **19**, 1396 (1951).

Part III

Disordered Alloys

Disordered Alloys and Their Surfaces: The Coherent Potential Approximation

I. Turek[1], J. Kudrnovský[2], and V. Drchal[2]

[1] Institute of Physics of Materials, Academy of Sciences of the Czech Republic,
Žižkova 22, CZ-616 62 Brno, Czech Republic
[2] Institute of Physics, Academy of Sciences of the Czech Republic,
Na Slovance 2, CZ-182 21 Praha 8, Czech Republic

Abstract. A recently developed ab initio approach to the electronic structure of substitutionally disordered alloys and their surfaces is reviewed. It is based on (i) the tight-binding linear muffin-tin orbital (TB-LMTO) method in the atomic sphere approximation which provides a physically transparent solution of the one-electron problem in metallic materials, (ii) the coherent potential approximation (CPA) for a mean-field treatment of the substitutional randomness, and (iii) the surface Green functions for a proper description of the true semi-infinite geometry of surfaces and interfaces. Theoretical formulation of fundamental electronic quantities, both site-diagonal (charge densities, densities of states) and site non-diagonal (the Bloch spectral functions) is presented. Transformation properties of the LMTO-CPA theory as well as specific problems of application of the local density approximation to random alloys are briefly discussed and basic algorithms employed in the numerical implementation of the formalism are described.

1 Introduction

Recent ab initio investigations of electronic properties of solids rely on the local spin-density approximation (LSDA) to the density-functional formalism and on a number of techniques solving the corresponding one-electron Schrödinger (Kohn-Sham) eigenvalue problem. These techniques comprise, e.g., the Korringa-Kohn-Rostoker (KKR) method [1,2], the linear muffin-tin orbital (LMTO) method [3,4], the linear augmented plane-wave (LAPW) method [3,5], or the optimized linear combination of atomic orbitals (LCAO) method [6]. They provide a reasonable description of the electronic structure for most of metallic solids even within the muffin-tin model [1,2] or the atomic sphere approximation (ASA) [3,4]. Full-potential versions of these techniques yield in principle an exact solution to the Schrödinger equation which is indispensable for accurate evaluation of total energies, forces, and other important quantities for perfect bulk solids (elemental metals, ordered alloys) as well as their defects (impurities, surfaces, grain boundaries).

Substitutionally disordered alloys (substitutional solid solutions) represent a broad class of systems where the above mentioned methods are only partially successful: their direct application requires large supercells simulating the randomness of real alloys. The coherent potential approximation (CPA) – introduced three decades ago [7] in terms of the Green functions – offered an effective-medium (mean-field) approach to the electronic structure of random alloys. Further development of the CPA was formulated first in a tight-binding picture (TB-CPA) [8,9] followed by the KKR-CPA theory [10,11] (for a review of both approaches see, e.g., [2]). In the early 1980's, the KKR-CPA became a theory of random alloys fully comparable to the existing charge selfconsistent techniques for non-random systems. The development of the TB-CPA continued towards an ab initio level which was motivated by a need for a physically simple description of the electronic structure of bulk alloys and their surfaces. This effort led to the LCAO-CPA [12,13] and the LMTO-CPA [14,15] methods which can be considered as alternatives to the KKR-CPA.

In this contribution a brief theoretical background of the LMTO-CPA method is given together with numerical techniques used in practice. The theory is developed within the TB-LMTO-ASA method [16,17] which results in an efficient unified scheme for the electronic structure of random and ordered bulk alloys, their surfaces and interfaces. Its full detailed description was presented in [15] while the numerical algorithms were reviewed in [18]. The paper is organized as follows. Section 2 summarizes the most important relations of the TB-LMTO-ASA method in terms of the Green functions. Section 3 is the central part of the paper: it introduces the concept of configurational averaging and describes the theoretical and numerical aspects of the LMTO-CPA method. Section 4 presents a short review of quantities and techniques for a treatment of layered systems (surfaces and interfaces). Section 5 deals with the application of the LSDA to random alloys. Finally, a brief survey of existing results and further extensions of the method is given in Sect. 6.

It should be noted that the LMTO-CPA formalism bears strong similarities with the KKR-CPA theory within the ASA, so that expressions for many quantities (e.g., densities of states, electronic charge densities) are fully analogous in both approaches. However, there are differences as well which arise from different Hilbert spaces and Hamiltonians: the KKR theory is based on the Hamiltonian $H = -\Delta + V(r)$ acting in the Hilbert space of functions $\psi(r)$, where r is a three-dimensional continuous variable, whereas the LMTO theory uses a local basis set with a finite number of orbitals per lattice site. The Hamiltonian is then a matrix quantity. Despite the fact that the spectra of both Hamiltonians are in principle identical (in a limited energy interval), some quantities (e.g., the Bloch spectral functions) become non-equivalent in the two approaches. From the point of view of the alloy theory, both formulations have their own merits: the Hilbert space of the KKR-CPA is explicitly non-random (independent of a particular alloy configuration), the TB (matrix) formulation of the LMTO-CPA offers, e.g., a simple perturbative treatment of relativistic effects (spin-orbit coupling) or an inclusion of many-body effects in terms of the intraatomic Coulomb and exchange integrals, etc.

2   Green Functions in the Atomic Sphere Approximation

The solution of the one-electron Schrödinger equation with a potential $V(r)$ can be equivalently formulated in terms of the one-electron Green function $G(r, r'; z)$ defined (with spin variables omitted) by [1,2]

$$
\begin{aligned}
[z + \Delta_r - V(r)] \, G(r, r'; z) &= \delta(r - r') \,, \\
[z + \Delta_{r'} - V(r')] \, G(r, r'; z) &= \delta(r - r') \,,
\end{aligned}
\tag{1}
$$

where z denotes a complex energy. The Green function $G(r, r'; z)$ is an analytic function of z with the exception of poles and/or branch cuts on the real energy axis. Within the ASA, the Green function for a closely packed solid can be written in the form [15,19,20]

$$
G(r + R, r' + R'; z) = - \delta_{RR'} \sum_L \varphi_{RL}(r^<, z) \, \tilde{\varphi}_{RL}(r^>, z)
$$

$$
+ \sum_{LL'} \varphi_{RL}(r, z) \, G_{RL,R'L'}(z) \, \varphi_{R'L'}(r', z) \,.
\tag{2}
$$

Here R, R' denote the lattice points (centers of the atomic spheres), L, L' are the angular momentum indices ($L = (\ell, m)$), the variables r, r' refer to positions of points inside the individual atomic spheres, and the symbol $r^<$ ($r^>$) denotes that of the vectors r, r' with the smaller (larger) modulus. The functions $\varphi_{RL}(r, z)$ and $\tilde{\varphi}_{RL}(r, z)$ are defined by

$$
\varphi_{RL}(r, z) = \varphi_{R\ell}(r, z) \, Y_L(\hat{r}) \,, \qquad \tilde{\varphi}_{RL}(r, z) = \tilde{\varphi}_{R\ell}(r, z) \, Y_L(\hat{r}) \,,
\tag{3}
$$

where $r = |r|$, $\hat{r} = r/r$, and $Y_L(\hat{r})$ denotes the real spherical harmonics. The radial amplitudes $\varphi_{R\ell}(r, z)$ and $\tilde{\varphi}_{R\ell}(r, z)$ are respectively regular and irregular solutions of the radial Schrödinger equation for the R-th atomic sphere of radius s_R and for the complex energy z. The regular solution is normalized to unity,

$$
\int_0^{s_R} \varphi_{R\ell}^2(r, z) \, r^2 \, dr = 1 \,,
\tag{4}
$$

while the irregular solution is unambiguously specified by a smooth matching at the sphere boundary ($r = s_R$) to the energy derivative of the regular solution $\dot{\varphi}_{R\ell}(r, z)$ (an overdot means energy derivative).

The Green function matrix $G_{RL,R'L'}(z)$ in (2) will be referred to as the physical Green function. It is given in terms of the potential functions $P_{R\ell}^0(z)$ and the canonical structure constants $S_{RL,R'L'}^0$ by

$$
G_{RL,R'L'}(z) = \lambda_{R\ell}^0(z) \, \delta_{RL,R'L'} + \mu_{R\ell}^0(z) \, g_{RL,R'L'}^0(z) \, \mu_{R'\ell'}^0(z) \,,
\tag{5}
$$

where the quantities on the r.h.s. are defined as

$$
\mu_{R\ell}^0(z) = \sqrt{\dot{P}_{R\ell}^0(z)} \,, \qquad \lambda_{R\ell}^0(z) = -\frac{1}{2} \frac{\ddot{P}_{R\ell}^0(z)}{\dot{P}_{R\ell}^0(z)} \,,
$$

$$
g_{RL,R'L'}^0(z) = \left\{ [P^0(z) - S^0]^{-1} \right\}_{RL,R'L'} \,.
\tag{6}
$$

In the last equation, the symbol $P^0(z)$ stands for a diagonal matrix of potential functions, $P^0_{RL,R'L'}(z) = P^0_{R\ell}(z)\,\delta_{RL,R'L'}$. The matrix $g^0_{RL,R'L'}(z)$ will be referred to as the auxiliary (or KKR-ASA) Green function. The superscript 0 of all quantities in (5, 6) denotes the canonical LMTO representation. The physical and auxiliary Green functions are connected by a trivial relation (5), the former one is directly related to the Green function in real space (2), the latter one is of a simpler form and thus better suited for numerical applications.

Let us now summarize the most important relations involved in the TB-LMTO theory [16,17]. The superscript α marks the corresponding representation specified by the screening constants $\alpha_{R\ell}$ (the trivial choice $\alpha_{R\ell} = 0$ corresponds to the canonical representation). The transformations of the screened potential functions $P^\alpha_{R\ell}(z)$ and the screened structure constants $S^\alpha_{RL,R'L'}$ from a particular representation α to some other representation β (specified by a different set of the screening constants $\beta_{R\ell}$) are given by

$$P^\beta_{R\ell}(z) \;=\; P^\alpha_{R\ell}(z)\,[\,1 + (\alpha_{R\ell} - \beta_{R\ell})\,P^\alpha_{R\ell}(z)\,]^{-1}\;,$$

$$S^\beta_{RL,R'L'} \;=\; \left\{ S^\alpha\,[\,1 + (\alpha - \beta)\,S^\alpha\,]^{-1} \right\}_{RL,R'L'}\;. \tag{7}$$

These relations serve simultaneously as definitions of the screened quantities from the canonical ones $\left(P^0(z), S^0\right)$. The second equation is written in a matrix notation with α, β being diagonal matrices of the form $\alpha_{RL,R'L'} = \alpha_{R\ell}\,\delta_{RL,R'L'}$. In analogy to (6), we define

$$\mu^\alpha_{R\ell}(z) \;=\; \sqrt{\dot{P}^\alpha_{R\ell}(z)}\;, \qquad\qquad \lambda^\alpha_{R\ell}(z) \;=\; -\frac{1}{2}\,\frac{\ddot{P}^\alpha_{R\ell}(z)}{\dot{P}^\alpha_{R\ell}(z)}\;,$$

$$g^\alpha_{RL,R'L'}(z) \;=\; \left\{[P^\alpha(z) - S^\alpha]^{-1}\right\}_{RL,R'L'}\;. \tag{8}$$

As a consequence, one can prove two important relations, namely

$$G_{RL,R'L'}(z) \;=\; \lambda^\alpha_{R\ell}(z)\,\delta_{RL,R'L'} \;+\; \mu^\alpha_{R\ell}(z)\,g^\alpha_{RL,R'L'}(z)\,\mu^\alpha_{R'\ell'}(z)\;, \tag{9}$$

and

$$g^\beta_{RL,R'L'}(z) \;=\; (\beta_{R\ell} - \alpha_{R\ell})\,\frac{P^\alpha_{R\ell}(z)}{P^\beta_{R\ell}(z)}\,\delta_{RL,R'L'}$$

$$+\; \frac{P^\alpha_{R\ell}(z)}{P^\beta_{R\ell}(z)}\,g^\alpha_{RL,R'L'}(z)\,\frac{P^\alpha_{R'\ell'}(z)}{P^\beta_{R'\ell'}(z)}\;. \tag{10}$$

The first relation (9) implies that the physical Green function is invariant with respect to the choice of the screening constants $\alpha_{R\ell}$, cf. (5). The second relation (10) means that the auxiliary Green functions in different representations are related to each other by a simple rescaling. Let us note that the first term on the r.h.s. of (9) does not contribute to calculated physical quantities in most cases. However, its presence is inevitable for correct analytic properties of the Green functions (2, 9) in the complex energy plane.

The only non-trivial step in a calculation of the Green function for a solid is the matrix inversion (8) defining the auxiliary Green function $g^\alpha_{RL,R'L'}(z)$. In the case of a bulk solid with three-dimensional translational symmetry, the lattice points R can be expressed in the form $R = B + T$ where B runs over a finite number of the basis vectors while T runs over the translation lattice vectors. The lattice Fourier transformation of the structure constant matrix leads to a k-dependent matrix quantity

$$S^\alpha_{BL,B'L'}(k) = \sum_T S^\alpha_{BL,(B'+T)L'} \exp(i\,k\cdot T) , \tag{11}$$

where k denotes a vector from the first Brillouin zone (BZ). The lattice Fourier transform of the auxiliary Green function, $g^\alpha_{BL,B'L'}(k,z)$, is given by the inverse of a finite-dimensional matrix:

$$g^\alpha_{BL,B'L'}(k,z) = \left\{ [P^\alpha(z) - S^\alpha(k)]^{-1} \right\}_{BL,B'L'} , \tag{12}$$

where $P^\alpha(z)$ denotes a diagonal matrix of the potential functions of the inequivalent atoms, $P^\alpha_{BL,B'L'}(z) = P^\alpha_{B\ell}(z)\delta_{BL,B'L'}$. The inverse Fourier transformation (a BZ-integration) yields then all elements of the auxiliary Green function as

$$g^\alpha_{BL,(B'+T)L'}(z) = \frac{1}{N} \sum_k g^\alpha_{BL,B'L'}(k,z) \exp(-i\,k\cdot T) , \tag{13}$$

where N is the number of cells in a large, but finite crystal with periodic boundary conditions.

Further, let us mention the link between the Green functions and the standard LMTO theory. By using parametrized forms of the potential functions $P^\alpha_{R\ell}(z)$ and the related quantities $\lambda^\alpha_{R\ell}(z)$ and $\mu^\alpha_{R\ell}(z)$ which are correct up to the second order in a limited energy region, we get

$$P^\alpha_{R\ell}(z) = [\,\Delta_{R\ell} + (\gamma_{R\ell} - \alpha_{R\ell})\,(z - C_{R\ell})\,]^{-1}\,(z - C_{R\ell}) ,$$

$$\mu^\alpha_{R\ell}(z) = [\,\Delta_{R\ell} + (\gamma_{R\ell} - \alpha_{R\ell})\,(z - C_{R\ell})\,]^{-1}\,\sqrt{\Delta_{R\ell}} ,$$

$$\lambda^\alpha_{R\ell}(z) = [\,\Delta_{R\ell} + (\gamma_{R\ell} - \alpha_{R\ell})\,(z - C_{R\ell})\,]^{-1}\,(\gamma_{R\ell} - \alpha_{R\ell}) , \tag{14}$$

where $C_{R\ell}$, $\Delta_{R\ell}$ and $\gamma_{R\ell}$ are the LMTO-ASA potential parameters [4,17]. The insertion of (14) with $\alpha_{R\ell} = 0$ into (5, 6) yields (in a matrix notation):

$$G(z) = (z - H)^{-1} , \qquad H = C + \sqrt{\Delta}\,S^0\,(1 - \gamma\,S^0)^{-1}\,\sqrt{\Delta} , \tag{15}$$

which means that the physical Green function is the resolvent of the second-order LMTO-ASA Hamiltonian H. It should be noted that the energy linearization of the LMTO method, which leads to (14) and to the Hamiltonian (15), is not of central importance for the Green function techniques discussed here, as they require matrix inversions rather than matrix diagonalizations.

Finally, let us sketch briefly the evaluation of basic physical observables. As a rule, they are directly related to a limit of the one-electron Green function

$G(r, r'; z)$ with respect to the upper complex halfplane, $z = E + i0$, where E denotes a real energy variable. At zero temperature, the electronic charge density $\varrho_R(r)$ inside the R-th atomic sphere (with spin index ignored) can be written as

$$\varrho_R(r) = -\frac{1}{\pi} \int_{-\infty}^{E_F} \text{Im } G(r + R, r + R; E + i0) \, dE$$

$$= \sum_{LL'} \int_{-\infty}^{E_F} \varphi_{RL}(r, E) \, n_{R,LL'}(E) \, \varphi_{RL'}(r, E) \, dE \, , \quad (16)$$

where E_F is the Fermi energy. The quantity $n_{R,LL'}(E)$ is the local density of states matrix which is given in terms of the site-diagonal block of the physical Green function

$$n_{R,LL'}(E) = -\frac{1}{\pi} \text{Im } G_{RL,RL'}(E + i0) \quad (17)$$

and which is closely related to RL-projected and local densities of states

$$n_{RL}(E) = n_{R,LL}(E) \, , \qquad n_R(E) = \sum_L n_{RL}(E) \, . \quad (18)$$

The total integrated density of states, $N(E)$, can be obtained from (8, 9, 17, 18). In a matrix notation, the result can be written as [21]

$$N(E) = \sum_{RL} \int_{-\infty}^{E} n_{RL}(\varepsilon) \, d\varepsilon$$

$$= \frac{1}{\pi} \text{Im } \left[\text{Tr } \log g^\alpha(E + i0) + \sum_{RL} \log \mu_{R\ell}^\alpha(E + i0) \right] , \quad (19)$$

where the symbol Tr means the trace over the composed RL-index. Note that it is the auxiliary (KKR-ASA) Green function which appears in the expression for the integrated density of states $N(E)$, in contrast to the physical Green functions entering the densities of states $n_R(E)$ and the charge densities $\varrho_R(r)$. Despite this fact, $N(E)$ (19) is representation-invariant as can be easily shown.

3   The Coherent Potential Approximation

Let us now consider the simplest model of a substitutionally disordered alloy. We assume several components (atomic species) labeled by a superscript Q ($Q = A, B, \ldots$) which occupy randomly the sites R of a given rigid lattice with probabilities c_R^Q satisfying the conditions

$$\sum_Q c_R^Q = 1 \, . \quad (20)$$

We neglect completely any correlations of occupations of different sites and assume that the one-electron potential inside the $\boldsymbol{R}$-th atomic sphere (and consequently the related quantities like the potential functions $P^\alpha_{R\ell}(z)$) depends solely on the occupation of this site. In order to express this model in a formal way, we introduce the random occupation index η^Q_R which takes on two values: $\eta^Q_R = 1$ if an atom of the species Q is at the site $\boldsymbol{R}$, and $\eta^Q_R = 0$ otherwise. Each configuration of the disordered alloy is thus uniquely specified by these occupation indices which obey the following trivial relations:

$$\sum_Q \eta^Q_R = 1 , \qquad \eta^Q_R \eta^{Q'}_R = \eta^Q_R \delta^{QQ'} , \tag{21}$$

which reflect the fact that a given site $\boldsymbol{R}$ cannot be empty or occupied by two different species simultaneously. Let us denote the configurational average of an arbitrary quantity as $\langle \dots \rangle$, then we get

$$\left\langle \eta^Q_R \right\rangle = c^Q_R , \quad \left\langle \eta^Q_R \eta^{Q'}_{R'} \right\rangle = c^Q_R \delta_{RR'} \delta^{QQ'} + c^Q_R c^{Q'}_{R'} (1 - \delta_{RR'}) , \tag{22}$$

where the second equation expresses the absence of correlations of the site occupations. The random potential functions $P^\alpha_{R\ell}(z)$ can be then written in a form

$$P^\alpha_{R\ell}(z) = \sum_Q \eta^Q_R P^{\alpha,Q}_{R\ell}(z) , \tag{23}$$

where $P^{\alpha,Q}_{R\ell}(z)$ denotes the non-random potential function of the atom Q occupying the site $\boldsymbol{R}$. Equation (23) represents an important assumption of the model; analogous relations are valid between the random quantities $\lambda^\alpha_{R\ell}(z)$, $\mu^\alpha_{R\ell}(z)$ (8) and their Q-dependent non-random counterparts $\lambda^{\alpha,Q}_{R\ell}(z)$, $\mu^{\alpha,Q}_{R\ell}(z)$. Let us further assume that the screening constants $\alpha_{R\ell}$ are non-random (configuration-independent). This implies that the structure constant matrix $S^\alpha_{RL,R'L'}$ is non-random. The basic problem is an (approximate) configurational averaging of the various one-electron quantities introduced in Sect. 2. In the following, we use a simplified notation with omitted angular momentum indices L, L' so that matrix quantities $X_{RL,R'L'}$ will be abbreviated as $X_{R,R'}$ (e.g., $X = S^\alpha, g^\alpha(z)$), while local (site-diagonal) quantities $W_{R,LL'}$ will be abbreviated by W_R (e.g., $W = P^\alpha(z), \lambda^\alpha(z), \mu^\alpha(z)$).

We start with the auxiliary Green function $g^\alpha_{R,R'}(z)$. Its configurational average $\bar{g}^\alpha_{R,R'}(z)$ can be formally written in a form (cf. (8))

$$\left\langle g^\alpha_{R,R'}(z) \right\rangle = \bar{g}^\alpha_{R,R'}(z) = \left\{ [\mathcal{P}^\alpha(z) - S^\alpha]^{-1} \right\}_{R,R'} \tag{24}$$

which is nothing but an implicit definition of a non-random matrix quantity $\mathcal{P}^\alpha_{R,R'}(z)$ – the so-called coherent potential function. The complete knowledge of the latter is equivalent to an exact configurational averaging in (24). Approximate alloy theories like the virtual crystal approximation, the average t-matrix approximation, and the single-site CPA are based on the neglect of all site non-diagonal blocks of $\mathcal{P}^{\alpha}_{\boldsymbol{R},\boldsymbol{R}'}(z)$:

$$\mathcal{P}^{\alpha}_{\boldsymbol{R},\boldsymbol{R}'}(z) = \mathcal{P}^{\alpha}_{\boldsymbol{R}}(z)\,\delta_{\boldsymbol{R}\boldsymbol{R}'}\,. \tag{25}$$

This assumption leads to a natural interpretation of the coherent potential functions: $\mathcal{P}^{\alpha}_{\boldsymbol{R},LL'}(z)$ describes the scattering properties of an effective atom at the lattice site $\boldsymbol{R}$. The average Green function (24) corresponds then to a non-random solid formed by the effective atoms placed at the rigid lattice sites.

There are several ways of introducing the single-site CPA [2,7,8]. Here we present the approach of [14,15]. The unknown coherent potential functions $\mathcal{P}^{\alpha}_{\boldsymbol{R}}(z)$ are determined in the following manner. Besides the solid with the effective atoms at all lattice sites, we consider a case with a particular site $\boldsymbol{R}$ occupied by a specified component Q while all other sites are occupied by the effective atoms. The auxiliary Green function in the former case is $\bar{g}^{\alpha}(z)$ (24), whereas that in the latter case will be denoted by $\bar{g}^{\alpha,(RQ)}(z)$. Since the two systems differ only by a perturbation $P^{\alpha,Q}_{\boldsymbol{R}}(z) - \mathcal{P}^{\alpha}_{\boldsymbol{R}}(z)$ which is localized on a single site, the two Green functions are related by

$$\bar{g}^{\alpha,(RQ)}_{\boldsymbol{R}',\boldsymbol{R}''}(z) = \bar{g}^{\alpha}_{\boldsymbol{R}',\boldsymbol{R}''}(z) - \bar{g}^{\alpha}_{\boldsymbol{R}',\boldsymbol{R}}(z)\,t^{\alpha,Q}_{\boldsymbol{R}}(z)\,\bar{g}^{\alpha}_{\boldsymbol{R},\boldsymbol{R}''}(z)\,. \tag{26}$$

The quantity $t^{\alpha,Q}_{\boldsymbol{R},LL'}(z)$ is the single-site t-matrix describing the scattering due to a Q-impurity in an effective medium formed by the effective atoms. It is explicitly given by

$$t^{\alpha,Q}_{\boldsymbol{R}}(z) = f^{\alpha,Q}_{\boldsymbol{R}}(z)\left[P^{\alpha,Q}_{\boldsymbol{R}}(z) - \mathcal{P}^{\alpha}_{\boldsymbol{R}}(z)\right]$$

$$= \left[P^{\alpha,Q}_{\boldsymbol{R}}(z) - \mathcal{P}^{\alpha}_{\boldsymbol{R}}(z)\right]\tilde{f}^{\alpha,Q}_{\boldsymbol{R}}(z)\,, \tag{27}$$

where

$$f^{\alpha,Q}_{\boldsymbol{R}}(z) = \left\{1 + \left[P^{\alpha,Q}_{\boldsymbol{R}}(z) - \mathcal{P}^{\alpha}_{\boldsymbol{R}}(z)\right]\bar{g}^{\alpha}_{\boldsymbol{R},\boldsymbol{R}}(z)\right\}^{-1}\,,$$

$$\tilde{f}^{\alpha,Q}_{\boldsymbol{R}}(z) = \left\{1 + \bar{g}^{\alpha}_{\boldsymbol{R},\boldsymbol{R}}(z)\left[P^{\alpha,Q}_{\boldsymbol{R}}(z) - \mathcal{P}^{\alpha}_{\boldsymbol{R}}(z)\right]\right\}^{-1}\,. \tag{28}$$

The CPA condition for the coherent potential functions can be now formulated as

$$\sum_{Q} c^{Q}_{\boldsymbol{R}}\,\bar{g}^{\alpha,(RQ)}_{\boldsymbol{R}',\boldsymbol{R}''}(z) = \bar{g}^{\alpha}_{\boldsymbol{R}',\boldsymbol{R}''}(z)\,, \tag{29}$$

which expresses the equivalence of the average Green function $\bar{g}^{\alpha}(z)$ and a concentration-weighted sum of the Green functions $\bar{g}^{\alpha,(RQ)}(z)$, see Fig. 1.

As it is obvious from (26), the relation (29) is equivalent to

$$\sum_{Q} c^{Q}_{\boldsymbol{R}}\,t^{\alpha,Q}_{\boldsymbol{R}}(z) = 0\,, \tag{30}$$

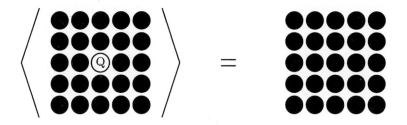

Fig. 1. The selfconsistency condition of the CPA

which is a condition for vanishing average scattering from the Q-impurities ($Q = A, B, \dots$) embedded in the effective medium.

Equation (30) represents the standard form of the CPA selfconsistency condition [2,7,8] which specifies implicitly the coherent potential functions $\mathcal{P}^\alpha_R(z)$. It should be noted that $\mathcal{P}^\alpha_{R,LL'}(z)$ are in general non-diagonal matrices in the L, L' indices, in contrast to the potential functions of the individual components ($P^{\alpha,Q}_{R,LL'}(z) = P^{\alpha,Q}_{R\ell}(z)\,\delta_{LL'}$). The CPA condition (30) has to be solved for all sites simultaneously as the single-site t-matrices (27, 28) involve the site-diagonal blocks of the full matrix inversion defining the average Green function (24). In practice, this can be done only if the whole lattice can be represented by a finite number of inequivalent sites. In the case of a bulk alloy with a crystal lattice and with a possible long-range order, the lattice sites can be written as $R = B + T$ (see the text before (11)), where B labels the inequivalent sites, and the alloy is specified by the concentrations c^Q_B and the component-dependent potential functions $P^{\alpha,Q}_{B\ell}(z)$. As a consequence, the coherent potential functions for all lattice sites reduce to a finite set of matrix quantities $\mathcal{P}^\alpha_{B,LL'}(z)$. In analogy to (12), the lattice Fourier transform of the average auxiliary Green function is given by

$$\bar{g}^\alpha_{BL,B'L'}(\boldsymbol{k}, z) = \left\{ [\mathcal{P}^\alpha(z) - S^\alpha(\boldsymbol{k})]^{-1} \right\}_{BL,B'L'} , \tag{31}$$

where the matrix $S^\alpha(\boldsymbol{k})$ is given by (11) and $\mathcal{P}^\alpha(z)$ denotes a matrix of the coherent potential functions of the inequivalent sites, $\mathcal{P}^\alpha_{BL,B'L'}(z) = \mathcal{P}^\alpha_{B,LL'}(z)\,\delta_{BB'}$. A subsequent BZ-integration yields the elements of the average auxiliary Green function (cf. (13))

$$\bar{g}^\alpha_{BL,(B'+T)L'}(z) = \frac{1}{N} \sum_{\boldsymbol{k}} \bar{g}^\alpha_{BL,B'L'}(\boldsymbol{k}, z)\, \exp(-\mathrm{i}\,\boldsymbol{k}\cdot\boldsymbol{T}) . \tag{32}$$

It should be noted that only the site-diagonal blocks ($B = B'$, $T = 0$) of $\bar{g}^\alpha(z)$ enter the CPA selfconsistency condition (30). The appearance of $\boldsymbol{k}$-dependent quantities in the description of random substitutional alloys reflects a well-known fact that the configurational averaging restores the translational symmetry (absent for individual configurations of the alloy).

Despite the fact that the CPA condition (30) represents a set of coupled non-linear equations for the complex matrix quantities $\mathcal{P}^{\alpha}_{\boldsymbol{R},LL'}(z)$, general theorems guarantee the existence of its unique solution which possesses the so-called Herglotz property. The latter means that (i) the coherent potential functions are analytic functions of z outside the real energy axis, and (ii) the imaginary part of the matrix $\mathcal{P}^{\alpha}_{\boldsymbol{R}}(z)$ is positive (negative) definite for $\mathrm{Im}\, z > 0$ ($\mathrm{Im}\, z < 0$).

3.1   Site-Diagonal Quantities

The calculation of average local quantities like the charge densities (16) or the local densities of states (18) requires a knowledge of additional quantities besides the site-diagonal blocks of the average Green function (24). One introduces so-called conditionally averaged local auxiliary Green functions $\bar{g}^{\alpha,Q}_{\boldsymbol{R},\boldsymbol{R}}(z)$ defined by

$$\bar{g}^{\alpha,Q}_{\boldsymbol{R},\boldsymbol{R}}(z) \;=\; \left(c^{Q}_{\boldsymbol{R}}\right)^{-1}\left\langle \eta^{Q}_{\boldsymbol{R}}\, g^{\alpha}_{\boldsymbol{R},\boldsymbol{R}}(z)\right\rangle . \tag{33}$$

This quantity corresponds to the site-diagonal $(\boldsymbol{R},\boldsymbol{R})$-th block of the Green function averaged under the condition that the site $\boldsymbol{R}$ is occupied by the atomic species Q. Within the CPA, $\bar{g}^{\alpha,Q}_{\boldsymbol{R},\boldsymbol{R}}(z)$ is equal to the $(\boldsymbol{R},\boldsymbol{R})$-th block of the Green function $\bar{g}^{\alpha,(\boldsymbol{R}Q)}(z)$ (26) corresponding to an $\boldsymbol{R}Q$-impurity in the effective medium:

$$\bar{g}^{\alpha,Q}_{\boldsymbol{R},\boldsymbol{R}}(z) \;=\; \bar{g}^{\alpha}_{\boldsymbol{R},\boldsymbol{R}}(z) - \bar{g}^{\alpha}_{\boldsymbol{R},\boldsymbol{R}}(z)\, t^{\alpha,Q}_{\boldsymbol{R}}(z)\, \bar{g}^{\alpha}_{\boldsymbol{R},\boldsymbol{R}}(z) . \tag{34}$$

An equivalent form of this result can be obtained with the help of (28):

$$\bar{g}^{\alpha,Q}_{\boldsymbol{R},\boldsymbol{R}}(z) \;=\; \bar{g}^{\alpha}_{\boldsymbol{R},\boldsymbol{R}}(z)\, f^{\alpha,Q}_{\boldsymbol{R}}(z) \;=\; \tilde{f}^{\alpha,Q}_{\boldsymbol{R}}(z)\, \bar{g}^{\alpha}_{\boldsymbol{R},\boldsymbol{R}}(z) . \tag{35}$$

It follows immediately from (34, 35) that the CPA selfconsistency condition (30) can be expressed in two other forms, namely,

$$\sum_{Q} c^{Q}_{\boldsymbol{R}}\, \bar{g}^{\alpha,Q}_{\boldsymbol{R},\boldsymbol{R}}(z) \;=\; \bar{g}^{\alpha}_{\boldsymbol{R},\boldsymbol{R}}(z) , \tag{36}$$

and

$$\sum_{Q} c^{Q}_{\boldsymbol{R}}\, f^{\alpha,Q}_{\boldsymbol{R}}(z) \;=\; \sum_{Q} c^{Q}_{\boldsymbol{R}}\, \tilde{f}^{\alpha,Q}_{\boldsymbol{R}}(z) \;=\; 1 . \tag{37}$$

The first of them can be easily interpreted: the concentration-weighted average of the Q-dependent conditionally averaged local Green functions is equal to the site-diagonal block of the average Green function.

Let us now discuss the averaging of local observables. We define the conditionally averaged local physical Green functions $\bar{G}^{Q}_{\boldsymbol{R},\boldsymbol{R}}(z)$ as

$$\bar{G}^{Q}_{\boldsymbol{R},\boldsymbol{R}}(z) \;=\; \left(c^{Q}_{\boldsymbol{R}}\right)^{-1}\left\langle \eta^{Q}_{\boldsymbol{R}}\, G_{\boldsymbol{R},\boldsymbol{R}}(z)\right\rangle . \tag{38}$$

Taking into account (9) and the simple configuration dependence of $P^\alpha_{R\ell}(z)$, $\lambda^\alpha_{R\ell}(z)$, and $\mu^\alpha_{R\ell}(z)$ (23), we obtain finally

$$\bar{G}^Q_{RL,RL'}(z) \;=\; \lambda^{\alpha,Q}_{R\ell}(z)\,\delta_{LL'} \;+\; \mu^{\alpha,Q}_{R\ell}(z)\,\bar{g}^{\alpha,Q}_{RL,RL'}(z)\,\mu^{\alpha,Q}_{R\ell'}(z)\;. \tag{39}$$

The expressions (16, 17, 18) can be modified to get the Q-resolved average quantities: the local density of states matrix

$$n^Q_{R,LL'}(E) \;=\; -\frac{1}{\pi}\,\mathrm{Im}\,\bar{G}^Q_{RL,RL'}(E+\mathrm{i}0)\;, \tag{40}$$

the densities of states

$$n^Q_{RL}(E) \;=\; n^Q_{R,LL}(E)\;, \qquad n^Q_R(E) \;=\; \sum_L n^Q_{RL}(E)\;, \tag{41}$$

and the charge densities

$$\varrho^Q_R(r) \;=\; \sum_{LL'}\int_{-\infty}^{E_F}\varphi^Q_{RL}(r,E)\,n^Q_{R,LL'}(E)\,\varphi^Q_{RL'}(r,E)\,dE\;. \tag{42}$$

One can also define average local quantities as concentration-weighted sums of the corresponding Q-resolved quantities, e.g.,

$$n_{RL}(E) \;=\; \sum_Q c^Q_R\,n^Q_{RL}(E)\;, \qquad n_R(E) \;=\; \sum_Q c^Q_R\,n^Q_R(E)\;, \tag{43}$$

which define average densities of states.

The CPA expression for the configuration average of the total integrated density of states $N(E)$ is not a simple generalization of (19). The final result is given by [2,14,22]

$$N(E) \;=\; \frac{1}{\pi}\,\mathrm{Im}\left[\,\mathrm{Tr}\,\log\,\bar{g}^\alpha(E+\mathrm{i}0) \;+\; \sum_{RQ} c^Q_R\,\mathrm{tr}\,\log\,f^{\alpha,Q}_R(E+\mathrm{i}0)\right.$$

$$\left.+\; \sum_{RQL} c^Q_R\,\log\,\mu^{\alpha,Q}_{R\ell}(E+\mathrm{i}0)\right]\;, \tag{44}$$

where the symbol tr means the trace over the angular momentum index L. Let us mention an important variational property of $N(E)$ (44), which is a direct consequence of the CPA selfconsistency [22]: $N(E)$ is stationary with respect to variations of the coherent potential functions $\delta P^\alpha_R(z)$. This property makes the CPA an excellent starting point for studies of alloy energetics within the generalized perturbation method [23,24].

In numerical implementations of the CPA as well as of other Green function techniques, complex energies are indispensable to obtain the limiting values at the real energy axis, cf. (40, 44). A useful approach to get the necessary limits $F(E+\mathrm{i}0)$ of a complex function $F(z)$ analytic in the upper halfplane is based on a relatively easy evaluation of the function $F(z)$ for Im $z > 0$ and a subsequent analytic continuation to the real axis [25]. This procedure is justified by the Riemann-Cauchy relations and it employs truncated Taylor expansions of the function $F(z)$. Suppose that $F(z)$ has to be evaluated for real energies on a dense equidistant mesh of energy points $E_n = E_0 + nh$, where h is an energy step and $n = 0, 1, \ldots, N$. Let us consider a discrete set of complex energy points

$$z_{n,m} = E_0 + nh + imh, \tag{45}$$

where $m = 0, 1, \ldots, M$, and let us abbreviate $F_{n,m} = F(z_{n,m})$. The first step of a continuation procedure is the calculation of $F_{n,M}$ for $-M \le n \le N + M$, i.e., for complex energies along a line parallel to the real axis. In each of the M following steps, the values of $F_{n,m}$ with m reduced by one are obtained from all previously calculated values. The simplest examples are given by relations:

$$F_{n,m-1} = 2 F_{n,m} + \frac{-1+i}{2} F_{n-1,m} + \frac{-1-i}{2} F_{n+1,m}, \tag{46}$$

which is based on a quadratic Taylor expansion, and

$$F_{n,m-1} = F_{n,m} + \frac{i}{2} F_{n-1,m} - \frac{i}{2} F_{n+1,m}, \tag{47}$$

which is based on a repeated linear Taylor expansion. There are many modifications of this procedure which employ higher-order expansions [15,25]. However, only the linear continuation (47) yields always strictly non-negative densities of states and the Bloch spectral functions. Typically, an energy increment of $h \sim 5$ mRy and $M \sim 2$ to 5 lead to a sufficiently large Im z for an initial calculation of $F(z)$. The continuation to the real axis according to (46, 47) represents then a negligible computational effort.

An example of average densities of states is presented in Fig. 2 for a spin-polarized random bcc $Fe_{0.7}V_{0.3}$ alloy. Due to the antiparallel magnetic moments of the Fe and V atoms, the spin-up electrons feel a much stronger disorder than the spin-down electrons. The different degree of disorder is nicely reflected in the shapes of the local densities of states: the spin-down densities for both components (Fig. 2b) resemble those for the pure elements in the bcc structure whereas the spin-up densities (Fig. 2a) are strongly modified due to alloying. Especially in the latter case, the CPA describes the electronic structure substantially better than other single-site theories (the virtual crystal approximation, the average t-matrix approximation).

3.2   Site Non-Diagonal Quantities

Let us now turn to the physical Green function $G_{R,R'}(z)$ and to its configurational average

$$\langle G_{R,R'}(z) \rangle = \bar{G}_{R,R'}(z), \tag{48}$$

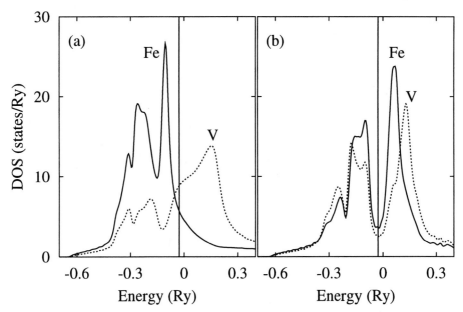

Fig. 2. Spin-polarized local densities of states for Fe (*full lines*) and V (*dotted lines*) atoms in the random bcc $Fe_{0.7}V_{0.3}$ alloy: (**a**) spin-up electrons, (**b**) spin-down electrons. The vertical lines denote the position of the Fermi energy

and let us treat separately its site-diagonal ($\boldsymbol{R} = \boldsymbol{R'}$) and site non-diagonal ($\boldsymbol{R} \neq \boldsymbol{R'}$) blocks. The former are given directly by

$$\bar{G}_{\boldsymbol{R},\boldsymbol{R}}(z) = \sum_{Q} c_{\boldsymbol{R}}^{Q}\, \bar{G}_{\boldsymbol{R},\boldsymbol{R}}^{Q}(z) , \tag{49}$$

where the conditionally averaged blocks $\bar{G}_{\boldsymbol{R},\boldsymbol{R}}^{Q}(z)$ can be expressed according to (39). The site non-diagonal blocks can be rewritten with the help of (9, 23) as

$$\bar{G}_{\boldsymbol{R},\boldsymbol{R'}}(z) = \sum_{QQ'} \mu_{\boldsymbol{R}}^{\alpha,Q}(z) \left\langle \eta_{\boldsymbol{R}}^{Q}\, g_{\boldsymbol{R},\boldsymbol{R'}}^{\alpha}(z)\, \eta_{\boldsymbol{R'}}^{Q'} \right\rangle \mu_{\boldsymbol{R'}}^{\alpha,Q'}(z) . \tag{50}$$

The configurational average on the r.h.s. of (50) represents (apart from a normalization) a more complicated case of a conditional average: it refers to the site non-diagonal ($\boldsymbol{R}, \boldsymbol{R'}$)-th block of the auxiliary Green function averaged under the condition that the two sites $\boldsymbol{R}, \boldsymbol{R'}$ are occupied by the atomic species Q, Q', respectively. The single-site CPA expression for this kind of conditional average is [10,11]

$$\left\langle \eta_{\boldsymbol{R}}^{Q}\, g_{\boldsymbol{R},\boldsymbol{R'}}^{\alpha}(z)\, \eta_{\boldsymbol{R'}}^{Q'} \right\rangle = c_{\boldsymbol{R}}^{Q}\, \tilde{f}_{\boldsymbol{R}}^{\alpha,Q}(z)\, \bar{g}_{\boldsymbol{R},\boldsymbol{R'}}^{\alpha}(z)\, c_{\boldsymbol{R'}}^{Q'}\, f_{\boldsymbol{R'}}^{\alpha,Q'}(z) . \tag{51}$$

Equation (51) can be derived for binary alloys by means of a simple algebraic technique [14] while for the multicomponent case one can use Green functions in an extended space [2,9]:

$$\hat{g}^\alpha_{RQL,R'Q'L'}(z) = \eta^Q_R \, g^\alpha_{RL,R'L'}(z) \, \eta^{Q'}_{R'} \, . \tag{52}$$

A single-site CPA averaging of (52) yields then the result (51) [2]. The final expression for the average physical Green function follows from (49, 50, 51) and can be compactly written as

$$\bar{G}_{R,R'}(z) = \bar{G}_{R,R}(z)\,\delta_{RR'} + \tilde{\mathcal{M}}^\alpha_R(z)\,\bar{g}^\alpha_{R,R'}(z)\,\mathcal{M}^\alpha_{R'}(z)\,(1-\delta_{RR'})$$

$$= \mathcal{L}^\alpha_R(z)\,\delta_{RR'} + \tilde{\mathcal{M}}^\alpha_R(z)\,\bar{g}^\alpha_{R,R'}(z)\,\mathcal{M}^\alpha_{R'}(z)\,, \tag{53}$$

where

$$\mathcal{M}^\alpha_R(z) = \sum_Q c^Q_R\,f^{\alpha,Q}_R(z)\,\mu^{\alpha,Q}_R(z)\,,$$

$$\tilde{\mathcal{M}}^\alpha_R(z) = \sum_Q c^Q_R\,\mu^{\alpha,Q}_R(z)\,\tilde{f}^{\alpha,Q}_R(z)\,, \tag{54}$$

and

$$\mathcal{L}^\alpha_R(z) = \bar{G}_{R,R}(z) - \tilde{\mathcal{M}}^\alpha_R(z)\,\bar{g}^\alpha_{R,R}(z)\,\mathcal{M}^\alpha_R(z)\,. \tag{55}$$

It should be noted that the final relation between $\bar{G}_{R,R'}(z)$ and $\bar{g}^\alpha_{R,R'}(z)$ (53) bears the same formal structure as (9) for the non-averaged Green functions.

The average physical Green function (53) can be now used to calculate the Bloch spectral functions. Let us consider again the case of a random bulk alloy with a crystal lattice and with a possible long-range order. The lattice sites can be written as $R = B + T$, where B labels the inequivalent sites and T runs over the translation vectors of the configurationally averaged system (see the text near (31, 32)). The Bloch spectral functions are defined in terms of the lattice Fourier transform of $\bar{G}_{R,R'}(z)$:

$$\mathcal{A}_{BL}(\boldsymbol{k},E) = -\frac{1}{\pi}\,\mathrm{Im}\,\bar{G}_{BL,BL}(\boldsymbol{k},E+\mathrm{i}0)\,,$$

$$\mathcal{A}_B(\boldsymbol{k},E) = \sum_L \mathcal{A}_{BL}(\boldsymbol{k},E)\,. \tag{56}$$

As follows from (53), the lattice Fourier transform of $\bar{G}_{R,R'}(z)$ can be reduced to that of $\bar{g}^\alpha_{R,R'}(z)$ which in turn is given by (31):

$$\bar{G}_{BL,BL}(\boldsymbol{k},z) = \sum_{L'L''} \tilde{\mathcal{M}}^\alpha_{B,LL'}(z)\,\bar{g}^\alpha_{BL',BL''}(\boldsymbol{k},z)\,\mathcal{M}^\alpha_{B,L''L}(z)$$

$$+ \mathcal{L}^\alpha_{B,LL}(z)\,. \tag{57}$$

Using (53) and elementary properties of lattice Fourier transformations, one can prove a relation between the Bloch spectral function $\mathcal{A}_{BL}(\boldsymbol{k},E)$ and the corresponding average density of states $n_{BL}(E)$ (43), namely,

$$n_{BL}(E) = \frac{1}{N}\sum_{\boldsymbol{k}} \mathcal{A}_{BL}(\boldsymbol{k},E)\,. \tag{58}$$

According to this sum rule, the Bloch spectral function reflects the contributions of different parts of the BZ to the resulting density of states of the configurationally averaged system. Let us note that in the case of non-random crystalline solids, the spectral functions for a given fixed k-vector reduce to sums of δ-functions located at the corresponding energy eigenvalues. Hence, the concept of the Bloch spectral functions substitutes energy bands in random alloys and can be used, e.g., for a definition of the Fermi surfaces. The latter are based on the k-dependence of the spectral functions (56) evaluated at a constant energy $(E = E_F)$.

3.3   Transformation Properties of the LMTO-CPA

The physical properties of a non-random system described by the TB-LMTO-ASA method do not depend on the choice of a particular LMTO representation α as expressed by (5, 9). In the context of random alloys, it is of fundamental importance to know whether this feature survives the approximate configuration averaging within the single-site CPA. The answer is positive [15] as will be shown below.

We assume that the representations α, β are specified by non-random screening constants $\alpha_{R\ell}, \beta_{R\ell}$, respectively. For simplicity, we will omit the energy arguments as well as the angular momentum indices L, L'. The transformation of the coherent potential functions is analogous to (7), namely,

$$\mathcal{P}_R^\beta = \mathcal{P}_R^\alpha \left[1 + (\alpha_R - \beta_R) \mathcal{P}_R^\alpha \right]^{-1} . \tag{59}$$

The transformations of the coherent potential functions (59) and of the non-random structure constants (7) lead to the following transformation of the average auxiliary Green functions:

$$\bar{g}_{R,R'}^\beta = (\beta_R - \alpha_R) \mathcal{P}_R^\alpha \left(\mathcal{P}_R^\beta \right)^{-1} \delta_{RR'}$$
$$+ \left(\mathcal{P}_R^\beta \right)^{-1} \mathcal{P}_R^\alpha \bar{g}_{R,R'}^\alpha \mathcal{P}_{R'}^\alpha \left(\mathcal{P}_{R'}^\beta \right)^{-1} \tag{60}$$

which is of the same structure as (10). The transformation of the perturbation related to a single Q-impurity embedded in the effective medium is given by

$$P_R^{\beta,Q} - P_R^\beta = P_R^{\beta,Q} \left(P_R^{\alpha,Q} \right)^{-1} \left(P_R^{\alpha,Q} - \mathcal{P}_R^\alpha \right) (\mathcal{P}_R^\alpha)^{-1} \mathcal{P}_R^\beta , \tag{61}$$

as can be easily derived from (7, 59). The transformation of the quantities (28) can be obtained with the help of (61) and of the site-diagonal blocks of (60). The result is

$$f_R^{\beta,Q} = \mathcal{P}_R^\beta (\mathcal{P}_R^\alpha)^{-1} f_R^{\alpha,Q} P_R^{\alpha,Q} \left(P_R^{\beta,Q} \right)^{-1} ,$$
$$\tilde{f}_R^{\beta,Q} = \left(P_R^{\beta,Q} \right)^{-1} P_R^{\alpha,Q} \tilde{f}_R^{\alpha,Q} (\mathcal{P}_R^\alpha)^{-1} \mathcal{P}_R^\beta , \tag{62}$$

which can be combined with (61) to get the transformation of the single-site t-matrices (27):

$$t_R^{\beta,Q} = \mathcal{P}_R^{\beta} \, (\mathcal{P}_R^{\alpha})^{-1} \, t_R^{\alpha,Q} \, (\mathcal{P}_R^{\alpha})^{-1} \, \mathcal{P}_R^{\beta} \, . \tag{63}$$

An immediate consequence of (63) is the simultaneous validity of the CPA self-consistency condition (30) in two different LMTO representations. This means that all CPA effective media are mutually equivalent irrespective of the particular LMTO representation used for the formulation and solution of the selfconsistency condition.

The transformations of other quantities can be derived from (59–63). This yields, e.g., for the conditionally averaged local auxiliary Green functions (35) a relation completely analogous to (10):

$$\bar{g}_{RL,RL'}^{\beta,Q}(z) = (\beta_{R\ell} - \alpha_{R\ell}) \frac{P_{R\ell}^{\alpha,Q}(z)}{P_{R\ell}^{\beta,Q}(z)} \delta_{LL'}$$

$$+ \frac{P_{R\ell}^{\alpha,Q}(z)}{P_{R\ell}^{\beta,Q}(z)} \, \bar{g}_{RL,RL'}^{\alpha,Q}(z) \, \frac{P_{R\ell'}^{\alpha,Q}(z)}{P_{R\ell'}^{\beta,Q}(z)} \, . \tag{64}$$

One can further show that CPA averages of the physical Green functions (38, 48) remain invariant with respect to different LMTO representations α, which in turn implies the invariance of all physical observables and proves a full compatibility of the single-site CPA with the TB-LMTO method.

3.4   Solution of the CPA Selfconsistency

It should be noted that although the above three forms of the CPA condition (30, 36, 37) are mathematically equivalent, not all of them are suitable for numerical applications. For this purpose we will introduce the so-called coherent interactor $\Omega_{R,LL'}^{\alpha}(z)$ [2,9,14] which is a local quantity defined implicitly in terms of the coherent potential function and the site-diagonal block of the average auxiliary Green function as

$$\bar{g}_{R,R}^{\alpha}(z) = [\, \mathcal{P}_R^{\alpha}(z) - \Omega_R^{\alpha}(z) \,]^{-1} \, , \tag{65}$$

or, explicitly, as

$$\Omega_R^{\alpha}(z) = \mathcal{P}_R^{\alpha}(z) - [\, \bar{g}_{R,R}^{\alpha}(z) \,]^{-1} \, . \tag{66}$$

The coherent interactor describes the effective coupling of a given site R to all other sites in the system. Using this definition, one can express the conditionally averaged local auxiliary Green function as

$$\bar{g}_{R,R}^{\alpha,Q}(z) = \left[\, P_R^{\alpha,Q}(z) - \Omega_R^{\alpha}(z) \,\right]^{-1} \, . \tag{67}$$

In the following, we describe a simple iterative scheme solving the CPA condition (36) using the coherent interactor and (65, 67). We assume that the energy z lies outside the real energy axis. For brevity, the energy argument z and the orbital indices L, L' will be omitted.

The algorithm starts from an input value $\Omega_R^{\alpha,(0)}$ which can be set either to zero or, e.g., to the converged coherent interactor for a neighboring energy argument. For a particular iteration leading from $\Omega_R^{\alpha,(n)}$ to the new value $\Omega_R^{\alpha,(n+1)}$, the procedure consists of three steps. First, the coherent potential function $\mathcal{P}_R^{\alpha,(n)}$ at each site R is set up in terms of $\Omega_R^{\alpha,(n)}$ and the potential functions $P_R^{\alpha,Q}$ and concentrations c_R^Q of all components Q according to the relation

$$\left[\mathcal{P}_R^{\alpha,(n)} - \Omega_R^{\alpha,(n)} \right]^{-1} = \sum_Q c_R^Q \left[P_R^{\alpha,Q} - \Omega_R^{\alpha,(n)} \right]^{-1}, \tag{68}$$

or, explicitly,

$$\mathcal{P}_R^{\alpha,(n)} = \left\{ \sum_Q c_R^Q \left[P_R^{\alpha,Q} - \Omega_R^{\alpha,(n)} \right]^{-1} \right\}^{-1} + \Omega_R^{\alpha,(n)}. \tag{69}$$

Second, these coherent potential functions are used to calculate the site-diagonal blocks $\bar{g}_{R,R}^{\alpha,(n)}$ of the average auxiliary Green function

$$\bar{g}_{R,R}^{\alpha,(n)} = \left\{ \left[\mathcal{P}^{\alpha,(n)} - S^\alpha \right]^{-1} \right\}_{R,R}. \tag{70}$$

Third, the new value of the coherent interactor $\Omega_R^{\alpha,(n+1)}$ at each site R is obtained from the relation

$$\left[\mathcal{P}_R^{\alpha,(n)} - \Omega_R^{\alpha,(n+1)} \right]^{-1} = \bar{g}_{R,R}^{\alpha,(n)}, \tag{71}$$

or, explicitly,

$$\Omega_R^{\alpha,(n+1)} = \mathcal{P}_R^{\alpha,(n)} - \left[\bar{g}_{R,R}^{\alpha,(n)} \right]^{-1}. \tag{72}$$

These three steps have to be repeated in order to obtain converged quantities Ω_R^α and $\mathcal{P}_R^\alpha$ at all sites. Steps (69, 70, 72) preserve the Herglotz property of the matrix quantities $\Omega_R^\alpha, \mathcal{P}_R^\alpha, \bar{g}_{R,R}^\alpha$. Convergence is achieved typically after 5 to 20 iterations depending on the alloy system and the complex energy variable.

Substantial acceleration of charge selfconsistent calculations for random systems can be achieved by repeated alternation of one CPA iteration and one update of one-electron potentials (see Sect. 5). In such case, the potential functions of all alloy components in (68, 69) are replaced by the n-dependent quantities $P_R^{\alpha,Q,(n)}$. The update of the one-electron potentials and the potential functions follows the CPA iteration (69, 70, 72) and is based on charge densities derived from the conditionally averaged local auxiliary Green functions

$$\bar{g}_{R,R}^{\alpha,Q} = \left[P_R^{\alpha,Q,(n)} - \Omega_R^{\alpha,(n+1)} \right]^{-1}. \tag{73}$$

In this way, the CPA selfconsistency is obtained simultaneously with the LSDA selfconsistency.

4   Surfaces and Interfaces

Applications of the single-site CPA to layered systems on lattices with two-dimensional (2D) translational symmetry require special approaches to calculate the Green function quantities involved. Below we summarize the most essential relations of a technique based on the concept of principal layers and the surface Green functions [26–28].

The approach rests on the use of the tight-binding LMTO representation β which provides the most localized structure constants $S^{\beta}_{RL,R'L'}$ [16,17], and on the representation invariance of the CPA (Sect. 3.3). The finite range of the tight-binding structure constants allows to introduce the principal layers in such a way that (i) each principal layer consists of a finite number of neighboring atomic layers, (ii) the whole lattice can be considered as a stacking of an infinite sequence of the principal layers labeled by an integer index p, see Fig. 3, and (iii) the structure constants $S^{\beta}_{RL,R'L'}$ couple only the neighboring principal layers. The sites R of a given system can be then written in a form $R \equiv (p, B, T_{\|})$, where p is the index of the principal layer, B denotes the corresponding basis vector (mostly an atomic layer) in the p-th principal layer, and $T_{\|}$ is a 2D translation vector such that $R = B + T_{\|}$. We assume for simplicity that each principal layer contains the same number n_B of the basis vectors B.

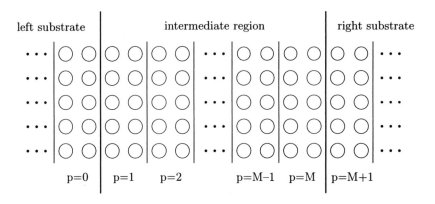

Fig. 3. Principal layers for a single interface of two semi-infinite systems

As a consequence of the 2D translational symmetry of the lattice, a 2D lattice Fourier transformation of the structure constant matrix leads to a $k_{\|}$-dependent matrix

$$S^{\beta}_{pBL,p'B'L'}(k_{\|}) = \sum_{T_{\|}} S^{\beta}_{pBL,p'(B'+T_{\|})L'} \exp(i\, k_{\|} \cdot T_{\|}) , \qquad (74)$$

where $k_{\|}$ denotes a vector in the 2D BZ. It should be noted that the tight-binding structure constants (74) vanish for $|p - p'| > 1$, i.e., they form a block tridiagonal matrix with respect to the principal-layer index. In the following, we will often omit the composed matrix index BL so that matrix quantities with elements $X_{pBL,p'B'L'}$ and $W_{p,BL,B'L'}$ will be respectively abbreviated as $X_{p,p'}$ and W_p. The dimension of the latter matrices is equal to $n_B(\ell_{max} + 1)^2$, where ℓ_{max} denotes the angular-momentum cutoff.

Let us consider the case of a single interface of two semi-infinite systems. Examples of this situation are: a surface of a bulk alloy (solid-vacuum interface), an epitaxial interface of two alloys (metals), a special grain boundary in a bulk metal, etc. The treatment of all these cases can be greatly simplified due to the fact that all inhomogeneities are confined to an intermediate region of a finite thickness (principal layers $1 \leq p \leq M$) placed between two semi-infinite substrates, see Fig. 3. The electronic properties (e.g., the coherent potential functions) of both unperturbed substrates are supposed to be known and the main interest then concentrates on the intermediate region. Let us assume that the configuration-independent properties of the layered alloy system exhibit the 2D translational symmetry of the underlying lattice. As a consequence, the coherent potential functions for all lattice sites reduce to pB-resolved quantities $\mathcal{P}^\beta_{pB,LL'}(z)$ which form matrices $\mathcal{P}^\beta_p(z)$ with elements $\mathcal{P}^\beta_{p,BL,B'L'}(z) = \mathcal{P}^\beta_{pB,LL'}(z)\,\delta_{BB'}$. The average auxiliary Green function (24) can be then calculated using the corresponding lattice Fourier transform $\bar{g}^\beta_{pBL,p'B'L'}(\mathbf{k}_\parallel, z)$.

The layer-diagonal $(p = p')$ blocks of the latter can be expressed as

$$\bar{g}^\beta_{p,p}(\mathbf{k}_\parallel, z) = \Big[\,\mathcal{P}^\beta_p(z) - S^\beta_{p,p}(\mathbf{k}_\parallel)$$
$$- \Gamma^{\beta,<}_p(\mathbf{k}_\parallel, z) - \Gamma^{\beta,>}_p(\mathbf{k}_\parallel, z)\,\Big]^{-1}, \qquad (75)$$

where the the first two terms in the bracket correspond to the isolated p-th layer while the so-called embedding potentials $\Gamma^{\beta,<}_p(\mathbf{k}_\parallel, z)$ and $\Gamma^{\beta,>}_p(\mathbf{k}_\parallel, z)$ reflect the influence of the two semi-infinite parts adjacent to the p-th principal layer – the superscript $<$ $(>)$ refers to the part consisting of all principal layers $p' < p$ $(p' > p)$. The embedding potentials for layers inside the intermediate region $(1 \leq p \leq M)$ can be calculated from recursion relations

$$\Gamma^{\beta,<}_p(\mathbf{k}_\parallel, z) = S^\beta_{p,p-1}(\mathbf{k}_\parallel)\,\Big[\,\mathcal{P}^\beta_{p-1}(z) - S^\beta_{p-1,p-1}(\mathbf{k}_\parallel)$$
$$- \Gamma^{\beta,<}_{p-1}(\mathbf{k}_\parallel, z)\,\Big]^{-1} S^\beta_{p-1,p}(\mathbf{k}_\parallel),$$

$$\Gamma^{\beta,>}_p(\mathbf{k}_\parallel, z) = S^\beta_{p,p+1}(\mathbf{k}_\parallel)\,\Big[\,\mathcal{P}^\beta_{p+1}(z) - S^\beta_{p+1,p+1}(\mathbf{k}_\parallel)$$
$$- \Gamma^{\beta,>}_{p+1}(\mathbf{k}_\parallel, z)\,\Big]^{-1} S^\beta_{p+1,p}(\mathbf{k}_\parallel), \qquad (76)$$

and from the starting values

$$\Gamma^{\beta,<}_1(\mathbf{k}_\parallel, z) = S^\beta_{1,0}(\mathbf{k}_\parallel)\,\mathcal{G}^\beta_{left}(\mathbf{k}_\parallel, z)\,S^\beta_{0,1}(\mathbf{k}_\parallel),$$

$$\Gamma^{\beta,>}_M(\mathbf{k}_\parallel, z) = S^\beta_{M,M+1}(\mathbf{k}_\parallel)\,\mathcal{G}^\beta_{right}(\mathbf{k}_\parallel, z)\,S^\beta_{M+1,M}(\mathbf{k}_\parallel). \qquad (77)$$

The matrix quantities $\mathcal{G}_{\text{left}}^{\beta}(\boldsymbol{k}_{\|}, z)$ and $\mathcal{G}_{\text{right}}^{\beta}(\boldsymbol{k}_{\|}, z)$ in (77) are the surface Green functions of the two semi-infinite substrates sandwiching the intermediate region. The calculation of the layer-diagonal blocks of the Green function according to (75, 76) is obviously an order-M procedure.

The surface Green function (SGF) is defined as a projection of the full Green function of a semi-infinite layered system onto its outer principal layer. For a semi-infinite system consisting of identical principal layers, one can apply the concept of removal invariance [2] to derive a closed condition for the SGF which reflects the true semi-infinite geometry of the system. In the case of the left substrate in Fig. 3, this condition is

$$
\mathcal{G}_{\text{left}}^{\beta}(\boldsymbol{k}_{\|}, z) = \Big[P_0^{\beta}(z) - S_{0,0}^{\beta}(\boldsymbol{k}_{\|})
$$
$$
- S_{0,-1}^{\beta}(\boldsymbol{k}_{\|}) \, \mathcal{G}_{\text{left}}^{\beta}(\boldsymbol{k}_{\|}, z) \, S_{-1,0}^{\beta}(\boldsymbol{k}_{\|}) \Big]^{-1}, \tag{78}
$$

whereas an analogous condition for the right substrate is omitted here for brevity. Both conditions are of the same form, namely

$$
\mathcal{G} = (D - A \, \mathcal{G} \, B)^{-1}, \tag{79}
$$

where $\mathcal{G}$ is the SGF and where the $\boldsymbol{k}_{\|}$- and z-arguments were suppressed. The most direct method to solve (79) is based on simple iterations [28]

$$
\mathcal{G}^{(n+1)} = (D - A \, \mathcal{G}^{(n)} \, B)^{-1} \tag{80}
$$

starting from an input value $\mathcal{G}^{(0)}$ which can be set either to zero or, e.g., to the converged SGF for a neighboring energy argument. The latter choice of $\mathcal{G}^{(0)}$ substantially reduces the number of necessary iteration steps, especially for complex energies close to the real axis. The iterative procedure (80) is easy to implement, leads always to the correct solution of (79) satisfying the Herglotz property, and has a direct physical meaning: $\mathcal{G}^{(n)}$ with the initial value $\mathcal{G}^{(0)} = 0$ corresponds to the SGF of a stacking of n identical principal layers. The number of steps to get a converged SGF depends on the imaginary part of the complex energy z, but in most applications several tens of iterations are sufficient. In the cases where an enhanced accuracy of the SGF and/or very small Im z (less than 10 mRy) are needed, the SGF can be more efficiently obtained by means of the renormalization-decimation technique [15,18,29]. The high efficiency of the latter method is due to an exponential increase of the thickness of an effective layer with the number of iterations, in contrast to the linear increase inherent to the simple procedure (80).

For evaluation of local physical observables as well as for the solution of the CPA condition, the site-diagonal blocks of the average auxiliary Green function are of central importance (see Sect. 3). They can be obtained by a 2D BZ-integration of (75) as

$$
\bar{g}_{pBL,pBL'}^{\beta}(z) = \frac{1}{N_{\|}} \sum_{\boldsymbol{k}_{\|}} \bar{g}_{pBL,pBL'}^{\beta}(\boldsymbol{k}_{\|}, z), \tag{81}
$$

where $N_\parallel$ is the number of $k_\parallel$-points sampling the 2D BZ. The layer-diagonal $k_\parallel$-dependent average auxiliary Green functions (75) enter also the corresponding Bloch spectral functions ($k_\parallel$-resolved densities of states). They are defined in analogy to (56) as

$$A_{pBL}(k_\parallel, E) = -\frac{1}{\pi} \operatorname{Im} \bar{G}_{pBL,pBL}(k_\parallel, E + \mathrm{i}0) , \qquad (82)$$

where the lattice Fourier transform of the average physical Green function is given by (cf. (57))

$$\bar{G}_{pBL,pBL}(k_\parallel, z) = \sum_{L'L''} \tilde{\mathcal{M}}^\beta_{pB,LL'}(z) \, \bar{g}^\beta_{pBL',pBL''}(k_\parallel, z) \, \mathcal{M}^\beta_{pB,L''L}(z)$$
$$+ \mathcal{L}^\beta_{pB,LL}(z) . \qquad (83)$$

It should be noted that the Bloch spectral functions (82) represent a suitable tool to study surface/interface states in disordered as well as ordered layered systems.

Figure 4 shows the local densities of states at the (001) surface of a random non-magnetic bcc $Fe_{0.15}V_{0.85}$ alloy. One can clearly see a rapid convergence of the layer-resolved densities to their bulk counterparts, which justifies numerically the concept of the intermediate region of a finite thickness. The bands in the top surface layer are narrower than the bulk ones and the pronounced minima in the middle of the bcc bulk bands are absent in the top surface layer. These effects can be ascribed to the reduced coordination of the surface atoms. As a consequence, both components exhibit a strong enhancement of the surface densities of states at the Fermi energy (see Fig. 4) which in turn can induce a surface magnetic instability of the non-magnetic bulk alloy.

5   Charge Selfconsistency for Random Alloys

The LSDA selfconsistency for substitutionally disordered systems within the CPA and the ASA is based on the average component-resolved charge densities (42). In the following formulas, we use atomic Rydberg units ($e^2 = 2$) and assume a spin-polarized non-relativistic system with a collinear spin structure. The spin-dependent charge densities inside the individual atomic spheres will be denoted $\varrho^Q_{R\sigma}(r)$ where $\sigma = \uparrow, \downarrow$ is the spin index. Related quantities are the spherically averaged spin-dependent densities

$$\tilde{\varrho}^Q_{R\sigma}(r) = \frac{1}{4\pi} \int \varrho^Q_{R\sigma}(r) \, d^2\hat{r} , \qquad (84)$$

and the total electronic charge densities

$$\varrho^Q_R(r) = \varrho^Q_{R\uparrow}(r) + \varrho^Q_{R\downarrow}(r) , \quad \tilde{\varrho}^Q_R(r) = \tilde{\varrho}^Q_{R\uparrow}(r) + \tilde{\varrho}^Q_{R\downarrow}(r) . \qquad (85)$$

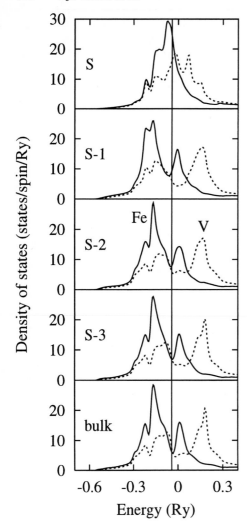

Fig. 4. Layer-resolved local densities of states for Fe (*full lines*) and V (*dotted lines*) atoms at the (001) surface of the random bcc Fe$_{0.15}$V$_{0.85}$ alloy. The top four layers and the bulk layer are denoted by S, S-1, S-2, S-3, and bulk, respectively. The vertical line denotes the position of the Fermi energy

The selfconsistent one-electron component-dependent ASA potentials are then given by

$$V_{R\sigma}^Q(r) = -2\,Z_R^Q\,r^{-1} + \int_{(R)} 2\,\tilde{\varrho}_R^Q(r')\,|\boldsymbol{r}-\boldsymbol{r}'|^{-1}\,d^3\boldsymbol{r}'$$

$$+ V_{\mathrm{xc},\sigma}\big(\tilde{\varrho}_{R\uparrow}^Q(r),\tilde{\varrho}_{R\downarrow}^Q(r)\big) + V_{\mathrm{Mad},Rs}\,, \qquad (86)$$

where the integration is carried out over the $\boldsymbol{R}$-th atomic sphere. The first term in (86) is the Coulomb potential due to the point-like nuclear charge Z_R^Q, the second term is the Hartree potential due to the spherically symmetric charge density $\tilde{\varrho}_R^Q(r)$, the third term represents the exchange-correlation contribution, and the last term is the Madelung contribution. The exchange-correlation term is evaluated according to a standard relation

$$V_{\mathrm{xc},\sigma}(\varrho_\uparrow, \varrho_\downarrow) = \frac{\partial}{\partial \varrho_\sigma} \left[(\varrho_\uparrow + \varrho_\downarrow) \, \varepsilon_{\mathrm{xc}}(\varrho_\uparrow, \varrho_\downarrow) \right] , \tag{87}$$

where $\varepsilon_{\mathrm{xc}}(\varrho_\uparrow, \varrho_\downarrow)$ is the exchange-correlation energy per particle of a spin-polarized homogeneous electron gas. The Madelung contribution $V_{\mathrm{Mad},Rs}$ in (86) is a special case (for $L = (\ell, m) = (0,0)$) of the multipole Madelung terms defined by

$$V_{\mathrm{Mad},RL} = \sum_{R'L'}{}' M_{RL,R'L'} \, \bar{q}_{R'L'} , \tag{88}$$

where the primed sum indicates exclusion of the term $R' = R$. The constants $M_{RL,R'L'}$ in (88) describe the electrostatic interactions between two multipoles located at the sites R, R' and the quantities $\bar{q}_{RL}$ are average multipole moments due to the total (electronic and nuclear) charge densities inside the atomic spheres,

$$\bar{q}_{RL} = \sum_Q c_R^Q \, q_{RL}^Q ,$$

$$q_{RL}^Q = \sqrt{\frac{4\pi}{2\ell+1}} \int_{(R)} r^\ell \, Y_L(\hat{r}) \, \varrho_R^Q(r) \, d^3r - Z_R^Q \, \delta_{\ell,0} . \tag{89}$$

Let us note that $\bar{q}_{Rs}$ and q_{Rs}^Q ($\ell = 0$ in (89)) refer to the net charges inside the R-th sphere. The summations in (88) for infinite lattices with two- or three-dimensional translational symmetry can be performed using the corresponding Ewald techniques [15,18,30]. For bulk systems the Madelung contribution (88) is often calculated only from the net charges $\bar{q}_{Rs}$, whereas for surfaces an inclusion of the dipole moments is inevitable, e.g., for a good description of the surface dipole barrier and the work function [30].

For calculations of charge densities, the energy dependence of the regular radial amplitude in (3) is replaced by a truncated Taylor expansion at an energy $E_{\nu,R\ell\sigma}^Q$ in the center of the occupied part of the valence band

$$\varphi_{R\ell\sigma}^Q(r, E) = \phi_{R\ell\sigma}^Q(r) + \dot{\phi}_{R\ell\sigma}^Q(r) \, (E - E_{\nu,R\ell\sigma}^Q)$$

$$+ \frac{1}{2} \ddot{\phi}_{R\ell\sigma}^Q(r) \, (E - E_{\nu,R\ell\sigma}^Q)^2 , \tag{90}$$

which results in a simple expression for the spherically averaged charge density (84)

$$\tilde{\varrho}_{R\sigma}^Q(r) = \frac{1}{4\pi} \sum_\ell \left\{ m_{R\ell\sigma}^{Q,0} \left(\phi_{R\ell\sigma}^Q(r) \right)^2 + 2 \, m_{R\ell\sigma}^{Q,1} \, \phi_{R\ell\sigma}^Q(r) \, \dot{\phi}_{R\ell\sigma}^Q(r) \right.$$

$$\left. + m_{R\ell\sigma}^{Q,2} \left[\left(\dot{\phi}_{R\ell\sigma}^Q(r) \right)^2 + \phi_{R\ell\sigma}^Q(r) \, \ddot{\phi}_{R\ell\sigma}^Q(r) \right] \right\}$$

$$+ \varrho_{R\sigma}^{Q,\mathrm{core}}(r) . \tag{91}$$

In (91) the last term denotes the core contribution while the quantities $m_{R\ell\sigma}^{Q,k}$ $(k = 0, 1, 2)$ represent the lowest energy moments of the $QR\ell\sigma$-projected valence densities of states (41)

$$m_{R\ell\sigma}^{Q,k} = \int_{E_B}^{E_F} \left(E - E_{\nu,R\ell\sigma}^Q\right)^k \sum_{m=-\ell}^{\ell} n_{RL\sigma}^Q(E) \, dE \,, \qquad (92)$$

where E_B denotes the bottom of the valence band. Similarly, the multipole moments q_{RL}^Q (89) reduce to several radial and energy integrations [15,30]. The latter are of the type $(k, k' = 0, 1, 2)$

$$m_{R,LL',\sigma}^{Q,kk'} = \int_{E_B}^{E_F} \left(E - E_{\nu,R\ell\sigma}^Q\right)^k n_{R,LL',\sigma}^Q(E) \left(E - E_{\nu,R\ell'\sigma}^Q\right)^{k'} dE \qquad (93)$$

representing thus the lowest energy moments of the local density of states matrix $n_{R,LL',\sigma}^Q(E)$ (40).

As follows from (40), the energy integrals (92, 93) over the occupied part of the valence spectrum can be generally formulated as

$$-\frac{1}{\pi} \int_{E_B}^{E_F} \text{Im} \, F(E + i0) \, dE = \frac{1}{2\pi i} \int_C F(z) \, dz \,. \qquad (94)$$

The function $F(z)$ is an analytic function of the complex energy variable z (except at poles and/or branch cuts lying on the real energy axis) which satisfies $F(z^*) = F^*(z)$. The r.h.s. integral in (94) is taken along a closed contour C intersecting the real energy axis at the Fermi level and enclosing the occupied valence band. Standard quadrature techniques lead to an approximation

$$\frac{1}{2\pi i} \int_C F(z) \, dz \approx \text{Re} \left[\sum_{n=1}^N w_n \, F(z_n) \right], \qquad (95)$$

which replaces the original integral along the real axis by a finite sum with N complex weights w_n and nodes $z_n \in C$. All nodes z_n can be chosen in the upper (or the lower) complex halfplane. They are usually taken along a semicircle contour with a denser mesh near the Fermi energy. Experience shows that a relatively modest number of nodes ($N \sim 10$ to 20) is sufficient to achieve desired accuracy in most charge selfconsistent calculations. Minor complications arise in selfconsistent bulk calculations in which the Fermi energy E_F is unknown and changes in each iteration (contrary to the case of surfaces where the value of E_F is fixed from a previous calculation of the bulk substrate). Fortunately, it is not necessary to locate the bulk Fermi level exactly in each iteration but merely to update its value so that the convergence of E_F proceeds simultaneously with that of the one-electron potentials.

Iterative procedures leading to selfconsistent one-electron potentials (or charge densities) have been recently systematically accelerated by means of quasi-Newton methods (like the Anderson and the second Broyden mixing scheme) [31]. These techniques are efficient also for random alloy systems where the LSDA-CPA self-consistency can be achieved by alternating updates of the one-electron potentials and the coherent interactors (Sect. 3.4). According to our experience, the full convergence in all-electron calculations can be obtained in 30 to 80 iterations for most systems.

The total energy for non-random systems within the ASA [17,20,30] can be directly generalized to the case with substitutional randomness. The final formula is given by a concentration-weighted sum of RQ-dependent terms

$$\mathcal{E} = \sum_{RQ} c_R^Q \, \mathcal{E}_R^Q \,, \tag{96}$$

where the individual contributions $\mathcal{E}_R^Q$ are explicitly given by

$$\mathcal{E}_R^Q = \sum_{\sigma j} \epsilon_{R\sigma j}^{Q,\text{core}} + \sum_{L\sigma} \int_{E_B}^{E_F} E \, n_{RL\sigma}^Q(E) \, dE$$

$$- \sum_{\sigma} \int_{(R)} \tilde{\varrho}_{R\sigma}^Q(r) \, V_{R\sigma}^Q(r) \, d^3r$$

$$+ \int_{(R)} \tilde{\varrho}_R^Q(r) \left[\varepsilon_{\text{xc}}\big(\tilde{\varrho}_{R\uparrow}^Q(r), \tilde{\varrho}_{R\downarrow}^Q(r)\big) - 2 \, Z_R^Q \, r^{-1} \right] d^3r$$

$$+ \int_{(R)} \int_{(R)} \tilde{\varrho}_R^Q(r) \, \tilde{\varrho}_R^Q(r') \, |r - r'|^{-1} \, d^3r \, d^3r'$$

$$+ \frac{1}{2} \sum_{L} q_{RL}^Q \, V_{\text{Mad},RL} \,. \tag{97}$$

The first term in (97) is the sum of core eigenvalues $\epsilon_{R\sigma j}^{Q,\text{core}}$ labeled by j, while the second term represents an energy contribution due to the valence densities of states $n_{RL\sigma}^Q(E)$ (41). It can be trivially expressed in terms of the moments $m_{R\ell\sigma}^{Q,k}$ (92).

Let us note that the above presented formulas for the one-electron potentials (86) and for the total-energy contributions (97) were derived under a complete neglect of any correlations (i) between the occupation of a particular site R and the charge densities inside the other atomic spheres, and (ii) between the charge densities inside different atomic spheres. These neglected correlations result then in the component-independent Madelung terms $V_{\text{Mad},RL}$ (88) due to the average multipole moments $\bar{q}_{R'L'}$ (89). This simple treatment is fully compatible with the mean-field nature of the single-site CPA.

However, it has been found in a number of applications of the CPA that the neglected charge correlations lead to substantial errors in the calculated total energies. Several schemes were suggested to remove this drawback. Let us consider for simplicity only the case of random binary bcc or fcc alloys. The condition of the overall charge neutrality together with the neglect of higher multipole moments leads to a vanishing mean-field Madelung contribution $V_{\text{Mad},s}$

to the one-electron potential (the site index $\boldsymbol{R}$ is omitted). The screened CPA [32] and the screened impurity model [33] lead to a component-dependent Madelung term

$$V_{\text{Mad},s}^{Q} = -2\, q_{s}^{Q}\, d_{\text{nn}}^{-1}\,, \tag{98}$$

where d_{nn} denotes the distance between the nearest neighboring sites of the lattice. This shift of the one-electron potentials follows from an assumption of a perfect screening of the net charge q_{s}^{Q} by compensating charges located on the nearest neighbors. The correction to the total alloy energy per lattice site is then given by

$$\Delta \mathcal{E}_{1} = \beta \sum_{Q} c^{Q}\, q_{s}^{Q}\, V_{\text{Mad},s}^{Q}\,, \tag{99}$$

where the prefactor β equals $1/2$ for the screened CPA [32,34] whereas for the screened impurity model the whole interval $1/2 \leq \beta \leq 1$ was considered [33,35,36]. Another approach employs an idea of neutral atomic spheres [37,38] where the sphere radii s^{Q} are changed (keeping the average atomic volume fixed) to achieve vanishing net charges ($q_{s}^{Q} = 0$) for both alloy components. All of these schemes improve considerably the calculated total energies for many alloy systems but a detailed assessment of their validity especially for alloy surfaces remains yet to be done.

6   Extensions and Applications of the LMTO-CPA

The non-relativistic TB-LMTO-CPA theory of substitutionally disordered alloys can be generalized to include properly all relativistic effects based on the Dirac equation. The relativistic theory in the non-magnetic case represents a straightforward modification of the non-relativistic counterpart [15,39] whereas for spin-polarized systems certain theoretical as well as technical problems appear [15,40]. Nevertheless, many of the theoretical concepts introduced above remain valid.

The energetics of metallic alloys and their surfaces with applications to ordering and segregation phenomena is usually studied in terms of effective interatomic interactions. They can be determined from ab initio electronic structure calculations using either the generalized perturbation method [23] or the Connolly-Williams inversion scheme [41]. In the context of the LMTO-CPA theory, the generalized perturbation method was described in [15,42] and reviewed in [43], while a modification of the Connolly-Williams approach was developed in [44].

Recent applications of the selfconsistent LMTO-CPA method cover a large area of the modern theory of alloys. The ground-state properties of non-magnetic bulk random alloys were investigated, e.g., in [33,35,38], while the Fermi surfaces and the electronic topological transitions were studied in [45]. Existing applications to magnetic bulk alloys include studies of the local magnetic moments

[40,46], various aspects of the Invar alloys [47,48], the structural stability [49,50], the ordering tendencies [49,51,52], and the Curie temperatures [43].

The electronic structure of surfaces of random alloys was investigated, e.g., in [39,53–55], studies of the surface segregation were published in [42–44,54,56,57], and calculations of the surface magnetic properties of random alloys were presented in [58,59].

Two-dimensional random alloys which can be formed at an epitaxial interface of two different metals represent another field of applicability of the LMTO-CPA method. The electronic structure of non-magnetic random overlayers on metallic substrates was calculated, e.g., in [55,60,61] while random magnetic overlayers on non-magnetic substrates were studied in [46,62,63]. The adlayer core-level shifts of random overlayers were calculated in [64], the ordering tendencies in surface non-magnetic alloys were analysed in [65], the interplay of magnetism and ordering was considered in [51], and the stability of metallic interfaces was investigated in [66].

The interlayer exchange coupling, encountered in epitaxial magnetic multilayers, is another quantity which can be influenced by substitutional disorder both in the magnetic layers and in the non-magnetic spacer. Applications of the LMTO-CPA to this problem can be found in the review [67] and references therein.

The formalism presented in this paper as well as the applications listed above are heavily based on the ASA. A development of a full-potential version of the LMTO-CPA is difficult due to the dependence of each LMTO on the occupation of all lattice sites. This complicated configuration dependence of the LMTO's can be removed in the so-called pure-L approximation for the TB-LMTO's [68] and the corresponding single-site CPA theory can be then derived. This was done in [69,70] together with applications to random fcc Li-Al and Ni-Pt bulk alloys.

Acknowledgements This work is a part of activities of the Center for Computational Materials Science sponsored by the Academy of Sciences of the Czech Republic. This research was supported by the Grant Agency of the Czech Republic (Project 202/97/0598), the Grant Agency of the Academy of Sciences of the Czech Republic (Project A1010829), and the Czech Ministry of Education, Youth, and Sports (in the framework of the COST Action P3 'Simulation of physical phenomena in technological applications').

References

1. P. Weinberger, Electron Scattering Theory for Ordered and Disordered Matter (Clarendon Press, Oxford, 1990).
2. A. Gonis, Green Functions for Ordered and Disordered Systems (North-Holland, Amsterdam, 1992).
3. O.K. Andersen, Phys. Rev. B **12**, 3060 (1975).
4. H.L. Skriver, The LMTO Method (Springer, Berlin, 1984).

5. D.J. Singh, Planewaves, Pseudopotentials and the LAPW Method (Kluwer Academic Publishers, Boston, 1994).
6. H. Eschrig, Optimized LCAO Method and the Electronic Structure of Extended Systems (Springer, Berlin, 1989).
7. P. Soven, Phys. Rev. **156**, 809 (1967).
8. B. Velický, S. Kirkpatrick, and H. Ehrenreich, Phys. Rev. **175**, 747 (1968).
9. J.A. Blackman, D.M. Esterling, and N.F. Berk, Phys. Rev. B **4**, 2412 (1971).
10. J.S. Faulkner and G.M. Stocks, Phys. Rev. B **21**, 3222 (1980).
11. J.S. Faulkner, Prog. Mater. Sci. **27**, 1 (1982).
12. R. Richter, H. Eschrig, and B. Velický, J. Phys. F: Met. Phys. **17**, 351 (1987).
13. K. Koepernik, B. Velický, R. Hayn, and H. Eschrig, Phys. Rev. B **55**, 5717 (1997).
14. J. Kudrnovský and V. Drchal, Phys. Rev. B **41**, 7515 (1990).
15. I. Turek, V. Drchal, J. Kudrnovský, M. Šob, and P. Weinberger, Electronic Structure of Disordered Alloys, Surfaces and Interfaces (Kluwer Academic Publishers, Boston, 1997).
16. O.K. Andersen and O. Jepsen, Phys. Rev. Lett. **53**, 2571 (1984).
17. O.K. Andersen, O. Jepsen, and M. Šob, in: Electronic Band Structure and Its Applications, edited by M. Yussouff (Springer, Berlin, 1987) p. 1.
18. V. Drchal, J. Kudrnovský, and I. Turek, Comput. Phys. Commun. **97**, 111 (1996).
19. C. Koenig and E. Daniel, J. Physique Lettres **42**, L 193 (1981).
20. O. Gunnarsson, O. Jepsen, and O.K. Andersen, Phys. Rev. B **27**, 7144 (1983).
21. V. Drchal, J. Kudrnovský, L. Udvardi, P. Weinberger, and A. Pasturel, Phys. Rev. B **45**, 14328 (1992).
22. F. Ducastelle, J. Phys. C: Solid State Phys. **8**, 3297 (1975).
23. F. Ducastelle and F. Gautier, J. Phys. F: Met. Phys. **6**, 2039 (1976).
24. F. Ducastelle, Order and Phase Stability (North-Holland, Amsterdam, 1991).
25. K.C. Hass, B. Velický, and H. Ehrenreich, Phys. Rev. B **29**, 3697 (1984).
26. B. Velický and J. Kudrnovský, Surf. Sci. **64**, 411 (1977).
27. J. Kudrnovský, P. Weinberger, and V. Drchal, Phys. Rev. B **44**, 6410 (1991).
28. B. Wenzien, J. Kudrnovský, V. Drchal, and M. Šob, J. Phys.: Condens. Matter **1**, 9893 (1989).
29. M.P. López Sancho, J.M. López Sancho, and J. Rubio, J. Phys. F: Metal Phys. **15**, 851 (1985).
30. H.L. Skriver and N.M. Rosengaard, Phys. Rev. B **43**, 9538 (1991).
31. V. Eyert, J. Comput. Phys. **124**, 271 (1996).
32. D.D. Johnson and F.J. Pinski, Phys. Rev. B **48**, 11553 (1993).
33. P.A. Korzhavyi, A.V. Ruban, I.A. Abrikosov, and H.L. Skriver, Phys. Rev. B **51**, 5773 (1995).
34. F.J. Pinski, J.B. Staunton, and D.D. Johnson, Phys. Rev. B **57**, 15177 (1998).
35. A.V. Ruban, I.A. Abrikosov, and H.L. Skriver, Phys. Rev. B **51**, 12958 (1995).
36. I.A. Abrikosov and B. Johansson, Phys. Rev. B **57**, 14164 (1998).
37. A. Gonis, P.E.A. Turchi, J. Kudrnovský, V. Drchal, and I. Turek, J. Phys.: Condens. Matter **8**, 7869 (1996).
38. A. Gonis, P.E.A. Turchi, J. Kudrnovský, V. Drchal, and I. Turek, J. Phys.: Condens. Matter **8**, 7883 (1996).
39. V. Drchal, J. Kudrnovský, and P. Weinberger, Phys. Rev. B **50**, 7903 (1994).
40. A.B. Shick, V. Drchal, J. Kudrnovský, and P. Weinberger, Phys. Rev. B **54**, 1610 (1996).
41. J.W.D. Connolly and A.R. Williams, Phys. Rev. B **27**, 5169 (1983).
42. V. Drchal, J. Kudrnovský, A. Pasturel, I. Turek, and P. Weinberger, Phys. Rev. B **54**, 8202 (1996).

43. V. Drchal, J. Kudrnovský, A. Pasturel, I. Turek, P. Weinberger, A. Gonis, and P.E.A. Turchi, in: Tight-Binding Approach to Computational Materials Science, edited by P.E.A. Turchi, A. Gonis, and L. Colombo, MRS Symp. Proc. Vol. **491** (Materials Research Society, Warrendale, 1998), p. 65.

44. A.V. Ruban, I.A. Abrikosov, D. Ya. Kats, D. Gorelikov, K.W. Jacobsen, and H.L. Skriver, Phys. Rev. B **49**, 11383 (1994).

45. N.V. Skorodumova, S.I. Simak, I.A. Abrikosov, B. Johansson, and Yu. Kh. Vekilov, Phys. Rev. B **57**, 14673 (1998).

46. I. Turek, J. Kudrnovský, V. Drchal, and P. Weinberger, Phys. Rev. B **49**, 3352 (1994).

47. I.A. Abrikosov, O. Eriksson, P. Söderlind, H.L. Skriver, and B. Johansson, Phys. Rev. B **51**, 1058 (1995).

48. R. Hayn and V. Drchal, Phys. Rev. B **58**, 4341 (1998).

49. I.A. Abrikosov, P. James, O. Eriksson, P. Söderlind, A.V. Ruban, H.L. Skriver, and B. Johansson, Phys. Rev. B **54**, 3380 (1996).

50. P. James, I.A. Abrikosov, O. Eriksson, and B. Johansson, in: Properties of Complex Inorganic Solids, edited by A. Gonis, A. Meike, and P.E.A. Turchi (Plenum Press, New York, 1997), p. 57.

51. J. Kudrnovský, I. Turek, A. Pasturel, R. Tetot, V. Drchal, and P. Weinberger, Phys. Rev. B **50**, 9603 (1994).

52. S.K. Bose, V. Drchal, J. Kudrnovský, O. Jepsen, and O.K. Andersen, Phys. Rev. B **55**, 8184 (1997).

53. J. Kudrnovský, I. Turek, V. Drchal, P. Weinberger, S.K. Bose, and A. Pasturel, Phys. Rev. B **47**, 16525 (1993).

54. I.A. Abrikosov and H.L. Skriver, Phys. Rev. B **47**, 16532 (1993).

55. J. Kudrnovský, V. Drchal, S.K. Bose, I. Turek, P. Weinberger, and A. Pasturel, Comput. Mater. Sci. **2**, 379 (1994).

56. I.A. Abrikosov, A.V. Ruban, H.L. Skriver, and B. Johansson, Phys. Rev. B **50**, 2039 (1994).

57. A. Christensen, A.V. Ruban, P. Stoltze, K.W. Jacobsen, H.L. Skriver, J.K. Nørskov, and F. Besenbacher, Phys. Rev. B **56**, 5822 (1997).

58. I. Turek, J. Kudrnovský, M. Šob, and V. Drchal, in: Stability of Materials, edited by A. Gonis, P.E.A. Turchi, and J. Kudrnovský (Plenum Press, New York, 1996), p. 431.

59. I. Turek, S. Blügel, and J. Kudrnovský, Phys. Rev. B **57**, R11065 (1998).

60. J. Kudrnovský, I. Turek, V. Drchal, P. Weinberger, N.E. Christensen, and S.K. Bose, Phys. Rev. B **46**, 4222 (1992).

61. M.V. Ganduglia-Pirovano, J. Kudrnovský, I. Turek, V. Drchal, and M.H. Cohen, Phys. Rev. B **48**, 1870 (1993).

62. J. Kudrnovský, I. Turek, V. Drchal, and P. Weinberger, Prog. Surf. Sci. **46**, 159 (1994).

63. I. Turek, J. Kudrnovský, M. Šob, V. Drchal, and P. Weinberger, Phys. Rev. Lett. **74**, 2551 (1995).

64. M.V. Ganduglia-Pirovano, J. Kudrnovský, and M. Scheffler, Phys. Rev. Lett. **78**, 1807 (1997).

65. J. Kudrnovský, S.K. Bose, and V. Drchal, Phys. Rev. Lett. **69**, 308 (1992).

66. A.M.N. Niklasson, I.A. Abrikosov, and B. Johansson, Phys. Rev. B **58**, 3613 (1998).

67. J. Kudrnovský, V. Drchal, I. Turek, P. Bruno, P. Dederichs, and P. Weinberger, these Proceedings.

68. O.K. Andersen, Z. Pawlowska, and O. Jepsen, Phys. Rev. B **34**, 5253 (1986).

69. Prabhakar P. Singh and A. Gonis, Phys. Rev. B **48**, 1989 (1993).
70. Prabhakar P. Singh and A. Gonis, Phys. Rev. B **48**, 2139 (1993).

Locally Self-Consistent Green's Function Method and Its Application in the Theory of Random Alloys

I. A. Abrikosov, P. A. Korzhavyi, and B. Johansson

Condensed Matter Theory Group, Physics Department,
Uppsala University, S-75121 Uppsala, Sweden

Abstract. A formulation of the order-N locally self-consistent Green's function, LSGF, method in conjunction with the linear muffin-tin orbital (LMTO) basis set is discussed. The method is particularly suitable for calculating the electronic structure of systems with an arbitrary distribution of atoms of different kinds on an underlying crystal lattice. We show that in the framework of the tight-binding representation it can be generalized to systems without ideal three-dimensional symmetry of the underlying lattice, like, for instance, alloys with local lattice relaxations or surface alloys. We also show that multipole corrections to the atomic sphere approximation can be easily incorporated into the formalism. Thus, the method represents a powerful tool for studing different problems within alloy theory.

1 Introduction

Recent research in solid state physics has shown a number of encouraging results for the investigation of physical properties of metallic alloys. In particular, the computational schemes which allow one to treat ordered, as well as random alloys, their surfaces and interfaces have been developed and applied with a great success. This has led to a much deepened understanding of the behavior of thermodynamic and magnetic properties, structural and phase stabilities, impurity, surface and segregation energies through the transition metal series [1–5]. On the other hand, first-principles investigations are still limited to certain ideal systems, like, for instance, completely ordered or completely random alloys, while for materials and problems of technological importance these studies are quite rare. A possible improvement of this circumstance consists in the development of more efficient computational schemes, for example, methods that scale linearly with increasing number of atoms in the system (order-N methods), thereby allowing a study of more realistic systems.

The problem is schematically illustrated in Fig. 1. In the framework of density functional theory (DFT) [6] our purpose is to solve the Kohn-Sham equations [7] for an infinite system of atoms. This set of effective one-electron equations is conventionally solved with a particular choice of basis functions. However, such an approach relies on the Bloch theorem, and thus requires ideal three-dimensional periodicity of the system at hand. If this is not the case, the periodicity is usually imposed artificially by considering only a finite part of the original system, the so called supercell, subject to periodic boundary conditions (Fig. 1a). This allows one to construct a Hamiltonian matrix which upon Fourier transformation to reciprocal k-space has a dimension proportional to the number of atoms N in the supercell. But due to the famous $O(N^3)$ scaling problem (i.e. the computational time increases as N^3 with the number of atoms N in the supercell) the size of the cell is often limited by the computer power rather than by the physical problem itself. On the other hand, to account for short range order effects in alloys [8,9] or to calculate interaction energies between point defects in a metal [10] one needs supercells with more than a hundred of atoms. For such big supercells conventional approaches are not at all efficient, and scaling properties of the computational technique must be improved.

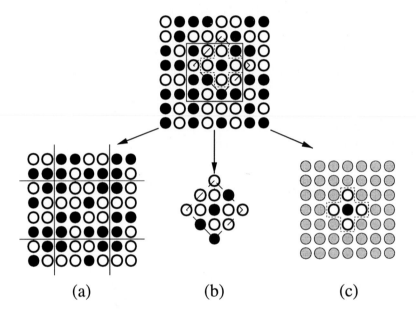

(a)   (b)   (c)

Fig. 1. Three ways to calculate electronic structure of infinite system composed of chemically nonequivalent atoms (*filled and open circles*) with an arbitrary degree of disorder. (**a**) In the framework of the supercell method one chooses some part of the original system (indicated by a *full line*), and repeats it periodically. (**b**) Conventional $O(N)$ methods are based on the direct solution of the electronic structure problem for a finite part of the original system (*dashed line*), the local interaction zone, centered at all the sites of this system. (**c**) Within the LSGF method the boundary conditions for the multiple scattering problem inside a LIZ are substantially improved by embedding the LIZ into a self-consistent effective medium (*gray circles*). As a result the size the LIZ is greatly reduced (*dotted line*)

Within the last decade considerable attention has been devoted to the development of so-called $O(N)$ methods. Most of them are based on the principle of nearsightedness [11], illustrated in Fig. 1b. Similar to the supercell method, one considers a finite subsystem of the original system, the so called local interaction zone (LIZ) [12,13]. But rather than introducing periodic boundary conditions, one solves the electronic structure problem for the particular LIZ, and then proceeds to the next LIZ, thus dividing the N-atom problem into N more or less independent problems for each LIZ. Such a procedure can be partly justified by the experience gained in the application of real-space cluster methods in electronic structure calculations that shows that for a large cluster the properties of an atom deep inside the cluster are very close to those given by band structure methods. Also, it guarantees linear scaling of the computational efforts with N.

Several $O(N)$ computational techniques differ from each other mainly by the methods of solving the electronic structure problem for the LIZ. For example, in Refs. [14–18] this is done by employing localized orbitals, Refs. [19–23] take advantages of the density matrix formulation, in Refs. [24,25] tigh-binding (TB) representation is used, while the techniques presented in Refs. [12,13,26,27] are based on a Green's function approach. However, if there are M atoms in the LIZ, then the computational effort required for an exact solution of the electronic structure problem inside a particular LIZ scales as M^3. Thus the total computational time scales as M^3N. One can see from Fig. 2 that there is a certain minimum number of atoms $\tilde{N}$ for which the $O(N)$ methods become more efficient than conventional $O(N^3)$ methods. Thus, the problem with $O(N)$ methods is not only to achieve the linear scaling of the computational efforts with increasing number of atoms, but also to minimize the size of the LIZ. Unfortunately, applications of the above mentioned linear scaling methods are very limited due to the fact that the size of their LIZ must in general be chosen quite large to give reliable description of the electronic properties, especially for metals [12,13].

The origin of this drawback can be understood as follows. From an illustration in Fig. 1b one can see that all the information about the system beyond the LIZ is totally neglected. Therefore, one needs to keep too many atoms directly in the LIZ in order not to loose essential physical information about the original system. Thus, there is a question if one can make the central atom of the LIZ more nearsighted. In Refs. [8,28] we have suggested that one can do this if one keeps at least some information about the system beyond the truncation region in the form of an effective medium (Fig. 1c). And, of course, the more information is kept, the better the convergence will be with respect to the size of the LIZ. Therefore, we have suggested to choose the effective medium that describes properties of the original system on average as accurate as possible. It is also clear that the symmetry of the effective medium may be much higher than the symmetry of the system under consideration. In Refs. [8,28] we have demonstrated how the computational effort of $O(N)$ calculations may be considerably reduced by embedding the LIZ in this effective medium (see Fig. 1c). Such an embedding may be established by means of the Dyson equation connecting the desired Green's function to the Green's function of the reference system.

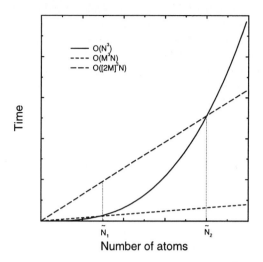

Fig. 2. Schematic representation of computational efforts required to solving the electronic structure problem by a conventional $O(N^3)$ method (*full line*), as well as by $O(N)$ methods with different number of atoms, M (*dashed line*) and $2M$ (*long-dashed line*), in the local interaction zone

In the present paper we first review the main ideas of the locally self-consistent Green's function (LSGF) technique proposed by Abrikosov *et al.* [8,28]. We then discuss some resent developments, in particular, multipole corrections to the atomic sphere approximation (ASA) [29,30] for the one-electron potential. We will next show how the tight-binding representation of the linear muffin-tin orbital method [31–35] can be incorporated in our Green's function technique. This will allow us to formulate general equations for systems without ideal three-dimensional symmetry of the underlying lattice, like, for instance, alloys with local lattice relaxations or surface alloys. Applications of the LSGF method to several problems of solid state physics will also be briefly discussed.

2   Locally Self-Consistent Green's Function Method

2.1   Assumptions and Definitions

We will formulate the locally self-consistent Greens function method in the framework of the density functional theory and the local spin density approximation (LSDA) [6,7] in conjunction with a linear muffin-tin orbitals basis in the atomic sphere approximation of Andersen [31–35] including multipole correction terms [29,30], ASA+M. In this section we will consider the problem of calculating the electronic structure of a system of N atoms illustrated in Fig. 1. Though this is not necessary, we assume that the original system is represented by a supercell subjected to periodic boundary conditions. This assumption allows us to calculate easily the long-ranged electrostatic contributions to the one-electron potential and energy, and to construct appropriate LIZ for atoms close to the boundary of the supercell. We allow for an arbitrary distribution of the atoms that may be of different types on the sites of the underlying crystal lattice. However, in this section we do not allow these lattice sites to deviate from their ideal positions, i.e. we neglect the so-called local relaxation effects. This problem will be considered in Sec. 3.2.

We will assign to each site of the supercell a corresponding atomic sphere (AS) centered at radius-vector $\mathbf{R}$. Inside each AS we will define the electron density $\rho_{\mathbf{R}}(\mathbf{r})$ and one-electron potential $v_{\mathbf{R}}(\mathbf{r})$. Also the potential function $P_{\mathbf{R}l}^{\alpha}(z)$

$$P_{\mathbf{R}l}^{\alpha}(z) = \frac{z - C_{\mathbf{R}l}}{(z - C_{\mathbf{R}l})(\gamma_{\mathbf{R}l} - \alpha_l) + \Delta_{\mathbf{R}l}} \tag{1}$$

for an arbitrary (complex) energy z and angular-momentum quantum number l may be assigned to each site. It is expressed by means of the band center C, bandwidth Δ, and the γ LMTO potential parameters calculated at an arbitrary energy $\varepsilon_{\nu \mathbf{R}l}$ in the energy range of interest [33]. Index α denotes the LMTO representation, and in the following discussion it will be used only when relevant. The so-called KKR-ASA Green's function matrix $\mathbf{g}$ is defined like

$$[\mathbf{P}(z) - \mathbf{S}(\mathbf{k})]\mathbf{g}(\mathbf{k}, z) = \mathbf{1}, \tag{2}$$

where $\mathbf{P}, \mathbf{S}$ and $\mathbf{g}$ are $(\mathbf{R}L, \mathbf{R}'L')$ matrices, L is the combined angular-momentum quantum numbers (l, m), and the structure constant matrix $\mathbf{S}$ contains all the information about the crystal structure. The real space KKR-ASA Green's function matrix $g_{\mathbf{R}L, \mathbf{R}'L'}(z)$ is obtained from $\mathbf{g}(\mathbf{k})$ by integration over the Brillouin zone

$$g_{\mathbf{R}L, \mathbf{R}'L'}(z) = (V_{BZ})^{-1} \int_{BZ} d\mathbf{k} \, e^{i\mathbf{k} \cdot (\mathbf{T} - \mathbf{T}')} g_{\mathbf{U}L, \mathbf{U}'L'}(\mathbf{k}, z), \tag{3}$$

where V_{BZ} is the volume of the Brillouin zone, $\mathbf{U}$ is a basis vector of the unit cell. It is connected to the lattice site $\mathbf{R}$ by a translation $\mathbf{T}$, i.e., $\mathbf{R} = \mathbf{U} + \mathbf{T}$. The on-site element $g_{\mathbf{R}L, \mathbf{R}L'}(z)$ is a key quantity which determines both the electron density and the density of states at the $\mathbf{R}$-th site. Therefore, our purpose is to calculate these matrix elements for all the sites of the supercell. In a conventional Green's function technique this requires an inversion of Eq. (2) at each k-point in the Brillouin zone and for each energy point z. This is essentially the operation that results in $O(N^3)$ scaling of the problem. In the following we will show how this problem can be solved without a direct inversion of Eq. (2) in the framework of the LSGF method.

2.2   Concept of Local Self-Consistency and Effective Medium

The concept of local self-consistency is based on the suggestion that the electron density and the density of states on a particular atom within a large condensed system may be obtained with sufficient accuracy by considering only the electronic multiple scattering processes in a finite spatial region centered at that atom. This concept has been applied by by Nicholson *et al.* [12,13] in the framework of the locally self-consistent multiple scattering (LSMS) method. Within this method each atom of a system is surrounded by few shells of nearest neighbors (see Fig. 1b), the local interaction zone, and the multiple scattering problem is solved in the real space for each LIZ independently of the other LIZs. The method was proven to have essentially $O(N)$ scaling properties with increasing supercell size. Nevertheless, it turns out that the minimal size of the LIZ should be about 100 atoms and even more to provide the accuracy of order 0.1 *mRy/atom* for the total energy. It is quite clear that the reliable description of the central site scattering properties will be achieved only when the size of the LIZ is so large, that an atom at this site becomes completely insensitive to what is happening beyond the LIZ. In particular, if one adds an extra atom at the LIZ's boundary, this should have no effect on the central site.

The concept of local self-consistency has been further developed by Abrikosov *et al.* [8,28] in the locally self-consistent Green's function method (LSGF), where the concept of an effective medium has been combined together with the LIZ concept. This resulted in the development of an efficient $O(N)$ technique. The combination of these two concepts can be most easily illustrated by the example of a substitutional alloy with an arbitrary degree of short and long range order.

In choosing the effective medium we tried to satisfy the following main criteria, namely its scattering properties as viewed by the central atom of the LIZ must be as close as possible to those of the original system at the shortest possible distance. Additionally, it has to be as simple (i.e. symmetric) as possible. In the case of the above mentioned substitutional alloy it is clear that a very good representation of the real system beyond the LIZ would be a completely random alloy. In particular, one can expect that the convergence with respect to the LIZ size will be achieved as soon as the central site becomes insensitive to the interchange of positions of two atoms at the LIZ's boundary, in contrast to a requirement of the complete insensitivity in conventional $O(N)$ schemes. Also, the random alloy has on the average the highest possible symmetry, and this will allow us to perform at least partly calculations in the reciprocal space of the effective lattice which in its own turn has a minimal number of atoms in the unit cell.

Now, there is a question of how to represent mathematically the effective medium in the form of a completely random alloy. Based on the experience gained in the study of alloys Abrikosov *et al.* [8,28] have suggested to do this in the framework of the multicomponent generalization of the coherent-potential approximation (CPA). In fact, the CPA effective medium fulfills both the above mentioned criteria. Despite the fact that the CPA is a very simple single-site approximation, it has been shown to give density of states for random alloys in very good agreement with experiment as well as with more accurate calculations that go beyond the single-site approximation [36–38]. In addition, the CPA Green's function decays, apart from an oscillating factor, exponentially as $R^{-1}e^{-R/l}$, where l is the mean free path [2].

However, conventionally the CPA describes scattering property of a binary alloy in terms of two potentials, for A and B alloy component. In reality situation is different, and in a N atom system, even though it is composed of just two chemically inequivalent atoms, A and B, all the local potentials are in general inequivalent due to the differences in local environments of any site. Therefore, one must define the effective medium to be used in the embedding of the LIZ as that given by the CPA for a multicomponent alloy. The number of components at each sublattice will be equal to the number of equivalent positions in the supercell formed from the underlying lattice, i.e., in the simplest case of a monoatomic underlying lattice each atom in the supercell is considered to be a component of an N-atom alloy. We thereby assume that the atoms are randomly distributed on their sublattices and neglect the fact that they occupy definite positions in the system. The difference between different atoms (or alloy components) will enter through their one-electron potentials. This method has recently been elevated to an idea of polymorphous CPA [38].

2.3 Computational Algorithm

In this section we will describe in details the complete self-consistent procedure of the LSGF method within the LMTO basis set as applied to the problem of calculating the electronic properties of a paramagnetic system of N atoms in a supercell subjected to periodic boundary conditions, such as illustrated in Fig. 1. Generalization to systems with collinear magnetic moments is straightforward, but there also exists a generalization of the method for systems with arbitrary orientations of local magnetic moments [39].

The principal scheme is similar to any other Green's function technique [34,40–43], but there are certain new steps that are specific for the LSGF method. We start with a guess for the charge density of all the atoms in the system, calculated as a renormalized atomic density or by means of a conventional CPA calculation for a random alloy with the same composition as in the supercell. The first problem is to construct an effective medium (all parameters that characterize the latter are denoted by a tilde). This means we must determine the potential function for the effective scatterers $\tilde{P}$ and the effective Green's function $\tilde{g}$. Here we remark, that in some cases it turns out to be more efficient to chose the effective medium in the form of multisublattice alloy. In particular, it has been demonstrated in Ref. [28] that for a partially ordered Ni(Ni$_{9.375}$Al$_{90.625}$) alloy the convergence with respect to the size of the LIZ is substantially enhanced by substituting the most symmetric bcc effective medium for an effective medium on B2 underlying lattice with two nonequivalent types of the effective atoms, one for each nonequivalent sublattice. Thus, one must solve the following system of coupled single-site equations for the $\tilde{U}$ sites in the unit cell of the underlying lattice [43,44]

$$\tilde{g}_{\tilde{U}\tilde{U}} = (\tilde{V}_{BZ})^{-1} \int_{BZ} d\mathbf{k} \, ([\tilde{\mathbf{P}} - \tilde{\mathbf{S}}(\mathbf{k})]^{-1})_{\tilde{U}\tilde{U}}$$
$$g_{\mathbf{R}}^0 = \tilde{g}_{\tilde{U}\tilde{U}} + \tilde{g}_{\tilde{U}\tilde{U}}(\tilde{P}_{\tilde{U}} - P_{\mathbf{R}})g_{\mathbf{R}}^0 \qquad (4)$$
$$\tilde{g}_{\tilde{U}\tilde{U}} = (N_{\tilde{U}})^{-1} \sum_{\mathbf{R}} g_{\mathbf{R}}^0,$$

where $\tilde{U}$ denotes a particular sublattice of the underlying lattice, $\mathbf{R} \in \tilde{U}$, i.e., $\mathbf{R} = \tilde{U} + \tilde{\mathbf{T}}$, $\tilde{V}_{BZ}$ is the volume of the Brillouin zone of the underlying unit cell and the integration is performed over the corresponding Brillouin zone, $\tilde{\mathbf{S}}$ is the structure constant matrix of the underlying lattice. In Eq. (4) $N_{\tilde{U}}$ is a number of atoms at the corresponding sublattice of the effective lattice, i.e. $N_{\tilde{U}} = N$ for a Bravais lattice, while, for instance, for the B2 lattice $N_{\tilde{U}} = N/2$. Eq. (4) must be solved for each sublattice of the effective lattice. In practice it is solved by iterations with an initial guess for the effective potential function in the following form:

$$\tilde{P}_{\tilde{U}}^{-1} = (N_{\tilde{U}})^{-1} \sum_{\mathbf{R} \in \tilde{U}} P_{\mathbf{R}}^{-1} \qquad (5)$$

using an efficient procedure described in details in Ref. [28].

As has been specified in Sec. 2.1, the key quantity to be calculated in order to get access to all the electronic properties of a system is on-site elements of the Green's function matrix, $g_{\mathbf{R}L,\mathbf{R}L'}(z)$. In the framework of the LSGF method we calculate them using the concept of the local self-consistency, i.e. for each site of the supercell separately. This site is considered as a central site of a corresponding LIZ embedded in the effective medium constructed by the procedure described above (Fig 1c). The Green's function for the LIZ can be found by solving corresponding Dyson equation

$$g_{\mathbf{R}\mathbf{R}} = \tilde{g}_{\mathbf{R}\mathbf{R}} + \sum_{\mathbf{R}'=1}^{M} \tilde{g}_{\mathbf{R}\mathbf{R}'}(\tilde{P}_{\mathbf{R}'} - P_{\mathbf{R}'})g_{\mathbf{R}'\mathbf{R}}, \qquad (6)$$

where the sum runs over the M atoms in the LIZ around site $\mathbf{R}$. Eq. (6) has to be solved for all the sites in the original system. Thus, the problem of solving the Kohn-Sham equations for an N-atom system is decomposed into N linked locally self-consistent problems for the LIZ associated with each atom in the system. Note also that in Eq.(6) $\tilde{g}_{\mathbf{R}\mathbf{R}'}$ is a matrix element of the complete (diagonal, as well as off-diagonal) Green's function matrix of the effective medium, and it is calculated from the corresponding Brilluoin zone integral, Eq. (3), where $\mathbf{U}$ now represents a basis vector of the underlying lattice $\tilde{U}$. At each iteration and for any energy point this Green's function must be calculated only once for the entire system, and shall not be updated when moving the LIZ from one site to another due to the invariance of all the sites of the underlying lattice that belong to the same sublattice $\tilde{U}$ with respect to a translation $\tilde{T}$. Moreover, due to the high symmetry of the effective medium calculating integral (3) is not a time-consuming operation.

We remark that in general $g_{\mathbf{RR}'}$ calculated by the Dyson equation (6) will not correspond to the one of a real system, especially close to the LIZ boundary. But the on-site element for the central atom will approach that of the real atom at $\mathbf{R}$ for a sufficiently large LIZ. In this sense $g_{\mathbf{RR}}$ will be locally self-consistent. This also distinguishes a solution of Eq. (6) from a solution of a single-site CPA Eq. (4), thus justifying our use of different notations for the corresponding on-site elements of the Green's function in these equations.

After the on-site elements of the Green's function have been calculated for the entire system, i.e. Eq. (6) has been solved for all the sites $\mathbf{R}$, the remaining procedure is completely similar to any conventional Green's function technique [34,40–43]. We proceed by transforming the KKR-ASA Green's function g into the Hamiltonian Green's function G:

$$G^{\gamma}_{\mathbf{R}L,\mathbf{R}L'}(z) = \frac{1}{z - V^{\alpha}_{\mathbf{R}l}}\delta_{LL'} + \frac{\sqrt{\Gamma^{\alpha}_{\mathbf{R}l}}}{z - V^{\alpha}_{\mathbf{R}l}}g^{\alpha}_{\mathbf{R}L,\mathbf{R}L'}(z)\frac{\sqrt{\Gamma^{\alpha}_{\mathbf{R}l'}}}{z - V^{\alpha}_{\mathbf{R}l'}}, \tag{7}$$

where the LMTO-representation dependent potential parameters $V^{\alpha}_{\mathbf{R}l}$, $\Gamma^{\alpha}_{\mathbf{R}l}$ are

$$V^{\alpha}_{\mathbf{R}l} = C_{\mathbf{R}l} - \frac{\Delta_{\mathbf{R}l}}{\gamma_{\mathbf{R}l} - \alpha_l} \tag{8}$$

$$\Gamma^{\alpha}_{\mathbf{R}l} = \frac{\Delta_{\mathbf{R}l}}{(\gamma_{\mathbf{R}l} - \alpha_l)^2}.$$

Then the moments of the state density can be calculated

$$m^{q'q''}_{\mathbf{R}L'L''} = \frac{1}{2\pi i}\oint dz(z - \varepsilon_{\nu\mathbf{R}l'})^{q'}G^{\gamma}_{\mathbf{R}L',\mathbf{R}L''}(z)(z - \varepsilon_{\nu\mathbf{R}l''})^{q''} \tag{9}$$

and using the second-order Taylor expansion of the partial wave, $\phi_{\mathbf{R},l}(\varepsilon;r)$:

$$\phi_{\mathbf{R}l}(\varepsilon;r) \approx \phi_{\nu\mathbf{R}l}(r) + (\varepsilon - \varepsilon_{\nu\mathbf{R}l})\,\dot{\phi}_{\nu\mathbf{R}l}(r) + \frac{1}{2}(\varepsilon - \varepsilon_{\nu\mathbf{R}l})^2\,\ddot{\phi}_{\nu\mathbf{R}l}(r), \tag{10}$$

the valence charge density $n^{v}_{\mathbf{R}}(r)$ in the corresponding atomic spheres can be found as the one-centre expansion

$$n^{v}_{\mathbf{R}}(r) = (4\pi)^{-1}\sum_{L}\{[\phi_{\nu\mathbf{R}l}(r)]^2 m^{00}_{\mathbf{R}LL} + 2[\phi_{\nu\mathbf{R}l}(r)\dot{\phi}_{\nu\mathbf{R}l}(r)]m^{10}_{\mathbf{R}LL}$$

$$+ [\dot{\phi}_{\nu\mathbf{R}l}(r)\dot{\phi}_{\nu\mathbf{R}l}(r) + \phi_{\nu\mathbf{R}l}(r)\ddot{\phi}_{\nu\mathbf{R}l}(r)]m^{20}_{\mathbf{R}LL}\}. \tag{11}$$

In Eq. (9) the contour must enclose the occuped valence states, while $\phi_{\nu Rl}$, $\dot{\phi}_{\nu Rl}$, and $\ddot{\phi}_{\nu Rl}$ in Eqs. (10) and (11) denote the partial wave and its first and second energy derivatives, respectively, evaluated at the energy $\varepsilon_{\nu Rl}$. An iteration is completed by solving Poissons equation for the electrostatic potential and adding the exchange-correlation potential. In addition, in contrast to the conventional treatment of an alloy problem in the framework of the CPA, adding corrections to the ASA becomes meaningful within the LSGF scheme. This will be described in the next section. The entire procedure is then repeated until self-consistency.

2.4   Multipole Corrections to the ASA

A supercell approach allows one to calculate the electrostatic contributions to the one-electron potential and energy exactly by performing the Madelung summation, because the atomic sphere charges in that case are calculated explicitly [12,13,8]. However, in the ASA, the electron charge is usually assumed to have spherical symmetry inside each atomic sphere. This approximation works quite well if all the atomic positions have high-symmetry local coordinations. When the local symmetry is broken, this approximation becomes less appropriate, and calculations based on the atomic sphere approximation sometimes lead to substantial errors when calculating physical properties of solids. For example, the vacancy formation energy is usually overestimated by as much as a factor of 2 [45–47], and the work function is about 50 % higher than experimental values [29,4].

The problem of Madelung contributions naturally emerges if one divides the crystal space into atomic or muffin-tin spheres and tries to describe the electrostatic interaction between the interior of a sphere and the rest of the crystal by adding a certain, constant throughout the sphere, electrostatic shift to the one-electron potential. However, if one considers, for example, an isolated vacancy in a metal matrix, then one sees that vacancy itself has a very symmetrical local coordination in the case of a cubic or hexagonal crystal structure of the metal matrix. But the local coordination of its close neighbors is very unsymmetrical because one atom has been removed to form the vacancy. On each of the vacancy neighbors a dipole moment of the electron charge should appear as a result of the fact that atomic density tails penetrate into this atomic sphere from all of its neighbors except the vacant site. An additional monopole potential shift on the vacant site must be thus induced by the dipole moments of the surrounding atoms. It is clear, that this contribution is omitted if the spherical approximation for the electron density is used.

This simple example illustrates the importance of taking into account multipole Madelung contributions to the potential and energy when dealing with a system in which the symmetry of the local configurations is low. Otherwise, the system can be adequately described by only the monopole Madelung terms. It is, of course, also important to examine the relative contributions of the multipoles higher than the dipole, but it is clear that the dipole term is dominant in the case of a vacancy as well as for surface problems.

Probably the most consistent way of dealing with this problem would be in the framework of a full-potential approach. However, as has been shown in several works [29,4,48,49,30], one can expect to be able to determine an accurate total energy of a system by still using a spherical potential provided the spherical approximation to the electron charge density is lifted. The monopole ($L = 0$) Madelung contribution to the one-electron potential is then evaluated using the monopole as well as multipole components of the valence electron charge, $Q_L^{\mathbf{R}}$, as:

$$V_0^{\mathbf{R}} = \frac{1}{S} \sum_{\mathbf{R}',L'} M_{0L'}^{\mathbf{R}\mathbf{R}'}\, Q_{L'}^{\mathbf{R}'}, \tag{12}$$

where $M_{LL'}^{\mathbf{R}\mathbf{R}'}$ is the multipole Madelung matrix which is equivalent to the conventional (unscreened) LMTO structure constants for the *entire* supercell, and S is the average Wigner-Seitz radius in all space.

On the other hand, the total energy is calculated including all possible electrostatic interactions between the multipole charges (monopole - monopole, monopole - dipole, dipole - dipole, etc.):

$$E_M = \frac{1}{2S} \sum_{\mathbf{R},L} Q_L^{\mathbf{R}} \sum_{\mathbf{R}',L'} M_{LL'}^{\mathbf{R}\mathbf{R}'}\, Q_{L'}^{\mathbf{R}'}. \tag{13}$$

In Eqs. (12) and (13) multipole moments of the charge density are calculated as integrals over the atomic sphere with the origin taken at $\mathbf{R}$:

$$\begin{aligned}
Q_L^{\mathbf{R}} &= \frac{\sqrt{4\pi}}{2l+1} \int_{\mathbf{R}} d\mathbf{r}\, Y_L^*(\hat{\mathbf{r}}) \left(\frac{r}{S_{\mathbf{R}}}\right)^l n_{\mathbf{R}}(\mathbf{r}) - Z_{\mathbf{R}}\delta_{l,0} \\
&= \frac{\sqrt{4\pi}}{2l+1} \sum_{L',L''} C_{L,L',L''} \int_0^{S_{\mathbf{R}}} dr \left(\frac{r}{S_{\mathbf{R}}}\right)^l r^2 f_{\mathbf{R}L'L''}(r) - Z_{\mathbf{R}}\delta_{l,0}. \tag{14}
\end{aligned}$$

Here $S_{\mathbf{R}}$, $n_{\mathbf{R}}(\mathbf{r})$ and $Z_{\mathbf{R}}$ are the atomic sphere radius, valence (nonspherical) electron density and valence number, respectively, of the atom at the site $\mathbf{R}$, $Y_L(\hat{\mathbf{r}})$ is a real harmonic, and $C_{L,L',L''}$ are real-harmonic Gaunt coefficients. Radial parts of the valence density, $f_{\mathbf{R}L'L''}(r)$, are obtained using the second-order Taylor expansion of the partial wave (10) as

$$\begin{aligned}
f_{\mathbf{R}L'L''} &= \phi_{\nu\mathbf{R}l'}\phi_{\nu\mathbf{R}l''} m_{\mathbf{R}L'L''}^{00} + \phi_{\nu\mathbf{R}l'}\dot{\phi}_{\nu\mathbf{R}l''} m_{\mathbf{R}L'L''}^{01} \\
&\quad + \dot{\phi}_{\nu\mathbf{R}l'}\phi_{\nu\mathbf{R}l''} m_{\mathbf{R}L'L''}^{10} + \dot{\phi}_{\nu\mathbf{R}l'}\dot{\phi}_{\nu\mathbf{R}l''} m_{\mathbf{R}L'L''}^{11} \\
&\quad + \frac{1}{2}\phi_{\nu\mathbf{R}l'}\ddot{\phi}_{\nu\mathbf{R}l''} m_{\mathbf{R}L'L''}^{02} + \frac{1}{2}\ddot{\phi}_{\nu\mathbf{R}l'}\phi_{\nu\mathbf{R}l''} m_{\mathbf{R}L'L''}^{20}. \tag{15}
\end{aligned}$$

In the last expression, the radial dependence of $f_{\mathbf{R}L'L''}(r)$ and the expansion coefficients $\phi_{\nu\mathbf{R}l}(r)$ has been omitted. In Eq. (15) the density of states moments,

$m^{q'q''}_{RL'L''}$, are given by Eq. (9). In this regard we remark that if the angular momentum cutoff for the Green's function is equal to l_{max}, then multipole moments of the charge density, Eq. (14), have nonzero components up to $2l_{max}$ due to the properties of Gaunt coefficients.

3   Taking Advantages of Tight-Binding LMTO Representation

All equations presented in Sec. 2 are formulated for a general LMTO representation α. Indeed, in the case of a bulk alloy without local lattice relaxations LSGF calculations can be carried out in any representation with the same computational efforts. However, when in addition to a substitutional disorder the ideal three-dimensional symmetry of an underlying lattice is also broken, a formulation of the method in terms of tight-binding (TB) LMTO basis set can give one a substantial advantage. Below we will consider two such cases, that is we will present generalization of the LSGF method to the case of (i) a surface alloy and (ii) an alloy with local distortions of atomic positions from the sites of ideal underlying lattice (local lattice relaxations).

3.1   Electronic Properties of Surface Alloys: the LSGF Method

The large interest in theoretical investigations of surfaces of transition metals and their alloys is motivated by its fundamental scientific value, as well as by the great practical importance of these systems. The properties of solid surfaces and interfaces between two metals play an important role in such phenomena as catalysis, chemisorbtion, adhesion and corrosion, just to mention a few examples. Recently there has been significant progress in *ab initio* calculations of the electronic structure of surfaces of ordered materials using first-principles techniques. It appears that one may calculate surface related properties such as surface tension and work function with a high degree of accuracy [29,4]. In this context, random metallic alloys represent a large class of important materials. As to now the surface properties of these systems have only been investigated in a few cases.

Let us consider for simplicity a system of N atoms with an ideal surface, i.e. without any imperfections, like, for example, relaxation, reconstruction, steps, etc. (generalization of the method for the latest case is trivial). The system is schematically illustrated in Fig. 3. In general, there are at least two ways of solving the electronic structure problem for this system in the framework of the LSGF method. Firstly, one can use a traditional approach (Fig. 3a), and model this surface by a supercell or a slab, thus coming back to a three-dimensional problem that is to be solved by a conventional LSGF method (see Fig. 3b) described in Sec. 2. Such a computational scheme will, of course, scale linearly with increasing number of atoms in the supercell, and in principle, it has been already successfully applied for several systems [28,50]. However, in doing so one has to construct an effective medium that represents on the average scattering properties of the alloy components, as well as takes into account a vacuum region modeled in the framework of LMTO-ASA method by empty spheres. Giving the fact that these properties differ considerably from each other, one will expect a substantial increase of the LIZ size.

At the same time, as has been demonstrated in Ref. [28], one of the advantages of the LSGF method is the possibility to vary the effective medium to suit a problem at hand. Thus, in a case of surface alloy the better choice of the latter will be the one with a layer dependent Green's function (Fig. 3c). This can be achieved in the framework of interface Green's function technique, proposed by Lambrecht and Andersen [51] and developed by Skriver and Rosengaard [29]. An implementation of this technique in conjunction with the CPA method has been described by Abrikosov and Skriver [43].

Similar to these techniques, a solution to the problem of calculating the electronic structure of a surface or an interface is separated into two parts. First, one must calculate the Green's function for the ideal bulk crystals on the both sides of the interface. In the case of a random alloy this has to be done by means of the LSGF method (Sec. 2) for a supercell which models the bulk alloy, i.e. without any presence of the surface or the interface. At the second stage one can construct an appropriate supercell to model a region of the alloy in the neighborhood of the interface. Note, that due to a local character of the perturbation induced by the surface or the interface, the size of this supercell along the direction perpendicular to the interface is not supposed to be too large. Proper boundary conditions will be ensured by using the bulk potential function from Eq. (4) for the effective medium outside the surface region obtained at the first stage. On the contrary, within the scheme illustrated in Fig. 3a and b, the whole problem must be solved for the same supercell which in this case should probably be quite large.

Following the main idea behind the LSGF method, we construct an effective medium for each layer Λ in the neighborhood of a surface as in the case of a multicomponent two-dimensional alloy:

$$
\begin{aligned}
\tilde{g}^{\beta}_{\tilde{\Omega}\tilde{\Omega}}(z) &= (\tilde{A}_{SBZ})^{-1} \int_{SBZ} d\mathbf{k}_{\parallel} \, \tilde{g}^{\beta}_{\tilde{\Omega}\tilde{\Omega}}(\mathbf{k}_{\parallel}, z) \\
g^{\beta 0}_{\mathbf{R}} &= \tilde{g}^{\beta}_{\tilde{\Omega}\tilde{\Omega}} + \tilde{g}^{\beta}_{\tilde{\Omega}\tilde{\Omega}}(\tilde{P}^{\beta}_{\tilde{\Omega}} - P^{\beta}_{\mathbf{R}}) g^{\beta 0}_{\mathbf{R}} \\
\tilde{g}^{\beta}_{\tilde{\Omega}\tilde{\Omega}} &= (N_{\tilde{\Omega}})^{-1} \sum_{\mathbf{R}} g^{\beta 0}_{\mathbf{R}},
\end{aligned}
\tag{16}
$$

where $\tilde{A}_{SBZ}$ is the area of the 2D surface Brilouin zone (SBZ), $N_{\tilde{\Omega}}$ and $P^{\beta}_{\mathbf{R}}$ the number of atoms and the potential function of real atoms, respectively, at the sites $\tilde{\Omega} = (\tilde{\Omega}_{\parallel}, \tilde{\Omega}_{\perp})$ that belong to the same sublattice of the 2D unit cell of the underlying surface lattice in the Λ layer (note that there can be more than one site $\tilde{\Omega}$ in the layer Λ), $\tilde{P}^{\beta}_{\tilde{\Omega}}$ the potential function of an effective scatterer at $\tilde{\Omega}$, $\mathbf{R} \in \tilde{\Omega}$, and β denotes the most localized, tight-binding LMTO-representation [32,33] which is essential to use in the calculation of the surface Green's function

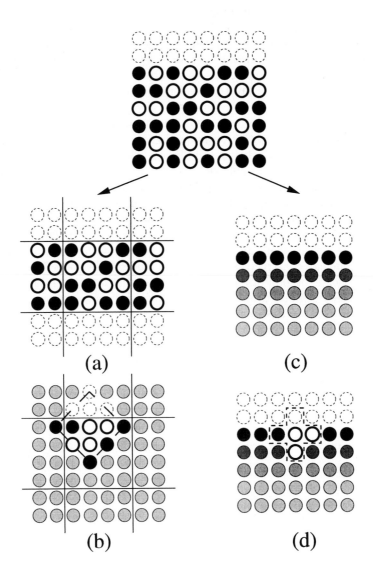

Fig. 3. Two ways of modeling surface alloy in the framework of the LSGF method. (a) One may employ a slab or a supercell approach, and repeat a part of the sample, including empty spheres (*thin dotted circles*) in all three directions. (b) An effective medium (*gray circles*) shall represent on the average scattering properties of real alloy components (*filled and open circles*), as well as empty spheres, and the embedding procedure can be carried out in a conventional manner. (c) By means of the interface Green's function technique in conjunction with the LSGF method the surface alloy is treated in its true semi-infinite geometry, and a layer-dependent effective medium is introduced. (d) Embedding procedure can be carried out, and the size of the LIZ is supposed to be smaller than in the case of the slab or the conventional supercell

$\tilde{g}^{\beta}_{\tilde{\Omega}\tilde{\Omega}}(\mathbf{k}_{\parallel}, z)$. As usual in the case of surfaces, a good quantum number is $\mathbf{k}_{\parallel} \in$ SBZ, and the dependence of the surface Green's function on $\mathbf{k}_{\perp}$ is integrated off by means of the principle layer technique [52]. In doing so one (i) calculates the ideal Green's functions for both sides of the interface using the unperturbed bulk potential functions of effective scatterers obtained for the supercell which models the *bulk* alloy, and (ii) solves a set of recurrent equations that glue the two sides of the interface together and take care of the perturbation of local potentials in the interface region. Note that this operation scales as $O(N_\Lambda)$, where N_Λ is a number of layers in the interface region.

When the layer dependent effective medium is constructed by Eq. (16), i.e. $\tilde{P}^{\beta}_{\tilde{\Omega}}$ is calculated for any sublattice $\tilde{\Omega}$ and for all layers Λ in the supercell that models *surface* alloy, the embedding procedure has to be carried out in a similar manner as for the bulk alloy. If we now denote sites of the LIZ as $\mathbf{R} = (\mathbf{R}_{\parallel}, \mathbf{R}_{\perp})$, then the off-diagonal effective medium Green's function for the LIZ is calculated as an integral over the SBZ

$$\tilde{g}^{\beta}_{\mathbf{R}\mathbf{R}'}(z) = (\tilde{A}_{SBZ})^{-1} \int_{SBZ} d\mathbf{k}_{\parallel}\, e^{i\mathbf{k}_{\parallel} \cdot [(\tilde{\Omega}_{\parallel} - \tilde{\Omega}'_{\parallel}) - (\mathbf{R}_{\parallel} - \mathbf{R}'_{\parallel})]} \tilde{g}^{\beta}_{\tilde{\Omega}\tilde{\Omega}'}(\mathbf{k}_{\parallel}, z), \qquad (17)$$

and the Green's function for the central site of the LIZ embedded into the effective medium (Fig. 3d) is obtained by solving Dyson equation (6). The remaining procedure is analogous to the one presented in Sec. 2.

3.2   Electronic Properties of Alloys with Local Lattice Relaxations

In general, alloys are composed of elements which are not only chemically nonequi-

ivalent, but also have different atomic sizes. As a result this size mismatch causes interatomic distances in the alloy to be different, i.e. it causes so-called local relaxation effects. The system can also be looked upon as one where local positions of all the atoms are moved from the sites of an ideal periodic underlying lattice (Fig. 4a). Though the recent systematic study of lattice relaxations around a single impurity in Cu by Papanikolaou *et al.* [53] has shown that their contribution to the impurity solution energy in general is small compared to the values calculated earlier without lattice relaxations [3], this effect can influence the results for the density of states and the total energy calculations in some cases of very large size mismatches [54], and it is important to be able to take the local relaxations into consideration.

Within the LSGF method in conjunction with the TB-LMTO basis set the following procedure can be suggested. Firstly, we construct an effective medium for the undistorted underlying lattice in essentially the same way as has been described in Sec. 2. Then we embed a LIZ that is composed of an atom at the particular site and several shells of its nearest neighbors and that also includes all the relaxations in this effective medium (Fig. 4b). Thus, the Dyson equation that has to be solved now shall include both, the perturbation due to chemical

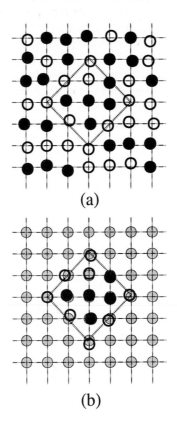

(a)

(b)

Fig. 4. Schematic representation of the LSGF method applied for an alloy with local lattice relaxations. (a) An original system composed of two chemically nonequivalent components (*filled and open circles*) can be looked upon as the system where local positions of all the atoms are moved from the sites of an ideal periodic underlying lattice (indicated by *thin lines*). (b) Effective atoms (*gray circles*) are placed at the sites of the undistorted underlying lattice, and the Dyson equation (18) for the LIZ (*dotted line*) embedded into this effective medium accounts for two kinds of perturbations, due to the potential part, as well as due to structural perturbations

disorder, as well as due to the fact that the atomic positions are shifted from the sites of the underlying lattice. In the framework of the TB-LMTO method this perturbation can be included in Eq. (6) through the difference between the real space structure constant matrices for the original system **S** and for the underlying lattice **S̃** as following

$$g(z) = [(\tilde{g}(z))^{-1} + (P(z) - \tilde{P}(z)) - (S - \tilde{S})]^{-1}, \qquad (18)$$

where bold symbols denotes $(RL, R'L')$ matrices, and $R, R' \in$ LIZ.

Note, that the only element needed from a solution of Eq. (18) is again the on-site element $g_{\mathbf{R}L,\mathbf{R}L'}(z)$ for the central site of the LIZ. Therefore, the advantages of the TB-LMTO representation shall show up in full strength here. Not only it allows one to calculate easily structure constant matrices for the effective and the original lattices in the real space, but also due to their localized character the difference between these lattices as it is seen from the central site of the LIZ vanishes exactly at very short distancies. As has been specified in Sec. 2 the above criterion is one of the major requirements to the effective medium that ensures the minimal size of the LIZ. Therefore, though the LIZ size necessary to carry out calculations for systems with local lattice relaxations will increase compared to the unrelaxed case, this increase is not supposed to be too large.

4   Summary

The order-N locally self-consistent Green's function method has been formulated in the framework of the LMTO method in the tight-binding representation. The atomic sphere approximation has been used, but we also have shown how multipole corrections to the ASA can be easily incorporated in our formalism. The LSGF method employs two basic concepts, the local interaction zone concept, that is the multiple scattering problem is solved in real space for each atom surrounded by few shells of nearest neighbors, and the effective medium concept that provides appropriate boundary conditions for the multiple scattering problem by embedding the LIZ into a self-consistent effective medium constructed in such a way as to describe as close as possible scattering properties of the original system beyond the LIZ boundary. The latter concept ensures a minimal size of the LIZ, and therefore enhances the efficiency of the LSGF method as compared to other $O(N)$ methods. The tight-binding representation of the LMTO method allows us to present generalization of the method to two important cases, a surface alloy, and an alloy with local lattice relaxations.

Applications of the LSGF method to several problems of alloy theory have shown that it is an excellent tool for studing systems with an arbitrary distribution of atoms on the sites of the underlying crystal lattice. The method allows one to include directly short range order effects when calculating the electronic structure and the total energy of random alloys. In particular, mixing energies of fcc Cu-Zn alloys calculated by the LSGF method for alloys with appropriate amount of short-range order were found to be in much better agreement with experiment than those where short-range order was neglected [8]. Moreover, the LSGF method takes into account all the local environment effects. Thus, it provides one with an atomic scale resolution when analyzing the electronic properties of materials, for example, spectral properties [37] or local magnetic moments [55].

The LSGF method gives one the opportunity to calculate the so-called effective cluster interactions which can be later applied in the framework of a statistical mechanical technique in order to study phase stabilities of alloys. In particular, one can extract *concentration dependent* effective cluster interactions.

This may be done by calculating the total energies of alloys with different sets of correlation functions but for some fixed concentrations of the alloy components and then by mapping these energies onto the corresponding cluster expansion. Since these calculations are fast one may perform a large number of them and thereby increase the accuracy of the interaction parameters obtained. The resulting effective cluster interactions will include all contributions to the total energy. Such calculations have been carried out with a great success by Simak et al. [9] for a multicomponent Cu_2NiZn alloy.

By means of the LSGF method one can also study properties of point defects and their clusters. In particular, in order to study a realistic metal-vacancy system, it is necessary to consider a large supercell in which the vacancies are well separated. To calculate vacancy-vacancy or vacancy-solute interaction energies in alloys or to study complex thermal defects in some intermetallics, very large supercells are necessary, so conventional methods of band structure calculations become extremely inefficient. Here our $O(N)$ technique has allowed us, for example, to perform a systematic study of the vacancy formation energy of the $3d$, $4d$, and $5d$ transition and noble metals, and to discuss its variation through a transition metal series, as well as the effects of crystal and magnetic structure [30], and the interaction of vacancies with other defects [10].

In conclusion, we find in number of applications and numerical tests that the LSGF method in conjunction with the TB-LMTO basis set leads to a reliable description of electronic properties of alloys. In general it yields results in excellent agreement with those obtained by alternative first-principles techniques, but becomes more efficient than the latter already for systems that contain 30 to 100 atoms in the unit cell. Thus, the LSGF method is a powerful technique for solving different problems of materials science theory.

Acknowledgements

This work has been partly supported by the Swedish Natural Science Research Council and by SKB AB, the Swedish Nuclear Fuel and Waste Management Company. The support by the TMR-network "Interface Magnetism" and by the Swedish Materials Consortium #9 are acknowledged. IAA is grateful to Dr. S. I. Simak for useful discussion of the manuscript. Collaboration with Dr. A. V. Ruban and Prof. H. L. Skriver in the development of the LSGF method is also acknowledged.

References

1. D. Pettifor, *Bonding and structure of molecules and solids* (Clarendon Press, Oxford, 1995).
2. F. Ducastelle, *Order and Phase Stability in Alloys* (North-Holland, Amsterdam, 1991).
3. B. Drittler, M. Weinert, R. Zeller, and P. H. Dederichs, Phys. Rev. B**39**, 930 (1989).
4. H. L. Skriver and N. M. Rosengaard, Phys. Rev. B **46** 7157 (1992).

5. A. Christensen, A. V. Ruban, P. Stoltze, K. W. Jacobsen, H. L. Skriver and J. K. Nørskov, Phys. Rev. B **56**, 5822 (1997).

6. P. Hohenberg and W. Kohn, Phys. Rev. **136B** 864 (1964).

7. W. Kohn and L.J. Sham, Phys. Rev. **140** A1133 (1965).

8. I. A. Abrikosov, A. M. N. Niklasson, S. I. Simak, B. Johansson, A. V. Ruban, and H. L. Skriver, Phys. Rev. Lett. **76**, 4203 (1996).

9. S. I. Simak, A. V.Ruban, I. A. Abrikosov, H. L. Skriver, and B. Johansson, Phys. Rev. Lett. **81**, 188 (1998).

10. P. A. Korzhavyi, I. A. Abrikosov, and B. Johansson, Acta Mater. **47**, 1417 (1999).

11. W. Kohn, Phys. Rev. Lett. **76**, 3168 (1996).

12. D. M. C. Nicholson, G. M. Stocks, Y. Wang, W. A. Shelton, Z. Szotek, and W. M. Temmerman, Phys. Rev. B **50**, 14686 (1994).

13. Y. Wang, G. M. Stocks, W. A. Shelton, D. M. C. Nicholson, Z. Szotek, and W. M. Temmerman, Phys. Rev. Lett. **75**, 2867 (1995).

14. W. Yang, Phys. Rev. Lett. **66**, 1438 (1991); T. Zhu, W. Pan, and W. Yang, Phys. Rev. B **53**, 12713 (1996).

15. G. Galli and M. Parrinello, Phys. Rev. Lett. **69**, 3547 (1992); F. Mauri, G. Galli, and R. Car, Phys. Rev. B **47**, 9973 (1993); F. Mauri and G. Galli, Phys. Rev. B **50**, 4316 (1994).

16. P. Ordejón, D. A. Drabold, M. P. Grumbach, and R. M. Martin, Phys. Rev. B **48**, 14646 (1993); P. Ordejón, D. A. Drabold, R. M. Martin, and M. P. Grumbach, Phys. Rev. B **51**, 1456 (1995); S. Itoh, P. Ordejón, D. A. Drabold, and R. M. Martin, Phys. Rev. B **53**, 2132 (1996), P. Ordejón, E. Artacho, and J. M. Soler, Phys. Rev. B **53**, R10441 (1996).

17. W. Kohn, Chem. Phys. Lett. **208**, 167 (1993).

18. E. B. Stechel, A. R. Williams, and P. J. Feibelman, Phys. Rev. B **49**, 10088 (1994); W. Hierse and E. B. Stechel, Phys. Rev. B **50** 17811 (1994).

19. X.-P. Li, R.W. Nunes, and D. Vanderbilt, Phys. Rev. B **47**, 10891 (1993).

20. M. S. Dow, Phys. Rev. B **47**, 10895 (1993).

21. S.-Y. Qiu, C. Z. Wang, K. M. Ho, and C. T. Chan, J. Phys.: Condens. Matter **6**, 9153 (1994).

22. E. Hernandez and M. J. Gillan, Phys. Rev. B **51** 10157 (1995); E. Hernandez, M. J. Gillan, and C. M. Goringe, Phys. Rev. B **53** 7147 (1996).

23. A. E. Carlsson, Phys. Rev. B **51**, 13935 (1995).

24. S. Goedecker and L. Colombo, Phys. Rev. Lett. **73**, 122 (1994); S. Goedecker and M. Teter, Phys. Rev. B **51**, 9455 (1995).

25. A. F. Voter, J. D. Kress, and R. N. Silver, Phys. Rev. B **53** 12733 (1996).

26. A. P. Horsfield, A. M. Bratkovsky, M. Fearn, D. G. Pettifor, and M. Aoki, Phys. Rev. B **53** 12694 (1996).

27. S. Baroni and P. Giannozzi, Europhys. Lett. **17**, 547 (1992).

28. I. A. Abrikosov, S. I. Simak, B. Johansson, A. V. Ruban, and H. L. Skriver, Phys. Rev. B **56**, 9319 (1997).

29. H. L. Skriver and N. M. Rosengaard, Phys. Rev. B **43** 9538 (1991).

30. P. A. Korzhavyi, I. A. Abrikosov, B. Johansson, A. V. Ruban, and H. L. Skriver, Phys. Rev. B **59**, 11693 (1999).

31. O. K. Andersen, Phys. Rev. B **12**, 3060 (1975).

32. O. K. Andersen and O. Jepsen, Phys. Rev. Lett. **53**, 2571 (1984).

33. O.K. Andersen, O. Jepsen, and D. Glötzel, in *Highlights of Condensed-Matter Theory* , edited by F. Bassani, F. Fumi, and M. P. Tosi (North Holland, New York, 1985).

34. O. Gunnarsson, O. Jepsen, and O. K. Andersen, Phys. Rev. B **27**, 7144 (1983).
35. H. L. Skriver, *The LMTO Method* (Springer-Verlag, Berlin, 1984).
36. J. S. Faulkner, Prog. Mater. Sci. **27**, 1 (1982).
37. I. A. Abrikosov and B. Johansson, Phys. Rev. B **57**, 14164 (1998).
38. J. S. Faulkner, N. Y. Moghadam, Y. Wang and G. M. Stocks, Phys. Rev. B **57**, 7653 (1998).
39. I. A. Abrikosov and B. Johansson, Philos. Mag. B **78**, 481 (1998).
40. R. Podloucky, R. Zeller, and P. H. Dederichs, Phys. Rev. B **22**, 5777 (1980); B. Drittler, M. Weinert, R. Zeller, and P. H. Dederichs, Phys. Rev. B**39**, 930 (1989).
41. C. Koenig, N. Stefanou, and J. M. Koch, Phys. Rev. B **33**, 5307 (1986).
42. D. D. Johnson, D. M. Nicholson, F. J. Pinski, B. L. Gyorffy, and G. M. Stocks, Phys. Rev. Lett. **56**, 2088 (1986); D. D. Johnson, D. M. Nicholson, F. J. Pinski, B. L. Gyorffy, and G. M. Stocks, Phys. Rev. B **41**, 9701 (1990).
43. I. A. Abrikosov and H. L. Skriver, Phys. Rev. B **47**, 16532 (1993).
44. A. V. Ruban, A. I. Abrikosov, and H. L. Skriver, Phys. Rev. B **51** 12958 (1995).
45. T. Beuerle, R. Pawellek, C. Elsässer, and M. Fähnle, J. Phys.: Condens. Matter **3**, 1957 (1991).
46. P. Braun, M. Fähnle, M. van Schilfgaarde, and O. Jepsen, Phys. Rev. B **44**, 845 (1991).
47. M. Sinder, D. Fuks, and J. Pelleg, Phys. Rev. B **50**, 2775 (1994).
48. B. Drittler, M. Weinert, R. Zeller, and P. H. Dederichs, Solid State Comm., **79**, 31 (1991).
49. P. H. Dederichs, B. Drittler, R. Zeller, Mater. Research Soc. Symp. Proc. **253**, 185 (1992).
50. A. V. Ruban, private communication.
51. W. Lambrecht and O. K. Andersen, Surface Sci. **178**, 256 (1986); Private communication
52. J. Kudrnovský, P. Weinberger, and V. Drchal, Phys. Rev. B **44**, 6410 (1991).
53. N. Papanikolaou, R. Zeller, P. H. Dederichs, and N. Stefanou, Phys. Rev. B **55**, 4157 (1997).
54. Z. W. Lu, S.-H. Wei, and A. Zunger, Phys. Rev. B **45**, 10314 (1992).
55. P. James, O. Eriksson, B. Johansson, and I. A. Abrikosov, Phys. Rev. B **59**, 419 (1999).

Part IV

Large-Scale Real-Space Calculations

Sparse Direct Methods: An Introduction

J. A. Scott

Department for Computation and Information, Rutherford Appleton Laboratory, Chilton, Didcot, Oxon OX11 0RA, England. (J.Scott@rl.ac.uk)

Abstract. The solution of large-scale linear systems lies at the heart of many computations in science, engineering, industry, and (more recently) finance. In this paper, we give a brief introduction to direct methods based on Gaussian elimination for the solution of such systems. We discuss the methods with reference to the sparse direct solvers that are available in the Harwell Subroutine Library. We briefly consider large sparse eigenvalue problems and show how the efficient solution of such problems depends upon the efficient solution of sparse linear systems.

1   Introduction

Sparse matrices arise in very many application areas, including such diverse fields as structural analysis, chemical engineering, surveying, and economics. A matrix is sparse if many of its coefficients are zero and there is an advantage in exploiting the zeros. In this paper, we present a brief introduction to direct methods for the solution of large sparse linear systems of equations

$$Ax = b. \tag{1}$$

We are concerned with methods that are based on Gaussian elimination. That is, we compute an LU factorization of a permutation of A

$$PAQ = LU,$$

where P and Q are permutation matrices and L and U are lower and upper triangular matrices, respectively. These factors are used to solve the system (1) through the forward elimination

$$Ly = Pb$$

followed by the back substitution

$$Uz = y.$$

The required solution x is then the permuted vector $x = Qz$. When A is symmetric positive definite, it is normal to use the Cholesky factorization

$$PAP^T = LL^T.$$

For more general symmetric matrices, the factorization

$$PAP^T = LDL^T,$$

is more appropriate. For a stable decomposition in the indefinite case, the matrix D is block diagonal with blocks of order 1 or 2, and L is unit lower triangular.

Several different approaches to Gaussian elimination for sparse matrices have been developed. They can each be divided into a number of phases:

1. Preordering to exploit structure eg preordering to block triangular form

$$PAQ = \begin{pmatrix} B_{11} & & & \\ B_{21} & B_{22} & & \\ B_{31} & B_{32} & B_{33} & \\ . & . & . & \\ B_{N1} & B_{N2} & B_{N3} & ... & B_{NN} \end{pmatrix}$$

   so that only the diagonal blocks B_{ii} need to be factorized.
2. Analyse - the sparsity pattern of A is analyzed to produce a suitable ordering and data structures for efficient factorization.
3. Factorize - the numerical factorization of A is performed.
4. Solve - the factors are used to solve one or more systems (1) using forward elimination and back substitution.

Some codes combine the analyse and factorize phases so that numerical values are available when the ordering is being generated. Phase 3 (or the combined phase 2 and 3) generally requires most computational time. If more than one matrix with the same sparsity pattern is to be factorized, the analyse phase only needs to be performed once. Thus in contrast to dense solvers, for sparse solvers, it is potentially much faster to perform subsequent factorizations.

In the following sections, we introduce three approaches: the general approach, frontal methods, and multifrontal methods. We illustrate these different approaches using software from the Harwell Subroutine Library (HSL) [1] and use numerical examples from a range of scientific and industrial applications. A useful reference is the book by Duff, Erisman, and Reid [2] and for an extensive list of references to recent developments in the area, we recommend the review by Duff [3].

We end this section by noting that the order of a matrix that is considered large is a function of time depending on both the development of dense and sparse codes and advances in computer architecture. However, by today's standards, A need not be very large for it to be worthwhile to exploit sparsity. This is illustrated in Table 1 (taken from Duff [4]), in which we compare the performance of the HSL sparse solver MA48 with that of the dense solver SGESV from LAPACK. The problems are all from the Harwell-Boeing Sparse Matrix Collection [5]. The experiments were performed on a single processor of a CRAY Y-MP vector supercomputer and the timings are given in seconds.

Table 1. A comparison of timings on the CRAY Y-MP for MA48 and SGESV on Harwell-Boeing matrices.

Identifier	Order	Number of entries	MA48	SGESV
FS 680 3	680	2646	0.06	0.96
PORES 2	1224	9613	0.54	4.54
BCSSTK27	1224	56126	2.07	4.55
NNC1374	1374	8606	0.70	6.19
WEST2021	2021	7353	0.21	18.88
ORANI678	2529	90158	1.17	36.37

2   The General Approach

The principal features of the general approach are:

- sparse data structures are used throughout
- numerical and sparsity pivoting are performed at the same time (phases 2 and 3 are combined).

The efficient implementation of techniques for handling sparse data structures is of crucial importance. The most common sparse data structure and the one used in most general-purpose codes holds the matrix by rows. All rows are stored in the same way with the real values and column indices in two arrays with a one-to-one correspondence between the arrays, so that the real value in position k, say, is in the column indicated by the entry in position k of the column index array. A sparse matrix can then be stored as a collection of sparse rows in two arrays; one integer, the other real, both of length nz, where nz is the number of entries in A. A third "pointer" array of length $n+1$ (n is the order of A) is used to identify the position in the first two arrays of the data structure for each row.

To illustrate this scheme, consider the following 4×4 sparse matrix:

$$A = \begin{pmatrix} 6 & 0 & 1 & 0 \\ 2 & 4 & 0 & -1 \\ 0 & 1 & 0 & 0 \\ 7 & 0 & 0 & -3 \end{pmatrix}$$

Storing this matrix as a collection of sparse row vectors we have

column index	1	3	2	4	1	2	1	4
value	6.	1.	4.	-1.	2.	1.	7.	-3.
row pointer	1	3	6	7	9			

Note that the column indices within each row need not be held in order. The advantages of the scheme are that it is a simple and compact method of storing the matrix and it is straightforward to access the matrix by rows. A disadvantage is that it is difficult to insert entries, which is needed in Gaussian elimination when a multiple of one row (the pivot row) of the matrix is added to the other (non-pivot) rows with different sparsity patterns.

In general, the matrix factors L and U are denser than A. For efficiency, in terms of both storage and floating-point operations (flops), it is essential to try and restrict the amount of "fill-in" (that is, the number of entries in L and U that correspond to zeros in A). The rows and columns of A need to be ordered to preserve sparsity. For example, consider the symmetric matrix with sparsity pattern

$$A = \begin{pmatrix} x & x & x & x & x \\ x & x & & & \\ x & & x & & \\ x & & & x & \\ x & & & & x \end{pmatrix}.$$

Choosing the pivots in order down the diagonal, the Cholesky factorization of A is LL^T, where L has the form

$$L = \begin{pmatrix} x & & & & \\ x & x & & & \\ x & x & x & & \\ x & x & x & x & \\ x & x & x & x & x \end{pmatrix}.$$

Thus all sparsity has been lost. However, reordering the rows of A in reverse order

$$PA = \hat{A} = \begin{pmatrix} x & & & & x \\ & x & & & x \\ & & x & & x \\ & & & x & x \\ x & x & x & x & x \end{pmatrix},$$

and $\hat{L}$ retains the sparsity of $\hat{A}$

$$\hat{L} = \begin{pmatrix} x & & & & \\ & x & & & \\ & & x & & \\ & & & x & \\ x & x & x & x & x \end{pmatrix}.$$

A simple but effective strategy for maintaining sparsity is due to Markowitz [6]. We motivate this strategy by considering the first step of Gaussian elimination for the matrix A partitioned in the form

$$A = A^{(0)} = \begin{pmatrix} \alpha & a_r^T \\ a_c & A_R^{(0)} \end{pmatrix}.$$

Assuming α is a suitable pivot choice, the first step in the matrix factorization is

$$A = \begin{pmatrix} 1 & 0 \\ a_c \alpha^{-1} & A^{(1)} \end{pmatrix} \begin{pmatrix} \alpha & a_r^T \\ 0 & I \end{pmatrix},$$

where

$$A^{(1)} = A_R^{(0)} - \frac{a_c a_r^T}{\alpha}. \tag{2}$$

Fill-in occurs when an entry in the rank-one matrix $a_c a_r^T$ is a zero in $A_R^{(0)}$. Clearly the dominant cost in the update (2) is that of forming the outer product $a_c a_r^T$. This cost is proportional to the product of the number of nonzeros in a_c and the number of nonzeros in a_r^T. At each stage of Gaussian elimination, Markowitz therefore selects as the pivot the nonzero entry of the remaining reduced submatrix that minimizes this product. That is, $a_{ij}^{(k)}$ is chosen as the pivot for the k-th stage to minimize

$$(r_i^{(k)} - 1)(c_j^{(k)} - 1),$$

where $r_i^{(k)}$ and $c_j^{(k)}$ denote, respectively, the number of nonzeros in row i and column j of $A^{(k)} = \{a_{ij}^{(k)}\}$.

For stability, pivot candidates must also satisfy some numerical criteria. In particular, assuming row and column interchanges have been performed to bring the pivot candidate selected by the Markowitz criteria to the diagonal, the pivot is only acceptable if it satisfies the inequality

$$|a_{kk}^{(k)}| \geq u \, |a_{ik}^{(k)}|, \quad i \geq k,$$

where u is a preset threshold parameter in the range $0 < u \leq 1$. The choice $u = 1$ corresponds to partial pivoting but, in general, this value is too restrictive and leads to a large number of pivots being rejected and to unnecessary fill-in. If u is chosen to be too small, instability can result. A common choice is $u = 0.1$. Experience has shown that this value usually provides a good compromise between maintaining stability and preserving sparsity.

In the Harwell Subroutine Library, the package MA48 of Duff and Reid [7] (and its complex counterpart ME48) is a general sparse solver which uses Markowitz pivoting. It uses the numerical values in the analyse phase and its default value for the threshold parameter is 0.1. The code offers a "fast factorization" for matrices with exactly the same sparsity pattern as one that has already been factorized. One of the ways in which the code achieves high performance, particularly on vector or super scalar machines, is by switching to full-matrix processing and using Level 3 BLAS [8] once the matrix is sufficiently dense. MA48 is frequently used as a benchmark against which new sparse solvers are judged.

The strength of the general approach is that it gives satisfactory performance for many matrix structures and is often the method of choice for very sparse unstructured problems. Some gains and simplifications are possible if A is symmetric. In particular, the Markowitz ordering is replaced by minimum degree ordering. At the kth stage the pivot is chosen to be $a_{ii}^{(k)}$, where

$$r_i^{(k)} = \min_l r_l^{(k)},$$

and $r_l^{(k)}$ is the number of nonzero entries in row l of $A^{(k)}$. If the matrix is additionally positive definite, numerical pivoting is not needed.

3   Frontal Methods

Frontal methods have their origins in the solution of finite-element problems for structural analysis, in which the matrix is symmetric and positive definite. The method can, however, be extended to general unsymmetric systems and need not be restricted to finite-element applications (see Duff [9]).

To describe the method, we assume that A is a sum of finite-element matrices

$$A = \sum_{l=1}^{m} A^{[l]} \tag{3}$$

where each element matrix $A^{[l]}$ has nonzeros only in a few rows and columns and corresponds to the matrix from element l. The main feature of the frontal method is that the contributions $A^{[l]}$ are assembled one at a time and the construction of the assembled coefficient matrix A is avoided by interleaving assembly and elimination operations. An assembly operation is of the form

$$a_{ij} \Leftarrow a_{ij} + a_{ij}^{[l]}, \tag{4}$$

where $a_{ij}^{[l]}$ is the (i,j)th nonzero entry of the element matrix $A^{[l]}$. A variable is *fully summed* if it is involved in no further sums of the form (4) and is *partially summed* if it has appeared in at least one of the elements assembled so far but is not yet fully summed. The Gaussian elimination operation

$$a_{ij} \Leftarrow a_{ij} - a_{il}[a_{ll}]^{-1} a_{lj} \tag{5}$$

may be performed as soon as all the terms in the triple product in (5) are fully summed. At any stage during the assembly and elimination processes, the fully and partially summed variables are held in a dense matrix, termed the *frontal matrix*. Assuming k variables are fully summed, the frontal matrix F can be partitioned in the form

$$F = \begin{pmatrix} F_{11} & F_{12} \\ F_{21} & F_{22} \end{pmatrix},$$

where F_{11} is a square matrix of order k. Pivots may be chosen from anywhere in F_{11}. For symmetric positive-definite systems, they can be taken from the diagonal in order but in the unsymmetric case, pivots must be chosen to satisfy a threshold criteria. Assuming k pivots can be chosen, F_{11} is factorized as $L_{11}U_{11}$. Then F_{12} and F_{21} are updated as

$$\hat{F}_{21} = F_{21}U_{11}^{-1} \quad \text{and} \quad \hat{F}_{12} = L_{11}^{-1}F_{12}$$

and finally the Schur complement

$$\hat{F}_{22} = F_{22} - \hat{F}_{21}\hat{F}_{12}$$

is formed. The factors L_{11} and U_{11}, as well as $\hat{F}_{12}$ and $\hat{F}_{21}$, are stored as parts of L and U, before further elements are assembled with the Schur complement to form a new frontal matrix.

The power of frontal schemes comes from the following observations:

- since the frontal matrix is held as a dense matrix, dense linear algebra kernels (in particular, the BLAS) can be used during the numerical factorization,
- the matrix factors need not be held in main memory, which allows large problems to be solved using only modest amounts of high-speed memory.

There are a number of frontal codes in the Harwell Subroutine Library: MA42 (Duff and Scott [10]) and its complex counterpart ME42 are for general unsymmetric systems and MA62 (Duff and Scott [11]) is for symmetric positive definite finite-element problems. Both codes use Level 3 BLAS in the innermost loop and optionally store the matrix factors in auxiliary storage. High level BLAS are also used in the solve phase, and this increases the efficiency when the codes are used to solve for multiple right-hand sides.

On a machine with fast Level 3 BLAS, the performance of the Harwell frontal solvers can be impressive. This is illustrated in Table 2. Here MA42 is used to solve a standard finite-element test problem on a single processor of a CRAY

Table 2. Performance (in Mflop/s) of MA42 on a standard test problem running on a CRAY Y-MP.

Dimension of element grid	16 × 16	32 × 32	48 × 48	64 × 64	96 × 96
Max order frontal matrix	195	355	515	675	995
Total order of problem	5445	21125	47045	83205	186245
Mflop/s	145	208	242	256	272

Y-MP, whose peak performance is 333 Mflop/s and on which the Level 3 BLAS matrix-matrix multiply routine SGEMM runs at 313 Mflop/s on sufficiently large matrices. It is important to realize that the Mflop rates given in Table 2 include all overheads for holding the factors in auxiliary storage.

The performance of frontal methods, in terms of both the number of floating-point operations and the storage requirements, is dependent upon the order in which the elements are assembled. Many of the proposed algorithms for element ordering are similar to those for profile reduction of assembled matrices (see, for example, Reid and Scott [12]). Routine MC63 by Scott [13] is a new element ordering routine in the Harwell Subroutine Library.

For non-element problems, the frontal method proceeds by assembling the rows of the matrix one at a time. In this case, efficiency depends on the ordering of the rows. For matrices with a symmetric or almost symmetric sparsity pattern, a profile reduction algorithm using the sparsity pattern of the Boolean sum of the patterns of A and A^T can be used. For highly unsymmetric problems, new algorithms for row ordering have recently been introduced by Scott [14], and implemented as routine MC62 in the Harwell Subroutine Library.

4   Multifrontal Methods

Although high Megaflop rates can be achieved by frontal solvers, there are several important deficiences with the method:

- for some problems, many more flops are performed than are needed by other methods
- the factors can be denser than those produced by other methods
- there is little scope for parallelism other than that which can be obtained within the higher level BLAS.

These problems can be at least partially overcome through the use of more than one front.

The multiple front approach partitions the underlying "domain" into subdomains, performs a frontal decomposition on each subdomain separately and then factorizes the remaining "interface" variables, perhaps by also using a frontal scheme. The strategy corresponds to a bordered block diagonal ordering of the matrix and can be nested. With judicious ordering within each subproblem, the amount of work required can be reduced and, since the factorizations of the subproblems are independent, there is much scope for parallelism. Preliminary results presented by Duff and Scott [15] are encouraging.

Multifrontal methods are a further extension of the frontal method. In place of a small number of frontal matrices corresponding to the number of subdomains used, many frontal matrices are used. Each is used for one or more pivot steps, and the resulting Schur complement is summed with other Schur complements to generate a further frontal matrix. To illustrate this idea, assume A is a sum of finite-element matrices (3). Assuming the elements are assembled in the natural order 1,2,..., the frontal method uses the summation

$$((..((\mathbf{A}^{[1]} + \mathbf{A}^{[2]}) + \mathbf{A}^{[3]}) + \mathbf{A}^{[4]}) + ...).$$

However, there are other ways in which the summation can be performed. One possible alternative is to sum the elements in pairs, and then sum the pairs in pairs, and so on

$$((\mathbf{A}^{[1]} + \mathbf{A}^{[2]}) + (\mathbf{A}^{[3]} + \mathbf{A}^{[4]})) + ((\mathbf{A}^{[5]} + \mathbf{A}^{[6]}) + (\mathbf{A}^{[7]} + \mathbf{A}^{[8]})) + ...$$

A judicious ordering of the summation can reduce the work involved in the factorization and the density of the resulting factors. This added freedom in the way in which the assemblies are organized gives the multifrontal method an advantage over frontal methods. By using a sparsity ordering technique for symmetric systems (usually based on minimum degree), the method can be used efficiently for any matrix whose sparsity pattern is symmetric or almost symmetric. The restriction to nearly symmetric patterns arises because the initial ordering is performed using the sparsity pattern of $A + A^T$. The approach can, however, be used on any system. If the matrix is very unsymmetric (that is for many entries $a_{ij} \neq 0$ but $a_{ji} = 0$), numerical pivoting in the factorization phase can significantly perturb the ordering given by the analyse phase. This is much reduced if the matrix is permuted to have a zero-free diagonal before the analyse phase and further gains can sometimes be obtained by permuting entries with large modulus to the diagonal. For further details, see Duff and Koster [16]. This enables the multifrontal method to perform well on a wide range of matrices.

As in the frontal method, multifrontal methods use dense matrices in the innermost loops. There is, however, more data movement than in the frontal scheme, and the innermost loops are not so dominant.

In the Harwell Subroutine Library, MA41 by Amestoy and Duff [17] is a multifrontal code for non-element matrices while the MA46 code of Damhaug and Reid [18] is designed for element problems. For symmetric problems, routines MA27 and MA47 (and complex versions ME27 and ME47) are available.

Although sparsity pivoting is usually separated from numerical pivoting, a recent variant of the multifrontal approach due to Davis and Duff [19] combines the analyse and factorize phases. This approach is implemented as subroutine MA38 in the Harwell Subroutine Library.

5   A Comparison of Codes

The performance of the codes MA42, MA41, MA48, and MA38 is compared in Table 3. The timings are in seconds on a Sun ULTRA-1 workstation. MA42 is used with the row reordering package MC62. The results show that no single code is clearly better than the others. The choice of code is dependent on the problem being solved. MA41 generally has the fastest factorize time for problems such as PORES 3, which have a nearly symmetric structure. For problems that are far from symmetric in structure (for example, LHR14C), MA38 is very competitive, while MA48 performs well when the matrix is very sparse (problem WEST2021). MA42 has the advantage of requiring much less in-core storage than the other codes and this reduction in main memory requirement can mean that it is feasible to solve problems with the frontal code which cannot be solved using the other codes.

Finally, we remark that the comparative behaviour of the codes in terms of timings is, to some extent, dependent on the computing environment. In particular, the performance of MA42 is impressive on vector machines and if out-of-core storage is used, its performance is significantly affected by the speed

Table 3. A comparison of HSL codes on unsymmetric assembled problems (Sun ULTRA-1).

Identifier	Order	No. of entries	Code	Factor time	Factor ops $(*10^6)$	Storage (Kwords)	
						In-core	Factor
BAYER04	20545	159082	MA42	10.1	123.8	32	2333
			MA41	14.9	159.0	2757	1928
			MA48	7.8	15.7	1130	926
			MA38	11.9	55.7	1557	1138
LHR07C	7337	156508	MA42	8.0	48.7	22	936
			MA41	11.2	151.6	2416	1387
			MA48	11.8	56.3	1553	1253
			MA38	9.1	33.9	1317	936
LHR14C	14270	307858	MA42	17.7	129.1	94	2041
			MA41	48.1	315.2	5025	3498
			MA48	24.4	88.8	2978	2399
			MA38	16.6	63.5	2201	1728
NNC1374	1374	8606	MA42	0.52	5.3	5	137
			MA41	0.58	9.2	209	154
			MA48	0.63	5.0	155	135
			MA38	1.03	5.6	184	129
PORES 3	532	3474	MA42	0.15	0.3	1	19
			MA41	0.06	0.2	33	16
			MA48	0.08	0.9	26	18
			MA38	0.11	0.2	34	20
WEST2021	2021	7353	MA42	0.32	1.32	2	82
			MA41	0.19	0.35	104	47
			MA48	0.10	0.05	44	25
			MA38	0.24	0.06	73	43

of the i/o. It is important to exploit machine characteristics, such as cache, for efficient implementation. Some experiments using other computing environments are reported on by Duff and Scott [20].

6   Computing the Inverse of a Sparse Matrix

Once the *LU* factors of a matrix have been computed they can be used to solve for any number of right-hand sides b. Some software, including MA42, allows the user to input multiple right-hand sides. BLAS 3 routines can then be exploited in the solve phase and this allows a single call with $k > 1$ right-hand sides to be significantly faster than k calls with a single right-hand side (see Duff and Scott [10]).

An important special case of more than one right-hand side is where the inverse A^{-1} is required, since this can be obtained by solving

$$AX = I$$

by taking columns of I as successive right-hand side vectors. If a sequence of problems with the same matrix but different right-hand sides b is to be solved, it is tempting to calculate the inverse and use it to form the product

$$x = A^{-1}b. \tag{6}$$

However, there is almost no occasion when it is appropriate to compute the inverse in order to solve a linear system. In general, the inverse of a sparse matrix is dense, whereas the L and U factors are usually sparse, so using the relation (6) can be many times more expensive than using the relation $LUx = b$.

There are many applications where only specified entries of the inverse of A are required. For example, the diagonal entries of A^{-1} may be needed. Since solving (6) computes the entries of A^{-1} a column at a time, the entire lower triangle of A^{-1} would have to be computed to obtain all the diagonal entries. This may be avoided using the algorithm of Erisman and Tinney [21], which allows advantage to be taken of sparsity. Let $Z = A^{-1}$ and suppose a sparse factorization of A

$$A = LDU$$

has been computed, where L and U are unit lower and upper triangular matrices, respectively, and D is diagonal. It can be shown that

$$Z = D^{-1}L^{-1} - (I - U)Z$$

and

$$Z = U^{-1}D-1 + Z(I - L).$$

Since $(I - L)$ and $(I - U)$ are strictly lower and upper triangular matrices, respectively, the following relations hold:

$$z_{ij} = [(I - U)Z]_{ij}, \quad i < j,$$

$$z_{ij} = [Z(I - L)]_{ij}, \quad i > j,$$

$$z_{ii} = d_i^{-1}i + [(I - U)Z]_{ii},$$

and

$$z_{ii} = d_i^{-1}i + [Z(I - L)]_{ii}.$$

Using the sparsity of L and U, these formulae provide a means of computing particular entries of Z from previously computed ones. Further details are given in the book by Duff, Erisman, and Reid [2].

7   Eigenvalue Problems

Solution methods for large sparse linear systems of equations are important in eigenvalue calculations. Large-scale generalized eigenvalue problems of the form

$$Ax = \lambda Bx \tag{7}$$

arise in many application areas, including structural dynamics, quantum chemistry, and computational fluid dynamics. In many cases, only a few eigenvalues are required (for example, the largest or smallest eigenvalues). Solution techniques involve iterative methods based on Krylov subspaces. Well-known approaches include subspace iteration, the Lanczos method and Arnoldi's method. The book by Saad [22] provides a useful introduction to numerical methods for large-scale eigenvalues.

Before applying a Lanczos or Arnoldi method, we transform (7) into a standard eigenvalue problem of the form

$$Tx = \theta x.$$

Iterative eigensolvers rapidly provide approximations to well-separated extremal eigenvalues. T should therefore be chosen so that the sought-after eigenvalues of (A, B) are transformed to well-separated extremal eigenvalues of T that are easily recoverable from the eigenvalues of T. Additionally, because the iterative eigensolvers involve matrix-vector products and for large problems, these can represent the dominant cost, we need to select T so that $y = Tv$ can be computed efficiently.

A frequently used transformation is the shift-invert transformation

$$T_{SI}(\sigma) = (A - \sigma B)^{-1} B.$$

The scalar σ is the *shift* or *pole*. Because eigenvalues close to σ are mapped away from the origin while those lying far from σ are mapped close to zero, T_{SI} is useful for computing eigenvalues of (A, B) lying close to σ. Performing matrix-vector products $y = T_{SI}v$ is equivalent to solving the linear system

$$(A - \sigma B)y = b, \tag{8}$$

where $b = Bv$. Thus the efficiency of the eigensolver depends on the efficiency with which linear systems can be solved.

In recent years, a number of software packages have been developed for large-scale eigenvalue problems, including EB12 and EB13 in the Harwell Subroutine Library (Duff and Scott [23], Scott [24]) for unsymmetric problems, and the very general ARPACK package of Lehoucq, Sorensen, and Yang [25]. These codes use a reverse communication interface so that, each time a matrix-vector product $y = Tv$ is required, control is returned to the user. This approach allows the user to exploit the sparsity and structure of the matrix and to take full advantage of parallelism and/or vectorization. Additionally, the user can incorporate different preconditioning techniques in a straightforward way. For shift-invert transformations, if a direct method of solution is used for the linear system (8), the LU factorization of $(A - \sigma B)$ need be performed only once for each shift.

8   Brief Summary

We have given a brief introduction to direct methods for solving large sparse systems of linear equations. Numerical examples have shown that there is no single method and no single code that is the best for all applications. The size of problem that can be solved using direct methods is constantly growing with the development of more sophisticated numerical techniques and advances in computer architecture. However, direct methods cannot be used for really large systems (particularly those for which the underlying problem is three-dimensional). In this case, iterative methods or techniques that combine elements of both direct and iterative methods will have to be used.

9   Availability of Software

All the codes highlighted in this paper are written in ANSI Fortran 77 and all except ARPACK are available through the Harwell Subroutine Library. Anybody interested in using any of the HSL codes should contact the HSL Manager: Scott Roberts, AEA Technology, Building 477 Harwell, Didcot, Oxfordshire OX11 0RA, England, tel. +44 (0) 1235 432682, fax +44 (0) 1235 432023, email Scott.Roberts@aeat.co.uk, who will provide licencing information. Academic licences are available at a nominal cost. Further information may also be found on the World Wide Web at http://www.dci.clrc.ac.uk/Activity/HSL.

The ARPACK code is in the public domain and may be accessed at http://www.caam.rice.edu/software/ARPACK/.

There is a limited amount of sparse matrix software that implements direct methods available within the public domain. Some codes can be obtained through netlib (http://www.netlib.org) and others from the Web pages of the researcher developing the code. The problem with the latter source is that, in general, there is no guarantee of quality control or of software maintenance and user support.

Further information on sources of software for sparse linear systems may be found in the recent report of Duff [4].

Acknowledgements

I am grateful to Iain Duff of the Rutherford Appleton Laboratory and Walter Temmerman of Daresbury Laboratory for helpful comments on a draft of this paper.

References

1. Harwell Subroutine Library, A Catalogue of Subroutines (Release 12), Advanced Computing Department, AEA Technology, Harwell Laboratory, Oxfordshire, England (1996).
2. I.S. Duff, A.M. Erisman, and J.K. Reid, Direct Methods for Sparse Matrices, Oxford University Press, England (1986).
3. I.S. Duff, in The State of the Art in Numerical Analysis, I.S. Duff and G.A. Watson, eds., Oxford University Press, England (1997).
4. I.S. Duff, Technical Report, RAL-TR-1998-054, Rutherford Appleton Laboratory (1998).
5. I.S. Duff, R.G. Grimes, and J.G. Lewis, Technical Report, RAL-TR-92-086, Rutherford Appleton Laboratory (1992).
6. H.M. Markowitz, Management Science, **3**, 255 (1957).
7. I.S. Duff and J.K Reid, ACM Trans. Math. Soft., **22**, 187 (1996).
8. J.J. Dongarra, J. DuCroz, I.S. Duff, and S. Hammarling, ACM Trans. Math. Soft., **16**, 1 (1990).
9. I.S. Duff, Report AERE R10079, Her Majesty's Stationery Office, London (1981).
10. I.S. Duff and J.A. Scott, ACM Trans. Math. Soft., **22**, 30 (1996).
11. I.S. Duff and J.A. Scott, Technical Report RAL-TR-97-012, Rutherford Appleton Laboratory (1997).
12. J.K. Reid and J.A. Scott, Technical Report RAL-TR-98-016, Rutherford Appleton Laboratory (1998).
13. J.A. Scott, Technical Report RAL-TR-1998-031, Rutherford Appleton Laboratory (1998).
14. J.A. Scott, Technical Report RAL-TR-1998-056, Rutherford Appleton Laboratory (1998).
15. I.S. Duff and J.A. Scott, in Proceedings of the Fifth SIAM Conference on Applied Linear Algebra, J.G. Lewis, ed., SIAM, Philadelphia (1994).
16. I.S. Duff, and J. Koster, Technical Report RAL-TR-97-059, Rutherford Appleton Laboratory (1997).
17. P.R. Amestoy and I.S. Duff, Inter. J. of Supercomputer Applics, **3**, 41 (1989).
18. A.C Damhaug and J.K. Reid, Technical Report RAL-TR-96-10, Rutherford Appleton Laboratory (1996).
19. T.A. Davis and I.S. Duff, SIAM J. Matrix Analysis and Applics, **18**, 140 (1997).
20. I.S. Duff and J.A. Scott, Technical Report RAL-TR-96-102 (revised), Rutherford Appleton Laboratory (1996).
21. A. Erisman and W.F. Tinney, Communications ACM **18**, 177 (1975).
22. Y. Saad, Numerical Methods for Large Eigenvalue Problems, Halsted Press (1992).

23. I.S. Duff and J.A. Scott, ACM Trans. Math. Soft., **19**, 137 (1993).
24. J.A. Scott, ACM Trans. Math. Soft., **21**, 432 (1995).
25. R.B. Lehoucq, D.C. Sorensen and C. Yang, ARPACK USERS GUIDE: Solution of Large Scale Eigenvalue Problems with Implicitly Restarted Arnoldi Methods. SIAM, Philadelphia, PA (1998).

Real-Space Tight-Binding LMTO Approach to Magnetic Anisotropy: Application to Nickel Films on Copper

D. Spišák and J. Hafner

Institut für Theoretische Physik and Center for Computational Materials Science, Technische Universität Wien, Wiedner Hauptstraße 8-10/136, A-1040 Vienna, Austria

Abstract. The basic ingredients of a real-space tight-binding linear-muffin-tin orbital (RS-TB-LMTO) approach to non-collinear magnetism and to torque-force calculations of the magnetic anisotropy are described. Applications to face-centered-tetragonal Ni films epitaxially grown on Cu(100) substrates are presented. The tetragonal distortion of the films is calculated using an *ab-initio* local-density technique, and the RS-TB-LMTO method is used for calculating the magnetic anisotropy in films with up to 7 Ni monolayers. The accuracy of the approach allows for a detailed analysis of second- and fourth-order anisotropy constants.

1 Introduction

The tight-binding linear muffin-tin orbital method [1] has proven to be a very efficient technique for investigating the electronic and magnetic properties of complex materials, both in its reciprocal- and real-space forms. Exemplary applications include disordered alloys [2,3], metallic glasses [4] and quasicrystalline alloys [5]. Of particular interest in the study of magnetism are systems where the magnetically ordered ground state cannot be described as a simple ferro-, antiferro-, or ferrimagnetic order with all moments aligned parallel or antiparallel to the global axis of magnetisation. In disordered systems the competition between ferro- and antiferromagnetic exchange interactions and/or fluctuating local anisotropies can lead to the formation of a non-collinear ground-state describable as a spin-glass, a spero-, speri-, or asperomagnet [6]. Non-collinear magnetic structures can also arise as a consequence of uncompensated magnetic interactions in ordered intermetallic compounds. The symmetry criteria for the formation of non-collinear spin structures have been discussed in Ref. [7].

Techniques for solving the Kohn–Sham equations of local-spin-density theory for a non-collinear magnet have been implemented in various standard electronic structure codes: the augmented spherical wave (ASW) method [8], the LMTO technique [9,10], and empirical tight-binding [11,12]. Applications include the helical magnetic structures of $\gamma-$Fe [9,13] and of YMn_2 [14], the non-collinear magnetism in Mn_3Sn [8], in metallic glasses [3], in quasicrystals [15] and in spin-glasses [2], to cite only a few examples.

The possibility to tilt the magnetic moment at a given site with respect to its equilibrium orientation opens the way to a calculation of Ising-, or Heisenberg-type exchange pair interactions, allowing even for a calculation of bilinear and biquadratic exchange couplings [16,17]. If spin-orbit coupling is included in the Hamiltonian, the magnetic anisotropy energy (MAE) can be calculated by a magnetic torque-force approach [18–20]. In both cases, the RS-TB-LMTO provides sufficient convergence whereas k-space calculations converge only when extremely fine grids are used for Brillouin-zone intergations [21,22]. In addition, they have the potential to make the underlying physical mechanism more transparent and to allow the investigation of even very complex systems.

In the present paper we first briefly review the fundamentals of the noncollinear spin-polarised RS-TB-LMTO technique and describe its application to the calculation of the magnetocrystalline anisotropy and to the exchange coupling constants. In the second part we present detailed investigations of the magnetic properties of the fcc Ni films epitaxially grown on Cu(001) surfaces. The Ni/Cu(001) system is unique because of the re-entrant character of the perpendicular magnetic anisotropy: with increasing film thickness, the magnetic anisotropy switches from in-plane to perpendicular at a thickness of about 7 monolayers, retaining the orientation of the magnetic moments normal to the film plane for a thickness of up to 60 Å [23,24]. In "normal" system the equilibrium between the spin-orbit driven anisotropy and the shape anisotropy leads to a single transition from perpendicular to in-plane with an increasing number of monolayers. It is believed that the re-entrant behaviour of Ni/Cu(001) films is largely strain-induced, driven by the lattice mismatch between film and substrate.

2   TB-LMTO Approach and Real-Space Recursion Formalism

Our approach to the self-consistent electronic-structure calculation is based on the two-center TB-LMTO Hamiltonian

$$
H^{\alpha}_{ils,i'l's'} = \left[\frac{1}{2}\delta_{ii'}\delta_{ll'}(c^{\alpha}_{ils} + c^{\alpha}_{i'l's'}) + \sqrt{d^{\alpha}_{ils}}\,S^{\alpha}_{il,i'l'}\,\sqrt{d^{\alpha}_{i'l's'}}\,\right]\delta_{ss'} -
$$
$$
- \frac{1}{2}\delta_{ii'}\delta_{ll'}\Delta_{il}\sigma^{z}_{ss'} = H^{\alpha,\mathrm{para}}_{ils,i'l's'} + H^{\alpha,\mathrm{exch}}_{ils,i'l's'} \quad , \tag{1}
$$

expressed in terms of the structure constants S^{α} and the potential parameters c^{α}, d^{α} which are evaluated in the screened most-localised representation [25]. The potential parameters depend on the solution of the radial Schrödinger equation at the energies ϵ_{ν} chosen usually at the center of the occupied part of the bands. Essentially, c^{α} describes the center of gravity of the bands whereas d^{α} measures the band width. The matrix element given by Eq. (1) refers to the interaction between atoms i, i', orbitals l, l' of the spin s, s'. The Pauli matrices will be denoted as σ^{x}, σ^{y} and σ^{z}.

The first term in of Eq. (1) describes the non-magnetic part of the band structure, the second spin-dependent term gives rise to the shifts of the bands with different spins in the opposite directions. The shift is controlled by the exchange splitting field

$$\Delta_{il} = c^{\alpha}_{il\downarrow} - c^{\alpha}_{il\uparrow} \quad . \tag{2}$$

Self-consistent calculations for various magnetic systems containing 3d- or 4d-metals reveal that the proportionality relation

$$\Delta_{il} = I_{il}m_{il} \quad , \tag{3}$$

between the exchange splitting field Δ_{il} and the magnetic moment m_{il} is satisfied very well for d-orbitals [4,26,27]. Therefore the non-selfconsistent studies based on the TB-LMTO formalism can use a fixed Stoner parameter of $I_2 = 0.95$ eV$/\mu_B$ as a very reasonable approximation.

Till now it was supposed that the magnetic moments are all aligned along the z axis. The generalization of the presented approach to the treatment of a non-collinear magnetic order consists in rewriting the exchange part of the Hamiltonian (1) to a rotation invariant form

$$H^{\alpha,\text{exch}}_{ils,i'l's'} = -\frac{1}{2}\delta_{ii'}\delta_{ll'}\Delta_{il}\sigma_{ss'} = -\frac{1}{2}\delta_{ii'}\delta_{ll'}I_{il}\Delta_{il}\mathbf{n_i}.\sigma_{ss'} \quad , \tag{4}$$

where $\sigma = \sigma^x\,\mathbf{x} + \sigma^y\,\mathbf{y} + \sigma^z\,\mathbf{z}$ is the vector of the Pauli matrices with $\mathbf{x}$, $\mathbf{y}$, $\mathbf{z}$ the unit vectors spanning a global coordinate space and $\mathbf{n_i} = \mathbf{m_i}/|\mathbf{m_i}|$. Each magnetic moment direction defined by polar angles φ_i and ϑ_i with respect to the global coordinate system defines the moments' local coordinate system, in which the exchange part of the Hamiltonian keeps the form of Eq. (1) but with $\sigma^z = \sigma^z_i$ referring to the local coordinate system. Because the paramagnetic part of the Hamiltonian is constructed in the global coordinate system, the on-site exchange part must be transformed correspondingly for each atom. Of course, the opposite procedure of the transformation of the paramagnetic matrix elements, namely the structure constants into the local bases would be equivalent. Taking $\mathbf{n_i} = \cos\varphi_i \sin\vartheta_i\,\mathbf{x} + \sin\varphi_i\sin\vartheta_i\,\mathbf{y} + \cos\vartheta_i\,\mathbf{z}$ we obtain for the transformed exchange splitting field on the ith site

$$-\frac{1}{2}I_{il}\Delta_{il}\mathbf{n_i}\sigma = -\frac{1}{2}I_{il}\Delta_{il}\begin{pmatrix} \cos\vartheta_i & \sin\vartheta_i\exp\left(-\mathrm{i}\varphi_i\right) \\ \sin\vartheta_i\exp\left(\mathrm{i}\varphi_i\right) & -\cos\vartheta_i \end{pmatrix}$$

$$= -\frac{1}{2}I_{il}\Delta_{il}D(\varphi_i,\vartheta_i)\sigma^z_i D^+(\varphi_i,\vartheta_i) \quad . \tag{5}$$

$D(\varphi_i,\vartheta_i)$ is the Wigner $s = \frac{1}{2}$ rotation matrix from the local coordination system to the global one

$$D(\varphi_i,\vartheta_i) = \begin{pmatrix} \cos\frac{\vartheta_i}{2}\exp\left(-\frac{\mathrm{i}}{2}\varphi_i\right) & -\sin\frac{\vartheta_i}{2}\exp\left(-\frac{\mathrm{i}}{2}\varphi_i\right) \\ \sin\frac{\vartheta_i}{2}\exp\left(\frac{\mathrm{i}}{2}\varphi_i\right) & \cos\frac{\vartheta_i}{2}\exp\left(\frac{\mathrm{i}}{2}\varphi_i\right) \end{pmatrix}. \tag{6}$$

The band-structure problem is solved using the real-space recursion method [28], which is very efficient in combination with the TB-LMTO Hamiltonian. Therefore complex systems with many degrees of freedom can be treated.

The determination of a non-collinear magnetic structure proceeds as follows. In the first step the ground state of the collinear magnetic arrangement is found for the Hamiltonian (1). The potential parameters c^α, d^α together with the Stoner parameters I_{i2} calculated from Eqs. (2) and (3) are used in the construction of the Hamiltonian for the non-collinear calculation with the exchange part due to Eq. (5), so that all the procedure is parameter-free. The starting magnetic moment vectors are distributed randomly or small random transversal components are added to the magnetic moments resulting from previous collinear calculation. Partial densities of states are obtained for the x-, y-, z-directions of a moment in its local basis by choosing as the starting recursion vector the normalized eigenvector of σ^x, σ^y, σ^z, respectively, for up and down spin directions. The starting recursion vector is then rotated to the global coordinate system by multiplying it with $D(\varphi_i, \vartheta_i)$ from Eq. (6). In general, the new magnetic moments obtained from the integrated projected densities of states will have transversal components with respect to the last moment directions. The partial densities of state along the direction of the magnetic moments are used in the update of the charge densities, potential parameters and exchange splitting fields at each step. The process continues in an iterative way until the transversal components are sufficiently small. The moment rotations are quite slow during the iteration process, therefore the new orientations are extrapolated from the old and new directions and besides random noise components are added in order to avoid running into nearest local minima. In the prediction of the new non-collinear structure we use a Broyden mixing scheme [29]. For further technical details of the non-collinear calculations we refer to the papers [10,19,30].

In some cases symmetry restrictions allow only a few special spin arrangements in a system. Then the aim is to find a ground state spin configuration. This kind of calculations can be done as described above but only the densities of state projected along the moments are necessary what results in the much faster calculation. To the group of models with fixed directions of magnetic moments belongs a determination of the magnetic anisotropy energy.

2.1   Magnetocrystalline Anisotropy

The spin-orbit coupling responsible for the magnetocrystalline anisotropy can be included into Hamiltonian given by Eq. (1) and Eq. (4) by adding an intra-atomic term

$$H^{\text{so}}_{ils,i'l's'} = \frac{1}{2}\delta_{ll'}\xi_{ils,i'l's'}(E)\left(\sigma\mathbf{l}\right)_{ils,i'l's'} \quad . \tag{7}$$

The matrix elements $\frac{1}{2}(\sigma\mathbf{l})_{ils,i'l's'}$ for the d-orbitals in the frame rotated to the magnetic moment direction can be found in Ref. [31]. The spin-orbit

Table 1. Spin-orbit coupling parameters in meV for iron, cobalt and nickel calculated at their experimental lattice constants for d-bands at the Fermi level E_F and at the middle of the occupied parts of the d-bands ϵ_ν. The results obtained by LSDA exchange-correlation potentials are compared with those obtained by GGA exchange-correlation potentials (in parentheses).

	$\xi_{d\uparrow}(E_F)$	$\xi_{d\downarrow}(E_F)$	$\xi_{d\uparrow}(\epsilon_\nu)$	$\xi_{d\downarrow}(\epsilon_\nu)$
Fe	72(72)	56(55)	58(58)	45(45)
Co	94(94)	78(78)	73(74)	63(63)
Ni	109(111)	103(103)	91(88)	84(84)

parameters $\xi_{ils,i'l's'}(E)$ are assumed to be non-negligible only between the d-orbitals centered on the same site and they are given in Ry units [32] as

$$\xi_{ils,i'ls'}(E) = \frac{2}{c^2}\delta_{ii'}\delta_{ll'}\int \phi_{ils}(E,r)\frac{dV(r)}{dr}\phi_{i'l's'}(E,r)r^2 dr \quad . \tag{8}$$

Here c is the velocity of light, $\phi_{ils}(E,r)$ are the radial partial waves calculated at the energy E and $V(r)$ is the one-electron interaction potential. The spin-orbit coupling parameters obtained for iron, cobalt and nickel are shown in Table 1. We have found that using the Barth-Hedin-Janak local-spin-density approximation (LSDA) [33,34] and the generalised gradient approximation (GGA) [35] results for the spin-orbit coupling differing less than 3 %.

Because the spin-orbit coupling constants are much smaller than the band width for 3d-metals the magnetic anisotropy energy is often evaluated as a difference of the sums of the single-particle eigenvalues for the opposite spin directions treating the spin-orbit term (7) as a perturbation. Even then the calculations in the **k**-space are very laborious [21]. Recently it has been demonstrated in several papers that the real-space approach makes the task of the MAE estimation possible in a non-perturbative fashion from the total ground state energies [19,20,36]. From practical reasons the inclusion of the spin-orbit coupling in the non-collinear calculations has the advantage of reducing somewhat the drift of the overall magnetic moment in the course of the iteration process.

The determination of the preferential magnetisation orientation is of much interest especially for thin magnetic films and multilayers, which possess a lowered symmetry. However, in the layered systems the other significant contribution, the magnetostatic shape anisotropy coming from the dipole-dipole interaction, must be taken into account. Because the shape anisotropy always prefers the in-plane magnetisation, it is responsible for the changing the orientation of the magnetisation to the plane at some critical thickness if the spin-orbit contribution to the MAE happens to support a perpendicular anisotropy. When the thickness of the magnetic film is reduced to a few monolayers, the contributions from all discrete dipole pairs have to be summed up explicitly

$$E^{\mathrm{dip}} = \frac{1}{c^2} \sum_{<i,j>} \frac{1}{r_{ij}^3} \left(\mathbf{m}_i . \mathbf{m}_j - 3\frac{\mathbf{m}_i . \mathbf{r}_{ij} \mathbf{m}_j . \mathbf{r}_{ij}}{r_{ij}^2} \right) \quad , \tag{9}$$

rather than to resort to the continuum approximation. The sums appearing in Eq. (9) converge slowly due to the long-range character of the dipole-dipole interaction but they can be efficiently evaluated in the reciprocal space [37]. For the cubic sc, bcc and fcc lattices with one atom type per layer the magnetostatic dipolar energy can be expressed in Ry units as

$$E^{\mathrm{dip}} = \frac{1}{c^2 a^3 n_{2D}^2} \sum_{<i,j>} m_i m_j (\cos \vartheta_i \cos \vartheta_j -$$
$$-\frac{1}{2} \cos (\varphi_i - \varphi_j) \sin \vartheta_i \sin \vartheta_j) M_{ij} \quad , \tag{10}$$

where n_{2D} means a number of atoms in one layer, a is the lattice parameter of the basic cubic cell and the M_{ij} are the dipolar Madelung constants tabulated in Table 2.

2.2   Exchange coupling constants

The modern spin-polarised band theory gives an accurate description of the magnetic ground state of most metals and alloys. The magnetic excitations from the ground state are described in terms of various spin models in which the strength of a pair interaction is controlled by a magnitude of the exchange coupling.

Recently we have derived expressions for the exchange pair coupling constants and some other related quantities within a real-space approach [16]. The exchange interaction between the ith and jth moments takes a form

$$J_{ij} = \frac{\Delta_i \Delta_j}{2\pi} \mathrm{Im} \int^{E_F} \mathrm{Tr} \, G_{ij}^{\uparrow\uparrow}(E) G_{ji}^{\downarrow\downarrow}(E) dE \quad , \tag{11}$$

Table 2. Dipolar Madelung constants for the sc, bcc and fcc lattice geometries provided the basic cubic cell of unit volume, z stands for the interlayer distance.

layer	z	sc	bcc	fcc
0	0	9.03362	9.03362	25.55094
1	$\frac{1}{2}$	—	4.17639	4.04301
2	1	-0.32746	-0.32746	-0.06402
3	$\frac{3}{2}$	—	0.01238	0.00072
4	2	-0.00055	-0.00055	0.00001

where Δ is the exchange splitting field defined by Eq. (2) and the off-site Green's functions $G_{ij}^{ss}(E)$ for the up and down spins are calculated by the recursion method using the bonding and antibonding combinations for the sites i and j.

The mean-field estimation of the Curie temperature is related to the on-site exchange coupling J_{ii} via

$$T_{C,i} = \frac{1}{3k_B} \left(\frac{1}{2} \Delta_i m_i - J_{ii} \right) \quad . \tag{12}$$

The critical temperature $T_{C,i}$ should be viewed as a measure of the local stability of the magnetic state of the ith atom surrounded by all other atoms.

Another quantity related to the exchange couplings and accessible to the experimental verification is the spin-wave stiffness constant

$$D_i = \sum_j J_{ij} r_{ij}^2 \quad . \tag{13}$$

Here r_{ij} are the lattice vectors. The results for the nearest and the next nearest exchange couplings, the Curie temperature and the stiffness constants for iron, cobalt and nickel are presented in Table 3. The stiffness constant for iron was calculated using the first 14 terms, for cobalt and for nickel the first 17 terms in Eq. (13) were taken into account. The overall agreement with the experimental data confirms that the spin models formulated originally for systems with localised magnetic moments can be still be used as a reasonable approximation for itinerant magnets.

3 Ni/Cu(001) Films

3.1 Atomic Structure

Recently we have witnessed extensive experimental [24,42–47] and theoretical investigations [48–50] of the Ni films grown on Cu. The Ni/(001)Cu system has an average lattice mismatch of only 2.6 % favoring a coherent growth of Ni on

Table 3. The nearest and the next nearest exchange couplings J_1, J_2, the experimental and the calculated values of Curie temperature T_C and the spin-wave stiffness constant D for iron, cobalt and nickel.

	J_1 (meV)	J_2 (meV)	T_C (K)	$T_C^{\exp}$ (K)	D (meVÅ^2)	$D^{\exp}$ (meVÅ^2)
Fe (bcc)	16.27	17.29	890	1044[a]	280	280[b]
Co (hcp)	25.05	4.11	1000 (β Co)	1388[c]	1900	580[b]
Ni (fcc)	4.62	0.20	290	627[a]	530	555[d]

[a] Ref. [38], [b] Ref. [39], [c] Ref. [40], [d] Ref. [41]

Table 4. The relaxed atomic structures of films with one to seven Ni layers on Cu(001) from first-principles calculations using the VASP package. The results are compared with the available experimental data taken at room temperature.

ML	1		2		3			5		7	fct Ni
$d_{12}(\%)$	-6.2	-7.0	-11.9	-2.4	-11.4	-3.5	-1.1	-10.8	1.6	-11.0	-11.1
$d_{23}(\%)$	0.2	-1.0	-1.9	-1.0	-7.4	-7.5	-3.9	-5.1	-6.1	-5.8	-6.2
$d_{34}(\%)$			0.5	-0.5	-3.2	0.4	-2.8	-5.8	-5.0	-5.9	-6.9
$d_{45}(\%)$					0.0	6.6	-1.1	-6.8	-5.6	-5.3	-7.0
$d_{56}(\%)$								-1.2		-6.0	
$d_{67}(\%)$								1.7		-6.2	
$d_{78}(\%)$										-0.7	
$d_{89}(\%)$										3.0	
Ref.	[45]		[45]		[45]	[46]		[46]			

Cu in an artificial tetragonally distorted face centered cubic (fct) structure up to a thickness of about 40 layers above which the growth continues in the fcc structure.

We investigated first the relaxation of films with one to seven Ni layers on Cu(001) using the spin-polarised version of the Vienna ab-initio simulation package (VASP). A detailed description of the VASP and its algorithms can be found in Ref. [51]. We used the Ceperley and Alder [52] local spin-density functional and the generalised gradient approximation corrections [35] to the exchange-correlation energy.

In the first step of the calculation the equilibrium lattice spacing of bulk Cu was found 3.637 Å, in close agreement with the measured value 3.61 Å. Then we performed a geometry optimisation of the Cu surface for a slab with 6 layers. We found a 3.6 % inward relaxation of the surface layer. The predicted relaxation agrees well with previous ab-initio calculations [53], but is somewhat lower than the relaxation found in experiment (1.2 % in Ref. [54], 2.4 % in Ref. [55]). The difference is mostly due to the fact that experiments have been performed at room temperature. Afterwards the slab was extended on one side with 1 to 7 Ni layers and during the relaxation the lateral lattice spacing was kept at the Cu bulk value. The inspection of the obtained layer relaxations summarised in Table 4 reveals clear trends. The surface layer undergoes a strong inward relaxation, the subsurface layers form a fct lattice with an axial ratio $c/a \approx 0.94$. The relaxation of the bulk fct Ni with the lattice spacing of the Cu bulk leads to a tetragonal distortion of $c/a \approx 0.93$. The rightmost column of Table 4 gives the structure of the fct Ni surface modeled as a slab of eight layers. Again the surface layer relaxes inward by a 11 % and the c/a ratio reaches the value 0.93 in the middle of the slab. The estimate within the continuum elasticity theory of coherent epitaxy-induced structural changes gives $c/a \approx 0.965$.

Table 5. Magnetic moments in relaxed Ni/Cu(001) films with up to seven monolayers of Ni and of fct Ni in μ_B. First two interface Cu layers are shown as well. The results within RS-TB-LMTO approach can be compared with the moments obtained by VASP package (right columns).

ML	1		2		3		5		7		fct Ni	
1	0.12	0.38	0.76	0.75	0.78	0.74	0.77	0.74	0.74	0.73		
2	0.00	0.01	0.48	0.52	0.55	0.59	0.58	0.68	0.58	0.68		
3	-0.00	-0.01	-0.01	-0.01	0.50	0.47	0.69	0.71	0.68	0.69		
4			-0.00	0.00	-0.01	0.00	0.62	0.69	0.64	0.68		
5					-0.01	0.00	0.48	0.58	0.68	0.71		
6							-0.01	0.01	0.64	0.70		
7							-0.01	0.00	0.47	0.58		
8									-0.01	-0.01		
9									-0.01	-0.01		
$\bar{m}$	0.12	0.38	0.62	0.63	0.61	0.60	0.63	0.68	0.63	0.68	0.66	0.61

For a single monolayer of Ni/Cu(001), the predicted relaxation is in very good agreement with low-energy electron-diffraction (LEED) experiments by Kim et al. [45]. For thicker layers, however, LEED experiments predict only a minimal inward relaxation (and for yet thicker layers even an outward relaxation) of the top layer [45,46], at an almost homogeneous tetragonal distortion of the deeper part of the film. The first-principle calculations, on the other side predict a large inward relaxation ot the toplayer, but agree with experiment concerning the tetragonal distortion of the interior of the film.

Recent ab-initio calculations (based on the same technique) of the structural, electronic and magnetic properties of all low index surfaces of Ni [56] lead to excellent agreement with experiment. Similar discrepancies between ab-initio calculation and the experiment as those observed for Ni/Cu(001) have been recorded for a number of transition metal surfaces, e.g. Rh(001) and attributed to an anomalously large perpendicular thermal expansion at the surface [57], limiting the comparison between the room-temperature experiment and the $T = 0$ K calculation. It must be left to the future work whether this mechanism also explains the discrepancy between calculations and experiment for the surface of Ni/Cu(001) films.

3.2 Magnetic Structure

Using the relaxed atomic structural models discussed in the previous paragraph we have calculated the magnetic structure and the magnetic anisotropy energy

Table 6. Spin-orbit coupling parameters for up (left columns) and down (right columns) partial waves in the relaxed Ni/Cu(001) films with up to seven layers of Ni in meV. The results for bulk fct Ni are shown in the right-most column.

ML	1	2	3	5	7	fct Ni
1	97.3 95.9	107.2 98.3	106.8 97.7	106.6 97.6	106.5 97.8	
2		105.7 102.6	111.6 104.5	111.3 103.8	111.8 104.3	
3			107.1 101.0	111.1 102.3	111.4 102.7	
4				110.8 102.9	110.7 102.6	
5				106.6 100.7	111.3 102.7	
6					111.4 103.2	
7					106.7 100.7	
						110.2 101.9

of the Ni/Cu(001) films within the framework of RS-TB-LMTO method. The real-space recursion technique was applied to cells with 1944 atoms for 1 Ni monolayer (ML) up to 2560 atoms for 7 Ni ML. Periodic boundary conditions in the lateral directions and the free boundary conditions normal to the layers were used. In all cases 20, 20, and 50 recursion levels were used for the s-, p-, d-orbitals, respectively. Because for the thicker Ni films we obtained a systematic inward relaxation 6 %, we studied also an infinite fct Ni crystal with the tetragonal distortion $c/a = 0.94$.

The layer-resolved and the average magnetic moments are shown in Table 5. As can be expected, the moments at the surface are enhanced (except for the monolayer) while the moments at the interface drop. This behaviour is observed independent of the film thickness. Although the total average magnetic moment approaches the bulk value gradually, even the system with 7 Ni ML is influenced so strongly by the surface that it is not possible to identify a subsurface region with steady bulk-like magnetic moments. We also note that the RS-TB-LMTO calculations agree well with k-space results obtained using VASP.

Due to the relatively low Curie temperature of bulk Ni, the thin films of Ni can be studied in a wide temperature range as a function of film thickness. The comprehensive collection of experimental data establishes a clear picture of magnetic anisotropies in Ni/Cu(001) films [24,43,44]. As the film thickness increases the magnetisation changes its direction from [100] to [110] between 6 and 7 Ni ML and between 7 and 8 Ni ML it switches continuously to the [001] direction. The surface and volume contributions to the second-order anisotropy constants K_2^s and K_2^v at zero temperature lie between -100 up to -180 μeV/atom and 40 up to 75 μeV/atom [44], respectively. The large error bars are due to the uncertainty in the extrapolation down to zero temperature. For the fourth-order in-plane and out-of-plane anisotropy constants very small values of $K_{4\parallel} \approx$

-1 μeV/atom and $K_{4\perp} \approx 0.2$ μeV/atom at $T/T_c = 0.8$ were reported for 7 Ni ML [44].

We calculated MAE using the force theorem [21], treating the spin-orbit coupling as a perturbation. As can be seen from Table 6 the spin-orbit coupling parameters are only slightly affected by the surface and interface. The layer-resolved and the total band and dipole-dipole contributions to the MAE are compiled in Table 7. The contributions of the surface layer to the MAE in all but monolayer films support an out-of-plane orientation of the magnetisation and the magnitude of these contributions saturates with the film thickness. The contribution of the Ni layer at the interface to the Cu substrate also prefers a perpendicular orientation of the magnetic moments (except for the 2 ML film), but does not show a systematic variation as the number of layers increases. Quite surprisingly, we also find a large contribution from the subsurface layer and the second Ni layer from interface, always preferring an in-plane orientation. The contributions from the inner layers in films with ≥ 5 ML are always smaller than the surface and interface contributions. As can be seen from Fig. 1, the 2 ML and 3 ML films show the perpendicular magnetic anisotropy, all other films we have investigated have an easy axis in the plane. The spin-reorientation transition between the 3 ML and 5 ML case is driven by the subsurface contribution to the MAE. For the Ni monolayer we predict $K_2 = -263$ μeV/atom in a fair agreement with the measured value about -157 μeV/atom [43], previous calculations giving -94 μeV/atom [48] or -690 μeV/atom [50]. For 2 Ni ML our result $K_2 =$

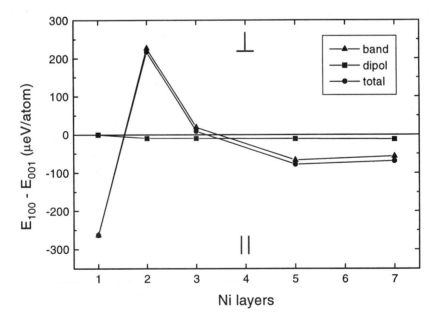

Fig. 1. Calcutated total magnetic anisotropy energy (circles) and its dipole-dipole (triangles) and band (squares) contributions for thin Ni films on Cu(001).

Table 7. Spin-orbit ($\triangle E_b$, left columns) and dipolar ($\triangle E_d$, right columns) contributions to the magnetic anisotropy energy $\triangle E = E_{100} - E_{001}$ in the relaxed Ni/Cu(001) films with up to seven Ni layers. In the last row the estimate of the fourth-order anisotropy constant $K_{4\parallel} = 4(E_{110} - E_{100})$ is given. The values for the layer-resolved contributions are in μeV, the averaged values are in μeV/atom.

ML	1		2		3		5		7		fct Ni	
1	-263	0	822	-14	1511	-15	2046	-15	2202	-14		
2			-369	-7	-2112	-10	-1036	-11	-1049	-11		
3					659	-6	355	-14	254	-13		
4							-1870	-11	-641	-12		
5							168	-6	-316	-13		
6									-1330	-12		
7									481	-6		
$\triangle \bar{E}_{b,d}$	-263	0	227	-10	19	-10	-67	-11	-57	-12	52	0
$\triangle(\bar{E}_b + \bar{E}_d)$	-263		217		9		-78		-69		52	
$K_{4\parallel}$	-331		103		24		-17		-6		-3	

227 μeV/atom compares very well with $K_2 = 300$ μeV/atom obtained by Wu and Freeman [50]. We note that the earlier calculations considered unrelaxed fct or fcc lattices.

For an infinite fct Ni crystal with the tetragonal distortion $c/a = 0.94$ we performed a series of calculations for models with different number of atoms. We obtained almost the same values for the MAE 50, 54, 52 μeV/atom for cells with 2048, 2916 and 6912 atoms, respectively. These values are only a bit lower than results 60 μeV [49] and 65 μeV [50] found in the k-space calculations.

The decreasing negative values of the MAE for 5 and 7 ML, together with the positive value for the infinite fct Ni indicate the possibility of an in-plane to perpendicular reorientation at a thickness > 7 ML. However, it must be left to future studies to locate this transition precisely.

The MAE for 2 Ni ML as a function of the tilt angle ϑ taken from the [001] direction is shown in Fig 2. The dependence on $\cos^2 \vartheta$ is almost linear. Our attempt to estimate the higher-order term yields a ratio between the fourth-order and the second-order contribution of 0.01. Here and below we use for the angular dependence of the MAE the expression [44]

$$E(\varphi, \vartheta) = E_0 - K_2 \cos^2 \vartheta - \frac{1}{2}K_{4\perp} \cos^4 \vartheta - \frac{1}{8}K_{4\parallel}(3 + \cos 4\varphi)\sin^4 \vartheta \quad . \quad (14)$$

Whether the reorientation of the magnetisation between the in-plane and out-of-plane orientations happens continuously or abruptly depends on the sign of

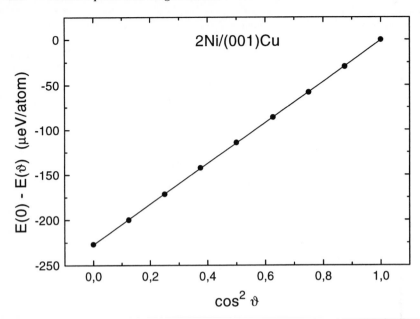

Fig. 2. Band energy difference plotted as a function of ϑ for 2 Ni ML films on Cu(001). The solid line represents the fit by a parabola.

$K_{4\parallel}$. For a stabilisation of a tilted magnetisation (and hence a continuous second or higher order spin-reorientation transition) a negative $K_{4\parallel}$ is needed. The value of the fourth-order in-plane anisotropy constant $K_{4\parallel}$ can be obtained from the variation of the MAE with $\cos 4\varphi$. In the last row of Table 7 we show our results for $K_{4\parallel}$. It can be concluded that its thickness dependence is rather complex and the $K_{4\parallel}$ changes sign between 3 and 5 ML. Because these values are typically as small as few μeV we carried out additional calculations for several intermediate angles between 0 and 45° for 2, 3 and 7 ML films (Fig. 3). From the scatter of the points around a linear fit versus $\cos 4\varphi$ we estimate that the confidence level in the numerical accuracy of our approach is better than 0.3 μeV/atom in all cases. Actually, the curves for 2 and 3 ML seem to be modulated systematically. Despite of the exceedingly small values of the fourth-order MAE the calculated values are in a reasonable agreement with experimentally observed trends.

In order to understand the influence of the surface and of the tetragonal distortion of the films on the magnetic anisotropy, a correlation between the number of holes in the Ni-d band (as observable in near-edge x-ray-absorption fine-structure (NEXAFS) experiments) and possible anisotropy of the d-band occupation has been evoked [47]. It has been argued that the number of holes in the Ni-d band is strongly reduced in the thinnest films, converging to a bulk-like value at a thickness of about 5 ML. In addition, an in-plane character of the d-holes irrespective of the thickness has been reported.

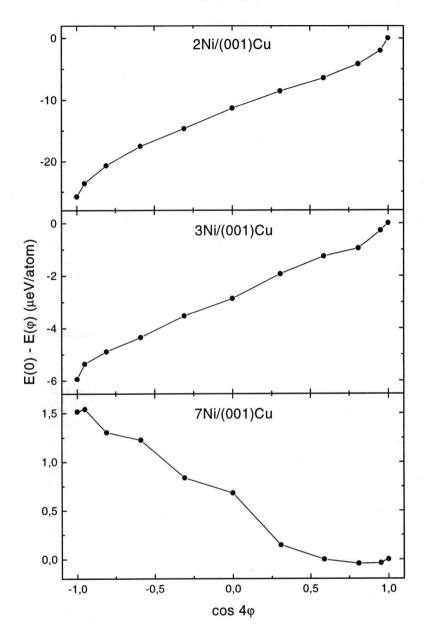

Fig. 3. Band energy difference plotted as a function of φ for 2, 3 and 7 Ni ML films on Cu(001).

In Table 8 we present the number of 3d-holes (unoccupied states) separated into holes in in-plane orbitals (xy, $x^2 - y^2$ for the (001) plane) and out-of-plane orbitals (yz, zx and $3z^2 - r^2$ for the (001) plane). It is obvious that the 3d-

Table 8. Numbers of 3d-holes with the in-plane ($d_\parallel$, left columns) and out-of-plane ($d_\perp$, right columns) symmetry for Ni atoms in one to seven Ni/Cu(001) films. The ratio $\bar{d}_\parallel/\bar{d}_\perp$ determines the character of the holes ($\bar{d}_\parallel/\bar{d}_\perp > 2/3$ means in-plane character, $\bar{d}_\parallel/\bar{d}_\perp < 2/3$ means out-of-plane character). The results for bulk fct Ni are shown in the rightmost column.

ML	1		2		3		5		7		fct Ni	
1	0.73	0.62	0.70	0.74	0.67	0.77	0.68	0.76	0.66	0.78		
2			0.53	0.88	0.51	0.93	0.51	0.93	0.51	0.93		
5					0.59	0.86	0.56	0.92	0.56	0.93		
6							0.55	0.92	0.56	0.92		
7							0.58	0.85	0.56	0.92		
									0.54	0.94		
									0.57	0.86		
$\bar{d}_{\parallel,\perp}$	0.73	0.62	0.61	0.81	0.59	0.85	0.58	0.88	0.57	0.90	0.55	0.91
$\bar{d}_\parallel + \bar{d}_\perp$	1.35		1.42		1.44		1.45		1.46		1.46	
$\bar{d}_\parallel/\bar{d}_\perp$	1.17		0.76		0.69		0.66		0.63		0.61	

band filling is reduced progressively as the film thickness increases and at the same time the hole character changes from the in-plane to out-of-plane between 4 and 5 Ni ML. The increase of the number of Ni holes with increasing film thickness compares well with the experimental observations reported in Ref. [47]. In addition we find that the ratio $d_\parallel/d_\perp$ is enhanced in the surface layer over the value $d_\parallel/d_\perp = 2/3$ corresponding to an isotropic distribution of the 3d-holes. Again this agrees with the conclusions derived from the NEXAFS experiments where the anisotropy has been attributed to the tetragonal distortion of the films. However, whether this conjecture is correct remains to be verified by reference calculations for undistorted films. The in-plane character of the 3d-holes means at the same time that the 3d-electrons have perpendicular character, and this agrees with the positive contributions of the surface layers to the MAE. In the deeper layers, the ratio $d_\parallel/d_\perp$ drops below 2/3 and is smallest in the subsurface layer and the second layer from the interface. Again this correlates well with the negative contributions to the MAE noted for these layers. Taking the average over the entire film, we find that the hole character changes between 3 and 5 ML what correlates with the reversal of the MAE. Altogether this analysis demonstrates that there are important changes in the partial electronic density of states near the Fermi level as a function of the film thickness whose evident correlations to the MAE deserve further investigation.

4   Conclusions

The RS-TB-LMTO formalism described in the first part of the paper, together with an *ab-initio* density functional approach to the reconstruction of the films, has been applied to study the variation of the magnetic anisotropy of Ni/Cu(001) films with increasing film thickness. The predicted tetragonal distortion of the deeper layers agrees with experimental observations, but there is disagreement concerning the obtained inward relaxation of the top layer – this is possibly related to the confrontation of the $T = 0$ K calculations with room-temperature experiments.

For the magnetic anisotropy, we predict a very complex behaviour: the change from in-plane (1 ML) to perpendicular (2, 3 ML) back to in-plane(5, 7 ML) and eventually again back to perpendicular for thicker layers (as long as the film remains tetragonally distorted). A detailed analysis reveals a competition between surface and interface contributions favouring a perpendicular orientation and subsurface and subinterface contributions favouring in-plane orientation of the magnetic moments. The correlations to a changing anisotropic population of the Ni-3d bands have been investigated and found to agree with the interpretation of NEXAFS experiments.

Acknowledgments

This work has been supported by the Austrian Ministery for Science and Transport within the project "Magnetism on the Nanometer Scale".

References

1. Andersen, O. K., Jepsen, O., Šob, M. in: Electronic Band Structure and its Applications, Ed. Yussouff, M., Springer, Berlin (1986).
2. Becker, Ch., Hafner, J., Lorenz, R. , J. Magn. Magn. Mat. **157-158**, 619 (1996).
3. Lorenz, R., Hafner, J., Jaswal, S. S., Sellmyer, D., J., Phys. Rev. Lett. **74**, 3688 (1995).
4. Turek, I., Becker, Ch., Hafner, J., J. Phys.: Condens. Matter **4**, 7257 (1992).
5. Hafner, J., Krajčí, M., Phys. Rev. B **57**, 2849 (1998).
6. Mattis, D., C., The Theory of Magnetism, Vols. I-II, Springer, Berlin (1985).
7. Sandratskii, L., M. , Adv. Phys **47**, 91 (1985).
8. Sticht, J., Höck, K.-H., Kübler, J. , J. Phys.: Condens. Matter **1**, 8155 (1989).
9. Mryasov, O. N., Liechtenstein, A. I., Sandratskii, L. M., Gubanov, V. A. J. Phys.: Condens. Matter **3**, 7683 (1991).
10. Lorenz, R., Hafner, J., J. Magn. Magn. Mat. **139**, 209 (1995).
11. Krey, U., Krompiewski, S., Krauss, U., J. Magn. Magn. Mat. **86**, 85 (1990).
12. Freyss M., Stoeffler, D., Dreyssé, H., , Phys. Rev. B **54**, R12677 (1996).
13. Uhl, M., Sandratskii, L., M., Kübler, J., J. Magn. Magn. Mat. **103**, 314 (1992).
14. Kübler, J., Sandratskii, L. M., Uhl, M., J. Magn. Magn. Mat. **104-107**, 695 (1992).
15. Smirnov, A., V., Bratkovsky, A., M., Phys. Rev. B **53**, 8515 (1996).
16. Spišák, D., Hafner, J., J. Magn. Magn. Mat. **168**, 257 (1997).

17. Spišák, D., Hafner, J., Phys. Rev. B **55** 8304 (1997), ibid. B **56**, 2646 (1997).
18. Lorenz, R., Hafner, J., J. Phys.: Condens. Matter **7**, L253 (1995).
19. Lorenz, R., Hafner, J., Phys. Rev. B **54**, 15 937 (1996).
20. Beiden, S. V., Temmerman, W., M., Szotek, Z., Gehring, G. A., Stocks, G., M., Wang, Y., Nicholson, D. M. C., Shelton,W. A., Ebert, H., Phys. Rev. B **57**, 14 247 (1998).
21. Daalderop, G. H. O., Kelly, P. J., Schuurmans, M. F. H., Phys. Rev. B **41**, 11919 (1990).
22. Hoermandinger, G., Weinberger, P. ,J. Phys.: Condens. Matter **4**, 2185(1992) .
23. Schulz, B., Baberschke, K., Phys. Rev. B **50**, 13 467 (1994).
24. Bochi, G., Ballentine, C. A., Inglefield, H. E., Thompson, C. V., O'Handley, R. C., Hug Hans J., Stiefel, B., Moser, A., Güntherodt, H.-J., Phys. Rev. B **52**, 7311 (1995).
25. Nowak, H. J., Andersen, O. K., Fujiwara, T., Jepsen, O., P. Vargas, P., Phys. Rev. B **44**, 3577 (1991) .
26. Becker, Ch., Hafner, J., Phys. Rev. B **50**, 3933 (1994).
27. Spišák, D., Becker, Ch., Hafner, J., Phys. Rev. B **51**, 11 616 (1995).
28. Haydock, R., Heine, V., Kelly, M. J. in: Solid State Physics, Advances in Research and Applications, Eds. Ehrenreich, H., Turnbull, D., Seitz, F., Vol **35** Academic Press, New York (1980).
29. Srivastava, G. P., J. Phys. A: Math. Gen. **17**, L317 (1984).
30. Lorenz, R., Hafner, J., Phys. Rev. B **58**, 5197 (1998).
31. Abate, E., Asdente, M., Phys. Rev. **140**, A1303 (1965).
32. Andersen, O. K., Phys. Rev. B **12**, 3060 (1975).
33. von Barth, U., Hedin, L. , J. Phys. C: Solid State Phys. **5**, 1629 (1972).
34. Janak, J. F., Solid State Commun. **25**, 53 (1978).
35. Perdew, J. P., Wang, Y., Phys. Rev. B **45**, 13 244 (1992). **41**, 11 919
36. Dorantes-Dávila, J., Pastor, G. M., Phys. Rev. Lett. **77**, 4450 (1996).
37. Tsymbal, E., J. Magn. Magn. Mat. **130**, L6 (1994).
38. Leger, J. M., Loriers-Susse, C., Vodar, B., Phys. Rev. B **6**, 4250 (1972).
39. Panthenet, R., J. Appl. Phys. **53**, 8187 (1982).
40. Cowan, D. L., Anderson, L. W., Phys. Rev. A **139**, 424 (1965).
41. Mook, H. A., Lynn, J. W., Nicklow, R. M., Phys. Rev. Lett. **30**, 556 (1973).
42. Huang, F., Kief, M. T., Mankey, G. J., Willis, R. F., Phys. Rev. B **49**, 3962 (1994).
43. Baberschke, K., Appl. Phys. A **62**, 417 (1996).
44. Farle, M., Mirwald-Schulz, B., Anisimov, A. N., Platow, W., Baberschke, K., Phys. Rev. B **55**, 3708 (1997).
45. Kim, S. H., Lee, K. S., Min, H. G., Seo, J., Hong, S. C., Rho, T. H., Kim, J.-S., Phys. Rev. B **55**, 7904 (1997).
46. Müller, S., Schulz, B., Kostka, G., Farle, M., Heinz, K., Baberschke, K., Surf. Science **364**, 235 (1996).
47. Srivastava, P., Haack, N., Wende, H., Chauvistré, R., Baberschke, K., Phys. Rev. B **56**, R4398 (1997).
48. Moos, T. H., Hübner, W., Bennemann, K. H., Solid State Commun. **98**, 639 (1996).
49. Hjortstam, O., Baberschke, K., Wills, J. M., Johansson, B., Eriksson, O., Phys. Rev. B **55**, 15026 (1997).
50. Wu, R., Chen, L., Freeman, A. J., J. Appl. Phys.**81**, 4417 (1997).
51. Kresse, G., Furthmüller, J., Comput. Mater. Sci. **6**, 15 (1996), Phys. Rev. B **54**, 11 169 (1996).
52. Ceperley D. M., Alder, B. J., Phys. Rev. Lett. **45**, 566 (1980).

53. Rodach, Th., Bohnen, K.-P., Ho, K. M., Surf. Sci. **286**, 66 (1993).
54. Lind, D., M., Dunning, F., B., Walters, G. K., Davis, H. L., Phys. Rev. B **35**, 9037 (1987).
55. Jiang, Q., T., Fenter, P., Gustafsson, T. Phys. Rev. B **44**, 5773 (1991).
56. Mittendorfer, F., Eichler, A., Hafner, J., Surf. Sci. (in press) (1998).
57. Cho, J. H., Scheffler, M., Phys. Rev. Lett. **78**, 1299 (1997).

Combining Real Space and Tight Binding Methods for Studying Large Metallic Systems

C. Cornea and D. Stoeffler

Institut de Physique et Chimie des Matériaux de Strasbourg (CNRS UMR 7504),
Groupe d'Etude des Matériaux Métalliques,
23, rue du Loess, F-67037 Strasbourg, France

Abstract. In this paper some problems experienced during studies combining real space and tight binding methods are addressed. These methods have been mainly used for studying the magnetic properties of thin films deposited on substrates and of multilayers taking into account interfacial imperfections. This paper is illustrated with calculations of the electronic structure of Fe/Cr multilayered systems which are particularly interesting. First, the use of d and spd tight binding parameterisations of the electronic structure for transition metals and its relation to the recursion technique is discussed. Second, some advantages of using real space cells for studying complex systems are presented. Finally, the application of these methods for systems presenting non-collinear magnetism is discussed.

1 Introduction

During the last ten years, the electronic structure of large and complex metallic systems has been extensively studied mainly due to the enhancement of the computer facilities. Powerful computers with large memories became available allowing to reach rapidly self consistency in the band structure calculations for cells containing up to a few hundred of heavy atoms. One of these kinds of systems, concerns the metallic multilayers presenting new magnetic properties like the Interlayer Magnetic Coupling (IMC) or the Giant Magneto Resistance (GMR) effect particularly interesting for applications. This paper deals with the use of the real space recursion technique for the study of the magnetic order in such metallic multilayers.

The multilayered $A_m B_n$ system built by alternating a m monolayers thick A layer with a n monolayers thick B layer consists in a long elemental chemical cell containing, in the simplest case, one non equivalent atom in the in plane cell and $m + n$ atoms in the growth direction perpendicular to the plane of the layers. Because these multilayers are periodic in the 3 directions of space, the band structure is usually calculated in the **k** space of the reciprocal lattice. However, since the aim of more complete studies is usually to determine the magnetic properties for thin overlayers during the growth of the multilayer, to include interfacial imperfections, to relate the growth mode and the magnetic behaviour, ... a real space technique is used in order to have the possibility to calculate the electronic structure of all these situations with the same method.

The aim of this paper is to discuss possible problems and solutions used during various studies of the Fe/Cr multilayered system. A d restricted tight binding modelling of the band structure has been first used mainly because it needs less computer time than a full *spd* description and allows to use an exact cluster for the recursion technique. More recently, due to physical lacks in some results, the description has been extended taking the *spd* hybridisation into account. However, in order to reduce the computer time, clusters smaller than the exact one are used for the calculation. In the first section, some possible choices for the shape of these "inexact" clusters are given and their use for large cells like the one of multilayers is discussed. In the second section, the advantages of using real space cells for studying such complex systems are discussed. Finally, in the last section, the application of these methods for systems presenting non-collinear magnetism is presented.

2   Tight Binding Parameterisation and Recursion Technique

2.1   The Recursion Technique

This method, proposed by Haydock and Heine [1–3], is well suited for the determination of the electronic structure when (i) the Hamiltonian H can be expressed in a finite basis of localized orbital $|i, \lambda\rangle$ of symmetry λ on the site i (like the one considered in the next subsection) and (ii) when the knowledge of the Green function $G_{i,\lambda}(z) = \langle i, \lambda | G(z) | i, \lambda \rangle = \langle i, \lambda | (z - H)^{-1} | i, \lambda \rangle$ elements is sufficient for the calculation of the band structure. For example, this method applies to situations presenting no symmetry like amorphous or disordered systems, around impurities or structural imperfections.

For each given site and symmetry (i, λ), a new basis $|n\}$ is built in order to have a tridiagonal matrix for the representation of H in this new basis. The basis $|n\}$ is recursively obtained starting from the $|i, \lambda\rangle$ basis function with the following expressions:

$$
\begin{aligned}
|0\} &= |i, \lambda\rangle \\
|1\} &= H|0\} - a_1^{i,\lambda}|0\} \\
|n+1\} &= H|n\} - a_{n+1}^{i,\lambda}|n\} - b_n^{i,\lambda}|n-1\}.
\end{aligned}
\tag{1}
$$

The sets $(a_n^{i,\lambda}, b_n^{i,\lambda})$ are called the recursion coefficients. They are easily obtained by calculations of simple scalar products:

$$
\begin{aligned}
a_{n+1}^{i,\lambda} &= \frac{\{n|H|n\}}{\{n|n\}} \\
b_{n+1}^{i,\lambda} &= \frac{\{n|H|n+1\}}{\{n|n\}} = \frac{\{n+1|n+1\}}{\{n|n\}}.
\end{aligned}
\tag{2}
$$

The orthonormalized recursion basis is then obtained by:

$$
|n\rangle = \frac{|n\}}{\sqrt{\{n|n\}}}
\tag{3}
$$

and the recursion coefficients correspond to:

$$a_{n+1}^{i,\lambda} = \langle n|H|n \rangle$$
$$\sqrt{b_{n+1}^{i,\lambda}} = \langle n|H|n+1 \rangle = \langle n+1|H|n \rangle. \tag{4}$$

The desired tridiagonal matrix representation of H is consequently obtained:

$$H = \begin{pmatrix} a_1^{i,\lambda} & \sqrt{b_1^{i,\lambda}} & & & 0 \\ \sqrt{b_1^{i,\lambda}} & a_2^{i,\lambda} & \sqrt{b_2^{i,\lambda}} & & \\ & \sqrt{b_2^{i,\lambda}} & a_3^{i,\lambda} & \sqrt{b_3^{i,\lambda}} & \\ & & \sqrt{b_3^{i,\lambda}} & a_4^{i,\lambda} & \ddots \\ 0 & & & \ddots & \ddots \end{pmatrix}. \tag{5}$$

Since $|i, \lambda\rangle = |0\rangle$, the Green function $G_{i,\lambda}(z)$ is equal to $\langle 0|(z - H)^{-1}|0\rangle$ which corresponds to the first element $[(z - H)^{-1}]_{00}$ of the inverse matrix of $(z - H)$. This particular element is easily obtained by considering the determinants $\det(z - H_n) = ||z - H_n||$ where H_n is the part of the H matrix limited to the elements $\{|n\rangle, |n + 1\rangle, |n + 2\rangle, ...\}$ of the recursion basis,

$$H_n = \begin{pmatrix} a_{n+1}^{i,\lambda} & \sqrt{b_{n+1}^{i,\lambda}} & & & 0 \\ \sqrt{b_{n+1}^{i,\lambda}} & a_{n+2}^{i,\lambda} & \sqrt{b_{n+2}^{i,\lambda}} & & \\ & \sqrt{b_{n+2}^{i,\lambda}} & a_{n+3}^{i,\lambda} & \sqrt{b_{n+3}^{i,\lambda}} & \\ & & \sqrt{b_{n+3}^{i,\lambda}} & a_{n+4}^{i,\lambda} & \ddots \\ 0 & & & \ddots & \ddots \end{pmatrix}. \tag{6}$$

The desired Green function is then equal to

$$\begin{aligned} G_{i,\lambda}(z) &= \frac{||z - H_1||}{||z - H_0||} \\ &= \frac{||z - H_1||}{(z - a_1^{i,\lambda})||z - H_1|| - b_1^{i,\lambda}||z - H_2||} \\ &= \frac{1}{z - a_1^{i,\lambda} - b_1^{i,\lambda}\frac{||z-H_2||}{||z-H_1||}} \end{aligned} \tag{7}$$

corresponding to the continuous fraction expansion

$$G_{i,\lambda}(z) = \cfrac{1}{z - a_1^{i,\lambda} - \cfrac{b_1^{i,\lambda}}{z-a_2^{i,\lambda} - \cfrac{b_2^{i,\lambda}}{z-a_3^{i,\lambda} - \cfrac{b_3^{i,\lambda}}{z-a_4^{i,\lambda} - \cfrac{b_4^{i,\lambda}}{\ddots}}}}}. \tag{8}$$

An analytical expression of the Green function is then obtained. However, an exact determination of $G_{i,\lambda}(z)$ requires an infinite number of recursion coefficients which is never the case in practical calculations. Usually, only a few levels (N pairs of (a_n, b_n) coefficients) of the continuous fraction are determined and the missing part of the fraction expansion is replaced by a terminator function $\Sigma_{i,\lambda}(z)$. The continuous fraction expansion becomes

$$G_{i,\lambda}(z) = \cfrac{1}{z - a_1^{i,\lambda} - \cfrac{b_1^{i,\lambda}}{z - a_2^{i,\lambda} - \cfrac{b_2^{i,\lambda}}{z - a_3^{i,\lambda} - \cfrac{b_3^{i,\lambda}}{\ddots \cfrac{}{z - a_N^{i,\lambda} - \cfrac{b_N^{i,\lambda}}{\Sigma_{i,\lambda}(z)}}}}}}. \tag{9}$$

For energy bands presenting no gap, the recursion coefficients have asymptotic limits corresponding to $(a_\infty^{i,\lambda}, b_\infty^{i,\lambda})$. The most easy way to determine $\Sigma_{i,\lambda}(z)$ is to assume that for $n > N$ the recursion coefficients are equal to their asymptotic values. The terminator function is then equal to:

$$\Sigma_{i,\lambda}(z) = z - a_\infty^{i,\lambda} - \cfrac{b_\infty^{i,\lambda}}{z - a_\infty^{i,\lambda} - \cfrac{b_\infty^{i,\lambda}}{z - a_\infty^{i,\lambda} - \cfrac{b_\infty^{i,\lambda}}{\ddots}}}$$

$$= z - a_\infty^{i,\lambda} - \frac{b_\infty^{i,\lambda}}{\Sigma_{i,\lambda}(z)}. \tag{10}$$

The solution of this equation gives the square root expression of the terminator function:

$$\Sigma_{i,\lambda}(z) = \frac{z - a_\infty^{i,\lambda} \pm \sqrt{(z - a_\infty^{i,\lambda})^2 - 4b_\infty^{i,\lambda}}}{2}. \tag{11}$$

The Beer-Pettifor method is used [4] to determine $(a_\infty^{i,\lambda}, b_\infty^{i,\lambda})$ which is based on the calculation of the band limits $(\epsilon_{min}, \epsilon_{max})$ given by $(a_\infty^{i,\lambda} - 2\sqrt{b_\infty^{i,\lambda}}, a_\infty^{i,\lambda} + 2\sqrt{b_\infty^{i,\lambda}})$. When the continuous fraction is truncated at the level N, the projected density of states (PDOS)

$$n_{i,\lambda}(\epsilon) = -\frac{1}{\pi}\text{Im}(G_{i,\lambda}(\epsilon + i0)) \tag{12}$$

corresponds to a sum of Dirac functions and the band limits can be identified by the energies of the lowest and highest Dirac functions. This method is extremely easy to use and, because it needs only to determine the diagonal representation of relatively small matrices, is also rapid and numerically stable. However, because the band limits correspond to Dirac functions, using the exact values obtained by this method gives usually diverging values for the PDOS near the band limits. Since the band width is underestimated by this method (due to the truncation of the continuous fraction), this problem can be nicely solved by enlarging the calculated band by 1%. One way to check the efficiency of this method for the determination of the terminator function is to verify that the PDOS are correctly normalized which is usually the case with an error smaller that 10^{-5}.

For each level of the recursion calculation (2), the hamiltonian H is applied on the n-th element $|n\}$ of the new basis. This adds to the expression of $|n+1\}$ all new "neighbours" linked through H of all sites included in the expression of $|n\}$. In other words, at each level of the recursion calculation, the contribution of the n-th shell of "neighbours" of the starting site $|i, \lambda\rangle$ is taken into account. Consequently, the number of sites needed for calculating N exact levels of the continuous fraction (or N exact pairs of recursion coefficients) is proportional to N^3 [5].

2.2   Tight Binding Hamiltonian

In the Linear Combination of Atomic Orbitals method, the one-electron wave function $|\Psi\rangle$ is expressed as a linear combination of localized atomic orbital $|i, \lambda\rangle$ on site i and spin-symmetry λ:

$$|\Psi\rangle = \sum_{i,\lambda} c_{i,\lambda} |i, \lambda\rangle. \tag{13}$$

Solving Schroedinger equation becomes the eigen-value problem

$$\sum_{i,\lambda} c_{i,\lambda}(\langle j, \mu|H|i, \lambda\rangle - \epsilon\langle j, \mu|i, \lambda\rangle) = 0$$

$$\sum_{i,\lambda} c_{i,\lambda}(H_{i,j}^{\lambda,\mu} - \epsilon S_{i,j}^{\lambda,\mu}) \qquad = 0. \tag{14}$$

The overlap matrix $S_{i,j}^{\lambda,\mu}$ plays an essential role when the atomic orbital can not be assumed as being orthogonal [6]. In this work, it is assumed that S is equal to identity $S_{i,j}^{\lambda,\mu} = \delta_{i,j}\delta_{\lambda,\mu}$. The hamiltonian can then be directly expressed in the atomic orbital basis

$$H = \sum_{(i,\lambda),(j,\mu)} |j, \mu\rangle H_{i,j}^{\lambda,\mu} \langle i, \lambda| \tag{15}$$

which can be split into intrasite and intersite terms

$$H = \sum_{i,\lambda,\mu} |i, \mu\rangle \epsilon_{i,\lambda,\mu} \langle i, \lambda| + \sum_{(i,\lambda),(j\neq i,\mu)} |j, \mu\rangle \beta_{i,j}^{\lambda,\mu} \langle i, \lambda|. \tag{16}$$

$\epsilon_{i,\lambda,\mu}$ are the on-site energy levels and $\beta_{i,j}^{\lambda,\mu}$ are the two-center hopping integrals linking sites i and j. In order to reduce the numbers of parameters of the hamiltonian, the on-site energy levels are assumed to be equal to

$$\epsilon_{i,\lambda,\mu} = \left(\epsilon_{i,\lambda}^0 + U_{i,l(\lambda)} \Delta N_{i,l(\lambda)} + \sigma_\lambda \frac{I_{i,l(\lambda)} M_{i,l(\lambda)}}{2} \right) \delta_{\lambda,\mu} \tag{17}$$

where $\epsilon_{i,\lambda}^0$, $U_{i,l(\lambda)}$, $\Delta N_{i,l(\lambda)}$ and $I_{i,l(\lambda)}$ are respectively the spin independent energy level reference, the effective Coulomb integral, the charge variation and the effective Exchange integral of site i, $M_{i,l(\lambda)}$ is the local magnetic moment, $l(\lambda)$ and σ_λ being respectively the orbital quantum number (corresponding to s, p or d states) and the spin of the spin-symmetry λ. The charge variation $\Delta N_{i,l(\lambda)}$ and the local magnetic moment $M_{i,l(\lambda)}$ are obtained from the PDOS by

$$\Delta N_{i,l} = \sum_\lambda \delta_{l(\lambda),l} \int^{\epsilon_F} n_{i,\lambda}(\epsilon)d\epsilon - N_{i,l}^0 \qquad (18)$$

where $N_{i,l}^0$ is the bulk l band occupation of site i and

$$M_{i,l} = \sum_\lambda \delta_{l(\lambda),l} \int^{\epsilon_F} (\delta_{\sigma_\lambda,+} n_{i,\lambda}(\epsilon) - \delta_{\sigma_\lambda,-} n_{i,\lambda}(\epsilon))d\epsilon \qquad (19)$$

Equations (17), (18) and (19) define the self consistency solution. The input charge variations and magnetic moments allow to calculate the input energy levels of the hamiltonian for which the output PDOS are determined and integrated to obtain the output charge variations and magnetic moments. Self consistency is reached when input and output quantities do not more differ significantly.

Usually, four energy levels ϵ_s^0, ϵ_p^0, $\epsilon_{E_g}^0$ and $\epsilon_{T_{2g}}^0$ are needed for each site and the hopping integrals $\beta_{i,j}^{\lambda,\mu}$ between each pair of sites are expressed in terms of ten simple Slater-Koster matrix elements [7] $ss\sigma$, $sp\sigma$, $sd\sigma$, $pp\sigma$, $pp\pi$, $pd\sigma$, $pd\pi$, $dd\sigma$, $dd\pi$ and $dd\delta$. It has been shown that a good description of the band structure can be obtained by limiting the hopping integrals to nearest neighbours pairs of sites. Various sets of hopping integrals for a given element can be found in the literature showing that the Slater-Koster matrix elements do not have an unique value [8].

As an illustration, the density of states (DOS) obtained using the Slater-Koster parameters deduced by fitting *ab initio* band structure calculations by Papaconstantopoulos [9] and those obtained using the TB-LMTO method by Andersen *et al* [10] are compared. Fig. 1 shows the DOS obtained for bulk non magnetic Chromium using these two sets of parameters. The d bands are very similar for both calculations. For the s and p DOS, if the bottom of the bands for both calculations are very similar, the DOS differ significantly for higher energies (larger than 15 eV). Due to the finite band width of tight binding DOS, the band structure shows non physical structures at high energies for all sets of parameters. However, with tight binding parameters taken from Papaconstantopoulos, the total band width (W) is approximately equal to 45 eV whereas with the second set of parameters W is approximately equal to 25 eV. Since the energy resolution of the DOS calculated with the recursion method is proportionnal to W divided by the number of levels, the larger W is, the smaller the energy resolution of the DOS - for a same number of recursion levels - is. Consequently, the d DOS - which is the most important for itinerant magnetism properties - exhibits more fine structures when W is smaller (Fig. 1.b) than for the other

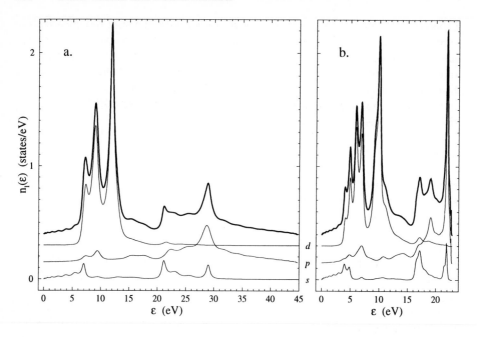

Fig. 1. *s*, *p*, *d* and total *(bold line)* densities of states obtained with 24 exact levels of the continuous fraction for bulk non magnetic Chromium with tight binding parameters taken from *a*. Papaconstantopoulos [9] and *b*. Andersen *et al* [10]

calculation (Fig. 1.a). It is then essential to choose tight binding parameters giving the smallest total band width in order to describe more precisely the *d* band of transition metals around the Fermi level. This is why, in the following, the tight binding parameters deduced by Andersen *et al* [10] are used.

2.3   Clusters for the Recursion Method

It has been shown previously that, for each additional level of the recursion fraction, the next shell of "neighbours" of the starting site |0} is taken into account. For the calculation of N_{exact} recursion levels, we have to built a cluster containing all sites which will contribute. This exact cluster correspond to all sites geometrically included in the N_{exact}-th shell of "neighbours". The shape of the exact cluster depends on the crystallographic structure and the cut-off r_c of the hopping integrals ($\beta_{i,j}^{\lambda,\mu} = 0$ when $|\mathbf{r}_i - \mathbf{r}_j| > r_c$) which can be approximated by a sphere centered on the starting site of radius $R = R^* = N_{\text{exact}}.r_c$ (this sphere contains the exact cluster). The number of sites N_{site} included in the sphere increases with N_{exact} like N_{exact}^3 and the computation time increases in the same way. This is why, in most cases smaller clusters are used. However, the recursion coefficients are affected by the size and the shape of these "inexact" clusters.

As an illustration, cubic and spherical "inexact" clusters are built and the

DOS obtained for various cluster sizes are compared. The minimal cubic cluster containing the exact cluster includes all sites i verifying $|x_i| < a/2$, $|y_i| < a/2$, $|z_i| < a/2$ with $a = a^* = 2N_{exact} \cdot r_c$. "Inexact" clusters are built with a smaller radius for spheres or a smaller edge for cubes. They are defined by the ratio $\rho = R/R^*$ for spherical clusters and $\rho = a/a^*$ for cubic ones.

Tables 1 and 2 present some characteristics of such "inexact" clusters and Figs. 2 and 3 display the DOS obtained for two of these examples. These results (Tables 1 and 2) shows that, as expected, for a similar number of sites in the clusters, the number of exact levels obtained during the calculation of 24 levels, is larger for a spherical cluster than for a cubic one. However, the comparison with the exact calculation (Fig. 1.b) shows that the DOS (Figs. 2 and 3) obtained with spherical clusters exhibit more non physical peaks (mainly near the bottom of the DOS) than the others; these peaks are mostly found in the s band. This can be easily understood because the spherical cluster is a much better approximation of the exact cluster than a cube and consequently the missing sites correspond to complete shells whereas, for cubic cluster, they correspond to fractions of shells. During the calculation of successive levels, the missing sites have a more progressive impact when cubic clusters are used than with spherical ones for which the impact occurs abruptly at a given level. In this work, cubic clusters are used, in order to keep the cubic symmetry of the considered crystals, with $\rho = 0.25$ and the DOS are determined with 24 levels in the continuous fraction. Of course, for other crystals, other "inexact" (non spherical) clusters have to be consider.

The previous considerations on the cluster shape have been done for bulk

Table 1. Some examples of "inexact" cubic clusters for various ratio ρ. The number of exact recursion levels and the computation time required for calculating 24 levels are displayed; the cluster built with $\rho = 1$ contains the exact cluster needed for the calculation of 24 exact recursion levels

Linear scale ratio ρ	Number of atoms	Number of exact levels	Computation time (s)
0.167	855	5	16
0.25	3925	11	38
0.5	29449	24[a]	238
1	228241	24	6106

[a] differences between calculated and exact coefficients smaller than 10^{-10}.

situation for which there is only one non equivalent atom in the unit cell. For cells containing a large number of non equivalent sites, there are two possible ways to build an "inexact" cluster (Fig. 4): (i) building a set of cubic clusters centered on each non equivalent site or (ii) building an unique cluster by joining

Table 2. Some examples of "inexact" spherical clusters for various ratio ρ. The number of exact recursion levels and the computation time required for calculating 24 levels are displayed; the cluster built with $\rho = 1$ contains the exact cluster needed for the calculation of 24 exact recursion levels

Linear scale ratio ρ	Number of atoms	Number of exact levels	Computation time (s)
0.195	869	6	16
0.325	3942	15	40
0.635	29627	24[a]	250
1	115633	24	1556

[a] differences between calculated and exact coefficients smaller than 10^{-10}.

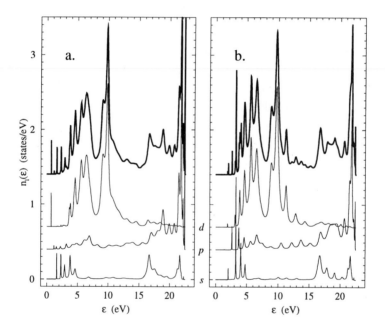

Fig. 2. s, p, d and total *(bold line)* densities of states obtained with 24 levels of the continuous fraction for bulk non magnetic Chromium for *a.* a cubic cluster with $\rho = 0.167$ and *b.* a spherical cluster with $\rho = 0.195$; both clusters contain approximately 850 sites

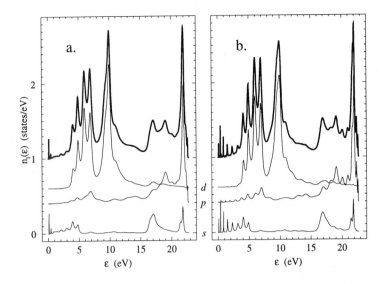

Fig. 3. s, p, d and total *(bold line)* densities of states obtained with 24 levels of the continuous fraction for bulk non magnetic Chromium for *a.* a cubic cluster with $\rho = 0.25$ and *b.* a spherical cluster with $\rho = 0.325$; both clusters contain approximately 3900 sites

together all clusters built previously. In the first construction, if the unit cell is longer than $a/2$, all non equivalent sites are not included in one cluster (Fig. 4.a). This is problematic when interactions between distant atoms are studied because one of the atoms is outside the cluster and no direct interactions are taken into account. The second cluster construction solves this problem by including all atoms of the unit cell in the cluster. However, this long cluster breaks the cubic symmetry and the different sites are no more equivalent from the point of view of the recursion calculation.

In order to have an idea of the fluctuations introduced by such a long cluster, the bulk antiferromagnetic Chromium situation is considered where all atoms are equivalent and the length n of the unit cell is artificially increased. With exact clusters, the result does not depend on the size of the cell. The results obtained with the "inexact" cluster are presented on Table 3. The local magnetic moment fluctuates only very slightly (fluctuations from site to site smaller than 10^{-5} μ_B) but it decreases when the unit cell is increased. The on-site energy shows more pronounced fluctuations from site to site. This shows clearly that the size and the shape of the cluster play an essential role on the calculated properties when the unit cell is varied. This is exactly what is usually done for multilayers where the properties are studied as a function of the layer thickness.

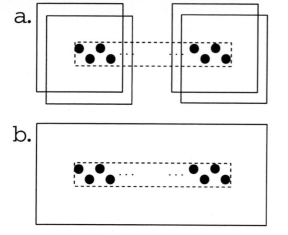

Fig. 4. Schematic representation of the two kind of clusters considered for long unit cells *(dashed line)* containing a large number of non equivalent sites *(filled circles)*: *a.* a set of cubic clusters centered on each non equivalent site are built, *b.* an unique cluster containing all clusters of the *(a)* situation is built

Table 3. Magnetic moment values (M) and energy on the sites i in the unit cell of length n which are not equivalent in the "inexact" unique cluster relative to the value obtained with a single atom in the unit cell $\Delta E(i) = E(i) - E(0,\ n = 1)$

	M (μ_B)	$\Delta E(0)$ (meV)	$\Delta E(1)$ (meV)	$\Delta E(2)$ (meV)	... ...
$n = 1$	0.6001	0			
$n = 2$	0.5868	−0.52			
$n = 3$	0.5796	−0.50	−0.69		
$n = 4$	0.5743	−0.68	−0.72		
$n = 5$	0.5731	−0.75	−0.75	−0.75	
$n = 6$	0.5727	−0.76	−0.76	−0.76	
$n = 11$	0.5725	−0.76	−0.76	−0.76	...

3   Periodic Versus Real Space Cells for Studying Bulk Magnetic Wall in Cr

The study of the Interlayer Magnetic Couplings (IMC) in Fe_mCr_n as a function of n requires the determination of the total energy from the electronic structure for various interlayer magnetic arrangement (IMA). Usually, they are restricted to collinear "ferromagnetic" (F) and "antiferromagnetic" (AF) IMA corresponding respectively to parallel and antiparallel magnetisation of successive Fe layers. The IMC are positive (respectively negative) as expected from the occurrence of a central magnetic defect in the Cr spacer when its thickness corresponds to an odd (even) number of atomic layers with an AF (F) interlayer arrangement [11]. For large spacer thickness, this defect becomes a bulk wall in the [001] direction in a layered antiferromagnetic crystal. This bulk wall resembles to a Bloch wall in a ferromagnet separating two domains in the limit of a very strong anisotropy giving collinear magnetism. In the present case, the bulk wall in a antiferromagnet corresponds to an antiphase in the layered antiferromagnetic order. It is easy to build such magnetic configurations using periodic or real space cells. The periodic cell consists in a Cr_1Cr_{2n-1} superlattice with an AF interlayer arrangement between the Cr_1 layers (Fig. 5.a). The real space cell consists in a block of $2n - 1$ atomic planes on which the wall is induced by a symmetry relative to the central atomic plane is applied (if this plane has a zero index, the symmetry corresponds to a magnetic moment on the i-th atomic plane given by $M_i = -M_{-i}$), this block being surrounded by blocks for which the magnetic moments are frozen in a AF bulk-like configuration (Fig. 5.b).

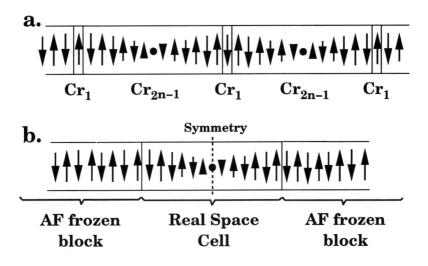

Fig. 5. Schematic representation of the cells used for calculating the bulk magnetic wall in AF Cr: *a.* a periodic cell built like a Cr_1Cr_{2n-1} superlattice with an AF interlayer arrangement, *b.* a real space cell, containing $2n - 1$ atomic planes with a symmetry applied relative to the central atomic plane reversing the magnetic moments, surrounded by bulk-like frozen AF Cr blocks

The calculation of the magnetic structure using a periodic cell is similar to the one of Fe_mCr_n superlattices: the Fermi level E_F is given by the global neutrality requirement:

$$\sum_{i,l} \Delta N_{i,l} = 0. \tag{20}$$

With the real space cell, the situation is more complex because the two semi-infinite bulk-like frozen AF Cr blocks fix E_F to the bulk value. The local energy levels $\epsilon_{i,l}$ are adjusted requiring the local neutrality:

$$\sum_l \Delta N_{i,l} = 0 \qquad (21)$$

which is a reasonable approximation since the charge variations in transition metals are usually small.

In this section, the study is restricted to a d description, because for collinear magnetism, the results obtained with this band structure description are more than qualitatively correct [12]. The characteristics (extent and energy) of the AF bulk wall can be determined with both cells using large thickness. In this paragraph, by increasing n, the two approaches (periodic and real space) are compared and the minimal value for n, for which the characteristics of the bulk wall do no more change significantly, is determined. With a periodic cell, the criterion is to recover the bulk value of the magnetic moment on the Cr_1 atomic plane and, with the real space cell, it is to have a continuous behaviour at the frontier between the cell and the "frozen" blocks. Fig. 6 presents these values a function of n. The criterion is more rapidly satisfied with a real space cell than with a periodic one. Moreover, all magnetic moments are found equal to zero for periodic cells with $n < 9$ whereas with this value for n, 90 % of the bulk moment is reached with the real space cell. This result is not very surprising since, with the real space cell, the moment at the frontier between the cell and the "frozen" blocks is strongly maintained by the proximity of a frozen bulk magnetic moment on one side and only slightly reduced by the magnetic defect on the other side. For periodic cells, the magnetic defect cancels all moments in too small cells. This illustrates the efficiency of the use of real space cells for studying non interacting magnetic configurations.

Using the real space approach, the characteristics of the bulk collinear magnetic wall in Cr can now be investigated. The insert of Fig. 7 shows the magnetic moment profile of the wall: its extent is found approximately equal to 40 atomic planes. This shows that a frustration in the Cr AF order has repercussions over a large range of planes making this spacer particularly suited for studies with the present light approach since a large number of Cr atoms are concerned. Finally, the asymptotic limit of the energy of this wall as a function of n seems to be $\gamma_{Cr} = 21$ meV per in plane atom (Fig. 7). This energy corresponds to the energy of the interlayer couplings obtained for large Cr thickness demonstrating that the coupling energy does not decrease in the collinear restriction when the spacer thickness increases in such Fe/Cr superlattices.

4   Non-Collinear Magnetism

4.1   Continuous Fraction Expansion and Non-Collinear Magnetism

All studies presented in the previous section have been realized in the collinear magnetism framework which saves computer time but represents a strong limitation as compared to experiments. The band structure non-collinear magnetism

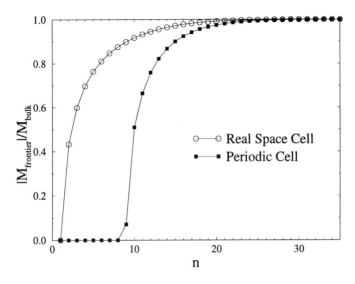

Fig. 6. Magnetic moments values relative to the bulk value on the Cr_1 atomic plane (*filled squares*) of the Cr_1Cr_{2n-1} periodic cell and on the atomic plane at the frontier between the cell and the "frozen" blocks (*open circles*)

allows to include the angular degree of freedom in the self consistent calculation [13,14]. The magnetic moment is a vector having 3 components M_x, M_y and M_z or can be described by its magnitude M_r and two spherical angles θ and ϕ. First, the input hamiltonian expression has to be modified in order to take into account the varying local spin quantization axis ζ_i (whose direction is given by the two usual spherical angles θ_i and ϕ_i) for each site i and, second, to determine the components $M_{r,i}$, $M_{\theta,i}$ and $M_{\phi,i}$ of the output magnetic moment.

The hamiltonian of (16) can be rewritten as a sum of a band H_{band} and an exchange H_{exch} hamiltonian where:

$$H_{band} = \left[\sum_{i,\lambda} |i, \lambda\rangle (\epsilon_{i,\lambda}^0 + U_{i,l(\lambda)} \Delta N_{i,l(\lambda)}) \langle i, \lambda| \right.$$

$$\left. + \sum_{(i,\lambda)\ (j \neq i,\mu)} |i, \lambda\rangle \beta_{i,j}^{\lambda,\mu} \langle j, \mu| \right] \begin{pmatrix} 1 & 0 \\ 0 & 1 \end{pmatrix} \qquad (22)$$

and

$$H_{exch} = \sum_{i,\lambda} |i, \lambda\rangle \frac{-I_{i,l(\lambda)} M_{i,l(\lambda)}}{2} \langle i, \lambda| \begin{pmatrix} \cos\theta_i & e^{-i\phi_i}\sin\theta_i \\ e^{i\phi_i}\sin\theta_i & -\cos\theta_i \end{pmatrix}. \qquad (23)$$

In these expressions, the spin part of the hamiltonian is represented by the 2×2 matrix which is site dependent only in H_{exch}. Expression (23) is obtained by applying a rotation on the σ_z Pauli matrix in order to align the local quantization

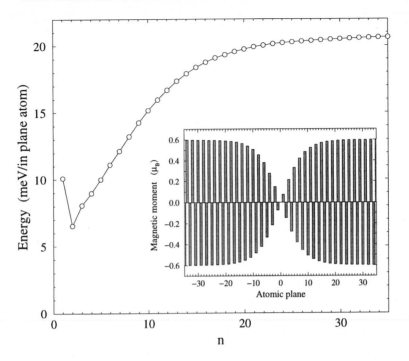

Fig. 7. Energy (in meV per in plane atom) of the bulk collinear magnetic wall as a function of the size n of the real space cell. The insert presents the magnetic moments profile of the wall

axis ζ_i with the global z axis.

The PDOS on site i for the symmetry λ is obtained with the recursion technique by setting the starting element $|0\rangle$ of the recursion basis equal to $|i,\lambda\rangle$. If the spin states are represented with the z quantization axis by a two elements vector, we have:

$$|i,\lambda,\sigma_\lambda = +\rangle = |i,\lambda'\rangle \begin{pmatrix} 1 \\ 0 \end{pmatrix} \quad , \quad |i,\lambda,\sigma_\lambda = -\rangle = |i,\lambda'\rangle \begin{pmatrix} 0 \\ 1 \end{pmatrix} \quad (24)$$

where λ' corresponds to the symmetry of the spin symmetry λ. The PDOS on an arbitrary axis ζ defined by the two spherical angles (Θ, Φ) in the spin space is obtained by starting with the following initial recursion basis element:

$$|i,\lambda,\sigma_\lambda = +\rangle_\zeta = |i,\lambda'\rangle \begin{pmatrix} e^{-i\Phi/2}\cos\frac{\Theta}{2} \\ e^{-ii\Phi/2}\sin\frac{\Theta}{2} \end{pmatrix}$$

$$|i,\lambda,\sigma_\lambda = -\rangle_\zeta = |i,\lambda'\rangle \begin{pmatrix} -e^{i\Phi/2}\sin\frac{\Theta}{2} \\ e^{i\Phi/2}\cos\frac{\Theta}{2} \end{pmatrix} . \quad (25)$$

For the determination of $M_{r,i}$, $M_{\theta,i}$ and $M_{\phi,i}$ using (19), the majority and minority spin states densities of states have to be calculated for all sites i and symmetries λ' for $\zeta = \zeta_i$ aligned with $\mathbf{u}_{r,i}$ ($\Theta = \theta_i, \Phi = \phi_i$), for ζ aligned with $\mathbf{u}_{\theta,i}$ ($\Theta = \theta_i + \pi/2, \Phi = \phi_i$) and for ζ aligned with $\mathbf{u}_{\phi,i}$ ($\Theta = \pi/2, \Phi = \phi_i + \pi/2$). The first step of the non-collinear study has been done using a d restricted

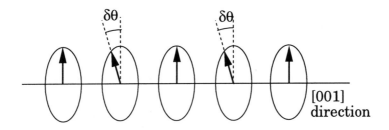

Fig. 8. Schematic representation of the slightly non-collinear magnetic configuration built by tilting the magnetic moment of one over two (001) atomic plane by a small angle $\delta\theta$

band structure description. In most cases studied, a discontinuous behaviour between collinear and slightly non-collinear calculations has been found. For example, a slightly non-collinear magnetic configuration by tilting the magnetic moment directions of half the (001) atomic planes by a small angle $\delta\theta$ is built as shown by Fig. 8. An antiferromagnetic order like in Cr is obtained for $\delta\theta = \pi$. When $\delta\theta$ is varied starting from zero (ferromagnetic order) and increased progressively, a nice continuous behaviour should be obtained. This is not obtained with a d band structure as shown by Fig. 9: the magnetic moment and the energy show rapid and large variations when $\delta\theta$ is varied from zero to 2°. On the contrary, the expected nice behaviour is obtained with a *spd* band structure as shown on the same figure. This peculiar behaviour comes from the not spin $(+)$ and $(-)$ mixed d densities of states in the ferromagnetic collinear configuration $(\delta\theta = 0)$: the $(+)$ and $(-)$ d densities of states are determined completely independently from a calculation where all 2×2 matrices of (25) are diagonal. Since it is assumed that the tight binding parameters are not spin dependent, the $(+)$ and $(-)$ densities of states are the same (they have the same band width W and are only split in energy by $I_{i,d}M_{i,d}$) but they have different band limits (Fig. 10) and the $(+)$ and $(-)$ spin recursion coefficients $a_n^{i,\lambda}$ have different $a_\infty^{i,\lambda}$ limits:

$$a_\infty^{i,d,+} = \frac{\epsilon_{min}^{i,d,+} + \epsilon_{max}^{i,d,+}}{2} = a_\infty^{i,d,-} - I_{i,d}M_{i,d}$$

$$b_\infty^{i,d,+} = b_\infty^{i,d,-} = \left(\frac{W}{4}\right)^2 \tag{26}$$

When, the magnetic configuration is non-collinear (even slightly non-collinear), the $(+)$ and $(-)$ spin states are mixed and the $(+)$ and $(-)$ densities of states have the same band limits (this is not exactly the case in Fig. 10 because only 8 recursion levels are used but larger band widths are obtained) and the recursion

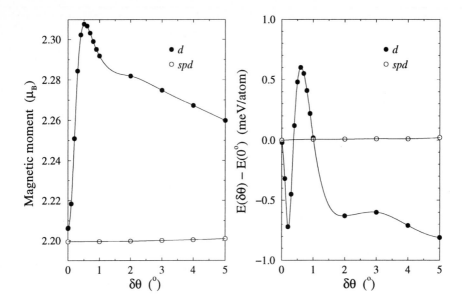

Fig. 9. Magnetic moments and energy variation as a function of the angle $\delta\theta$ of the magnetic configuration represented in Fig. 8 obtained with a d (*filled circles*) and with a spd (*open circles*) band structure

coefficients $(a_n^{i,\lambda}, b_n^{i,\lambda})$ should have the same $(a_\infty^{i,\lambda}, b_\infty^{i,\lambda})$ limit:

$$a_\infty^{i,d,+} = a_\infty^{i,d,-}$$

$$b_\infty^{i,d,+} = b_\infty^{i,d,-} = \left(\frac{W + I_{i,d}M_{i,d}}{4}\right)^2. \tag{27}$$

This explains the large differences in the densities of states represented by Fig. 10. The most significant changes are mainly noticeable at the top of the majority spin band where a large peak occurs around one eV even for a very small $\delta\theta$ value. Such an unphysical behaviour is of course not obtained with an spd hamiltonian. In this case, the d band which carries the magnetism is hybridised with the s and p bands having a large band width and consequently the recursion coefficients limits $a_\infty^{i,\lambda}$ are nearly insensitive to changes in the magnetic configuration.

For non-collinear studies, a restricted d hamiltonian has to be used very carefully in order to avoid unphysical results related to numerical problems in the continuous fraction expansion. All these problems are nicely solved when a spd hamiltonian is used.

4.2 Angular Dependence of the Interlayer Magnetic Couplings in Fe/Cr Multilayers

The interlayer magnetic couplings discussed previously can now be studied in the non-collinear framework and their angular dependence can be investigated.

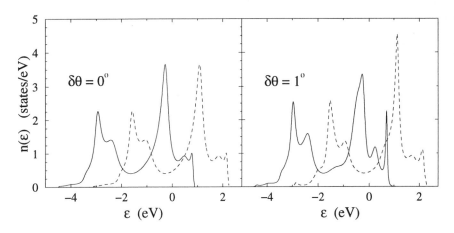

Fig. 10. Densities of states for bulk Fe obtained considering the magnetic configuration of Fig. 8 with $\delta\theta = 0°$ (ferromagnetic order) and $\delta\theta = 1°$ (slightly non-collinear order). The majority (respectively minority) spin bands are represented by solid (dashed) lines

This is done by fixing the angle $\Delta\theta$ between the directions of the inner magnetic moments of successive ferromagnetic layers during the self-consistent calculation as displayed by Fig. 11. During the calculation, all not fixed magnetic moments

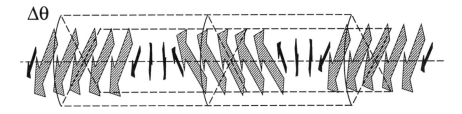

Fig. 11. Schematic representation of the magnetic configuration considered for the calculation of the interlayer magnetic couplings. Each vector corresponds to the magnetic moment of all atoms in the (001) atomic plane which are equivalent

are free to rotate and self-consistency is assumed to be achieved when the output perpendicular components $M_{\theta,i}$ and $M_{\phi,i}$ on all these sites i are nearly equal to zero. In this paper, the angular variation is restricted to θ in order to reduce the computer time but also because it has been checked that all magnetic moments vectors are in the plane defined by the two fixed magnetic moments.

Because the d states are the most essential for the magnetism, the effective exchange integrals I_s and I_p are usually equal to zero. However, if this is usually arbitrarily assumed, setting $I_s = I_p = 0$ is now required because we obtain self-consistency by an iterative way for non-collinear solutions. This comes from the expression of the output magnetic moment in terms of s, p and d contributions in the input $\mathbf{u}_{i,r}^{(\text{in})}$ and $\mathbf{u}_{i,\theta}^{(\text{in})}$ spherical basis:

$$M_{i,r}^{(\text{out})} = M_{i,r,s}^{(\text{out})} + M_{i,r,p}^{(\text{out})} + M_{i,r,d}^{(\text{out})}$$
$$M_{i,\theta}^{(\text{out})} = M_{i,\theta,s}^{(\text{out})} + M_{i,\theta,p}^{(\text{out})} + M_{i,\theta,d}^{(\text{out})}$$
$$\mathbf{M}_i^{(\text{out})} = M_{i,r}^{(\text{out})} \mathbf{u}_{i,r}^{(\text{in})} + M_{i,\theta}^{(\text{out})} \mathbf{u}_{i,\theta}^{(\text{in})}. \tag{28}$$

The magnetic moment used as input for the next iteration is obtained by varying θ_i in order to align the local quantization axis with the direction of the output magnetic moment given by:

$$\mathbf{M}_i^{(\text{next in})} = M_{i,r}^{(\text{next in})} \mathbf{u}_{i,r}^{(\text{out})}$$
$$= \mathbf{M}_i^{(\text{out})} \tag{29}$$

in the output $\mathbf{u}_{i,r}^{(\text{out})}$ and $\mathbf{u}_{i,\theta}^{(\text{out})}$ spherical basis. The s, p and d decomposition of the next input magnetic moment has to be obtained:

$$M_{i,r}^{(\text{next in})} = M_{i,r,s}^{(\text{next in})} + M_{i,r,p}^{(\text{next in})} + M_{i,r,d}^{(\text{next in})} \tag{30}$$

which is equal to the magnitude of the output magnetic moment vector

$$\sqrt{(M_{i,r,s}^{(\text{out})} + M_{i,r,p}^{(\text{out})} + M_{i,r,d}^{(\text{out})})^2 + (M_{i,\theta,s}^{(\text{out})} + M_{i,\theta,p}^{(\text{out})} + M_{i,\theta,d}^{(\text{out})})^2}. \tag{31}$$

However, this s, p and d decomposition is lost when the magnitude of $\mathbf{M}_i^{(\text{out})}$ given by (31) is calculated. If $I_s = I_p = 0$, the results are the same whatever the values of $M_{i,r,s}^{(\text{in})}$ and $M_{i,r,p}^{(\text{in})}$ are (they do not contribute to the exchange field) and we have to do the self-consistent calculation only for the d component of the magnetic moments:

$$M_{i,r,d}^{(\text{next in})} = \sqrt{(M_{i,r,d}^{(\text{out})})^2 + (M_{i,\theta,d}^{(\text{out})})^2}. \tag{32}$$

The calculation is assumed to be converged when

$$\text{Max}_i\{|M_{i,r,d}^{(\text{out})} - M_{i,r,d}^{(\text{in})}|\} < \epsilon$$
$$\text{Max}_i\{|M_{i,\theta,d}^{(\text{out})}|\} \qquad < \epsilon$$
$$\text{Max}_i\{|E_{\text{tot}}^{(\text{out})} - E_{\text{tot}}^{(\text{in})}|\} \quad < \epsilon' \tag{33}$$

with $\epsilon = 5 \times 10^{-5} \ \mu_B$ and $\epsilon' = 10^{-5}$ eV.

Fig. 12 presents the interlayer magnetic couplings obtained with the d and the two spd tight binding parameters as a function of $\Delta\theta$ for Fe_5Cr_4 and Fe_5Cr_5 superlattices. Around the energy minimum, the coupling energy follows a parabolic expression $C_+(\Delta\theta - \pi)^2$ for $n = 4$ and $C_+(\Delta\theta)^2$ for $n = 5$ as predicted by a phenomenological model [15]. The couplings obtained with the d tight binding

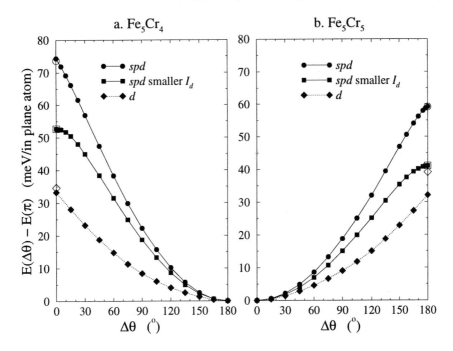

Fig. 12. Interlayer magnetic couplings $\Delta E(\Delta\theta) - E(\pi \ or \ 0)$ obtained with the d (*filled diamonds*) and the *spd* tight binding parameters without (*filled circles*) and with (*filled squares*) a smaller interfacial I_d as a function of $\Delta\theta$ for Fe_5Cr_4 and Fe_5Cr_5 superlattices. The open symbols correspond to the collinear solution

parameters follow the parabolic function over the whole range of $\Delta\theta$ considered. The collinear solution obtained for the frustrated configuration does not correspond to the solution obtained for $\Delta\theta = 0$ $(n = 4)$ and π $(n = 5)$ and has a higher energy. This is not very surprising since, with these parameters, the Fe and Cr magnetic moments do not vary significantly when $\Delta\theta$ varies and the local angles θ_i vary nearly linearly with $\Delta\theta$ as previously reported [16]. Such a *nearly constant magnetic moment magnitude* behaviour corresponds exactly to the phenomenological model which assumes a helical configuration in a Heisenberg model for the antiferromagnetic spacer. The behaviour of the coupling energies for *spd* tight binding parameters is completely different: they follow the parabolic function only over half the range of $\Delta\theta$ considered. This is particularly pronounced when the interfacial Cr I_d is reduced from 0.96 eV to 0.90 eV (in order to have a better agreement with *ab initio* calculations) where the coupling energies show a maximum at $\Delta\theta = 0$ $(n = 4)$ and π $(n = 5)$. Moreover, the frustrated collinear solution energies are nearly degenerate with the ones of the corresponding solution obtained with the non-collinear calculations. For example, for Fe_5Cr_4 superlattices, during the decrease of $\Delta\theta$ from π to 0 (i) the Fe magnetic moments have a nearly constant magnitude (ii) the magnitude of the magnetic moments on the Cr atoms decreases strongly when $\Delta\theta$ reaches 0 and the magnetic moment of the central Cr atomic planes nearly vanishes, (iii) the local angles vary linearly when $\Delta\theta$ decreases from 180° down to approximately 60°; for smaller $\Delta\theta$ values they vary very rapidly and reach values corresponding to the frustrated collinear magnetic configuration for $\Delta\theta = 0$.

This result explains qualitatively the different behaviours experimentally obtained for FeCo/Mn superlattices where the parabolic function applies [17] and for Fe/Cr superlattices where the saturation is better reproduced with a $J_1.\cos(\Delta\theta) + J_2.\cos^2(\Delta\theta)$ expression for the coupling energy [18].

4.3   Step Induced Non-Collinear Magnetism

In the previous paragraph, the non-collinear character is induced by the variation of $\Delta\theta$. This is similar, for $n = 4$, to the situation where an increasing external magnetic field is applied on the multilayer. Non-collinear magnetic configurations can also be obtained when interfacial imperfections frustrate the natural magnetic order in the multilayer.

This is illustrated by Fig. 13 for interfacial atomic steps. Because Cr is anti-

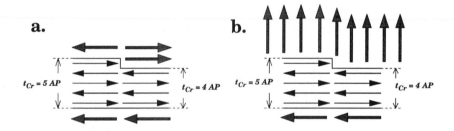

Fig. 13. Schematic representation of the frustration induced by an interfacial atomic step in an Fe/Cr/Fe sandwich: the Cr thickness t_{Cr} variation from 4 to 5 atomic planes (AP) *a.* splits the second Fe layer into domains of opposite magnetisation when the interfacial coupling is preserved or *b.* induces a 90° interlayer arrangement when the interfacial coupling is partially frustrated

ferromagnetic, the sign of the interfacial Cr magnetic moment changes from one (001) atomic plane to the next. If the antiparallel interfacial Fe-Cr coupling and the Cr antiferromagnetic order (for small Cr thickness) are preserved, the interlayer magnetic coupling changes from AF to F when the Cr thickness t_{Cr} varies from $n = 4$ to $n = 5$ atomic planes at the atomic step (Fig. 13.a). The second Fe layer is then split into domains of opposite magnetisation and the domain walls correspond to the steps [19]. However, if we allow non-collinear magnetism, the second Fe layer can be nearly monodomain if its magnetisation is perpendicular to the one of the first Fe layer as shown by Fig. 13.b where we have assumed that only the interfacial Fe-Cr coupling is partially frustrated. This behaviour, resulting from the competition between the strong Fe ferromagnetism and the fluctuations of the interlayer coupling, is usually invoked for explaining the occurrence of 90° interlayer arrangements but does not result from an intrinsic biquadratic interlayer coupling. The aim of this part is to use our method to determine explicitly the magnetic moment map for such configurations.

For these calculations, all magnetic moments are free to rotate until self-consistency is achieved. However, because we do not include the spin-orbit coupling, a global rotation can be applied on all moments without changing the solution (there are no privileged directions for the magnetisation). Consequently, when all moments are effectively free to rotate, the calculation can never converge because a global rotation can occur during each iteration and θ_i can never reach its asymptotic value. Fixing arbitrary one moment (on site 0 for example) and allowing all others to rotate is not the best way to obtain the self-consistent map: the torque applied on the fixed site by all other sites is usually very large and a large number of iterations is needed in order to reduce significantly $M_{0,\theta,d}$ and to satisfy the convergence criteria (33) on all sites. A better way is to applied a global rotation at each iteration in order to keep the moment on a given site (the site 0) fixed in direction. The value for θ_i at the next input is then

$$\theta_i^{(\text{next in})} = \theta_i^{(\text{out})} - \theta_0^{(\text{out})} \tag{34}$$

and the angular self-consistency is obtained when

$$\text{Max}_i\{|M_{i,\theta,d}^{(\text{out})} - M_{i,r,d}^{(\text{out})}.\theta_0^{(\text{out})}|\} < \epsilon. \tag{35}$$

In this case, the final map is obtained when the magnetic moments do no more rotate relatively each others even if they continue to rotate globally from one iteration to the next.

Superlattices with atomic steps at one Fe/Cr interface like in Fig. 13 are modelled by periodic superlattices having a perfectly flat Fe layer separated from a rough second Fe layer by the Cr spacer with varying thickness. In the cases considered in this paper, $i.e.$ atomic steps along the [010] direction with flat terraces having all the same size of 5 atomic rows, the cell corresponds to the lateral juxtaposition of 5 $Fe_5/Cr_5/Fe_5/Cr_5$ and 5 $Fe_5/Cr_4/Fe_7/Cr_4$ cells (see Fig. 14). The total real space cell contains 110 non equivalent sites. The calculations have been done using the d restricted parameters and the two sets of spd parameters previously used. The magnetic moment map obtained with the d restricted parameters of Fig. 14 shows clearly that (i) the rough Fe layer is structured in domains of opposite magnetisation corresponding exactly to the terraces, (ii) only the Fe atoms at the border line of the domains have a local magnetic moment perpendicular to the others, and (iii) the Cr spacer layer displays only a slight non-collinear character. This result does not correspond to the expected 90° interlayer arrangement and presents a strongly reduced magnetisation of the 'rough' Fe layer. On the contrary, with the spd parameters, the 'rough' Fe layer is (i) only slightly structured in magnetic domains having their magnetisation making an angle of approximately 80° and not 180°, (ii) the Cr spacer layer presents a more pronounced non-collinear character, and (iii) the magnetisation of the 'rough' layer, which is only slightly reduced, is preferentially perpendicular to the one of the flat Fe layer. The two sets of spd parameters used in this work give nearly the same result. This behaviour corresponds more to

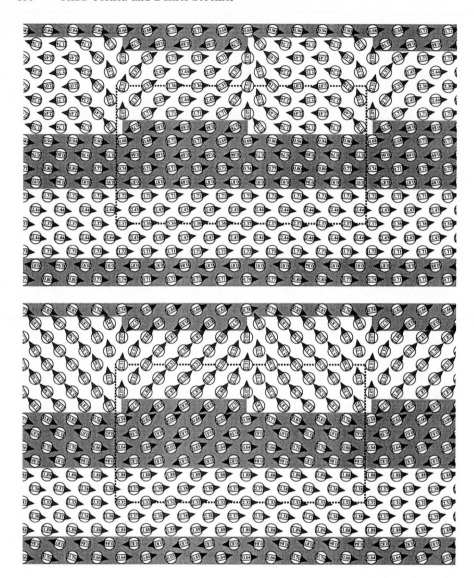

Fig. 14. Magnetic moment maps of a $Fe_{5.5}Cr_{4.5}$ superlattice having a perfectly flat 5 atomic planes thick Fe layer (*bottom layer*) and a rough Fe layer whose thickness varies from 5 to 7 atomic planes (*top layer*) separated by a Cr spacer layer whose thickness varies from 4 to 5 atomic planes obtained with the d restricted parameters (upper map) and with the *spd* parameters and the reduced Cr interfacial I_d (lower map). The arrow gives the direction of the local magnetic moment whose magnitude is given. The Fe sites have a grey arrow head and the Cr sites have a black one and a grey background

the expected one.

5   Summary

It has been shown that real space methods like the recursion technique allow to study the magnetic properties of complex systems in a much larger variety of configurations than most of the other approaches. However, the use of a tight binding description of the band structure limits the confidence in the results and it has been exhibited that d restricted and spd parameters give significantly different results when non-collinear magnetic solutions are allowed. This is why, more accurate band structure description consistent with a real space approach are needed and the TB-LMTO method is a possible candidate.

Acknowledgements: The calculations presented in this work have been realized using facilities at the Institut de Physique et Chimie des Matériaux de Strasbourg, on the T3E parallel computer at the Institut du Développement et des Ressources en Informatique Scientifique (IDRIS) of the CNRS, on the SP2 parallel computer of the Centre National Universitaire Sud de Calcul (CNUSC) and on the parallel computer Origin 2000 at the Institut du Calcul Parallèle de Strasbourg (ICPS) of the Université Louis Pasteur. The authors acknowledge the TMR 'Interface Magnetism' Network (contract FMR-CT96-0089) for support.

References

1. R. Haydock, V. Heine, and M.J. Kelly, Journal of Physic C **5**, 2845 (1972).
2. R. Haydock, in Springer Series in Solid-State Sciences *The Recursion Method and Its Applications* published by Springer-Verlag (Berlin) **58**, 8 (1985).
3. V. Heine, in NATO ASI Series B: Physics *The Electronic Structure of Complex Systems* published by Plenum Press (New York) **113**, 761 (1984).
4. N. Beer, and D.G. Pettifor, in NATO ASI Series B: Physics *The Electronic Structure of Complex Systems* published by Plenum Press (New York) **113**, 769 (1984).
5. H. Dreyssé, and R. Riedinger, J. Physique **48**, 915-920 (1987).
6. R. Riedinger, M. Habar, L. Stauffer, and Dreyssé H., Léonard P., Nath Mukherjee M., Phys. Rev. B **39**, 2442 (1989).
7. J.C. Slater, and G.F. Koster, Physical Review **94**, 1498 (1954).
8. for example see in MRS Symp. Proc. **491** (1998).
9. Papaconstantopoulos D. A., *Handbook of the Band Structure of Elemental Solids* published by Plenum Press (New York) (1986).
10. O.K. Andersen, O. Jepsen, and D. Gloetzel, in *Highlights of Condensed Matter Theory* edited by Bassani F., Fumi F., Tosi M. P. published by North Holland (Amsterdam), page 59 (1985).
11. D. Stoeffler, and F. Gautier, Progress of Theoretical Physics, Supplement No. **101**, 139 (1990).
12. D. Stoeffler, and F. Gautier, J. Magn. Magn. Mater. **147**, 260 (1995).
13. D. Stoeffler, and C. Cornea, Computational Materials Science **10**, 217 (1998).

14. C. Cornea, and D. Stoeffler, Computational Materials Science **10**, 245; 249 (1998).
15. J.C. Slonczewski, J. Magn. Magn. Mater. **150**, 13 (1995).
16. M. Freyss, D. Stoeffler, and H. Dreyssé, Phys. Rev. B **54**, 12677 (1996).
17. M.E. Filipowski, J.J. Krebs, G.A. Prinz, and C.J. Gutierrez, Physical Review Letters **75**, 1847 (1995).
18. B. Heinrich *et al*, J. Magn. Magn. Mater., **140-144**, 545 (1995).
19. D. Stoeffler, F. Gautier, J. Magn. Magn. Mater. **156**, 114-116 (1996).

Printing: Weihert-Druck GmbH, Darmstadt
Binding: Buchbinderei Schäffer, Grünstadt

Lecture Notes in Physics

Monographs

For information about Vols. 1–19
please contact your bookseller or Springer-Verlag